ANNUAL REVIEW OF
NAN🌐 RESEARCH

VOLUME 2

ANNUAL REVIEW OF NANO RESEARCH

Series Editors: Guozhong Cao (*University of Washington, USA*)
C Jeffrey Brinker (*University of New Mexico &*
Sandia National Laboratories, USA)

Vol. 1: ISBN-13 978-981-270-564-8
ISBN-10 981-270-564-3
ISBN-13 978-981-270-600-3 (pbk)
ISBN-10 981-270-600-3 (pbk)

Vol. 2: ISBN-13 978-981-279-022-4
ISBN-10 981-279-022-5
ISBN-13 978-981-279-023-1 (pbk)
ISBN-10 981-279-023-3 (pbk)

ANNUAL REVIEW OF NAN●RESEARCH

VOLUME 2

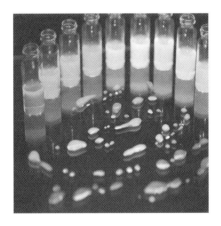

EDITORS

GUOZHONG CAO
University of Washington, USA

C. JEFFREY BRINKER
University of New Mexico and Sandia National Laboratories, USA

World Scientific

NEW JERSEY · LONDON · SINGAPORE · BEIJING · SHANGHAI · HONG KONG · TAIPEI · CHENNAI

Published by

World Scientific Publishing Co. Pte. Ltd.

5 Toh Tuck Link, Singapore 596224

USA office: 27 Warren Street, Suite 401-402, Hackensack, NJ 07601

UK office: 57 Shelton Street, Covent Garden, London WC2H 9HE

British Library Cataloguing-in-Publication Data
A catalogue record for this book is available from the British Library.

ANNUAL REVIEW OF NANO RESEARCH, Vol. 2

ISBN-13 978-981-279-022-4
ISBN-10 981-279-022-5
ISBN-13 978-981-279-023-1 (pbk)
ISBN-10 981-279-023-3 (pbk)

Printed in Singapore by Mainland Press Pte Ltd

CONTENTS

**Chapter 4.   Silicon Nanocrystal Assemblies: Universal
              Spin-Flip Activators?   159**
Dmitri Kovalev and Minoru Fujii

Chapter 5. DNA-Templated Nanowires: Context, Fabrication, Properties and Applications **217**
Qun Gu and Donald T. Haynie

**Chapter 7.   One- and Two-Dimensional Assemblies of
                Nanoparticles: Mechanisms of Formation
                and Functionality   345**
Nicholas A. Kotov and Zhiyong Tang

Chapter 10. Environmental Application of Nanotechnology   439
             G. Ali Mansoori, Tahereh Rohani Bastami,
             Ali Ahmadpour and Zarrin Eshaghi

**Chapter 11. Nanostructured Ionic and Mixed
            Conducting Oxides   495**
Xin Guo and Sangtae Kim

PREFACE

Annual Review of Nano Research publishes excellent review articles in selected topic areas authored by those who are authorities in their own subfields of nanotechnology with two vital aims: (1) to present a comprehensive and coherent distilling of the state-of-the-art experimental results and understanding of theories detailed from the otherwise segmented and scattered literature, and (2) to offer critical opinions regarding the challenges, promises, and possible future directions of nano research.

The second volume of *Annual Review of Nano Research* includes 13 articles offering a concise review detailing recent advancements in a few selected subfields in nanotechnology. The first topic to be focused upon in this volume is the electronic and optical properties of nanostructured materials and their applications. The second featured subfield is the recent advancement in the synthesis and fabrication of nanomaterials or nanostructures. Applications of nanostructures and nanomaterials for environmental and/or energy conversion and storage purposes are the focus in this volume.

Dr. Jeff Zhang has devoted many long hours in the editing and formatting of all the review articles published within this volume. Mr. Yeow-Hwa Quek from World Scientific Publishing was responsible for much of the coordination necessary to make the publication of this volume possible.

Guozhong Cao
Seattle, WA

C. Jeffrey Brinker
Albuquerque, NM

CONTRIBUTING AUTHORS

Ahmadpour, Ali
  * University of Illinois at Chicago, USA

Bastami, Tahereh R.
  * University of Illinois at Chicago, USA

Cao, Guozhong
  * University of Washington, USA

Cooper, Andrew I.
  * University of Liverpool, United Kingdom

Eshaghi, Zarrin
  * University of Illinois at Chicago, USA

Fujii, Minoru
  * Kobe University, Japan

Grant, Christian D.
  * Livermore National Laboratories, USA

Gu, Qun
  * Pacific Nanotechnology, Inc., USA

Guo, Xin
  * Research Center Jülich, Germany

Haynie, Donald T.
  * Central Michigan University, USA

Kim, Sangtae
  * University of California at Davis, USA

Kotov, Nicholas A.
  * University of Michigan, USA

Kovalev, Dmitri
  * University of Bath, United Kingdom

Léonard, Alexandre
  * The University of Namur (FUNDP), Belgium

Li, Jinghong
  * Tsinghua University, China

Liu, Jun
  * Pacific Northwest National Laboratory, USA

Mansoori, G. Ali
  * University of Illinois at Chicage, USA

Norris, Jordan
  * Western Kentucky University, USA

Rumbles, Garry
  * National Renewable Energy Laboratory, USA

Scholes, Gregory D.
  * University of Toronto, Canada

Su, Bao-Lian
  * The University of Namur (FUNDP), Belgium

Tang, Zhiyong
  * National Center for Nanoscience and Technology, China

Vantomme, Aurélien
  * The University of Namur (FUNDP), Belgium

Wang, Ying
  * University of Washington, USA
  * Northwestern University, USA

Wang, Zhouping
  * Jiangnan University, China

Wood, C. D.
  * University of Liverpool, United Kingdom

Zeng, Tingying
  * Western Kentucky University, USA

Zhang, Jin Z.
  * University of California, USA

Zhang, Qifeng
  * University of Washington, USA

CHAPTER 1

OPTICAL AND DYNAMIC PROPERTIES OF UNDOPED AND DOPED SEMICONDUCTOR NANOSTRUCTURES

Jin Z. Zhang[1]* and Christian D. Grant[2]

*[1]Department of Chemistry and Biochemistry, University of California, Santa Cruz, CA 95064 USA [2]Lawrence Livermore National Laboratories, Livermore, CA 94550 USA *Corresponding author, email: zhang@chemistry.ucsc.edu*

This chapter provides an overview of some recent research activities on the study of optical and dynamic properties of semiconductor nanomaterials. The emphasis is on unique aspects of these properties in nanostructures as compared to bulk materials. Linear, including absorption and luminescence, and nonlinear optical as well as dynamic properties of semiconductor nanoparticles are discussed with focus on their dependence on particle size, shape, and surface characteristics. Both doped and undoped semiconductor nanomaterials are highlighted and contrasted to illustrate the use of doping to effectively alter and probe nanomaterial properties. Some emerging applications of optical nanomaterials are discussed towards the end of the chapter, including solar energy conversion, optical sensing of chemicals and biochemicals, solid state lighting, photocatalysis, and photoelectrochemistry.

1. Introduction

Nanomaterials are the cornerstones of nanoscience and nanotechnology and are anticipated to play an important role in future economy, technology, and human life in general. The strong interests in nanomaterials stem from their unique physical and chemical properties and functionalities that often differ significantly from their corresponding bulk counterparts. Many of these unique properties are extremely promising for emerging technological applications, including nanoelectronics, nanophotonics, biomedicine, information storage, communication, energy conversion, catalysis, environmental protection, and space exploration.

One of the most fascinating and useful aspects of nanomaterials is their optical properties, including linear and non-linear absorption, photoluminescence, electroluminescence, and light scattering. For instance, semiconductor nanomaterials with spatial features on the order of a few nanometers exhibit dramatic size dependence of optical properties due to the quantum confinement effect [1, 2]. Shape and interaction between particles can also play an important role [3-5]. Therefore, their optical properties can be varied for different applications by controlling the size and shape of the nanostructures.

Since the surface-to-volume ratio (1/R scaling for spherical nanoparticles with radius R) is exceedingly large for nanomaterials, typically a million-fold increase compared to bulk, many of their properties, including optical, are extremely sensitive to surface characteristics [6]. As a result, one could also manipulate or modify the surface to influence and control their properties. Understanding of the surface properties of nanomaterials at the atomic level is still quite primitive at the present time.

While static studies, e.g. microscopy and XRD, provide important information about crystalline structure, size, shape, and surface, dynamic studies of charge carriers can provide complementary information that cannot be easily obtained from steady-state or time-integrated studies [7]. For example, the lifetime of charge carriers and their corresponding relaxation pathways determined from time-resolved studies can help gain insight into the effects of bandgap trap states that are due to surface or internal defects.

In the rest of this article, we will provide an overview of some recent research activities in the study of optical and dynamic properties of semiconductor nanomaterials. We will briefly discuss synthesis and structural characterization in order to make the article more self-contained. While we draw specific examples mostly from our own work, we attempt to cover as much relevant work as possible within the limited space. Even though we try our best to provide a balanced presentation of all the work cited, the viewpoints expressed in this article clearly reflects primarily our own interpretation and understanding.

2. Synthesis of Semiconductor Nanomaterials

Semiconductor nanostructures are synthesized by either chemical or physical methods. In a typical chemical synthesis, reactants are mixed in an appropriate solvent to produce the nanostructured product of interest. The result of the synthesis depends strongly on a number of factors such as concentration, temperature, mixing rate, or pH if in aqueous solution [8-15]. Surfactant or capping molecules are often used to stabilize the nanoparticles and can even direct particle growth along a particular crystal plane into that of a rod or other structure [16]. Truly bare or naked nanoparticles are not thermodynamically stable because of high surface tension and dangling chemical bonds on the surface. Impurities either from starting materials or introduced from some other source during synthesis can have deleterious effects by either profoundly altering their optical, structural, or chemical properties or even preventing the formation of the desired nanostructure. In light of this, extreme care should be taken to ensure that high purity reactants are used and that synthetic technique is as clean as possible.

Physical methods usually involve deposition onto appropriate substrate of the desired material from a source that is evaporated by heat or other type of energy such as light. Most techniques of nanomaterials synthesis are a combination of chemical and physical methods, such as CVD (chemical vapor deposition) or MOCVD (metal organic chemical vapor deposition) [17-25]. In CVD, a precursor, often diluted in a carrier gas or gasses, is delivered into a reaction chamber at approximately ambient temperatures. As it passes over or comes into contact with a heated substrate, it reacts or decomposes to form a solid phase that is deposited onto the substrate. The substrate temperature is critical and can influence what reaction takes place. The crystal structure of the substrate surface, along with other experimental parameters, determines what nanostructures can be generated. In MOCVD, atoms to be incorporated in a crystal of interest are combined with complex organic gas molecules and passed over a hot semiconductor wafer. The heat decomposes the molecules and deposits the desired atoms onto the substrate's surface. By controlling the composition of the gas, one can vary the properties of the crystal on an atomic scale. The crystal structure of the fabricated materials is dictated by the crystal structure of the substrate.

A special technique for synthesizing nanostructures, especially 1-D structures such as nanowires, is based on the VLS (vapor-liquid-solid) mechanism first discovered in the mid-1960s [26-29]. The mechanism consists of small metal particle catalysts deposited on a substrate. The substrate is then heated and vapor (e.g. Si, ZnO, GaN) of the material of choice is introduced. The vapor diffuses into the metal until a saturated solution is generated and the material of choice precipitates forming nanowires. There are several modern examples using VLS to grow many different types of nanowires or other one-dimensional nanostructures [30-34]. One such variation is where a laser ablates a substrate containing a metal/semiconductor mixture to create a semiconductor/metal molten alloy [35-37]. The resulting nanowires undergo VLS growth. Nanowires made by the laser assisted catalytic growth have lengths up to several μm [38-41].

The above synthetic techniques are generally considered bottom-up approaches where atoms and molecules are brought together to produce larger nanostructures. An opposite approach is top-down where large bulk scale structures are fabricated into smaller nanostructures. Lithographic techniques such as e-beam or photo-lithography are examples that allow creation of nanostructures on the micron and nanometer scales, easily down to tens of nanometers [42]. Such techniques lend conveniently for mass production of high quality and high purity structures critical in microelectronics and computer industry. It is currently a challenge to create structures on a few nm scale using typical lithographic methods. There is urgent need for developing new technologies that can meet this challenge, especially for the microelectronics and computer industry. The combination of top-down and bottom-up approaches may hold the key to solving this problem in the future.

There are a number of review articles and books that devote a significant amount of detail on nanomaterial synthesis [3, 5, 43-46], and we refer the reader to these resources since this chapter focuses more on the optical properties of semiconductor nanomaterials and their applications.

3. Structural Characterization

Structural determination and understanding are an important and integral part of nanomaterials research. Since the nanostructures are too small to be visualized with a conventional optical microscope, it is essential to use appropriate tools to characterize their structure in detail at the molecular or atomic level. This is important not only for understanding their fundamental properties but also for exploring their functional and technical performance in technological applications. There are a number of powerful experimental techniques that can be used to characterize structural and surface properties of nanomaterials either directly or indirectly, e.g. XRD (X-ray diffraction), STM (scanning tunneling microscopy), AFM (atomic force microscopy), SEM (scanning electron microscopy), TEM (transmission electron microscopy), XAS (X-ray absorption spectroscopy) such as EXAFS (extended X-ray absorption fine structure) and EXANES(extended X-ray absorption near edge structure), EDX (energy dispersive X-ray), XPS (X-ray photoelectron spectroscopy), IR (infrared), Raman, and DLS (dynamic light scattering) [43, 44, 47-49]. Some of these techniques are more surface sensitive than others. Some of the techniques are directly element-specific while others are not. The choice of technique depends strongly on the information being sought about the material.

X-ray diffraction (XRD) is a popular and powerful technique for determining crystal structure of crystalline materials. Diffraction patterns at wide-angles are directly related to the atomic structure of the nanocrystals, while the pattern in the small-angle region yields information about the ordered assembly of nanocrystals, e.g. superlattices [3, 50, 51]. By examining the diffraction pattern, one can identify the crystalline phase of the material. Small angle scattering is useful for evaluating the average interparticle distance while wide-angle diffraction is useful for refining the atomic structure of nanoclusters [Alivisatos, 1996, 933]. The widths of the diffraction lines are closely related to the size, size distribution, defects, and strain in nanocrystals.

As the size of the nanocrystal decreases, the line width is broadened due to loss of long range order relative to the bulk. This XRD line width can be used to estimate the size of the particle by using the Debye-Scherrer formula. However, this line broadening results in inaccuracies in the quantitative structural analysis of nanocrystals smaller than ~ 1 nm.

Scanning probe microscopy (SPM) represents a group of techniques, including scanning tunneling microscopy (STM), atomic force microscopy (AFM), and chemical force microscopy, that have been extensively applied to characterize nanostructures [47, 52]. A common characteristic of these techniques is that an atom sharp tip scans across the specimen surface and the images are formed by either measuring the current flowing through the tip or the force acting on the tip. SPM can be operated in a variety of environmental conditions, in a variety of different liquids or gases, allowing direct imaging of inorganic surfaces and organic molecules. It allows viewing and manipulation of objects on the nanoscale and its invention is a major milestone in nanotechnology.

STM is based on the quantum tunneling effect [53]. The wave function of the electrons in a solid extends into the vacuum and decay exponentially. If a tip is brought sufficiently close to the solid surface, the overlap of the electron wave functions of the tip with that of the solid results in the tunneling of the electrons from the solid to the tip when a small electric voltage is applied. Images are obtained by detecting the tunneling current when the bias voltage is fixed while the tip is scanned across the surface, because the magnitude of the tunneling current is very sensitive to the gap distance between the tip and the surface. Based on current-voltage curves measured experimentally, the surface electronic structure can also be derived. Therefore, STM is both an imaging as well as a spectroscopy technique. STM works primarily for conductive specimens or for samples on conducting substrates.

For non-conductive nanomaterials, atomic force microscopy (AFM) is a better choice [52, 54]. AFM operates in an analogous mechanism except the signal is the force between the tip and the solid surface. The interaction between two atoms is repulsive at short-range and attractive at long-range. The force acting on the tip reflects the distance from the tip atom(s) to the surface atom, thus images can be formed by detecting the force while the tip is scanned across the specimen. A more generalized application of AFM is scanning force microscopy, which can measure magnetic, electrostatic, frictional, or molecular interaction forces allowing for nanomechanical measurements.

Scanning electron microscopy (SEM) is a powerful and popular technique for imaging the surfaces of almost any material with a resolution down to about 1 nm [48, 49]. The image resolution offered by SEM depends not only on the property of the electron probe, but also on the interaction of the electron probe with the specimen. Interaction of an incident electron beam with the specimen produces secondary electrons, with energies typically smaller than 50 eV, the emission efficiency of which sensitively depends on surface geometry, surface chemical characteristics and bulk chemical composition [55].

Transmission electron microscopy (TEM) is a high spatial resolution structural and chemical characterization tool [56]. A modern TEM has the capability to directly image atoms in crystalline specimens at resolutions close to 0.1 nm, smaller than interatomic distance. An electron beam can also be focused to a diameter smaller than ~ 0.3 nm, allowing quantitative chemical analysis from a single nanocrystal. This type of analysis is extremely important for characterizing materials at a length scale from atoms to hundreds of nanometers. TEM can be used to characterize nanomaterials to gain information about particle size, shape, crystallinity, and interparticle interaction [48, 57].

X-ray based spectroscopies are useful in determining the chemical composition of materials. These techniques include X-ray absorption spectroscopy (XAS) such as extended X-ray absorption fine structure (EXAFS) and X-ray absorption near edge structure (XANES), X-ray Fluorescence spectroscopy (XRF), energy dispersive X-ray spectroscopy (EDX), and X-ray photoelectron spectroscopy (XPS) [58, 59]. They are mostly based on detecting and analyzing radiation absorbed or emitted from a sample after excitation with X-rays, with the exception that electrons are analyzed in XPS. The spectroscopic features are characteristic of specific elements and thereby can be used for sample elemental analysis. This is fundamentally because each element of the periodic table has a unique electronic structure and, thus, a unique response to electromagnetic radiation such as X-rays.

XAS is an element-specific probe of the local structure of atoms or ions in a sample. Interpretation of XAS spectra commonly uses standards with known structures, but can also be accomplished using theory to derive material structure. In either case, the species of the material is determined based on its unique local structure. X-ray absorption spectroscopy results form the absorption of a high energy X-ray by an atom in a sample. This absorption occurs at a defined energy corresponding to the binding energy of the electron in the material. The ejected electron interacts with the surrounding atoms to produce the spectrum that is observed. Occasionally, the electron can be excited into vacant bound electronic states near the valence or conduction bands. As a result, distinct absorptions will result at these energies. Often these features are diagnostic of coordination. XAS is commonly divided into two spectral region. The first is the X-ray absorption near edge structure or the XANES spectral region [59]. The XANES technique is sensitive to the valence state and speciation of the element of interest, and consequently is often used as a method to determine oxidation state and coordination environment of materials. XANES spectra are commonly compared to standards to determine which species are present in an unknown sample. XANES is sensitive to bonding environment as well as oxidation state and thereby it is capable of discriminating species of similar formal oxidation state but different coordination. The high energy region relative to XANES of the X-ray absorption spectrum is termed the extended X-ray absorption fine structure or EXAFS region. EXAFS yields a wealth of information, including the identity of neighboring atoms, their distance from the excited atom, the number of atoms in nearest neighbor shell, and the degree of disorder in the particular atomic shell. These distances and coordination numbers are diagnostic of a specific mineral or adsorbate-mineral interaction; consequently, the data are useful to identify and quantify major crystal phases, adsorption complexes, and crystallinity.

X-ray fluorescence (XRF) is a technique used to determine elemental composition in a material. The technique is based on irradiating a sample with either a lab based X-ray source (X-ray tube) or monochromatic radiation such as that obtained from a synchrotron. The emitted X-rays are characteristic of the element contained in the material.

In contrast, energy dispersive X-ray spectroscopy (EDS or EDX) is usually based on direct sample excitation with an electron beam (as in an SEM) with subsequent detection of an emitted X-ray. In either case, the information obtained from either XRF or EDX is equivalent in that it is chemically specific.

XPS is based on the measurement of photoelectrons following X-ray excitation of a sample. It is a quantitative spectroscopic technique that measures the chemical composition, redox state, and electronic state of the elements within a material. XPS spectra are obtained by irradiating a material with a beam of X-rays while simultaneously measuring the kinetic energy and number of electrons that escape from the top 1 to 10 nm of the material being analyzed. Thus, XPS is a surface sensitive analytic technique and it requires ultra high vacuum (UHV) conditions [58].

Optical spectroscopy such as IR and Raman provide more direct structure information while UV-visible electronic absorption and photoluminescence (PL) provide indirect structural information. For example, higher crystallinity and large particle size result in sharper Raman peaks and strong Raman signal. Disorder or high density of defects are reflected in low PL yield and trap state emission [7, 43]. Dynamic light scattering (DLS) can provide a measure of the overall size of nanoparticles in solution, usually when the size is larger than a few nm. In general optical spectroscopy is sensitive to structural properties but cannot provide a direct probe of the structural details.

4. Optical Properties

Semiconductor nanoparticles or quantum dots (QDs) have rich optical properties that strongly depend on size, especially when the particle size is less than the exciton Bohr radius of the material. Exciton Bohr radii are typically on the order of a few nm for semiconductors like CdSe, but are smaller for metal oxides like TiO_2. Their optical properties are also very sensitive to the surface characteristics and, to a lesser degree, of shape of the nanoparticles. For example, the photoluminescence spectrum and quantum yield can be altered by orders of magnitude by surface modification of the nanoparticles [60]. This fact can be used to advantage for specific applications of interest. Another factor affecting the optical properties is the interaction between nanoparticles or between the nanoparticles and their embedding environment [3, 4]. Interaction between nanoparticles typically leads to lower PL quantum yield and red-shifted PL spectrum due to shortened charge carrier lifetime [61]. Interaction with the environment is more complex and depends strongly on the chemical and physical nature of environment medium.

Optical properties are commonly characterized using spectroscopic techniques including UV-visible and photoluminescence spectroscopy, which both yield information about the electronic structure of nanoparticles. Related optical techniques such as Raman and IR provide information about the crystal structure such as phonon or vibrational frequencies and crystal phases. There are also a number of other more specialized optical techniques, often laser-based, that have been used to characterize the linear or non-linear optical properties of nanomaterials, such as second harmonic generation (SHG) [62-64], sum-frequency generation (SFG) [65, 66], and four-wave mixing [67].

4.1. Linear Optical Absorption and Emission

A striking optical signature of nanoparticles or quantum dots (QDs) is the strong size dependence of the absorption and photoluminescence (PL) (Figure 1 right) especially when the particle size is comparable to the exciton Bohr radius. An experimental manifestation of the size dependence is the blue shift of the UV-visible and PL spectra with decreasing particle size. This behavior is due to what is termed quantum confinement. The quantum confinement effect may be qualitatively understood using the particle-in-a-box model from quantum mechanics. In other words, a smaller box yields larger energy gaps between electronic states than does a larger box. For spherical particles, a quantification of quantum confinement is embodied in equation 1 [1, 2],

$$E_{g,effective}(R) = E_g(\infty) + \frac{\hbar^2\pi^2}{2R^2}(\frac{1}{m_e} + \frac{1}{m_h}) - \frac{1.8e^2}{\varepsilon R} \qquad (1)$$

where $E_g(\infty)$ is the bulk bandgap, m_e and m_h are the effective masses of the electron and hole, and ε is the bulk optical dielectric constant or relative permittivity. The second term on the right hand side shows that the effective bandgap is inversely proportional to R^2 and increases as size decreases. On the other hand, the third term shows that the bandgap energy decreases with decreasing R due to increased Columbic interaction. However, since the second term becomes dominant with small R, the effective bandgap is expected to increase with decreasing R, especially when R is small. This effect is illustrated schematically in Figure 1 (left). The effect of solvent or embedding environment is neglected in this form of the equation, but the effect of solvation is typically small compared to quantum confinement.

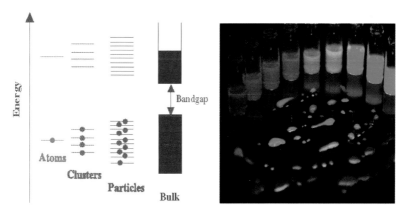

Figure 1. (left) Illustration of quantum confinement effect in different systems ranging from atoms to bulk materials. (Right) Photos of CdTe QDs with different sizes under UV illumination, ranging from 6 nm (red) to 2.5 nm (green) in size [73].

The quantum size confinement effect becomes significant particularly when the particle size becomes comparable to or smaller than the Bohr exciton radius, α_B, which is given by:

$$\alpha_B = \frac{\varepsilon_0 \varepsilon h^2}{\pi \mu e^2} \qquad (2)$$

where ε_0 and ε are the permittivity of vacuum and relative permittivity of the semiconductor, μ is the reduced mass of the electron and hole,

$m_e m_h/(m_e+m_h)$, and e the electron charge. For instance, the Bohr radius of CdS is around 2.4 nm [68] and particles with radius smaller or comparable to 2.4 nm show strong quantum confinement effects, as indicated by a significant blue-shift of their optical absorption relative to that of bulk [69-71]. Likewise, the absorption spectra of CdSe nanoparticles (NPs) show a dramatic blue-shift with decreasing particle size [72]. The emission spectra usually show a similar blue shift with decreasing size. Figure 1 (right) displays different sized CdTe nanoparticles exhibiting different PL center wavelengths with larger particles (left) showing redder luminescence.

The UV-visible absorption measured as a function of wavelength reflects the strength of the electronic transition between the valence (VB) and conduction bands (CB). The transition from the valence to the conduction band is the solid state analog to the HOMO-LUMO electronic transition in molecules. In the case of direct bandgap transitions, typically a strong excitonic band with a well-defined peak is observed at the low energy side of the spectrum. The excitonic state is located slightly below the bottom of the conduction band. The energy difference between the bottom of the CB and the excitonic state is the electron-hole binding energy, which is typically a few to a few hundred meV. Thus, the peak position of the excitonic absorption band provides an estimate of the bandgap of the nanoparticle. The bandgap energy increases with decreasing particle size, resulting in a blue-shift of the absorption spectrum as well as the excitonic peak. In contrast, indirect bandgap transitions lack an excitonic peak and the spectrum usually features a gradually and smoothly increasing absorption with decreasing wavelength. A well-known example is Si [7, 74]. Quantum confinement in indirect bandgap materials is less easily observable due to the lack of sharp or well-defined spectral peaks or bands. The intensity of the absorbance for QDs follows Beer's law. In this case, QDs can be considered as large molecules. Each QD typically contains a few hundred to a few thousands atoms and the absorption oscillator strength for one QD is proportional to the number of atoms in each QD [75, 76]. An experimental study by Yu *et al.* determining the molar absorptivity of CdS, CdSe, and CdTe as a function of size bears this out quite well [77].

In PL spectroscopy, photoemission is measured following excitation of the sample with a fixed wavelength of light. Photoluminescence reflects the electronic transition from the excited state, usually the excitonic state but also could be trap states, to the ground state, the valence band. Since PL is a "zero-background" experiment, it is much more sensitive, by approximately 1000 times, than UV-visible absorption measurements [78]. Thus PL provides a sensitive probe of bandgap states that UV-visible spectroscopy is much less sensitive to. For a typical nanoparticle sample, PL can be generally divided into bandedge emission, including excitonic emission, and trap state emission. If the size distribution is very narrow, bandedge luminescence is often characterized by a small Stokes shift from the excitonic absorption band along with a narrow bandwidth which usually means there is a narrow energy distribution of emitting states. In contrast, trap states are typically located within the semiconductor bandgap and hence their emission is usually red shifted relative to bandedge emission. In addition, trap state PL is often characterized by a large bandwidth reflecting a broad energy distribution of emitting states. The ratio between the two types of emission is determined by the density and distribution of trap states. Strong trap state emission indicates a high density of trap states and efficient electron and/or hole trapping.

It is possible to prepare high quality samples that have mostly bandedge emission when the surface is well passivated. For example, TOPO (tri-*n*-octylphosphine oxide) capped CdSe show mostly bandedge emission and weak trap state emission, which is an indication of a high quality sample [8, 79, 80]. Luminescence can also be enhanced by surface modification [81-86] or using core/shell structures [12, 87-89]. Many nanoparticles, including CdSe, CdS, ZnS, have been found to show strong photoluminescence [90]. Other nanoparticles have generally been found to be weakly luminescent or non-luminescent at room temperature, e.g. PbS [91], PbI_2 [92], CuS [93], Ag_2S [94]. The low luminescence can be due to either the indirect nature of the semiconductor or a high density of internal and/or surface trap states that quench the luminescence. Luminescence usually increases at lower temperature due to suppression of electron-phonon interactions and thereby increases the excited electronic state lifetime. Controlling the surface by removing surface trap states can lead to significant enhancement of luminescence as well as of the ratio of bandedge over trap state emission [81-86]. Surface modification often involves capping the particle surface with organic, inorganic, biological molecules, or even ions that reduce the amount of trap states that quench luminescence. This scheme likely removes surface trap states, enhances luminescence, and is important for many applications that require highly luminescent nanoparticles, e.g. lasers, LEDs, fluorescence imaging, and optical sensing.

One common issue encountered is PL quenching in solution over time. The reason for quenching varies and may be influenced by such factors as pH, the presence of O_2, CO, or other gas molecules, or even room light [95, 96]. More specifically, pH is one critical factor to consider if the particles need to be in aqueous solution. There is indirect evidence that acidic conditions may result in dissolution of the oxide or hydroxide layer present on the surface of the nanoparticle that serves to stabilize the QD's luminescence. When the protecting layer is dissolved under acidic conditions, there is an increase in surface trap or defect states that quench the PL [60]. Whatever the true reason for the PL quenching, the luminescence intensity decay over time presents a problem for applications like biological labeling or imaging. To address the problem, different approaches have been considered and used, primarily in terms of stabilizing the surface by using a protecting layer of another material, e.g. polymer, large bandgap semiconductor like ZnS, or insulator such as silica and polymers [97, 98]. One interesting example is SiO_2 coated CdTe nanoparticles [73]. As shown in Figure 2, the PL of CdTe QDs lacking a silica coating is quenched within 200 s when dispersed in a tris-borate EDTA (TBE) buffer solution (blue curve). TBE is a commonly used buffer in molecular biology involving nucleic acids, so determining the PL stability of QDs in this relevant buffer is important for biological applications. It should be pointed out that in SiO_2 coated CdTe the PL intensity will not decay or decreases only slightly if they are dissolved in water. With only a partial layer of silica coating, the PL is better stabilized (red curve) and the intensity lasts slightly longer than uncoated CdTe (blue curve) in TBE buffer. However, when a 2-5 nm shell of silica coats the CdTe QD surface, the

PL (green curve) persists longer than the uncoated or only partially coated CdTe. Attempts have been made recently in our lab to put even thicker layers of silica with the hope that the PL will be more stable for even longer. PL stability in biologically relevant buffers is essential for many PL based applications such as biomarker detection [73].

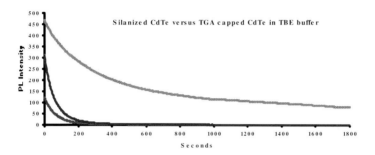

Figure 2. Effect of silica coating on the PL intensity in TBE buffer of CdTe nanoparticles. Adapted with permission from ref. [73]. Uncoated particles are shown in blue, partial silica coated in red, and with a 2-5 nm silica shell in green.

In addition to PL emission spectroscopy, another very useful PL experimental technique is photoluminescence excitation (PLE). This involves varying the wavelength of excitation while monitoring the PL intensity at a fixed wavelength. In the simplest case of a single emitting species (or state), the PLE is identical to the absorption. However, when there are several species present (e.g. different sized QDs) or a single species that exists in different forms in the ground state, the PLE and absorption bands are no longer superimposable. This technique can yield information about the nature of the emitting state or species. Specifically, in the case of ZnSe:Mn or ZnS:Mn QDs by monitoring the emission from the Mn dopant, the PLE band is identical to the absorption band indicating that the emission from Mn is due to energy transfer through excitation of the host crystal. Comparison of absorption of PLE often provides useful information on the types of states that are contributing to the PL.

There are two practical problems that are often encountered in PL measurements: Raman scattering and high order Rayleigh scattering. Raman scattering from solvent molecules can show up as relatively strong signal in PL spectroscopy, especially when the PL intensity is low. For nanoparticles, PL speaks are generally broad for ensemble samples while Raman peaks are usually narrow. A simple diagnostic to verify that a peak is due to Raman scattering is by changing the excitation wavelength and observe if the peak shifts accordingly. If the observed peak is from Raman scattering, it will shift by the same amount in frequency as the change in excitation wavelength, while there will be no shift if the emission is due to true PL.

Another potential artifact is high order Rayleigh scattering that occur at multiples of the excitation wavelength, λ. For example, if $\lambda=400$ nm is the excitation wavelength, due to the basic grating diffraction equation, $10^{-6}nk\lambda=sin\alpha+sin\beta$ (where n is the groove density of the grating, k is the diffraction order, α the angle of incidence, and β is the angle of diffraction), apparent "peaks" at $n\times400$ nm can show up on the PL spectrum, e.g. 400 nm, 800 nm and 1200 nm corresponding to $k=1,2,$ and 3 respectively, if the spectrometer scans cover these regions. Such apparent peaks do not correspond to real light at these wavelengths but are simply a grating effect from the 400 nm Rayleigh scattering. One indication is their narrow line widths. To determine this experimentally, one can use short or long pass optical filters to check if the observed peaks are from the sample or artifact from the instrument. For instance, if a peak at 800 nm does not disappear when a filter that blocks 800 nm light is placed in front of the detector, it is most likely that this peak is a second order Rayleigh scattering from the 400 nm excitation light. Of course, the first order 400 nm is usually blocked by a filter. But there is usually still 400 nm light leaking through the filter. Usually it is a good idea to try to avoid observing the first order excitation light directly by starting the PL spectral scan to the red of the excitation line. Of course the choice of PL scan range depends on the emission properties of the nanomaterial under consideration. Confounding mistakes of this sort due to Raman and Rayleigh scattering have appeared in the literature more often than expected.

4.2. Non-Linear Optical Absorption and Emission

Similar to bulk materials, nanomaterials exhibit non-linear optical properties such as multiphoton absorption or emission, harmonic generation, up- or down-conversion. Nanoparticles have interesting non-linear optical properties at high excitation intensities, including absorption saturation, shift of transient bleach, third and second harmonic generation, and up-conversion luminescence. The most commonly observed non-linear effect in semiconductor nanoparticles is absorption saturation and transient bleach shift at high intensities [82, 86, 99-104]. Similar non-linear absorption have been observed for quantum wires of GaAs [105, 106] and porous Si [107, 108]. These non-linear optical properties have been considered potentially useful for optical limiting and switching applications [109].

Another non-linear optical phenomenon is harmonic generation, mostly based on the third-order nonlinear optical properties of semiconductor nanoparticles [110-113]. The third order non-linearity is also responsible for phenomena such as the Kerr effect and degenerate four wave mixing (DFWM) [114]. For instance, the third order non-linear susceptibility, $\chi^{(3)}$ (~5.6x10^{-12} esu) for PbS nanoparticles has been determined using time-resolved optical Kerr effect spectroscopy and it was found to be dependent on surface modification [113]. Third order non-linearity of porous silicon has been measured with the Z-scan technique and found to be significantly enhanced over crystalline silicon [109]. DFWM studies of thin films containing CdS nanoparticles found a large $\chi^{(3)}$ value, ~10^{-7} esu, around the excitonic resonance at room temperature [115].

Only a few studies have been carried out on second-order nonlinear optical properties since it is usually believed that the centrosymmetry or near centrosymmetry of the spherical nanoparticles reduces their firs-order hyperpolarizability (β) to zero or near zero. Using hyper-Rayleigh scattering, second harmonic generation in CdSe nanocrystals has been observed [116]. The first hyperpolarizability β per nanocrystal was found to be dependent on particle size, decreasing with size down to about 1.3 nm in radius and then increasing with further size reduction. These results are explained in terms of surface and bulk-like contributions. Similar technique has been used for CdS nanoparticles for which the β-value per particle (4 nm mean diameter) was found to be on the order of 10^{-27} esu, which is quite high for solution species [117]. Second harmonic generation has also been observed for magnetic cobalt ferrite ($CoFe_2O_4$) colloidal particles when oriented with a magnetic field [118]. The nonlinear optical properties of nanoparticles are found to be strongly influenced by the surface.

As discussed earlier, the optical properties of isolated nanoparticles can be very different from those of assembled nanoparticle films. This is true for both linear and non-linear optical properties. Theoretical calculations on nonlinear optical properties of nanoparticle superlattice solids have shown that an ideal resonant state for a nonlinear optical process is the one that has large volume and narrow line width [119-121]. The calculations also showed that nonlinear optical responses could be enhanced greatly with a decrease in interparticle separation distance.

Anti-Stokes photoluminescence or photoluminescence up-conversion is another interesting non-linear optical phenomenon. In contrast to Stokes emission, the photon energy of the luminescence output is higher than the excitation photon energy. This effect has been previously reported for both doped [122, 123] and high purity bulk semiconductors [124, 125]. For bulk semiconductors, the energy up-conversion is usually achieved by (i) an Auger recombination process, (ii) anti-Stokes Raman scattering mediated by thermally populated phonons, or (iii) two-photon absorption [126, 127]. Luminescence up-conversion has been observed in semiconductor heterojunctions and quantum wells [127-143] and has been explained based on either Auger recombination [131, 136, 144] or two-photon absorption [137]. Long-lived intermediate states have been suggested to be essential for luminescence up-conversion in some heterostructures such as $GaAs/Al_xGa_{1-x}As$ [136]. For semiconductor nanoparticles or quantum dots with confinement in three dimensions, luminescence up-conversion has only recently been reported for CdS [145], InP [126, 146], CdSe [126], InAs/GaAs [147], and Er^{3+}-doped $BaTiO_3$ [148]. Surface states have been proposed to play an important role in the up-conversion in nanoparticles such as InP and CdSe [126].

Luminescence up-conversion in ZnS:Mn nanoparticles and bulk has been observed [149]. When 767 nm excitation was used, Mn^{2+} emission near 620 nm was observed with intensity increasing almost quadratically with excitation intensity. The red shift of Mn^{2+} emission from that usually observed at 580 nm to 620 nm has been proposed to be caused by the difference in particle size. However, a more likely explanation could be the local environment of the Mn^{2+} ion rather than particle size. Comparison with 383.5 nm excitation showed similar luminescence spectrum and decay kinetics, indicating that the up-converted luminescence with 767 nm excitation is due to a two-photon process. The observation of fluorescence up-conversion in Mn^{2+}-doped ZnS opens up some new and interesting possibilities for applications in optoelectronics, e.g. as infrared phosphors. There remain some unanswered questions, especially in terms of some intriguing temperature dependence of the up-converted luminescence [150]. It was found that the up-conversion luminescence of ZnS:Mn nanoparticles first decreases and then increases with increasing temperature. This is in contrast to bulk ZnS:Mn in which the luminescence intensity decreases monotonically with increasing temperature due to increasing electron-phonon interaction. The increase in luminescence intensity with increasing temperature for nanoparticles was attributed tentatively to involvement of surface trap states. With increasing temperature, surface trap states can be thermally activated, resulting in increased energy transfer to the excited state of Mn^{2+} and thereby increased luminescence. This factor apparently is significant enough to overcome the increased electron-phonon coupling with increasing temperature that usually results in decreased luminescence [150].

Raman scattering could also perhaps be considered as a non-linear optical phenomenon since it involves two photons and inelastic scattering. Raman scattering is a powerful technique for studying molecules with specificity. For nanomaterials, Raman scattering can be used to study vibrational or phonon modes, electron-phonon coupling, as well as symmetries of excited electronic states. Raman spectra of nanoparticles have been studied in a number of cases, including CdS [151-156], CdSe [157-159], ZnS [154], InP [160], Si [161-164], and Ge [165-171]. Resonance Raman spectra of GaAs [172] and CdZnSe/ZnSe

[173] quantum wires have also been determined. For CdS nanocrystals, resonance Raman spectrum reveals that the lowest electronic excited state is coupled strongly to the lattice and the coupling decreases with decreasing nanocrystal size [153]. Raman spectra of composite films of Ge and ZnO nanoparticles revealed a 300 cm^{-1} Ge–Ge transverse optical (TO) vibrational band of Ge nanocrystals, which shifted towards lower frequencies on decreasing the size of Ge nanocrystals due to phonon confinement in smaller crystallites [170]. For 4.5 nm nanocrystals of CdSe, the coupling between the lowest electronic excited state and the LO phonons is found to be 20 times weaker than in the bulk solid [157]. For CdZnSe/ZnSe quantum wires, resonance Raman spectroscopy revealed that the ZnSe-like LO phonon position depends on the Cd content as well as excitation wavelength due to relative intensity changes of the peak contributions of the wire edges and of the wire center [173].

4.3. Other Relevant Optical Properties: Chemiluminescence and Electroluminescence

Besides optical absorption and emission, nanomaterials have other interesting optical properties such as chemiluminescence (CL) and electroluminescence (EL) that are of interest for technological applications such as chemical sensing and biochemical detection. For example, CL has been observed for CdTe nanoparticles [174, 175]. CL in CdTe nanoparticles capped with thioglycolic acid (TGA) was induced by direct chemical oxidation in aqueous solution using hydrogen peroxide and potassium permanganate under basic conditions [175]. The oxidized CL of CdTe NCs displayed size-dependent effect and its intensity increased along with increasing the sizes of the nanoparticles. Electron and hole injection into the CdTe nanoparticles through radicals such as O_2^- and OH· are proposed to be responsible for the strong CL observed.

Electroluminescence has been reported in various nanoparticles including Si [176], ZnO [177], and CdSe/CdS core/shells [178]. With a semiconductor polymer poly(N-vinylcarbazole) (PVK) doped with CdSe/CdS core–shell semiconductor quantum dots (QDs), white light emission was observed and attributed to the incomplete energy and charge transfer from PVK to CdSe/CdS core–shell QDs.

5. Charge Carrier Dynamics

5.1. Ultrafast Time-Resolved Laser Techniques

Study of charge carrier relaxation in nanoparticles provides complementary information to steady-state experiments that may not be readily or easily accessible using the time-integrated techniques already discussed. This charge carrier dynamical information can lead to a deeper understanding of nanomaterial fundamental properties including but not limited to optical properties. Time-resolved laser spectroscopy is a powerful technique for probing charge carrier dynamics in nanomaterials [6, 7]. Two common techniques are transient absorption (TA) and time-resolved luminescence. In transient operation measurement, a short laser pulse excites the sample of interest, namely a pump-pulse, and a second short laser pulse (probe pulse) is used to interrogate an excited population of charge carriers, e.g. electrons. The probe pulse is delayed in time with respect to the pump or excitation pulse. Changes in the detected signal (transmission in the case of transient absorption) of the probe pulse with this time delay contains information of the dynamics or lifetime of the excited carriers being probed [7]. The assignment of the observed signal is usually not trivial and often control experiments are combined with other information such as theory to help make the appropriate determination of origin of the transient absorption signal. This is also partly due to the fact that the probe pulse initiates an electronic transition between two excited states that are often not well characterized, especially the higher-lying electronic state. Nonetheless, transient absorption is versatile and provides high time resolution since the instrument response is determined only by the cross correlation of the pump and probe pulses, which are usually very short temporally (easily down to a few tens of fs with current technology).

In time-resolved luminescence measurements, the excitation mechanism is the same as in transient absorption. The difference is in the monitoring of the excited population. Instead of monitoring the excited state population with a second short laser pulse, the time profile of the photoluminescence is monitored. If the PL is monitored directly with a photodetector, such as photomultiplier tube (PMT), photodiode, or charge-coupled device (CCD), the time resolution is limited by the detector, which is often much longer (ps or ns) than the excitation laser pulse. For example Time Correlated Single-Photon Counting (TCSPC) can have an instrument resolution down to 25 ps and with instrumental deconvolution lifetimes of a few ps are reliably obtained. One way to take full advantage of short laser pulses is to use a technique called luminescence up-conversion. In this method, a second short laser pulse is mixed with the PL in a non-linear crystal to generate a new up-converted or higher energy photon which is then directed into a spectrometer and detected. The width or time-profile of the up-converted pulse is mainly determined by the second laser pulse used for the up-conversion while the energy of the photons in the up-converted pulse is the sum of the energies of photons from the PL and second up-converting laser pulse. By changing the time delay between the second laser pulse with respect to the excitation pulse, a time profile of the PL kinetics is obtained. In this case, the time resolution is much higher as it is determined by the cross overlap of the second up-converting and first pump pulse. While the PL up-conversion technique provides high time resolution for PL dynamics measurement, it is often involved and challenging since the up-converted signal is typically small due to the low PL intensity and the non-linear nature of up-conversion [179].

5.2. Linear Dynamic Properties: Relaxation, Trapping, and Recombination

Figure 3 provides a summary of possible dynamic processes involved in charge carrier relaxation in nanoparticles. If we ignore non-linear dynamic processes for now (to be discussed in the next section), the mechanisms are relatively simple and straightforward. Electronic relaxation in nanoparticles is similar to that observed in bulk solids, except with the important complication involving trap states.

For simplicity, let us first ignore trap states and suppose that there is at most one exciton or electron-hole pair per nanoparticle. In this simple case, above bandgap excitation produces an exciton or an electron in the CB and hole in the VB bound to each other by Columbic attraction. If the electron and/or hole have excess kinetic energy, they will first relax to the bandedge (the electron to the bottom of the CB and the hole to the top of the VB) through electron-phonon interactions on the tens to hundreds of fs time scale. Subsequently, the relaxed electron and hole at the bandedge can recombine radiatively, producing PL, or non-radiatively, usually producing heat. In a perfect crystal with few or no defects and hence a very low trap state density, radiative recombination dominates. The PL quantum yield in this case is very high, near 90% or more.

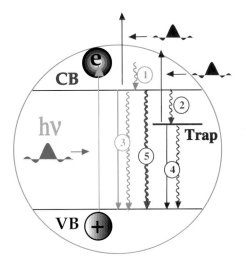

Figure 3. Illustration of a pump-probe approach for lifetime measurements. Different electronic relaxation pathways in a nanoparticle with trap states are illustrated: 1) electron relaxation through electron-phonon coupling with the conduction band (likewise for the hole in the valence band) following excitation across the bandgap; 2) trapping of electrons into trap states due to defects or surface states; 3) radiative and non-radiative bandedge electron-hole (or exciton) recombination; 4) radiative and non-radiative trapped electron-hole (or slightly relaxed exciton) recombination; and 5) non-linear and non-radiative exciton-exciton annihilation.

However, in a crystal with defects or a QD with a relatively high density of bandgap trap states due to surface or internal defects, trapping of electrons or holes into these states becomes significant. In many cases, trapping of charge carriers is faster than bandedge radiative recombination and therefore trapping dominates. Typically, the higher the density of trap states, the more likely and faster trapping will take place. Trapping lowers the bandedge PL yield and often the overall PL yield of the sample.

Following trapping, charge carriers (electrons, holes, or both) can either undergo further trapping (e.g. from shallow to deeper traps, not illustrated in Figure 3) or recombine radiatively or non-radiatively, as illustrated in Figure 3. This recombination process is quite similar in nature to bandedge carrier recombination processes. If this step is radiative, PL from trap states can be observed, red-shifted relative to bandedge PL since the trapping process causes non-radiative energy loss. Therefore, bandedge or trap state PL can be differentiated. If time-resolved PL is used, one can determine the lifetime of the bandedge states versus trap states. Since PL is a very sensitive technique, it is very useful for probing and understanding optical, dynamic, and electronic structure of nanomaterials.

With low excitation intensity, most nanoparticles will only have one exciton per particle, and no non-linear processes, such as exciton-exciton annihilation or Auger recombination, should occur. In this case, the charge carrier relaxation dynamics are relatively simple. Following relaxation within the conduction band for the electron and valence band for the hole, the carriers at or near the bandedge will recombine either radiatively or non-radiatively. The observed lifetime, τ_{ob}, is related to the radiative, τ_r, and non-radiative, τ_{nr}, lifetimes, by the following equation:

$$1/\tau_{ob}=1/\tau_r+1/\tau_{nr} \tag{3}$$

The lifetimes are related to the PL quantum yield, Φ_{PL}, by:

$$\Phi_{PL}= \tau_{ob}/\tau_r. \tag{4}$$

There is sometimes confusion between radiative lifetime (τ_r) and observed lifetime (τ_{ob}). These two are strictly speaking equal only in the limit that τ_{nr} is very long or the PL quantum yield is nearly 100%, as in a perfect single crystal, according to equations 3 and 4. Any lifetime measured experimentally based on time resolved PL or TA measurements is just the observed lifetime τ_{ob}, which contains contributions from both τ_r and τ_{nr}.

In nanoparticles, oftentimes there is a high density of trap states that lead to charge carrier trapping before or following relaxation in the CB or VB. The trapped carriers can recombine radiatively or non-radiatively, similar to bandedge carrier recombination. Trapping is typically a non-radiative process and contributes to τ_{nr} in equation 3. Also, there is usually a distribution of trap states and trapping can therefore be through several stages, e.g. through first shallow traps and then deep traps, thus complicating the kinetics. The shallow traps usually have shorter observed lifetimes than deep traps. The nature of trap states depends on the chemical nature, crystal structure, and details of surface characteristics. Surface-related trap states can be manipulated and this is often reflected sensitively in PL changes (spectral position and intensity). For example, based on time-resolved studies of a number of systems, the following general observations have been made: i) a high density of trap states corresponds to overall low PL yield; ii) a high density of trap states corresponds to relatively strong PL from trap states and weak or no PL from bandedge states; iii) a high density of trap states corresponds to short observed lifetime of charge carriers or the exciton; iv) a high density of trap states corresponds to a higher threshold for non-linear processes since it is harder to saturate all of the trap states.

The last statement implies that nanomaterials are better non-linear optical materials in the sense that they can tolerate a higher density of optical and possibly other radiation. This could be very useful for radiation protection applications.

In the scenario where the nanoparticles have no trap states within the bandgap at all, it is essentially a small perfect single crystal. In this ideal case, which is challenging to achieve experimentally, the behavior of the exciton or charge carrier is similar to that in bulk single crystals with the difference of spatial confinement. This would in principle allow for study of the pure spatial confinement effect without any influence from trap or surface states. However, this is not easy to achieve in reality due to bandgap states that are challenging to remove completely.

5.3. Non-Linear Dynamic Properties

Non-linear behavior occurs when there are multiple excitons generated in the same spatial region at the same time where there is strong interaction between the excitons. This is typically reflected as a dynamic process that depends non-linearly (e.g. quadratically or even higher order) on the excitation light intensity [6, 7, 71, 86, 180-182].

There have been various explanations for the observation of non-linear dynamical behavior in nanomaterials, including higher order kinetics, Auger recombination, and exciton-exciton annihilation (illustrated in Figure 4). It is challenging to assign an exact mechanism from only experimental data. All these models can explain the observations reasonably well. We have favored the exciton-exciton annihilation model since Auger recombination involves ionization and most time-resolved studies do not provide direct evidence for charge ejection. In the exciton-exciton annihilation model, high excitation laser intensity for the pump pulse produces multiple excitons per particle that can interact and annihilate, resulting in one exciton doubly excited and another one de-excited. If the rate of trapping is faster than the rate of exciton-exciton annihilation, which is often the case, trapping will reduce the probability of exciton-exciton annihilation. However, when trap states are saturated, exciton-exciton annihilation will take place. Therefore, nanoparticles with a higher density of trap states have a higher threshold for observing exciton-exciton annihilation or require higher pump laser intensities to observe this non-linear process. This behavior has been clearly demonstrated in CdS nanoparticles [60, 86].

The comparisons can be subtle and require careful attention when different sized or shaped nanoparticles are considered. This is partly because there are several factors, some competing, that need to be accounted for while making a comparison [183]. For example, when particles of different sizes but the same number of excitons per particle are compared, the smaller particles show a stronger non-linear effect or, conversely, a lower excitation threshold for observing the non-linear process. This is because smaller particles have stronger spatial confinement and lower density of states per particle that both facilitate exciton-exciton annihilation. On the other hand, when two samples of the same material such as CdS with the same nominal optical density or concentration but different particle sizes are compared under the same excitation intensity, the larger particles show stronger non-linear effect. This is apparently due to a larger number of excitons per particle for the larger particles. This indicates that the volume factor dominates over the effect of trap states. In other words, larger particles have a larger molar absorptivity (see the section on linear optical properties) and thus absorb more photons to create more excitons for a given laser pulse. The observation is opposite to what is expected for a larger number of trap states per particle, which for the larger particle should raise the threshold and thereby suppress exciton-exciton annihilation. These are illustrated schematically in Figure 5.

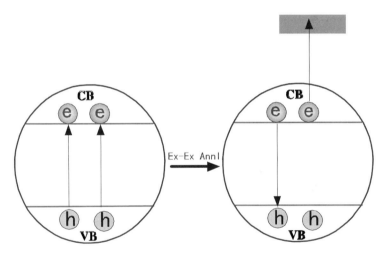

Figure 4. Illustration of non-linear and non-radiative exciton-exciton annihilation that results in the non-radiative de-excitation or recombination of one exciton and excitation of the other exciton to higher energy. The excited exciton will eventually relax radiatively or non-radiatively.

To further support the above argument, we have found that particles with similar volume but different shapes and thereby a different density/distribution of trap states show different thresholds for non-linear effects. For non-spherical particles, the PL yield is much lower compared to that of spherical particles, indicating a higher density of trap states for non-spherical particles. Based on the model discussed above, we should expect non-spherical particles to show a higher threshold for or weaker effect of exciton-exciton annihilation since it is harder to achieve trap state saturation. This is completely consistent with the experimental observation of stronger non-linear effect for the spherical particles [183].

It should be pointed out that our observations are qualitatively consistent with that made by Klimov *et al.* on CdSe nanoparticles [182]. However, the time constant for non-linear decay is much faster in our observation (a few ps) than that reported by Klimov *et al.* (as long as hundreds of ps). It is unclear if this difference is due to differences in the systems studied, experimental conditions, or even simply due to differences in data analysis and interpretation. Further studies are needed to better understand this issue.

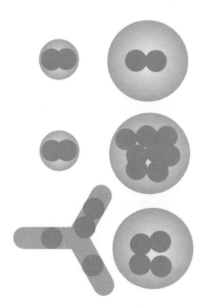

Figure 5. Illustration of the size and shape dependence of exciton-exciton annihilation.

Very recently, it has been suggested that multiple excitons can be generated using a single photon in small bandgap semiconductors such as PbSe and PbS [184-189]. This is potentially useful for solar energy conversion and other applications. Even though this is theoretically possible and there is preliminary supporting experimental evidence, it is unclear what the probability or efficiency is in practice. Given that electronic relaxation is typically very fast (less than 100 fs), this multi-exciton generation process with one photon is likely to be inefficient or has a very small cross section unless the energy levels involved can be carefully and intelligently designed to enhance the process. Impact ionization is a scheme proposed for enhancing multiple exciton generation (MEG) [184-188]. It is yet to be realized in practical device applications. Further research is needed to verify the feasibility of this approach and strategies to realize or enhance it.

5.4. Charge Transfer Dynamics Involving Nanoparticles

Beside dynamic processes of charge carrier relaxation in nanoparticles, other important dynamic processes of interest include charge transfer from nanoparticles to other species such as surface attached molecules or from molecules to nanoparticles (often referred to as charge injection). For example, dye molecules have been used to sensitize metal oxide nanoparticles such as TiO_2 and ZnO for potential solar cell applications [190-195]. In this case, one important process is electron injection from the dye molecule to the nanoparticle. The rate of injection is one critical factor in determining the solar conversion efficiency. The rate itself depends on a number of factors including the relative electronic energy levels of the dye and nanoparticles, strength of interaction between them, optical properties of the dye. Time-resolved studies have been successfully used to determine the rate of charge injection and its dependence on various factors such as distance and coupling strength [196]. In general, the injection rate has been found to be very fast, on the time scale of 100 fs or less [191].

Charge transfer from nanoparticles to molecules near or on the nanoparticles has also been studied using time-resolved techniques. Study of such processes is important for understanding photochemical and photocatalytic reactions when nanoparticles serve as photocatalysts or catalysts. For instance, electron transfer dynamics from CdS and CdSe NPs to electron acceptors, e.g. viologen derivatives, adsorbed on the particle surface have been studied using transient absorption, transient bleach and time-resolved fluorescence [197, 198]. Electron transfer was found to take place on the time scale of 200-300 fs and competes effectively with trapping and electron-hole recombination. These results are important to understanding interfacial charge transfer involved in photocatalysis and photoelectrochemistry applications.

6. Doped Semiconductor Nanomaterials

Doping is a powerful and effective way to alter the electronic and optical properties of a semiconductor. Doping is essential in the semiconductor industry since most semiconductors including silicon are essentially insulators without doping at room temperature.

Similar to bulk materials, doping has been used for semiconductor nanomaterials [45, 46]. There are some unique challenges with doping nanomaterials. For example, when the size is very small, one dopant ion per nanostructure can make a major difference in the properties of the nanostructure The addition of the dopant can introduce electronic and/or structural defects into the pristine nanomaterial that can be advantageous or deleterious. It is therefore critical to attempt to dope the nanostructures uniformly, i.e. same number of dopant ions per nanostructure (e.g. Figure 6). There are further complications to this issue beyond just the number of dopants. For example, the location of the dopant on the surface versus the interior affects the optical or electrical properties differently. Another issue is the interaction among the dopants when the dopant concentration per particle is high, e.g. two or more dopants in close proximity. This will remain a challenging and interesting issue for years to come, particularly when spatial features become smaller and the importance of the dopant becomes more critical.

Recently, there have been some reports of uniform doping using either growth or nucleation doping techniques by decoupling the doping and growth processes. Briefly, in nucleation doping reaction conditions are controlled in such a way along with judicious choice of reactants that a nucleus of dopants such as MnSe can be created followed by shell growth of ZnSe effectively confining the dopants to the center of the particle. Alternatively, also in a similar manner growth doping starts by creating a small ZnSe host crystal and then the doping atom may be introduced to either controllably dope the surface or the interior [199, 200].

Almost all studies of doped semiconductor nanostructures have been performed on ensemble averaged samples, i.e. the sample contains particles with a distribution of dopant per particle. Doping typically follows a Poisson distribution. The measured results need to be interpreted in such a manner. Several good review articles on doped semiconductor nanomaterials have appeared recently [5, 45, 46]. In this article, we will show a few examples to highlight the complexity and uniqueness of doped semiconductor nanoparticles.

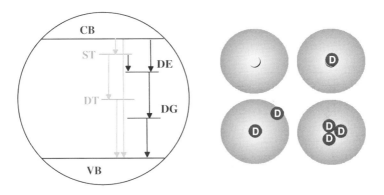

Figure 6. Left: Schematic illustration of energy levels of shallow trap (ST), deep trap (DT), dopant excited state (DE) and dopant ground state (DG) in a doped semiconductor nanoparticle with respect to the band edges of the valence band (VB) and conduction band (CB). Right: Illustration of nanoparticles with different numbers of dopant ions per particles as well as different locations of the dopant ions in the nanoparticles.

One of the most extensively studied doped nanoparticles is Mn doped ZnS, which is of interest for applications as a phosphor material [201-207]. Related systems studied recently include ZnSe:Mn that clearly show a strong correlation between optical emission and the location of the Mn dopant ion [208]. It has been found that at 1% Mn doping (measured in term of starting reactant materials, but actually determined for the product particles), the ZnSe bandedge PL is significantly quenched, by over a factor of 100 in peak intensity in comparison (Figure 7). It is clear that with roughly one Mn^{2+} ion per nanoparticle, significant PL quenching of the host ZnSe occurs. This makes sense since the time-integrated PL measured is conducted at low excitation intensity and is a single electron transition event. One Mn ion per particle can affect the relaxation pathways dramatically. Surprisingly, however, no PL from the Mn dopant ion was observed at this 1% doping level as one would expect. It was found that at this particular doping level, the Mn ions are primarily on the surface of the ZnSe nanoparticles and are consequently non-emissive.

This conclusion was reached after using a combination of PL, ESR, and XAFS on many different samples with varying doping levels. XAFS was able to provide a direct measure of the location and coordination environment of the different ions including Zn, Se, and Mn [208]. At 6% Mn doping, the ZnSe bandedge PL was further quenched by a factor of 800 relative to the undoped material with the characteristic 580 nm Mn dopant emission observed as expected. At this doping level, ESR clearly shows two different environments for Mn ions with XAFS data indicating two different coordination sites and symmetry, octahedron versus tetrahedron [208]. It was suggested that the tetrahedron Mn site is in the interior of the ZnSe nanoparticle with Mn substituting for the Zn cations while the octahedral Mn site is located at or near the ZnSe nanoparticle surface. While the interior Mn is emissive, the surface site is non-emissive, as illustrated in Figure 8, possibly because the extra ligands on the surface can potentially interact with either the capping agent or the solvent quenching the Mn emission. If this model is correct, it suggests that one should avoid having surface Mn in applications such as nanophosphors where Mn emission is desired. One needs to either remove surface Mn ions or find ways to encapsulate the Mn ions perhaps via a shell of a wide bandgap semiconductor such as ZnS into the lattice so they become optically emissive. Growth or nucleation growth, as discussed earlier, may be a way help solve this problem of encapsulation of dopants [199, 200].

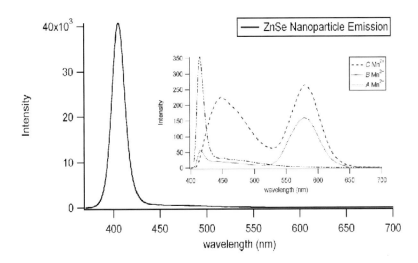

Figure 7. Photoluminescence spectra of ZnSe:Mn with different Mn doping levels: 1, 1, 5 and 10% (Reproduced with permission from ref. [208]).

Even though metal oxides are often considered insulators, in the context of this discussion, doped metal oxides share many similar properties to semiconductors as they can be considered large bandgap semiconductors. One very interesting doped metal oxide is ZnO nanoparticles doped with various ions such as Co^{2+} [209], Mn^{2+} [210, 211], and Cu^{2+} [212]. As high as 35% of Co can be doped into $Co_xZn_{1-x}O$ thin films without phase segregation [209]. As another example, TiO_2 nanoparticles and nanotubes have been doped with non-metal ions such as N to extend their photoresponse to the visible region and improve photoactivity [213-216]. Both TiO_2:N and ZnO:N have found success in narrowing the bandgap and increasing light harvesting efficiency [217, 218]. In addition, research has been conducted on the photoelectrochemical properties and photocatalytic activity for solar energy applications. Several different synthesis protocols have been developed to produce TiO_2:N. The usual doping process involves using ammonia as a nitrogen source by sol-gel, thermal, or hydrothermal chemical methods [219, 220].

Figure 8. Schematic illustration of two different Mn^{2+} sites in ZnSe nanoparticles with different coordination symmetry and optical properties: tetrahedral and emission at 580 nm for the interior site while octahedral and non-emissive for the surface site.

Charge carrier dynamics in doped semiconductor nanomaterials have also been studied with emphasis often placed on the carrier lifetime in the excited electronic state of the dopant ions. Such lifetimes can vary significantly from dopant to dopant, ranging from ns to ms. Some of the interesting and unresolved issues include how trap states mediate the rate of energy or charge transfer from host to dopant. It can be expected that the trap state could play an important role in energy or charge transfer processes especially when the trap states are located in between the donor states of the host and the acceptor states of the dopant. These issues require further investigation.

In some cases, the study can be complicated by host trap states, especially when there is spectral overlap between trap state and dopant transitions. For example, in the well-studied case of ZnS:Mn, there had been some controversy as well as some confusion about the carrier lifetime and related PL yield. In 1994, it was first reported that the PL lifetime of Mn^{2+} in ZnS:Mn nanoparticles was significantly shorter (~20 ns) than that in the bulk and had a greater luminescence efficiency [221-223]. The observed ns decays were five orders of magnitude shorter than the bulk luminescence lifetime (1.8 ms) [224]. This was explained using rehybridization between the s-p conduction band of ZnS and the 3d states of the Mn^{2+} because of quantum confinement. However, subsequent studies have shown that the Mn^{2+} PL lifetime in ZnS nanoparticles is the same as the bulk (1.8 ms) [225-228]. In our study of the PL kinetics of ZnS:Mn nanoparticles monitored at 580 nm, we observed a slow 1.8 ms decay that is similar to the Mn^{2+} emission lifetime in bulk ZnS as well as fast ns and μs decays that are also present in undoped ZnS particles and thereby attributed to trap state emission [227], as shown in Figure 9.

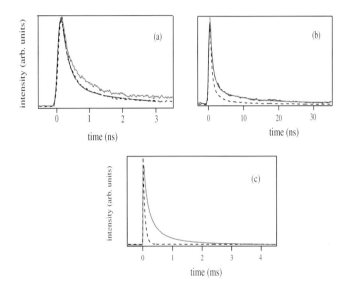

Figure 9. PL delay kinetics monitored at 580 nm of Mn-doped (solid) and undoped (dashed) ZnS nanoparticles on different time scales. Adapted with permission from ref. [227]. The major difference on long time scales is due to Mn doping.

Although these recent studies have consistently shown that the Mn^{2+} PL lifetime is similar in nanoparticles as in bulk ZnS:Mn, there are still active discussion over whether the PL quantum yield is higher in nanoparticles than in bulk. Recent studies [201-206], including a theoretical study [207], made claims of enhancement that seem to support the original claim of enhanced PL [222]. For instance, a study on the PL and EL properties of ZnS:Mn nanoparticles has found that the PL efficiency increased with decreasing particle size [229]. However, most of the reported PL or EL quantum yields have not been quantitatively compared between nanoparticles and bulk. Quantitative and calibrated measurements of the PL quantum yield are essential to establish if there is truly an enhancement in nanoparticles.

Another time resolved study that has come out of our lab is a comparative study between same diameter ZnSe:Mn with ZnSe using picosecond PL in combination with femtosecond transient absorption [230]. Briefly, it was shown that the electronic relaxation as determined from fs spectroscopy was multi-exponential and overall faster in Mn doped ZnSe particles relative to undoped ZnSe. Also of interest is that PL relaxation was also multi-exponential in nature and that in the doped material the overall PL decayed more rapidly then the undoped ZnSe. In addition on all time scales the relaxation was faster in ZnSe:Mn than in ZnSe. The difference in lifetimes on all time scales was attributed to a mechanism of energy transfer from host to dopant either mediated by trap states and another without (directly from the bandedge states). From these results the energy transfer possibly occurs on a time scale of tens to hundreds of ps which appears to be shorter than that reported in a study of ZnS:Mn by Chung where they found a rise time for the Mn^{2+} luminescence of 700 ps [228].

7. Applications of Optical Properties

While this chapter focuses on fundamental optical and dynamic properties of nanomaterials, we wish to briefly discuss some of the relevant emerging technological applications that span a wide range of fields from chemical sensing, photocatalysis, photoelectrochemistry, solar energy conversion, to biomedical detection and therapy. Most of these applications take advantage of some or all of the following unique features of nanomaterials: i) nanoscale sizes are comparable to carrier scattering lengths, this significantly reduces the scattering rate, thus increasing the carrier collection efficiency; ii) nanoparticles have strong optical absorption coefficient due to increased oscillator strength; iii) by varying the size, nanoparticle bandgap can be tuned to absorb in a particular wavelength region, possibly covering the entire solar spectrum; iv) nanoparticle-based devices can be built on flexible substrates;

v) nanoparticle-based devices can be lightweight and easy to make at potentially low cost; vi) nanomaterials have large surface-to-volume ratio and the surface can be modified and tailored for specific applications; vii) nanomaterials lend themselves conveniently for integration and assembly into larger and more sophisticated systems.

Of course, nanomaterials also have limitations when it comes to specific applications. For example, the large surface area of nanomaterials makes them vulnerable for surface defects and trap states that could have adverse effect on optical and other properties. Large particle surface areas potentially make nanoparticles more reactive due to dangling bonds. Even in carefully prepared, high quality samples, the density of trap states tend to be much higher than corresponding bulk materials and this results in low mobility or conductivity of charge carriers, which is undesirable for applications requiring good charge carrier transport. Some of these limitations can be overcome by careful design and engineering of the overall devices structures.

A few application examples will be given below in connection to the optical properties of nanomaterials.

7.1. Energy Conversion: Photovoltaics and Photoelectrochemistry

Solar energy conversion into electricity or chemical energy such as hydrogen represents one of the most promising applications of optical properties of nanomaterials [190, 192-195, 231]. For example, dye-sensitized solar cells have attracted significant attention since the initial report in 1991 of a power conversion efficiency of 12% [190]. Other variations of solar cells based on nanomaterials have also been demonstrated. While there are still issues related to efficiency, lifetime, and cost, they look very encouraging. While 0-D nanomaterials offer the largest surface-to-volume ratio that is often desired for solar cell applications, charge carrier transport is usually poor or mobility is low due to trapping of carrier by surface or other defect states and the need of carrier hopping for conduction. In this regard, 1-D or 2-D nanomaterials should offer better transport properties over 0-D nanomaterials. Of course, the surface-to-volume ratio is somewhat compromised.

There is growing recent interest in using various 1-D structures for solar cell applications, including CdSe, dye-sensitized TiO_2 and ZnO nanorods, nanowires, and nanotubes [232-239]. In many cases, improved photovoltaic performance has been found compared to 0-D nanoparticle films. However, most of these nanostructures are not as well aligned or ordered as one would like. The order or alignment of the 1-D structures should further improve their performance. Fabrication of well-ordered 1-D nanostructure arrays usually requires more sophisticated fabrication or synthesis techniques such as glancing angle deposition (GLAD) [240, 241]. Figure 10 shows an example of an idealized solar cell structure based on 1-D nanostructures.

Related to solar energy conversion directly into electricity is photocatalysis and photoelectrochemistry (PEC) that converts solar energy into chemical fuels such as hydrogen. One example is PEC conversion of water into hydrogen which is of strong current interest. In PEC, light illuminates one electrode (photoanode) that generates an electron and hole pair. The hole reacts with water molecules to produce O_2 while the electron is transported to another electrode (cathode, usually a metal such as Pt) where it reacts with protons (H^+) to produce H_2 gas. For most materials, the process requires an external bias in the 0 to 1 V range to assist in the conversion process and can be supplied by a battery or even a photovoltaic cell. The overall H_2 generation efficiency depends on a number of factors, most importantly the structural, chemical, and energetic characteristics of the cathode material that are often metal oxides or other semiconductors. The electrode materials can be bulk materials, thin films, or nanomaterials. Similar to solar cells, the issues of carrier transport versus surface area are often factors to consider when assessing their performances. 1-D materials again might hold some promise due to the combination of good electrical transport properties and large surface areas. It is important to develop inexpensive techniques for producing ordered or aligned 1-D structures.

7.2. Photochemistry and Photocatalysis

Nanomaterials have played a critical role in many important chemical reactions as reactants, catalysts, or photocatalysts. In relation to the optical properties that are of interest to this chapter, we will discuss briefly photochemical and photocatalytic reactions involving nanomaterials as reactants or photocatalysts. Their reactivities are often altered or enhanced due to size dependent changes in their redox potentials and high density of active surface states associated with a very large surface-to-volume ratio.

It has also been demonstrated that photooxidation of some small molecules on semiconductor nanoparticles can lead to the formation of biologically important molecules such as amino acids, peptide oligomers, and nucleic acids [242, 243]. In addition to photooxidation, photoreduction based on semiconductor nanoparticles have also been explored for synthesis of organic molecules [244, 245]. For example, photoinduced reduction of *p*-dinitrobezene and its derivatives on TiO_2 particles in the presence of a primary alcohol has been found to lead to the formation of benzimidoles with high yields [246].

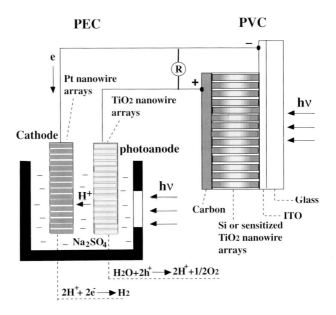

Figure 10. Schematic of an integrated photovoltaic cell (PVC) with a photoelectrochemical cell (PEC) for hydrogen generation from water splitting using 1-D nanostructures for both the PVC and PEC.

Another important area of application for semiconductor nanoparticles is in photoelectrochemical reactions. Similar to photochemical reactivities, the photoelectrochemical properties of nanoparticles are often quite different from those of their bulk counterparts. For example, structurally controlled generation of photocurrents has been demonstrated for double-stranded DNA-cross-linked CdS nanoparticle arrays upon irradiation with light [247]. The electrostatic binding of $[Ru(NH_3)_6]^{3+}$ to ds-DNA units provides tunneling routes for the conduction band electrons and thus results in enhanced photocurrents. This could be useful for DNA sensing applications. Photoelectrochemical behavior has been demonstrated in a number of semiconductor nanoparticle films, including CdS and CdSe [248-250], ZnO [251], TiO_2 [252-255], Mn-doped ZnS [256, 257], WO_3 [258], SnO_2/TiO_2 composite [259], and TiO_2/In_2O_3 composite [260]. The large bandgap semiconductors, e.g. TiO_2 and ZnO, often require sensitization with dye molecules so that photoresponse can be extended to the visible region of the spectrum [191, 252, 256].

Photocatalysis based on semiconductors plays an important role in chemical reactions of small inorganic, large organic, and biological molecules. The photocatalytic reactivities are strongly dependent on the nature and properties of the photocatalysts, including pH of the solution, particle size, and surface characteristics [261]. These properties are sensitive to preparation methods [261]. Impurities or dopants can significantly affect these properties as well as reactivities. For example, it has been shown that selectively doped nanoparticles have a much greater photoreactivity as measured by their quantum efficiency for oxidation and reduction than their undoped counterparts [262]. A systematic study of the effects of over 20 different metal ion dopants on the photochemical reactivity of TiO_2 colloids with respect to both chloroform oxidation and carbon tetrachloride reduction has been conducted [262, 263]. A maximum enhancement of 18-fold for CCl_4 reduction and 15-fold for $CHCl_3$ oxidation in quantum efficiency for Fe(III)-doped TiO_2 colloids have been observed [264]. Recent studies have shown that the surface photovoltage spectra (SPS) of TiO_2 and ZnO nanoparticles can be an effective method for evaluating the photocatalytic activity of semiconductor materials since it can provide a rapid, non-destructive monitor of the semiconductor surface properties such as surface band bending, surface and bulk carrier recombination and surface states [265]. It has been demonstrated that the weaker the surface photovoltage signal is, the higher the photocatalytic activity is in the case of nanosized semiconductor photocatalysts.

Photocatalytic oxidation of organic and biological molecules is of great interest for environmental applications, especially in the destruction of hazardous wastes. The ideal outcome is complete mineralization of the organic or biological compounds, including aliphatic and aromatic chlorinated hydrocarbons, into small inorganic, non- or less- hazardous molecules, such as CO_2, H_2O, HCl, HBr, SO_4^{2-}, and NO_3^-. Photocatalysts include various metal oxide semiconductors, such as TiO_2, in both bulk and particulate forms. Compounds that have been degraded by semiconductor photocatalysis include alkanes, haloalkanes, aliphatic alcohols, carboxylic acids, alkenes, aromatics, haloaromatics, polymers, surfactants, herbicides, pesticides and dyes, as summarized in an excellent review article by Hoffmann [264]. It has been found in many cases that the colloidal particles show new or improved photocatalytic reactivities over their bulk counterparts.

One of the most important areas of application of photocatalytic reactions is removal or destruction of contaminates in water treatment or purification [266-268]. Major pollutants in waste waters are organic compounds. Small quantities of toxic and precious metal ions or complexes are usually also present. As discussed above, semiconductor nanoparticles, most often TiO_2, offer an attractive system for degrading both organic and inorganic pollutants in water. Water treatment based on photocatalysis provides an important alternative to other advanced oxidation technologies such as $UV-H_2O_2$ and $UV-O_3$ designed for environmental remediation by oxidative mineralization. The photocatalytic mineralization of organic compounds in aqueous media typically proceeds through the formation of a series of intermediates of progressively higher oxygen to carbon ratios. For example, photodegradation of phenols yields hydroquinone, catechol and benzoquinone as the major intermediates that are eventually oxidized quantitatively to carbon dioxide and water [269].

In general, the details of the surface morphology, crystal structure, and chemical composition critically influence the photochemical and photocatalytic performance of the photocatalysts [270-274]. Therefore, these parameters need to be carefully controlled and evaluated when comparing photocatalytic activities of different materials.

7.3. Chemical and Biological Sensing

The unique optical properties of nanomaterials lend them conveniently for various optical sensing and detection of chemicals and biological samples. The optical luminescence from QDs is often used as a signature or probe [8, 97, 275-282]. For example, luminescent CdSe-ZnS core/shell quantum dot (QD) bioconjugates have been designed to detect proteolytic activity of enzymes by fluorescence resonance energy transfer (FRET) [283]. A modular peptide structure was developed for controlling the distance between the donors and acceptors. The high sensitivity of PL allows detection of single molecules and complex systems such as viruses [284].

7.4. Photonics and Solid State Lighting

Nanomaterials have found promising applications in photonics such as laser, LEDs, solid state lighting, and phosphors. In many of these applications, doping plays an important role, especially in phosphor materials. Lasers and LEDs based on thin films with thicknesses on the nanometer scale (2-D nanomaterials) have long been demonstrated and are currently used in commercial products. 1-D and 0-D nanomaterials have shown lasing properties based on optical pumping and demonstrated promising for technical applications [285, 286]. Lasing based on electrical pumping is a current challenge and if successful would represent a major technological breakthrough. The challenge is partly related to a higher density of surface trap states of 0-D and 1-D nanomaterials compared to their 2-D counter parts.

For laser applications, it is in principle possible to build lasers with different wavelengths by simply changing the particle size. There are two practical problems with this idea. First, the spectrum of most nanoparticles is usually quite broad due to homogeneous and inhomogeneous broadening. Second, the high density of trap states leads to fast relaxation of the excited charge carrier, making it difficult to create the population inversion necessary for lasing. When the surface of the particles is clean and has few defects, the idea of lasing can indeed be realized. This has been demonstrated mostly for nanoparticles self-assembled in clean environments based on physical methods, e.g. MBE (molecular beam epitaxy) [287-290] or MOCVD (metal organic chemical vapor deposition) [291]. Examples of quantum dot lasers include InGaAs [287], InAs [289], AlInAs [288, 290], and InP [292]. Stimulated emission has also been observed in GaN quantum dots by optical pumping [291]. The lasing action or stimulated emission has been observed mostly at low temperature [290, 292]. However, some room temperature lasing has also been achieved [287, 288].

Lasing action has been observed in colloidal nanoparticles of CdSe based on wet chemistry synthesis and optical pumping [286, 293]. It was found that, despite highly efficient intrinsic nonradiative Auger recombination, large optical gain can be developed at the wavelength of the emitting transition for close-packed solids of CdSe quantum dots. Narrow band stimulated emission with a pronounced gain threshold at wavelengths tunable with size of the nanocrystal was observed. This work demonstrates the feasibility of nanocrystal quantum dot lasers based on wet chemistry techniques. Whether real laser devices can be built based on these types of nanoparticles remains to be seen. Also, it is unclear if electrical pumping of such lasers can be realized. Likewise, nanoparticles can be potentially used for laser amplification and such an application has yet to be explored. Nanoparticles such as TiO_2 have also been used to enhance stimulated emission for conjugated polymers based on multiple reflection effect [294].

Room temperature ultraviolet lasing in ZnO nanowire arrays has been demonstrated [285]. The ZnO nanowires grown on sapphire substrates were synthesized with a simple vapor transport and condensation process. The nanowires form a natural laser cavity with diameters varying from 20 to 150 nm and lengths up to 10 μm. Under optical excitation at 266 nm, surface-emitting lasing action was observed at 385 nm with emission line width less than 0.3 nm. Such miniaturized lasers could have many interesting applications ranging from optical storage to integrated optical communication devices.

Nanoparticles have been used for LED application in two ways. First, they are used to enhance light emission of LED devices with other materials, e.g. conjugated polymers, as the active media. The role of the nanoparticles, such as TiO_2, is not completely clear but thought to enhance either charge injection or transport [295]. In some cases the presence of semiconductor nanocrystals in carrier-transporting polymers has been found to not only enhance the photoinduced charge generation efficiency but also extends the sensitivity range of the polymers, while the polymer matrix is responsible for charge transport [296]. This type of polymer/nanocrystal composite materials can have improved properties over the individual constituent components and may have interesting applications. Second, the nanoparticles are used as the active material for light generation directly [88, 297-299]. In this case, the electron and hole are injected directly into the CB and VB, respectively, of the NPs and the recombination of the electron and hole results in light emission. Several studies have been reported with the goal to optimize injection and charge transport in such device structures using CdS [299] and CdSe nanoparticles [297, 298].

Since the mobility of the charge carriers is usually much lower than in bulk single crystals, charge transport is one of the major limitations in efficient light generation in such devices. For example, photoconductivity and electric field induced photoluminescence quenching studies of close-packed CdSe quantum dot solids suggest that photoexcited, quantum confined excitons are ionized by the applied electric field with a rate dependent on both the size and surface passivation of the quantum dots [300, 301]. Separation of electron-hole pairs confined to the core of the dot requires significantly more energy than separation of carriers trapped at the surface and occurs through tunneling processes. New nanostructures such nanowires [285, 302], nanorods [16, 72, 303, 304], and nanobelts [305], may provide some interesting alternatives with better transport properties than nanoparticles. Devices such as LEDs and solar cells based on such nanostructures are expected to be developed in the next few years.

Solid state lighting is an area of fast growing interest. Approximately 30% of the United States electricity is consumed by lighting, an industry that is largely dominated by relatively old technologies such as the incandescent and fluorescent light bulb. New innovations in lower cost and higher efficiency solid state lighting are expected to significantly reduce our dependence on fossil fuels. A solid state lighting technology that is already compatible with low cost manufacturing is AC powder electroluminescence (ACPEL). Discovered in 1936 [306], powder electroluminescence utilizes emission from ~40-50 micron sized doped-ZnS:Cu,Cl phosphor particles and requires relatively low applied electric fields (10^4 V/cm) compared to DC EL which requires electric fields near 10^6 V/cm. More recently, microencapsulation technology has been successfully applied to ZnS:Cu,Cl powder phosphors so that the emissive particles can be deposited on plastic substrates under open air, non clean-room conditions using low cost, large area print-based manufacturing. Consequently, ACPEL lights are one of the least expensive large area solid state lighting technologies, and the characteristic blue-green light can now be found in many products. Furthermore, fluorescent energy conversion and doping can readily be used to convert the blue-green light to the white light preferred for normal everyday lighting. Using a variety of other dopants (I, Br, Al, Mn, Pr, Tm etc.) other colors can also be obtained [307, 308].

The mechanism for light emission from ZnS:Cu,Cl particles is thought to be due to localized electron and hole injection near Cu_xS inclusions which requires an alternating (AC) field to enable the frequency-dependent electron-hole recombination. While this process is highly efficient, the lifetime and power efficiency of ACPEL lights are heretofore too low to provide a replacement for white lights. The power efficiency is limited by the large ZnS:Cu,Cl particle size (> 20 μm) over which the electric field is dropped, resulting in the need for higher voltages, ~120 V, to achieve the brightness needed for solid state lighting. Research focused on phosphor nanoparticles has attempted to address the voltage issue; however, these systems typically have significantly reduced lifetime and quantum efficiency due to poor charge transport and/or trapping. For example, an EXAFS study has been conducted to understand how the degradation depends on the local microstructure about Cu in doped-ZnS:Cu,Cl and have found that the degradation process is reversible through modest application of elevated temperatures (~200 °C) [309]. Note that although the particle size is large (typically 20-50 μm), the active regions within each particle are near Cu_xS precipitates a few nm in size; hence it is essential to understand the Cu_xS nanoparticle precipitates within the ZnS host and how they interact with other optically active centers. Nanoparticle and nanorod systems exhibiting AC EL have been reported recently and will be studied for comparison [308, 310].

7.5. Single Molecule and Single Nanoparticle Spectroscopy

Most spectroscopy studies of nanoparticles have been carried out on ensembles of a large number of particles. The properties measured are thus ensemble averages of the properties of individual particles. Due to heterogeneous distributions in size, shape, environment, and surface properties, the spectrum measured is thus inhomogeneously broadened. This results in loss of spectral information. For instance, it has been predicted by theory that nanocrystallites should have a spectrum of discrete, atomic-like energy states [1, 2]. However, transition line widths observed experimentally appear significantly broader than expected, even though the discrete nature of the excited states has been verified [311, 312]. This is true even when size-selective optical techniques are used to extract homogeneous line widths [311-317].

One way to solve the above problem is to make particles with truly uniform size, surface, environment, and shape. However, this is almost impossible or at least very difficult, especially with regard to the surface. Another approach to remove the heterogeneity is to conduct the measurement on one single particle. This approach is similar to that used in the field of single-molecular spectroscopy [318, 319]. A number of single nanoparticle studies have been reported on semiconductor nanoparticles, including CdS [320] and CdSe [321-327]. Compared to ensemble averaged samples, single particle spectroscopy studies of CdSe nanoparticles revealed several new features, including fluorescence blinking, ultra narrow transition line width, a range of phonon couplings between individual particles, and spectral diffusion of the emission spectrum over a wide range of energies [327, 328].

One interesting observation in measuring the emission of a single nanoparticle is an intermittent on-off behavior in emission intensity under CW light excitation [323]. The intermittency observed was attributed to an Auger photoionization mechanism that leads to ejection of one charge (electron or hole) outside the particle. A "dark exciton" state was assigned to such ionized nanocrystal in which the emission is quenched because of excess charge. Similar studies on single CdSe/CdS core/shell nanocrystals as a function of temperature and excitation intensity and the observations are consistent with a darkening mechanism that is a combination of Auger photoionization and thermal trapping of charge [329].

Interestingly, single particle spectroscopy also reveals non-linear optical properties of single nanoparticles. For instance, in the low temperature near-field absorption spectroscopy study of InGaAs single quantum dots, the absorption change was found to depend non-linearly on the excitation intensity [330]. This non-linearity was suggested to originate from state filling of the ground state.

Single nanoparticle spectroscopy should be a powerful technique for studying properties of doped semiconductor nanoparticles in terms of the uniformity of doping, location and state of the dopant ions, and interaction between the dopant and host. To date, only one report on doped semiconductor nanoparticles, Te-doped CdSe, has been reported [331]. The photoluminescence spectra from single Te-doped CdSe (6 nm in diameter, passivated by ZnS) measured at room temperature was analyzed in comparison with those from undoped CdSe. No difference was observed in the emission linewidths of individual particles of doped and undoped samples, though emissions from doped samples were found to have wider ensemble bands due to increased inhomogeneity because of doping. Several dopant emissions showed irreversible blue shifts with repeated measurements, which significantly exceeded those observed for the undoped ones. The shifts were attributed to instability of the dopant in a nanocrystal upon excitation.

8. Concluding Remarks

While significant progress as been made in the last decade in the synthesis, characterization, and exploration of applications of nanomaterials, there is still much room for further study in all these areas. On the synthesis front, better control and assembly of nanostructures and functionality is important, including size, shape, surface, impurity, and interaction between structures. For better characterization, we need to have more and higher precision experimental tools for atomic level studies. Single particle studies represent a good step in this regard but still lack atomic resolution usually. In the ideal scenario, one would like to have high resolution in space, time, and energy that are atom or element specific. This is needed for a complete understanding of structural, optical, electronic, magnetic, and dynamic properties.

For applications, one key frontier issue is integration of different nanostructured components for desired functionality. This involves understanding, engineering, and controlling interfaces between the different components. Therefore, it is important to be able to design and fabricate interfaces with atomic precision since it is the atomic details at the interfaces that usually play a critical role in the properties and functionalities of the interface and assembled or integrated systems. This may require synergetic multi- and inter- disciplinary approaches involving collaborative and collective efforts from scientists and engineers in different areas and fields.

Acknowledgements

This work is partially supported by the BES Division of the US Department of Energy, National Science Foundation, NASA-UARC, UC-MEXUS, and UCSC (JZZ). The work at Lawrence Livermore National Laboratory was performed under the auspices of the University of California under DOE contract No. W-7405-Eng-48 (CDG).

References

1.   A.L. Efros and A.L. Efros, Fizika i Tekhnika Poluprovodnikov 16, 1209 (1982).
2.   L.E. Brus, J. Chem. Phys. 80, 4403 (1984).

3. C.B. Murray, C.R. Kagan and M.G. Bawendi, Ann. Rev. Materials Science 30, 545 (2000).
4. N. Zaitseva, Z.R. Dai, F.R. Leon and D. Krol, J. Am. Chem. Soc. 127, 10221 (2005).
5. C. Burda, X.B. Chen, R. Narayanan and M.A. El-Sayed, Chem. Rev. 105, 1025 (2005).
6. J.Z. Zhang, Acc. Chem. Res. 30, 423 (1997).
7. J.Z. Zhang, J. Phys. Chem. B 104, 7239 (2000).
8. C.B. Murray, D.J. Norris and M.G. Bawendi, J. Am. Chem. Soc. 115, 8706 (1993).
9. C.B. Murray, M. Nirmal, D.J. Norris and M.G. Bawendi, Z Phys D Atom Mol Cl 26, S231 (1993).
10. A.P. Alivisatos, Science 271, 933 (1996).
11. A.A. Guzelian, J.E.B. Katari, A.V. Kadavanich, U. Banin, K. Hamad, E. Juban, A.P. Alivisatos, R.H. Wolters, C.C. Arnold and J.R. Heath, J. Phys. Chem. 100, 7212 (1996).
12. X.G. Peng, M.C. Schlamp, A.V. Kadavanich and A.P. Alivisatos, J. Am. Chem. Soc. 119, 7019 (1997).
13. X.G. Peng, Chem-Eur J 8, 335 (2002).
14. J. Joo, H.B. Na, T. Yu, J.H. Yu, Y.W. Kim, F. Wu, J.Z. Zhang and T. Hyeon, J. Am. Chem. Soc. 125, 11100 (2003).
15. A.Y. Nazzal, X.Y. Wang, L.H. Qu, W. Yu, Y.J. Wang, X.G. Peng and M. Xiao, J. Phys. Chem. B 108, 5507 (2004).
16. L. Manna, E.C. Scher and A.P. Alivisatos, J. Am. Chem. Soc. 122, 12700 (2000).
17. J. Wang, M. Nozaki, M. Lachab, R.S.Q. Fareed, Y. Ishikawa, T. Wang, Y. Naoi and S. Sakai, J Cryst Growth 200, 85 (1999).
18. S.Y. Lu and S.W. Chen, J. Am. Ceram. Soc. 83, 709 (2000).
19. J.W. Wang, Y.P. Sun, Z.H. Liang, H.P. Xu, C.M. Fan and X.M. Chen, Rare Metal Mat Eng 33, 478 (2004).
20. Y.J. Li, P.B. Shi, R. Duan, B.R. Zhang, Y.P. Qiao, G.G. Qin and L. Huang, J. Infrared Millimeter Waves 23, 176 (2004).
21. M. Sacilotti, J. Decobert, H. Sik, G. Post, C. Dumas, P. Viste and G. Patriarche, J. Cryst. Growth 272, 198 (2004).
22. H.T. Lin, J.L. Huang, S.C. Wang and C.F. Lin, J. Alloy Compd. 417, 214 (2006).
23. D.Y. He, X.Q. Wang, Q. Chen, J.S. Li, M. Yin, A.V. Karabutov and A.G. Kazanskii, Chinese Sci. Bull. 51, 510 (2006).
24. Z.Z. Ye, J.Y. Huang, W.Z. Xu, J. Zhou and Z.L. Wang, Solid State Commun. 141, 464 (2007).
25. J.Y. Wang and T. Ito, Diamond Related Materials 16, 589 (2007).
26. R.L. Barns and W.C. Ellis, J. Appl. Phys. 36, 2296 (1965).
27. R.S. Wagner and W.C. Ellis, Transa. Metallurgical Soc. Aime 233, 1053 (1965).
28. J.P. Sitarik and W.C. Ellis, J. Appl. Phys. 37, 2399 (1966).
29. R.S. Wagner and C.J. Doherty, J. Electrochem. Soc. 115, 93 (1968).

30. P.X. Gao and Z.L. Wang, J. Phys. Chem. B 108, 7534 (2004).
31. Y. Ding, P.X. Gao and Z.L. Wang, J. Am. Chem. Soc. 126, 2066 (2004).
32. F.M. Kolb, H. Hofmeister, R. Scholz, M. Zacharias, U. Gosele, D.D. Ma and S.T. Lee, J. Electrochem. Soc. 151, G472 (2004).
33. M.C. Yang, J. Shieh, T.S. Ko, H.L. Chen and T.C. Chu, Jpn. J. Appl. Phys. 1 44, 5791 (2005).
34. J.L. Taraci, T. Clement, J.W. Dailey, J. Drucker and S.T. Picraux, Nuclear Instru. Meth. Phys. Res. B 242, 205 (2006).
35. A.M. Morales and C.M. Lieber, Science 279, 208 (1998).
36. Z.W. Pan, Z.R. Dai, C. Ma and Z.L. Wang, J. Am. Chem. Soc. 124, 1817 (2002).
37. J.R. Morber, Y. Ding, M.S. Haluska, Y. Li, P. Liu, Z.L. Wang and R.L. Snyder, J. Phys. Chem. B 110, 21672 (2006).
38. X.F. Duan and C.M. Lieber, J. Am. Chem. Soc. 122, 188 (2000).
39. Y.Q. Chen, K. Zhang, B. Miao, B. Wang and J.G. Hou, Chem. Phys. Lett. 358, 396 (2002).
40. J.C. Johnson, H.J. Choi, K.P. Knutsen, R.D. Schaller, P.D. Yang and R.J. Saykally, Nat. Mater. 1, 106 (2002).
41. C.Y. Meng, B.L. Shih and S.C. Lee, J. Nanoparticle Research 9, 657 (2007).
42. M. Gentili, C. Giovannella and S. Selci, eds. Nanolithography: A Borderland between STM, EB, IB, and X-Ray Lithographies, Kluwer Academic Publishers, New York, 1994.
43. J.Z. Zhang, Z.L. Wang, J. Liu, S. Chen and G.-Y. Liu, Self-assembled Nanostructures. Nanoscale Science and Technology, New York, Kluwer Academic/Plenum Publishers, 2003.
44. G. Cao, Nanostructures & Nanomaterials: Synthesis, Properties & Applications, London, Imperial College Press, 2004.
45. W. Chen, J.Z. Zhang and A. Joly, J. Nanosci.Nanotech. 4, 919 (2004).
46. J.D. Bryan and D.R. Gamelin, Prog. Inorg. Chem. 54, 47 (2005).
47. G.Y. Liu, S. Xu and Y.L. Qian, Acc. Chem. Res. 33, 457 (2000).
48. Z.L. Wang, J. Phys. Chem. B 104, 1153 (2000).
49. Z.L. Wang, P. Poncharal and W.A. de Heer, Microscopy and Microanalysis 6, 224 (2000).
50. R.L. Whetten, J.T. Khoury, M.M. Alvarez, S. Murthy, I. Vezmar, Z.L. Wang, P.W. Stephens, C.L. Cleveland, W.D. Luedtke and U. Landman, Advan. Mater. 8, 428 (1996).
51. C.B. Murray, C.R. Kagan and M.G. Bawendi, Science 270, 1335 (1995).
52. G. Binnig, C.F. Quate and C. Gerber, Phys. Rev. Lett. 56, 930 (1986).
53. G. Binnig, H. Rohrer, C. Gerber and E. Weibel, Appl. Phys. Lett. 40, 178 (1982).
54. H.J. Guntherodt, D. Anselmetti and E. Meyer, eds. Forces in scanning probe methods, NATO ASI Series. Series E, Applied sciences. Vol. 286, Kluwer Academic, 1995.

55. J.I. Goldstein, D.E. Newbury, P. Echlin, D.C. Joy, J. Romig, A.D., C.E. Lyman, C. Fiori and E. Lifshin, Scanning Electron Microscopy and X-ray Microanalysis. 2nd ed, New York, Plenum Press, 1992.

56. Z.L. Wang, ed. Characterization of Nanophase Materials. Wiley-VCH, New York, 2000.

57. Z.L. Wang, Advan. Mater. 10, 13 (1998).

58. T.L. Barr, Modern ESCA: The Principles and Practice of X-Ray Photoelectron Spectroscopy, New York, CRC Press, 1994.

59. D.C. Koningsberger and R. Prins, eds. X-Ray Absorption: Principles, Applications, Techniques of EXAFS, SEXAFS and XANES, Wiley-Interscience, New York, 1988.

60. F. Wu, J.Z. Zhang, R. Kho and R.K. Mehra, Chem. Phys. Lett. 330, 237 (2000).

61. H. Dollefeld, H. Weller and A. Eychmuller, Nano Letters 1, 267 (2001).

62. A.M. Malvezzi, M. Patrini, A. Stella, P. Tognini, P. Cheyssac and R. Kofman, Eur. Phys. J. D 16, 321 (2001).

63. A. Stella, S. Achilli, M. Allione, A.M. Malvezzi, M. Patrini and R. Kofman, Microelectron J. 34, 619 (2003).

64. Z. Gui, X. Wang, J. Liu, S.S. Yan, Y.Y. Ding, Z.Z. Wang and Y. Hu, J. Solid State Chem. 179, 1984 (2006).

65. C.Y. Wang, H. Groenzin and M.J. Shultz, Langmuir 19, 7330 (2003).

66. C.Y. Wang, H. Groenzin and M.J. Shultz, J. Phys. Chem. B 108, 265 (2004).

67. K.C. Jena, P.B. Bisht, M.M. Shaijumon and S. Ramaprabhu, Opt Commun. 273, 153 (2007).

68. M. Gratzel, Heterogeneous Photochemical Electron Transfer, Boca Raton, CRC Press, 1989.

69. V.L. Colvin, A.N. Goldstein and A.P. Alivisatos, J. Am. Chem. Soc. 114, 5221 (1992).

70. D. Duonghong, J.J. Ramsden and M. Gratzel, J. Am. Chem. Soc. 104, 2977 (1982).

71. J.Z. Zhang, R.H. O'Neil and T.W. Roberti, J. Phys. Chem. 98, 3859 (1994).

72. Z. Adam and X. Peng, J. Am. Chem. Soc. 123, 183 (2001).

73. A. Wolcott, D. Gerion, M. Visconte, J. Sun, A. Schwartzberg, S.W. Chen and J.Z. Zhang, J. Phys. Chem. B 110, 5779 (2006).

74. R.A. Bley, S.M. Kauzlarich, J.E. Davis and H.W.H. Lee, Chem. Mater. 8, 1881 (1996).

75. T. Takagahara, Phys. Rev. B. 36, 9293 (1987).

76. Y. Kayanuma, Phys. Rev. B. 38, 9797 (1988).

77. W.W. Yu, L.H. Qu, W.Z. Guo and X.G. Peng, Chem. Mater. 15, 2854 (2003).

78. B. Valeur, Molecular Fluorescence: Principles and Applications, New York, Wiley-VCH, 2001.

79. L.R. Becerra, C.B. Murray, R.G. Griffin and M.G. Bawendi, J. Chem. Phys. 100, 3297 (1994).

80. J.E.B. Katari, V.L. Colvin and A.P. Alivisatos, J. Phys. Chem. 98, 4109 (1994).

81. L. Spanhel, M. Haase, H. Weller and A. Henglein, J. Am. Chem. Soc. 109, 5649 (1987).

82. P.V. Kamat and N.M. Dimitrijevic, J. Phys. Chem. 93, 4259 (1989).

83. P.V. Kamat, M.D. Vanwijngaarden and S. Hotchandani, Israel J. Chem. 33, 47 (1993).

84. C. Luangdilok and D. Meisel, Israel J. Chem. 33, 53 (1993).

85. M.Y. Gao, S. Kirstein, H. Mohwald, A.L. Rogach, A. Kornowski, A. Eychmuller and H. Weller, J. Phys. Chem. B 102, 8360 (1998).

86. T.W. Roberti, N.J. Cherepy and J.Z. Zhang, J. Chem. Phys. 108, 2143 (1998).

87. M.A. Hines and P. Guyot-Sionnest, J. Phys. Chem. 100, 468 (1996).

88. M.C. Schlamp, X.G. Peng and A.P. Alivisatos, J. Appl. Phys. 82, 5837 (1997).

89. P. Palinginis and W. Hailin, Appl. Phys. Lett. 78, 1541 (2001).

90. M.A. Hines and P. Guyot-Sionnest, J. Phys. Chem. B 102, 3655 (1998).

91. A.A. Patel, F.X. Wu, J.Z. Zhang, C.L. Torres-Martinez, R.K. Mehra, Y. Yang and S.H. Risbud, J. Phys. Chem. B 104, 11598 (2000).

92. A. Sengupta, B. Jiang, K.C. Mandal and J.Z. Zhang, J. Phys. Chem. B 103, 3128 (1999).

93. M.C. Brelle, C.L. Torres-Martinez, J.C. McNulty, R.K. Mehra and J.Z. Zhang, Pure Appl. Chem. 72, 101 (2000).

94. M.C. Brelle, J.Z. Zhang, L. Nguyen and R.K. Mehra, J. Phys. Chem. A 103, 10194 (1999).

95. S.R. Cordero, P.J. Carson, R.A. Estabrook, G.F. Strouse and S.K. Buratto, J. Phys. Chem. B 104, 12137 (2000).

96. X.H. Gao, W.C.W. Chan and S.M. Nie, J. Biomed. Optics 7, 532 (2002).

97. D. Gerion, F. Pinaud, S.C. Williams, W.J. Parak, D. Zanchet, S. Weiss and A.P. Alivisatos, J. Phys. Chem. B 105, 8861 (2001).

98. X.H. Gao and S.M. Nie, Analyt. Chem. 76, 2406 (2004).

99. N.M. Dimitrijevic and P.V. Kamat, J. Phys. Chem. 91, 2096 (1987).

100. Y. Wang, A. Suna, J. McHugh, E.F. Hilinski, P.A. Lucas and R.D. Johnson, J. Chem. Phys. 92, 6927 (1990).

101. T. Vossmeyer, L. Katsikas, M. Giersig, I.G. Popovic, K. Diesner, A. Chemseddine, A. Eychmuller and H. Weller, J. Phys. Chem. 98, 7665 (1994).

102. A. Henglein, A. Kumar, E. Janata and H. Weller, Chem. Phys. Lett. 132, 133 (1986).

103. K.I. Kang, A.D. Kepner, S.V. Gaponenko, S.W. Koch, Y.Z. Hu and N. Peyghambarian, Phys. Rev. B 48, 15449 (1993).

104. V. Klimov, S. Hunsche and H. Kurz, Phys. Rev. B 50, 8110 (1994).

105. V. Dneprovskii, N. Gushina, O. Pavlov, V. Poborchii, I. Salamatina and E. Zhukov, Physics Letters A 204, 59 (1995).

106. N.V. Gushchina, V.S. Dneprovskii, E.A. Zhukov, O.V. Pavlov, V.V. Poborchii and I.A. Salamatina, Jetp. Lett.-Engl. Tr. 61, 507 (1995).

107. V. Dneprovskii, A. Eev, N. Gushina, D. Okorokov, V. Panov, V. Karavanskii, A. Maslov, V. Sokolov and E. Dovidenko, Phys. Status Solidi B 188, 297 (1995).

108. V. Dneprovskii, N. Gushina, D. Okorokov, V. Karavanskii and E. Dovidenko, Superlattice Microst. 17, 41 (1995).

109. F.Z. Henari, K. Morgenstern, W.J. Blau, V.A. Karavanskii and V.S. Dneprovskii, Appl. Phys. Lett. 67, 323 (1995).

110. B.L. Yu, C.S. Zhu and F.X. Gan, J. Appl. Phys. 82, 4532 (1997).

111. Y. Wang, Acc. Chem. Res. 24, 133 (1991).

112. T. Dannhauser, M. O'Neil, K. Johanseon, D. Whitter and G. McLendon, J. Phys. Chem. 90, 6074 (1986).

113. X.C. Ai, L. Guo, Y.H. Zou, Q.S. Li and H.S. Zhu, Mater. Lett. 38, 131 (1999).

114. Y.R. Shen, The Principles of Nonlinear Optics, New York, J. Wiley, 1984.

115. T. Yamaki, K. Asai, K. Ishigure, K. Sano and K. Ema, Synthet. Metal 103, 2690 (1999).

116. M. Jacobsohn and U. Banin, J. Phys. Chem. B 104, 1 (2000).

117. Z. Yu, F. Dgang, W. Xin, L. Juzheng and L. Zuhong, Colloids Surfaces A 181, 145 (2001).

118. J. Lenglet, A. Bourdon, J.C. Bacri, R. Perzynski and G. Demouchy, Phys. Rev. B 53, 14941 (1996).

119. T. Takagahara, Solid State Commun. 78, 279 (1991).

120. T. Takagahara, Surface Sci. 267, 310 (1992).

121. Y. Kayanuma, J. Phys. Soc. Japan 62, 346 (1993).

122. B. Clerjaud, F. Gendron and C. Porte, Appl. Phys. Lett. 38, 212 (1981).

123. Y. Mita, in Phosphor Handbook, S.a.Y. Shionoya, W.M., Editor, CRC Press, New York, 1999.

124. L.G. Quagliano and H. Nather, Appl. Phys. Lett. 45, 555 (1984).

125. E.J. Johnson, J. Kafalas, R.W. Davies and W.A. Dyes, Appl. Phys. Lett. 40, 993 (1982).

126. E. Poles, D.C. Selmarten, O.I. Micic and A.J. Nozik, Appl. Phys. Lett. 75, 971 (1999)

127. Y.H. Cho, D.S. Kim, B.D. Choe, H. Lim, J.I. Lee and D. Kim, Phys. Rev. B 56, R4375 (1997).

128. M. Potemski, R. Stepniewski, J.C. Maan, G. Martinez, P. Wyder and B. Etienne, Phys. Rev. Lett. 66, 2239 (1991).

129. P. Vagos, P. Boucaud, F.H. Julien, J.M. Lourtioz and R. Planel, Phys. Rev. Lett. 70, 1018 (1993).

130. W. Seidel, A. Titkov, J.P. Andre, P. Voisin and M. Voos, Phys, Rev, Lett, 73, 2356 (1994).

131. F. Driessen, H.M. Cheong, A. Mascarenhas, S.K. Deb, P.R. Hageman, G.J. Bauhuis and L.J. Giling, Phys. Rev. B 54, R5263 (1996).

132. R. Hellmann, A. Euteneuer, S.G. Hense, J. Feldmann, P. Thomas, E.O. Gobel, D.R. Yakovlev, A. Waag and G. Landwehr, Phys. Rev. B 51, 18053 (1995).

133. Z.P. Su, K.L. Teo, P.Y. Yu and K. Uchida, Solid State Commun. 99, 933 (1996).

134. J. Zeman, G. Martinez, P.Y. Yu and K. Uchida, Phys. Rev. B 55, 13428 (1997).

135. L. Schrottke, H.T. Grahn and K. Fujiwara, Phys. Rev. B 56, 15553 (1997).

136. H.M. Cheong, B. Fluegel, M.C. Hanna and A. Mascarenhas, Phys. Rev. B 58, R4254 (1998).

137. Z. Chine, B. Piriou, M. Oueslati, T. Boufaden and B. El Jani, J. Luminescence 82, 81 (1999).

138. T. Kita, T. Nishino, C. Geng, F. Scholz and H. Schweizer, Phys. Rev. B 59, 15358 (1999).

139. S.C. Hohng and D.S. Kim, Appl. Phys. Lett. 75, 3620 (1999).

140. L. Schrottke, R. Hey and H.T. Grahn, Phys. Rev. B 60, 16635 (1999).

141. W. Heimbrodt, M. Happ and F. Henneberger, Phys. Rev. B 60, R16326 (1999).

142. T. Kita, T. Nishino, C. Geng, F. Scholz and H. Schweizer, J. Luminescence 87-9, 269 (2000).

143. A. Satake, Y. Masumoto, T. Miyajima, T. Asatsuma and T. Hino, Phys. Rev. B 61, 12654 (2000).

144. G.G. Zegrya and V.A. Kharchenko, Zhurnal Eksperimentalnoi I Teoreticheskoi Fiziki 101, 327 (1992).

145. S.A. Blanton, M.A. Hines, M.E. Schmidt and P. Guyotsionnest, J. Luminescence 70, 253 (1996).

146. I.V. Ignatiev, I.E. Kozin, H.W. Ren, S. Sugou and Y. Matsumoto, Phys. Rev. B 60, R14001 (1999).

147. P.P. Paskov, P.O. Holtz, B. Monemar, J.M. Garcia, W.V. Schoenfeld and P.M. Petroff, Appl. Phys. Lett. 77, 812 (2000).

148. H.X. Zhang, C.H. Kam, Y. Zhou, X.Q. Han, S. Buddhudu and Y.L. Lam, Opt. Mater. 15, 47 (2000).

149. W. Chen, A.G. Joly and J.Z. Zhang, Phys Rev B 6404, 1202 (2001).

150. A.G. Joly, W. Chen, J. Roark and J.Z. Zhang, J. Nanosci. Nanotech. 1, 295 (2001).

151. J.J. Shiang, A.N. Goldstein and A.P. Alivisatos, J. Chem. Phys. 92, 3232 (1990).

152. G. Scamarcio, M. Lugara and D. Manno, Phys. Rev. B 45, 13792 (1992).

153. J.J. Shiang, S.H. Risbud and A.P. Alivisatos, J. Chem. Phys. 98, 8432 (1993).

154. M. Abdulkhadar and B. Thomas, Nanostruct. Mater. 5, 289 (1995).

155. A.A. Sirenko, V.I. Belitsky, T. Ruf, M. Cardona, A.I. Ekimov and C. TralleroGiner, Phys. Rev. B 58, 2077 (1998).

156. V.G. Melehin and V.D. Petrikov, Physics of Low-Dimensional Structures 9-10, 73 (1999).

157. A.P. Alivisatos, T.D. Harris, P.J. Carroll, M.L. Steigerwald and L.E. Brus, J. Chem. Phys. 90, 3463 (1989).

158. J.J. Shiang, I.M. Craig and A.P. Alivisatos, Z. Phys. D Atom Mol. Cl. 26, 358 (1993).

159. V. Spagnolo, G. Scamarcio, M. Lugara and G.C. Righini, Superlattice Microst. 16, 51 (1994).

160. J.J. Shiang, R.H. Wolters and J.R. Heath, J. Chem. Phys. 106, 8981 (1997).
161. V.A. Volodin, M.D. Efremov, V.A. Gritsenko and S.A. Kochubei, Appl. Phys. Lett. 73, 1212 (1998).
162. M.D. Efremov, V.V. Bolotov, V.A. Volodin and S.A. Kochubei, Solid State Commun. 108, 645 (1998).
163. W.F.A. Besling, A. Goossens and J. Schoonman, J. Phys. IV. 9, 545 (1999).
164. G.H. Li, K. Ding, Y. Chen, H.X. Han and Z.P. Wang, J. Appl. Phys. 88, 1439 (2000).
165. J.R. Heath, J.J. Shiang and A.P. Alivisatos, J. Chem. Phys. 101, 1607 (1994).
166. Y.Y. Wang, Y.H. Yang, Y.P. Guo, J.S. Yue and R.J. Gan, Mater. Lett. 29, 159 (1996).
167. W.K. Choi, V. Ng, S.P. Ng, H.H. Thio, Z.X. Shen and W.S. Li, J. Appl. Phys. 86, 1398 (1999).
168. A.V. Kolobov, Y. Maeda and K. Tanaka, J. Appl. Phys. 88, 3285 (2000).
169. Y.W. Ho, V. Ng, W.K. Choi, S.P. Ng, T. Osipowicz, H.L. Seng, W.W. Tjui and K. Li, Scripta Materialia 44, 1291 (2001).
170. U. Pal and J.G. Serrano, Appl. Surf. Sci. 246, 23 (2005).
171. Y. Batra, D. Kabiraj and D. Kanjilal, Eur. Phys. J.-Appl. Phys. 38, 27 (2007).
172. R. Rinaldi, R. Cingolani, M. Ferrara, A.C. Maciel, J. Ryan, U. Marti, D. Martin, F. Moriergemoud and F.K. Reinhart, Appl. Phys. Lett. 64, 3587 (1994).
173. B. Schreder, A. Materny, W. Kiefer, G. Bacher, A. Forchel and G. Landwehr, J. Raman Spectroscopy 31, 959 (2000).
174. Y. Bae, N. Myung and A.J. Bard, Nano Letters 4, 1153 (2004).
175. Z.P. Wang, J. Li, B. Liu, J.Q. Hu, X. Yao and J.H. Li, J. Phys. Chem. B 109, 23304 (2005).
176. R.K. Ligman, L. Mangolini, U.R. Kortshagen and S.A. Campbell, Appl. Phys. Lett. 90, 061116 (2007).
177. C.Y. Lee, Y.T. Haung, W.F. Su and C.F. Lin, Appl. Phys. Lett. 89, 231116 (2006).
178. Y. Xuan, D.C. Pan, N. Zhao, X.L. Ji and D.G. Ma, Nanotechnology 17, 4966 (2006).
179. D.F. Underwood, T. Kippeny and S.J. Rosenthal, J. Phys. Chem. B 105, 436 (2001).
180. J.J. Cavaleri, D.E. Skinner, D.P. Colombo and R.M. Bowman, J. Chem. Phys. 103, 5378 (1995).
181. D.E. Skinner, D.P. Colombo, J.J. Cavaleri and R.M. Bowman, J. Phys. Chem. 99, 7853 (1995).
182. V.I. Klimov, A.A. Mikhailovsky, D.W. McBranch, C.A. Leatherdale and M.G. Bawendi, Science 287, 1011 (2000).
183. F.X. Wu, J.H. Yu, J. Joo, T. Hyeon and J.Z. Zhang, Opt. Mater. 29, 858 (2007).
184. J.M. Luther, M.C. Beard, Q. Song, M. Law, R.J. Ellingson and A.J. Nozik, Nano Letters 7, 1779 (2007).
185. A. Luque, A. Marti and A.J. Nozik, Mrs Bulletin 32, 236 (2007).
186. G. Allan and C. Delerue, Phys. Rev. B 73, 205423 (2006).

187. J.E. Murphy, M.C. Beard, A.G. Norman, S.P. Ahrenkiel, J.C. Johnson, P.R. Yu, O.I. Micic, R.J. Ellingson and A.J. Nozik, J. Am. Chem. Soc. 128, 3241 (2006).

188. R.J. Ellingson, M.C. Beard, J.C. Johnson, P.R. Yu, O.I. Micic, A.J. Nozik, A. Shabaev and A.L. Efros, Nano Letters 5, 865 (2005).

189. V.I. Klimov, J. Phys. Chem. B 110, 16827 (2006).

190. B. Oregan and M. Gratzel, Nature 353, 737 (1991).

191. N.J. Cherepy, G.P. Smestad, M. Gratzel and J.Z. Zhang, J. Phys. Chem. B 101, 9342 (1997).

192. O. Khaselev and J.A. Turner, Science 280, 425 (1998).

193. J.A. Turner, Science 285, 687 (1999).

194. C.A. Parsons, M.W. Peterson, B.R. Thacker, J.A. Turner and A.J. Nozik, J. Phys. Chem. 94, 3381 (1990).

195. M. Gratzel, J. Photochem. Photobiol. C Photochem. Rev. 4, 145 (2003).

196. N.A. Anderson and T. Lian, Ann. Rev. Phys. Chem. 56, 491 (2005).

197. S. Logunov, T. Green, S. Marguet and M.A. El-Sayed, J. Phys. Chem. A 102, 5652 (1998).

198. C. Burda, T.C. Green, S. Link and M.A. El-Sayed, J. Phys. Chem. B 103, 1783 (1999).

199. W.Q. Peng, S.C. Qu, G.W. Cong, X.Q. Zhang and Z.G. Wang, J. Cryst. Growth 282, 179 (2005).

200. N. Pradhan and X.G. Peng, J. Am. Chem. Soc. 129, 3339 (2007).

201. G. Counio, T. Gacoin and J.P. Boilot, J. Phys. Chem. B 102, 5257 (1998).

202. J.Q. Yu, H.M. Liu, Y.Y. Wang, F.E. Fernandez and W.Y. Jia, J. Luminescence 76-7, 252 (1998).

203. A.D. Dinsmore, D.S. Hsu, H.F. Gray, S.B. Qadri, Y. Tian and B.R. Ratna, Appl. Phys. Lett. 75, 802 (1999).

204. W. Chen, R. Sammynaiken and Y.N. Huang, J. Appl. Phys. 88, 5188 (2000).

205. W. Chen, A.G. Joly and J.Z. Zhang, Phys. Rev. B 64, 41202 (2001).

206. M. Konishi, T. Isobe and M. Senna, J. Luminescence 93, 1 (2001).

207. K. Yan, C.K. Duan, Y. Ma, S.D. Xia and J.C. Krupa, Phys. Rev. B 58, 13585 (1998).

208. T.J. Norman, D. Magana, T. Wilson, C. Burns, J.Z. Zhang, D. Cao and F. Bridges, J. Phys. Chem. B 107, 6309 (2003).

209. A.C. Tuan, J.D. Bryan, A.B. Pakhomov, V. Shutthanandan, S. Thevuthasan, D.E. McCready, D. Gaspar, M.H. Engelhard, J.W. Rogers, K. Krishnan, D.R. Gamelin and S.A. Chambers, Phys. Rev. B 70, 054424 (2004).

210. P.V. Radovanovic, N.S. Norberg, K.E. McNally and D.R. Gamelin, J. Am. Chem. Soc. 124, 15192 (2002).

211. R. Viswanatha, S. Sapra, S. Sen Gupta, B. Satpati, P.V. Satyam, B.N. Dev and D.D. Sarma, J. Phys. Chem. B 108, 6303 (2004).

212. R. Viswanatha, S. Chakraborty, S. Basu and D.D. Sarma, J. Phys. Chem. B 110, 22310 (2006).

213. R. Asahi, T. Morikawa, T. Ohwaki, K. Aoki and Y. Taga, Science 293, 269 (2001).
214. C. Burda, Y.B. Lou, X.B. Chen, A.C.S. Samia, J. Stout and J.L. Gole, Nano Letters 3, 1049 (2003).
215. L.H. Huang, Z.X. Sun and Y.L. Liu, J. Ceramic Soc. Japan 115, 28 (2007).
216. J.W. Wang, W. Zhu, Y.Q. Zhang and S.X. Liu, J. Phys. Chem. C 111, 1010 (2007).
217. M. Sathish, B. Viswanathan, R.P. Viswanath and C.S. Gopinath, Chem. Mater. 17, 6349 (2005).
218. S. Moribe, T. Ikoma, K. Akiyama, Q.W. Zhang, F. Saito and S. Tero-Kubota, Chem. Phys. Lett. 436, 373 (2007).
219. H.Y. Chen, A. Nambu, W. Wen, J. Graciani, Z. Zhong, J.C. Hanson, E. Fujita and J.A. Rodriguez, J.Phys. Chem. C 111, 1366 (2007).
220. S. Yin, Y. Aita, M. Komatsu and T. Sato, J. European Ceramic Soc. 26, 2735 (2006).
221. R.N. Bhargava, D. Gallagher, X. Hong and A. Nurmikko, Phys. Rev. Lett. 72, 416 (1994).
222. R.N. Bhargava, D. Gallagher and T. Welker, J.Luminescence 60-1, 275 (1994).
223. R.N. Bhargava, J. Luminescence 70, 85 (1996).
224. H.E. Gumlich, J. Luminescence 23, 73 (1981).
225. A.A. Bol and A. Meijerink, Phys. Rev. B 58, R15997 (1998).
226. N. Murase, R. Jagannathan, Y. Kanematsu, M. Watanabe, A. Kurita, K. Hirata, T. Yazawa and T. Kushida, J. Phys. Chem. B 103, 754 (1999).
227. B.A. Smith, J.Z. Zhang, A. Joly and J. Liu, Phys. Rev. B 62, 2021 (2000).
228. J.H. Chung, C.S. Ah and D.-J. Jang, J. Phys. Chem. B 105, 4128 (2001).
229. T. Toyama, D. Adachi, M. Fujii, Y. Nakano and H. Okamoto, J. Non-Crystalline Solids 299, 1111 (2002).
230. E.M. Olano, C.D. Grant, T.J. Norman, E.W. Castner and J.Z. Zhang, J. Nanosci. Nanotech. 5, 1492 (2005).
231. A. Wolcott, T.R. Kuykendall, W. Chen, S.W. Chen and J.Z. Zhang, J. Phys. Chem. B 110, 25288 (2006).
232. S. Uchida, R. Chiba, M. Tomiha, N. Masaki and M. Shirai, Electrochem. 70, 418 (2002).
233. W.U. Huynh, J.J. Dittmer, W.C. Libby, G.L. Whiting and A.P. Alivisatos, Adv. Funct. Mater. 13, 73 (2003).
234. P. Gould, Materials Today 9, 18 (2006).
235. M. Law, L.E. Greene, A. Radenovic, T. Kuykendall, J. Liphardt and P.D. Yang, J. Phys. Chem. B 110, 22652 (2006).
236. K.S. Kim, Y.S. Kang, J.H. Lee, Y.J. Shin, N.G. Park, K.S. Ryu and S.H. Chang, B Kor. Chem. Soc. 27, 295 (2006).
237. Y.B. Liu, B.X. Zhou, B.T. Xiong, J. Bai and L.H. Li, Chinese Sci. Bull. 52, 1585 (2007).
238. T.Y. Lee, P.S. Alegaonkar and J.B. Yoo, Thin Solid Films 515, 5131 (2007).

239. E. Joanni, R. Savu, M.D. Goes, P.R. Bueno, J.N. de Freitas, A.F. Nogueira, E. Longo and J.A. Varela, Scripta Materialia 57, 277 (2007).
240. K. Robbie, J.C. Sit and M.J. Brett, J. Vacuum Science Technology B 16, 1115 (1998).
241. Y.P. Zhao, D.X. Ye, G.C. Wang and T.M. Lu, Nano Letters 2, 351 (2002).
242. W.W. Dunn, Y. Aikawa and BardmA.J., J. Am. Chem. Soc. 6893 (1981).
243. H. Harada, T. Ueda and T. Sakata, J. Phys. Chem. 93, 1542 (1989).
244. L.F. Lin and R.R. Kuntz, Langmuir 8, 870 (1992).
245. P. Zuman and Z. Fijalek, J. Electroanal. Chem. Interfacial Electrochem. 296, 583 (1990).
246. H.Y. Wang, R.E. Partch and Y.Z. Li, J. Org. Chem. 62, 5222 (1997).
247. I. Willner, F. Patolsky and J. Wasserman, Angew. Chem. Int. Edit 40, 1861 (2001).
248. C. Nasr, P.V. Kamat and S. Hotchandani, J. Electroanal. Chem. 420, 201 (1997).
249. S.G. Hickey, D.J. Riley and E.J. Tull, J. Phys. Chem. B 104, 7623 (2000).
250. G. Hodes, Israel Journal of Chemistry 33, 95 (1993).
251. K. Keis, L. Vayssieres, H. Rensmo, S.E. Lindquist and A. Hagfeldt, J. Electrochem. Soc.y 148, A149 (2001).
252. L. Zhang, M.Z. Yang, E.Q. Gao, X.B. Qiao, Y.Z. Hao, Y.Q. Wang, S.M. Cai, F.S. Meng and H. Tian, Chem. J. Chinese Universities-Chinese 21, 1543 (2000).
253. Y. Ren, Z. Zhang, E. Gao, S. Fang and S. Cai, J. Appl. Electrochem. 31, 445 (2001).
254. X.M. Qian, D.Q. Qin, Q. Song, Y.B. Bai, T.J. Li, X.Y. Tang, E.K. Wang and S.J. Dong, Thin Solid Films 385, 152 (2001).
255. M.S. Liu, M.Z. Yang, Y.Z. Hao, S.M. Cai and Y.F. Li, Acta Chim. Sinica 59, 377 (2001).
256. T. Yoshida, K. Terada, D. Schlettwein, T. Oekermann, T. Sugiura and H. Minoura, Advan. Mater. 12, 1214 (2000).
257. J.F. Suyver, R. Bakker, A. Meijerink and J.J. Kelly, Phys. Status Solidi B 224, 307 (2001).
258. C. Santato, M. Ulmann and J. Augustynski, J. Phys. Chem. B 105, 936 (2001).
259. K. Vinodgopal, I. Bedja and P.V. Kamat, Chem. Mater. 8, 2180 (1996).
260. S.K. Poznyak, D.V. Talapin and A.I. Kulak, J. Phys. Chem. B 105, 4816 (2001).
261. M.A. Fox and M.T. Dulay, Chem. Rev. 93, 341 (1993).
262. W.Y. Choi, A. Termin and M.R. Hoffmann, J. Phys. Chem. 98, 13669 (1994).
263. W.Y. Choi, A. Termin and M.R. Hoffmann, Angew. Chem., Int. Ed. Engl. 33, 1091 (1994).
264. M.R. Hoffmann, S.T. Martin, W.Y. Choi and D.W. Bahnemann, Chem. Rev. 95, 69 (1995).
265. L.Q. Jing, X.J. Sun, J. Shang, W.M. Cai, Z.L. Xu, Y.G. Du and H.G. Fu, Sol, Energ, Mat, Sol, Cell 79, 133 (2003).
266. D.F. Ollis and H. Al-Ekabi, Photocatalytic purification and treatment of water and air : proceedings of the 1st International Conference on TiO₂ Photocatalytic Purification and Treatment of Water and Air, London, Ontario, Canada, 8-13

November, 1992. Trace metals in the environment ; 3, Amsterdam, New York, Elsevier, 1993.

267. A. Mills, R.H. Davies and D. Worsley, Chem. Soc. Rev. 22, 417 (1993).

268. N. Serpone and R.F. Khairutdinov, in Semiconductor Nanoclusters-Physical, Chemical, and Catalytic Aspects, P.V. Kamat and D. Meisel, Editors, Elsevier, New York, 1997.

269. K. Okamoto, Y. Yamamoto, H. Tanaka, M. Tanaka and A. Itaya, B. Chem. Soc. Japan 58, 2015 (1985).

270. C.H. Kwon, H.M. Shin, J.H. Kim, W.S. Choi and K.H. Yoon, Mate.r Chem. Phys. 86, 78 (2004).

271. Y. Nemoto and T. Hirai, B. Chem. Soc. Jpn. 77, 1033 (2004).

272. Y. Jiang, P. Zhang, Z.W. Liu and F. Xu, Mater. Chem. Phys. 99, 498 (2006).

273. Z.X. Wang, S.W. Ding and M.H. Zhang, Chinese J. Inorg. Chem. 21, 437 (2005).

274. Y.C. Lee and S. Cheng, J. Chinese Chem. Soc. 53, 1355 (2006).

275. H. Mattoussi, J.M. Mauro, E.R. Goldman, G.P. Anderson, V.C. Sundar, F.V. Mikulec and M.G. Bawendi, J. Am. Chem. Soc. 122, 12142 (2000).

276. W.C.W. Chan and S.M. Nie, Science 281, 2016 (1998).

277. M. Bruchez, M. Moronne, P. Gin, S. Weiss and A.P. Alivisatos, Science 281, 2013 (1998).

278. N. Gaponik, D.V. Talapin, A.L. Rogach, K. Hoppe, E.V. Shevchenko, A. Kornowski, A. Eychmuller and H. Weller, J. Phys. Chem. B 106, 7177 (2002).

279. B.O. Dabbousi, J. RodriguezViejo, F.V. Mikulec, J.R. Heine, H. Mattoussi, R. Ober, K.F. Jensen and M.G. Bawendi, J. Phys. Chem. B 101, 9463 (1997).

280. E.R. Goldman, E.D. Balighian, M.K. Kuno, S. Labrenz, P.T. Tran, G.P. Anderson, J.M. Mauro and H. Mattoussi, Phys. Status Solidi B 229, 407 (2002).

281. S. Kim, Y.T. Lim, E.G. Soltesz, A.M. De Grand, J. Lee, A. Nakayama, J.A. Parker, T. Mihaljevic, R.G. Laurence, D.M. Dor, L.H. Cohn, M.G. Bawendi and J.V. Frangioni, Nat. Biotechnol. 22, 93 (2004).

282. E.R. Goldman, A.R. Clapp, G.P. Anderson, H.T. Uyeda, J.M. Mauro, I.L. Medintz and II. Mattoussi, Analyt. Chem. 76, 684 (2004).

283. I.L. Medintz, A.R. Clapp, F.M. Brunel, T. Tiefenbrunn, H.T. Uyeda, E.L. Chang, J.R. Deschamps, P.E. Dawson and H. Mattoussi, Nat. Mater. 5, 581 (2006).

284. A. Agrawal, C.Y. Zhang, T. Byassee, R.A. Tripp and S.M. Nie, Analyt. Chem. 78, 1061 (2006).

285. M.H. Huang, S. Mao, H. Feick, H.Q. Yan, Y.Y. Wu, H. Kind, E. Weber, R. Russo and P.D. Yang, Science 292, 1897 (2001).

286. V.I. Klimov, S.A. Ivanov, J. Nanda, M. Achermann, I. Bezel, J.A. McGuire and A. Piryatinski, Nature 447, 441 (2007).

287. R. Mirin, A. Gossard and J. Bowers, Electronics Letters 32, 1732 (1996).

288. S. Fafard, K. Hinzer, A.J. Springthorpe, Y. Feng, J. McCaffrey, S. Charbonneau and E.M. Griswold, Mat. Sci. Eng. B-Solid 51, 114 (1998).

289. K. Hinzer, C.N. Allen, J. Lapointe, D. Picard, Z.R. Wasilewski, S. Fafard and A.J.S. Thorpe, J. Vac. Sci. Technol. A 18, 578 (2000).

290. K. Hinzer, J. Lapointe, Y. Feng, A. Delage, S. Fafard, A.J. SpringThorpe and E.M. Griswold, J. Appl. Phys. 87, 1496 (2000).

291. S. Tanaka, H. Hirayama, Y. Aoyagi, Y. Narukawa, Y. Kawakami and S. Fujita, Appl. Phys. Lett. 71, 1299 (1997).

292. M.K. Zundel, K. Eberl, N.Y. Jin-Phillipp, F. Phillipp, T. Riedl, E. Fehrenbacher and A. Hangleiter, J. Cryst. Growth. 202, 1121 (1999).

293. V.I. Klimov, A.A. Mikhailovsky, S. Xu, A. Malko, J.A. Hollingsworth, C.A. Leatherdale, H.J. Eisler and M.G. Bawendi, Science 290, 314 (2000).

294. F. Hide, B.J. Schwartz, M.A. Diazgarcia and A.J. Heeger, Chem. Phys. Lett. 256, 424 (1996).

295. S.A. Carter, J.C. Scott and P.J. Brock, Appl. Phys. Lett. 71, 1145 (1997).

296. T.S. Ahmadi, Z.L. Wang, T.C. Green, A. Henglein, and M.A. El-Sayed, Science, 1924 (1996).

297. V.L. Colvin, M.C. Schlamp and A.P. Alivisatos, Nature 370, 354 (1994).

298. N.C. Greenham, X.G. Peng and A.P. Alivisatos, Phys. Rev. B 54, 17628 (1996).

299. S. Nakamura, K. Kitamura, H. Umeya, A. Jia, M. Kobayashi, A. Yoshikawa, M. Shimotomai and K. Takahashi, Electronics Letters 34, 2435 (1998).

300. C.A. Leatherdale, C.R. Kagan, N.Y. Morgan, S.A. Empedocles, M.A. Kastner and M.G. Bawendi, Phys. Rev. B 62, 2669 (2000).

301. H. Mattoussi, A.W. Cumming, C.B. Murray, M.G. Bawendi and R. Ober, Phys. Rev. B 58, 7850 (1998).

302. L. Brus, J. Phys. Chem. 98, 3575 (1994).

303. Z.L. Wang, M.B. Mohamed, S. Link and M.A. El-Sayed, Surface Sci. 440, L809 (1999).

304. W.U. Huynh, X.G. Peng and A.P. Alivisatos, Advan. Mater. 11, 923 (1999).

305. Z.W. Pan, Z.R. Dai and Z.L. Wang, Science 291, 1947 (2001).

306. Y.A. Ono, Annual Review of Materials Science 27, 283 (1997).

307. S. Tanaka, H. Kobayashi and H. Sasakura, in *Phosphor Handbook*, S. Shionoya and W.M. Yen, Editors, CRC Press, New York, 1999.

308. K. Manzoor, S.R. Vadera, N. Kumar and T.R.N. Kutty, Appl. Phys. Lett. 84, 284 (2004).

309. M. Warkentin, F. Bridges, S.A. Carter and M. Anderson, Phys. Rev. B 75, 075301 (2007).

310. K. Manzoor, V. Aditya, S.R. Vadera, N. Kumar and T.R.N. Kutty, Solid State Commun. 135, 16 (2005).

311. D.J. Norris and M.G. Bawendi, J. Chem. Phys. 103, 5260 (1995).

312. D.J. Norris and M.G. Bawendi, Phys. Rev. B 53, 16338 (1996).

313. A.P. Alivisatos, A.L. Harris, N.J. Levinos, M.L. Steigerwald and L.E. Brus, J. Chem. Phys. 89, 4001 (1988).

314. D.M. Mittleman, R.W. Schoenlein, J.J. Shiang, V.L. Colvin, A.P. Alivisatos and C.V. Shank, Phys. Rev. B 49, 14435 (1994).

315. U. Woggon, S. Gaponenko, W. Langbein, A. Uhrig and C. Klingshirn, Phys. Rev. B 47, 3684 (1993).

316. H. Giessen, B. Fluegel, G. Mohs, N. Peyghambarian, J.R. Sprague, O.I. Micic and A.J. Nozik, Appl. Phys. Lett. 68, 304 (1996).

317. V. Jungnickel and F. Henneberger, J. Luminescence 70, 238 (1996).

318. W.E. Moerner, Science 265, 46 (1994).

319. T. Basche, W.E. Moerner, M. Orrit and U.P. Wild, eds. Single-Molecule Optical Detection, Imaging and Spectroscopy. VCH, Weinheim, Cambridge, 1997.

320. J. Tittel, W. Gohde, F. Koberling, T. Basche, A. Kornowski, H. Weller and A. Eychmuller, J. Phys. Chem. B 101, 3013 (1997).

321. S.A. Blanton, A. Dehestani, P.C. Lin and P. Guyot-Sionnest, Chem. Phys. Lett. 229, 317 (1994).

322. S.A. Empedocles, D.J. Norris and M.G. Bawendi, Phys. Rev. Lett. 77, 3873 (1996).

323. M. Nirmal, B.O. Dabbousi, M.G. Bawendi, J.J. Macklin, J.K. Trautman, T.D. Harris and L.E. Brus, Nature 383, 802 (1996).

324. S.A. Blanton, M.A. Hines and P. Guyot-Sionnest, Appl. Phys. Lett. 69, 3905 (1996).

325. S.A. Empedocles and M.G. Bawendi, Science 278, 2114 (1997).

326. S.A. Empedocles, R. Neuhauser, K. Shimizu and M.G. Bawendi, Advan. Mater. 11, 1243 (1999).

327. S.A. Empedocles and M.G. Bawendi, J. Phys. Chem. B 103, 1826 (1999).

328. S. Empedocles and M. Bawendi, Acc. chem. Res. 32, 389 (1999).

329. U. Banin, M. Bruchez, A.P. Alivisatos, T. Ha, S. Weiss and D.S. Chemla, J. Chem. Phys. 110, 1195 (1999).

330. T. Matsumoto, M. Ohtsu, K. Matsuda, T. Saiki, H. Saito and K. Nishi, Appl. Phys. Lett. 75, 3246 (1999).

331. N. Murase, Chem. Phys. Lett. 368, 76 (2003).

CHAPTER 2

NANOSTRUCTURE PRESENTED CHEMILUMINESCENCE AND ELECTROCHEMILUMINESCENCE

Zhouping Wang [1] and Jinghong Li [2]*

*1 State Key Laboratory of Food Science and Technology, School of Food Science and Technology, Jiangnan University, Wuxi 214122, China; 2 Department of Chemistry, MOE Key Laboratory of Bioorganic Phosphorus Chemistry & Chemical Biology, Tsinghua University, Beijing 100084, China *E-mail: jhli@mail.tsinghua.edu.cn*

This paper reviews the newly advancement of nanomaterials and nanostructures applied in chemiluminescence (CL) and electrochemiluminescence (ECL). Numerous reports have demonstrated that nanomaterials and nanostructures possess high surface/volume ratio and the quantum size effect, which results in their properties of catalysis, luminescence, absorption and the others. And these properties have been utilized for CL and ECL analysis generating various excellent performances. Among these nanostructures, quantum dots (semiconductor nanocrystals) can be employed as luminophor in CL and ECL. Metal nanostructures and metal oxide nanostructures usually catalyzed the related CL and ECL reaction. Upon these nanostructure presented CL and ECL reactions, considerable procedures have been designed and proposed for analytical application.

1. Introduction

1.1. Nanomaterials and nanostructure

Nanomaterials (nanocrystalline materials) are materials possessing particles sizes on the order of a billionth of a meter, nanometer. When the size of particles in the scale of nanometer (1~100nm), it would perform some novel properties, such as quantum-size effect, small-size effect, surface effect, and macroscopic-quantum-tunnel effect. Nanomaterials have attracted widespread attention since the 1990s because of their specific features that differ from bulk materials. Recently, novel nanometer-scale materials provide analytical chemistry with various opportunities. [1-3]

Nanostructures are the ordered system of one-dimension, two-dimension or three-dimension constructed or assembled with nanometer-scale unit in certain pattern, which basically include nanosphere, nanorod, nanowire, nanobelt and nanotube. They manifest extremely fascinating and useful properties, which can be exploited for a variety of structural and non-structural applications. [4, 5]

In recent years, much attention has been paid to nanomaterials due to novel optical and electronic properties, which mainly come from the high surface/volume ratio and the quantum size effect. [6]

1.2. Chemiluminescence and electrochemiluminescence

Chemiluminescence (CL) is the generation of electromagnetic radiation as light by the release of energy from a chemical reaction. While the light can, in principle, be emitted in the ultraviolet, visible or infrared region, among which the emitting visible light is the most common. They are also the most interesting and useful. [7-10] Chemiluminescence takes its place among other spectroscopic techniques because of its inherent sensitivity and selectivity. It requires no excitation source (as does fluorescence and phosphorescence), only a single light detector such as a photomultiplier tube, no monochromator and often not even a filter. Although not as widely applicable as excitation spectroscopy, the detection limits for chemiluminescent methods can be 10 to 100 times lower than other luminescence techniques. Most chemiluminescence methods involve only a few chemical components to actually generate light. In many CL systems, a "fuel" is chemically oxidized to produce an excited state product. Due to the advantages of high sensitivity, wide linear range, and simple instrument, CL method has been extensively applied in clinic diagnostics, immunoassay, DNA hybridization, environmental monitor, or used as detector after separation, such as HPLC, capillary electrophoresis and micro-fluidic chip, or coupled with flow-injection analysis for automatization analysis. [11-19]

Electrochemiluminescence (ECL, also called electrogenerated chemiluminescence) involves the generation of species at electrode surfaces that then undergo electron-transfer reactions to form excited states that emit light. For example, application of a voltage to an electrode in the presence of an ECL luminophore such as $Ru(bpy)_3^{2+}$ (where bpy = 2,2'-bipyridine) results in light emission and allows detection of the emitter at very low concentrations ($\leq 10^{-11}$ M). [20] By employing ECL-active species as labels on biological molecules, ECL has found application in immunoassays and DNA analyses. [21-23] Commercial systems have been developed that use ECL to detect many clinically important analytes with high sensitivity and selectivity. [22-30]

It is also important to distinguish ECL from CL. Both involve the production of light by species that undergo highly energetic electron-transfer reactions. However, luminescence in CL is initiated and controlled by the mixing of reagents and careful manipulation of fluid flow. In ECL, luminescence is initiated and controlled by switching an electrode voltage.

2. Nanostructure presented chemiluminescence

2.1. Nanostructure as catalyst in chemiluminescence

Recent researches indicated that metal nanoparticles (such as gold nanoparticles, Pt nanoparticles), and metal oxide nanoparticles (such as TiO_2, Al_2O_3), due to their high catalytic activity, are the most useful nanostructure presented in liquid or air-phase CL analysis.

2.1.1. Liquid-phase chemiluminescence

Metal ions (such as Au^{3+}, Pt^{2+}, Co^{2+}, Cu^{2+}) often were used to sensitize CL reaction as catalysts in lipid phase. Usually, these CL reactions occurred in atomic or molecular level. If these metal atoms aggregated and form nanoparticles in a very ordered pattern, the catalysis activity would change in a large degree. The excellent catalysis of metal nanoparticles has appeared in many organic and inorganic reactions. But for CL reaction, until recent years, there began present some reports dealt with the catalysis of metal nanoparticles.

2.1.1.1. Non-labeled liquid-phase catalytic reaction

Non-labeled liquid-phase catalytic reaction is the basic manner to study the catalysis behavior of nanostructure in CL reaction. Cui's group took great attention in metal nanoparticles-catalyzed CL reaction. They observed that the reaction of gold nanoparticles with a potassium periodate-sodium hydroxide-carbonate system underwent CL with three emission bands at 380-390, 430-450, and 490-500 nm, respectively. [31] The light intensity increased linearly with the concentration of the gold nanoparticles, and the CL intensity increased dramatically when the citrate ions on the nanoparticle surface were replaced by SCN⁻. The shape, size, and oxidation state of gold nanoparticles after the chemiluminescent reaction were characterized by UV-visible absorption spectrometry, transmission electron microscopy (TEM), and X-ray photoelectron spectrometry (XPS). Gold nanoparticles are supposed to function as a nanosized platform for the observed chemiluminescent reactions without shape change before and after CL reaction (Figure 1).

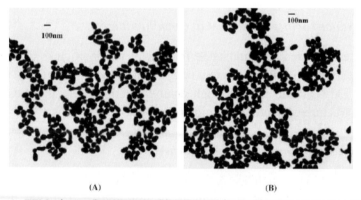

(A)   (B)

Figure 1. TEM photos for 68-nm gold nanoparticles before (A) and after (B) the addition of 2.35×10^{-4} mol/L KSCN. Reprinted with permission from [Ref. 31], H. Cui *et al.* J. Phys. Chem. B., 109, 3099(2005), Copyright @ American Chemical Society.

Scheme 1. Possible Mechanism for the Chemiluminescence Involving Carbon Dioxide Dimer and Singlet Oxygen Molecular Pair. Reprinted with permission from [Ref. 31], H. Cui *et al*. J. Phys. Chem. B., 109, 3099(2005), Copyright @ American Chemical Society.

$$IO_4^- + O_2 + 2\,OH^- \longrightarrow 2\,O_2^{\bullet-} + IO_3^- + H_2O \qquad (1)^{24}$$

$$O_2^{\bullet-} + CO_3^{2-} \longrightarrow CO_3^{\bullet-} + O_2^{2-} \qquad (2)^{24}$$

$$4\,O_2^{\bullet-} + 4\,H_2O \xrightarrow{\text{nanogold}} (O_2)_2{}^* + 2\,H_2O_2 + 4\,OH^- \qquad (3)$$

$$(O_2)_2{}^* \longrightarrow 2\,O_2 + h\nu \qquad (\lambda = 490 \sim 500 \text{ nm}) \qquad (4)$$

$$2\,CO_3^{\bullet-} \xrightarrow{\text{nanogold}} (CO_2)_2{}^* + O_2^{2-} \qquad (5)$$

$$(CO_2)_2{}^* \longrightarrow 2\,CO_2 + h\nu \qquad (\lambda = 430 \sim 450 \text{ nm}) \qquad (6)$$

Scheme 2. Possible Mechanism for the Chemiluminescence Involving the Oxidation of Surface Gold Atoms. Reprinted with permission from [Ref. 31], H. Cui *et al*. J. Phys. Chem. B., 109, 3099(2005), Copyright @ American Chemical Society.

$$2\,SCN^- + IO_4^- \longrightarrow (SCN)_2^{\bullet-} + IO_3 \qquad (7)$$

$$Au_n + O_2^{\bullet-} \longrightarrow Au_{n-1}\left[Au^I(O_2)\right]^{-*} \longrightarrow Au_{n-1}\left[Au^I(O_2)\right]^- + h\nu \qquad (8)$$

$$Au_n + CO_3^{\bullet-} \longrightarrow Au_{n-1}\left[Au^I(CO_3)\right]^{-*} \longrightarrow Au_{n-1}\left[Au^I(CO_3)\right]^- + h\nu \qquad (9)$$

$$Au_n + (SCN)_2^{\bullet-} \longrightarrow Au_{n-1}\left[Au^I(SCN)_2\right]^{-*} \longrightarrow Au_{n-1}\left[Au^I(SCN)_2\right]^- + h\nu \qquad (10)$$

$$(\lambda 1 = 380 \sim 390 \text{ nm}; \lambda 2 = 430 \sim 450 \text{ nm})$$

$$Au_{n-1}Au^I \xrightarrow{[O]} Au_{n-1}Au^{III} \qquad (11)$$

A chemiluminescent mechanism has been proposed in which the interaction between free $CO_3^{\bullet-}$ and $O_2^{\bullet-}$ radicals generated by a KIO_4-$NaOH$-Na_2CO_3 system and gold nanoparticles results in the formation of emissive intermediate gold(I) complexes, carbon dioxide dimers, and singlet oxygen molecular pairs on the surface of the gold nanoparticles (Schemes 1 and 2). [31]

Meanwhile, gold nanoparticles of different sizes were found to enhance the chemiluminescence (CL) of the luminol-H_2O_2 system, and the most intensive CL signals were obtained with 38-nm-diameter gold nanoparticles (Figure 2). [32] UV-visible spectra, X-ray photoelectron

Wang et al.

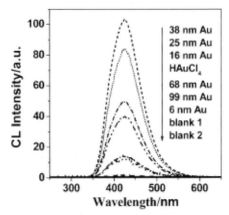

Figure 2. Chemiluminescence spectra for luminol-H_2O_2-gold colloids system.: $HAuCl_4$, 1×10^{-4} g/mL; blank 1, 2×10^{-4} g/mL $Na_3C_6O_7$; blank 2, 5.5×10^{-5} g/mL $Na_3C_6O_7$, 1.125×10^{-6} mol/L $NaBH_4$. Conditions: luminol, 2×10^{-4} mol/L in 0.01 mol/L NaOH; H_2O_2, 0.01 mol/L. Reprinted with permission from [Ref. 32], Z. F. Zhang *et al.* Anal. Chem. 77, 3324 (2005). Copyright @ American Chemical Society.

spectra, and transmission electron microscopy studies were carried out before and after the CL reaction to investigate the CL enhancement mechanism. The CL enhancement by gold nanoparticles of the luminol-H_2O_2 system was supposed to originate from the catalysis of gold nanoparticles, which facilitated the radical generation and electron-transfer processes taking place on the surface of the gold nanoparticles (Scheme 3). The effects of the reactant concentrations, the size of the gold nanoparticles and some organic compounds were also investigated. Organic compounds containing OH, NH_2, and SH groups were observed to inhibit the CL signal of the luminol-H_2O_2-gold colloids system, which made it applicable for the determination of such compounds.

In acid media, nanogold was also found catalyzing CL reaction. It was found that potassium permanganate ($KMnO_4$) could react with gold nanoparticles in a strong acid medium to generate particle size-dependent CL. [33] For gold nanoparticles with the size of 2.6 or 6.0 nm, the reaction was fast and could produce the excited state Mn(II)* with light emission around 640 nm. For gold nanoparticles larger than 6.0 nm, no light emission was observed due to a much slower reaction rate. The CL intensity was found to increase linearly with the concentration of 2.6 nm

Scheme 3. Possible Mechanism for the Luminol-H_2O_2-Gold Colloids CL System. Reprinted with permission from [Ref. 32], Z. F. Zhang *et al*. Anal. Chem. 77, 3324 (2005). Copyright @ American Chemical Society.

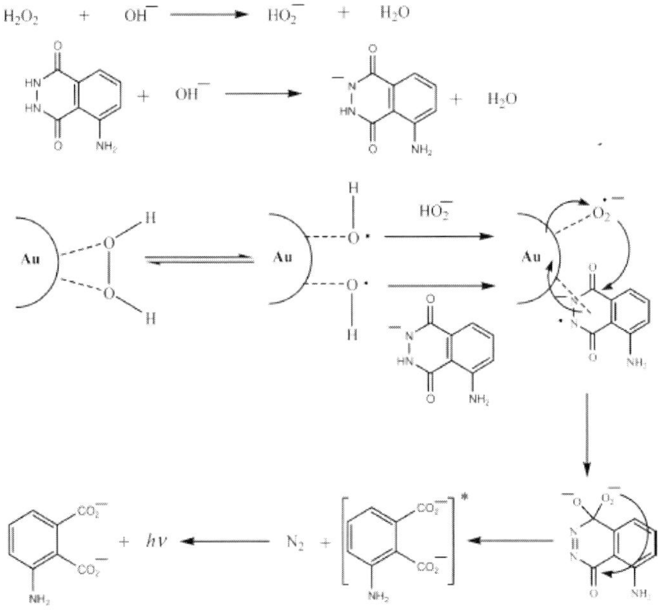

gold nanoparticles. The effects of the acid medium, concentration of $KMnO_4$ and presence of N_2 and O_2 were investigated. UV-Vis absorption spectra and X-ray photoelectron spectra (XPS) measured before and after the CL reaction were analyzed. A CL mechanism has been proposed suggesting that the potassium permanganate was reduced by gold nanoparticles in the strong acid medium to the excited state Mn(II)*, yielding light emission. The results bestow new light on the size-dependent chemical reactivities of the gold nanoparticles and on nanoparticle-induced CL.

Size-dependent effect is the basic property of nanoparticles. Besides above mentioned literatures, Cui's another research work also demonstrated the size-dependent effect of gold nanoparticles-catalyzed CL reaction. [34] They observed that gold nanoparticles of small size (<5 nm) could inhibit the CL of the luminol-ferricyanide system, whereas gold nanoparticles of large size (>10 nm) could enhance this CL, and the most intensive CL signals were obtained with 25-nm-diameter gold nanoparticles. Cui et al. examined the CL inhibition and enhancement mechanism by means of the studies of UV-visible spectra, CL spectra, X-ray photoelectron spectra, effects of concentrations of luminol and ferricyanide solution, and fluorescence quenching efficiency of gold colloids. The results indicated that the luminophor was identified as the excited-state 3-aminophthalate anion. The CL inhibition by gold nanoparticles of small size was supposed to originate from the competitive consumption of ferricyanide by gold nanoparticles and the relatively high quenching efficiency of the luminophor by gold nanoparticles. In contrast, the CL enhancement by gold nanoparticles of large size was ascribed to the catalysis of gold nanoparticles in the electron-transfer process during the luminol CL reaction and the relatively low quenching efficiency of the luminophor by gold nanoparticles. This work demonstrates that gold nanoparticles have the size-dependent inhibition and enhancement in the CL reaction, proposing a perspective for the investigation of new and efficient nanosized inhibitors and enhancers in CL reactions for analytical purposes.

Nanogold with different shapes also displays distinct catalytic activity on CL reaction. Li and co-worker tested the catalyzed CL efficiency of luminol CL system in the presence of different shaped nanogold, including gold nanosphere, hexagonal gold nanoparticles, tadpole gold nanoparticles and irregular shaped nanogold. [35] The results reveal that nanogold with more polar sites (tadpole gold nanoparticles and irregular shaped nanogold) performed more strong catalytic activity on luminol CL reaction. And the catalytic efficiency of tadpole gold nanoparticles and irregular shaped nanogold was more than 100-folds to spherical gold nanoparticles.

Upon the high catalytic activity of nanogold on luminol CL reaction, researchers have designed procedure to promote the analytical performance of luminol CL reaction in biochemical analysis. For instance, based on the enhancement of CL of luminol-hydrogen peroxide-gold nanoparticles system by fluoroquinolones (FQs), a novel and rapid CL method was reported for the determination of FQs derivatives. Under the optimum conditions, the CL intensity was proportional to the concentration of FQs derivative in solution. This proposed method has been applied to detect FQs derivatives in human urine successfully. [36]

Platinum nanoparticles are another kind of noble metal nanoparticles used for CL reaction. Platinum colloids prepared by the reduction of hexachloroplatinic acid with citrate in the presence of different stabilizers were also found to enhance the CL of the luminol-H_2O_2 system, and the most intensive CL signals were obtained with citrate-protected Pt colloids synthesized with citrate as both a reductant and a stabilizer. Light emission was intense and reproducible. Transmission electron microscopy and X-ray photoelectron spectroscopy studies were conducted before and after the CL reaction to investigate the possible CL enhancement mechanism. It is suggested that this CL enhancement was attributed to the catalysis of platinum nanoparticles, which could accelerate the electron-transfer process and facilitate the CL radical generation in aqueous solution. The effects of Pt colloids prepared by the hydroborate reduction were also investigated. The application of the luminol-H_2O_2-Pt colloids system was exploited for the determination of compounds such as uric acid, ascorbic acid, phenols and amino acids. [37]

Willner's group recently reported that a photoisomerizable monolayer consisting of carboxypropyl nitrospiropyran (*1a*) linked to an aminopropyl siloxane layer associated with an indium tin oxide surface was used to photoswitch the electrocatalyzed reduction of H_2O_2 in the presence of Pt nanoparticles (NPs) or to photostimulate the generation of CL in the presence of Pt NPs, H_2O_2, and luminol. Photoisomerization of *1a* to the protonated mcrocyaninc, *1b*, and layer resulted in the electrostatic attraction of the negatively charged Pt NPs to the surface. This facilitated the electrocatalyzed reduction of H_2O_2 or the catalyzed generation of chemiluminescence in the presence of H_2O_2/luminol. Further photo isomerization of the *1b* monolayer resulted in the formation of the nitrospiropyran layer that allowed the washing off of the surface-bound Pt NPs. The resulting *1a*-modified surface was inactive toward the reduction of H_2O_2 or toward the generation of CL. [38]

More recently, Cui et al. further investigated lucigenin CL behavior in the presence of noble metal nanoparticles including Ag, Au, and Pt nanoparticles. They found that these noble metal nanoparticles in the presence of adsorbates such as iodide ion, cysteine, mercaptoacetic acid, mercaptopropionic acid, and thiourea could reduce lucigenin (bis-N-methylacridinium) to produce CL. Lucigenin-Ag-KI system was chosen as a model to study the CL process. Absorption spectra and X-ray photoelectron spectra showed that when the Ag colloid was mixed with KI, Ag nanoparticles were covered by adsorbed iodide ions. X-ray diffraction patterns and fluorescence spectra indicated that Ag nanoparticles were oxidized to AgI and lucigenin was converted to N-methylacridone in the CL reaction. The addition of superoxide dismutase could inhibit the CL. According to Nernst's equation, the presence of iodide ions decreased the oxidation potential of Ag nanoparticles. As a result, lucigenin was rapidly reduced by Ag nanoparticle to a monocation radical, which reacted with oxygen to generate a superoxide anion; then the superoxide anion reacted with the monocation radical to produce CL. Other adsorbates such as cysteine, mercaptoacetic acid, mercaptopropionic acid, and thiourea that could decrease the oxidation potential of Ag nanoparticles could also induce the CL reaction. [39]

Another report still involved in silver nanoparticles presented CL reaction. Mixtures of silver (I) and citrate that were used to produce silver nanoparticles evoked intense chemiluminescence with tris(2,2'-bipyridyl)ruthenium(II) and cerium(IV), which can be exploited for the determination of citrate ions and other analytes over a wide concentration range. However, the CL reaction seems more related to citrate ions capped on the surface of silver nanoparticles rather than the state of nanoparticles. [40]

2.1.1.2. Nanostructure-labeled chemiluminescent bioassay

Preparation of nanostructure-biomolecules conjugates coupled with specific recognition reaction, such as immune affinity and DNA hybridization, is the important approach for the application of nanostructure-presented CL. The special surface properties or the surface modification procedures make most nanostructure compatible to bond with biomolecules via the forces of surface adsorption, electrostatic adsorption, covalent binding and the others.

Originally, nanostructure, mainly including gold nanoparticles and silver nanoparticles, are used as label to bind with anti-IgG or oligonucleotide in CL analysis. However, the catalytic activity of the metal nanoparticles on CL reaction was not yet found and the labeled nanoparticles often were stripped to form the corresponding metal ions, which then catalyzed the classic CL reactions, such as luminol-H_2O_2-Au^{3+} and Ag^+-Mn^{2+}-$K_2S_2O_8$-H_3PO_4-luminol CL systems. The remarkable merits of the proposed nanoparticles labeling and stripping CL detection are the labeled stability and the relative high sensitivity.

Lu et al. developed a novel, sensitive CL immunoassay by taking advantage of a magnetic separation/mixing process and the amplification feature of colloidal gold label. [41] First, the sandwich-type complex was formed in this protocol by the primary antibody immobilized on the surface of magnetic beads, the antigen in the sample, and the second antibody labeled with colloidal gold. Second, a large number of Au^{3+} ions from each gold particle anchored on the surface of magnetic beads were released after oxidative gold metal dissolution and then quantitatively determined by a simple and sensitive Au^{3+}-catalyzed luminol CL reaction. Third, this protocol was evaluated for a noncompetitive immunoassay of a human immunoglobulin G (Figure 3), and a concentration as low as 3.1×10^{-12} M was determined, which was competitive with colloidal gold-based anodic stripping voltammetry, colorimetric ELISA, or immunoassays based on fluorescent europium chelate labels. The high performance of this protocol was related to the sensitive CL determination of Au^{3+} ion (detection limit of 2×10^{-10} M), which was 25 times higher than that by ASV at a single use carbon-based screen-printed electrode. Based on the similar principle, Li et al. also developed a nanogold-labeling and stripping CL detection method for IgG in the absence of magnetic beads. [42]

Silver nanoparticles labeling and stripping CL have also been developed for ultrasensitive detection of DNA hybridization. [43] The assay relied on a sandwich-type DNA hybridization in which the DNA targets were first hybridized to the captured oligonucleotide probes immobilized on polystyrene microwells and then the silver nanoparticles

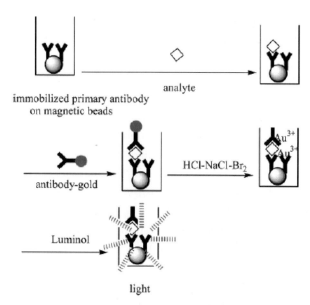

immobilized primary antibody
on magnetic beads

antibody-gold

Luminol

light

Figure 3. Representation of the noncompetitive CL immunoassay by using magnetic beads and colloidal gold label. Scheme 3. Possible Mechanism for the Luminol-H_2O_2-Gold Colloids CL System. Reprinted with permission from [Ref. 41], A. Fan *et al.* Anal. Chem. 77, 3238 (2005). Copyright @ American Chemical Society.

modified with alkylthiol-capped oligonucleotides were used as probes to monitor the presence of the specific target DNA. After being anchored on the hybrids, silver nanoparticles were dissolved to Ag^+ in HNO_3 solution and sensitively determined by a coupling CL reaction system (Ag^+-Mn^{2+}-$K_2S_2O_8$-H_3PO_4-luminol). The combination of the remarkable sensitivity of the CL method with the large number of Ag^+ released from each hybrid allowed the detection of specific sequence DNA targets at levels as low as 5 fM. The sensitivity increased 6 orders of magnitude greater than that of the gold nanoparticle-based colorimetric method and was comparable to that of surface enhanced Raman spectroscopy, which is one of the most sensitive detection approaches available to the nanoparticle-based detection for DNA hybridization. Moreover, the perfectly complementary DNA targets and the single-base mismatched DNA strands can be evidently differentiated through controlling the temperature, which indicates that the proposed CL assay offers great promise for single nucleotide polymorphism analysis.

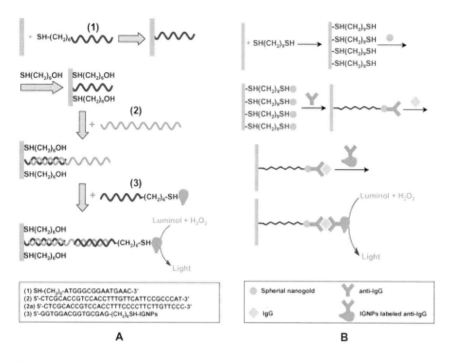

Figure 4. Scheme of in situ amplified chemiluminescence detection of DNA (A) and immunoassay of IgG (B) using IGNPs as label. Reprinted with permission from [Ref. 35], Z.P. Wang, et al. Clin. Chem. 52, 1958, (2006), Copyright @ American Association for Clinical Chemistry.

Consequently, researchers observed the ultra-high catalytic activity of metal nanoparticles (such as gold and platinum nanoparticles), then nanostructure-labeling and *in situ* catalytic amplified CL detection was proposed. Li et al. synthesized specially shaped, irregular gold nanoparticles (IGNPs), and observed their catalytic efficiency on luminol CL to be 100-fold greater than that of spherical Au-NPs. Using the IGNPs-functionalized DNA oligomers and the IGNPs-modified anti-IgG as in situ chemiluminescent probes, they established sandwich-type analytic methods for rapid, simple, selective, and sensitive sequence-specific DNA detection and for human plasma IgG immunoassay, respectively (Figure 4). [35] They used 12 clinical human plasma samples to examine the precision and accuracy of the proposed method for IgG content determination.

Zhang et al. designed a novel gold nanoparticles based protein immobilization method for CL imaging assay of H_2O_2 and recombinant human interleukin-6(rHu IL-6). In this study, gold nanoparticles not only were used as solid immobilized materials on poly (methyl methacrylate) (PMMA) plates to enlarge the loading capability of biomolecules, but also effectively enhanced the CL intensity of luminol-H_2O_2 and the detection sensitivity of analyte. [44]

2.1.2. Air-phase and aerosol chemiluminescence

Air-phase and aerosol CL provides another vital CL reaction approach, and extend CL analytical method covering gaseous and volatile substrates.

CL resulting from the interaction between gases and solid surfaces has been studied for decades. The phenomenon was observed during the catalytic oxidation of carbon monoxide on a thoria surface, [45] and was called cataluminescence (CTL). Application of the luminescence phenomenon has been developed as a means of revealing intermediate stages in adsorption and catalysis. [46] In recent years, numerous organic vapor CL sensors were developed based on this phenomenon, as shown in Table 1. The expanding availability of nanostructures has attracted widespread attention in the use of catalysis because of their high surface areas, high activity and good selectivity.

Table 1. Nanostructure presented air-phase chemiluminescence sensors.

Nanostructures	Analyte	Limit of detection (ppm)	References
γ-Al_2O_3	Ethanol	1 or less	[47,49,50]
	Butanol	1 or less	[48,49]
	Acetone	1 or less	[48,49]
	Butanoic acid	1 or less	[48,49]
	Dimethylbenzene	20	[48,49]
γ-Al_2O_3:Dy^{3+}	Iso-Butane	0.2	[51]
	Fragrance substances	0.1–1	[52]
$SrCO_3$	Ethanol	6–3750	[55]
ZrO_2	Air	0.6 µg/mL	[56]
$LaCoO_3$	NH_3	14.2	[57]
TiO_2	CCl_4	0.15–150	[58]
SnO_2	H_2S		[59]

McCord et al. [53] found that porous silicon treated with nitric acid or persulfate could result in intense CL. Recently, Zhang's group all through addressed to the research work of nanostructure-catalyzed air-phase or aerosol chemiluminescence. [60] Nanomaterials, including nanosized MgO, ZrO_2, TiO_2, Al_2O_3, Y_2O_3, $LaCoO_3$:Sr^{2+} and $SrCO_3$, were investigated and CL could be detected on seven of them, [54] while organic vapor was passing through. The response of organic vapors containing the groups of -Cl-, -P-, -O-, -S-, -N-, and -H-, were systematically examined. The results showed that acetone, [48, 49] gaseous ethanol, [55] NH_3, [57] chlorinated volatile organic compounds (CH_2Cl_2, $CHCl_3$, and CCl_4), [58] H_2S [59] could be catalyzed to produce CL signals on the surface of different nanomaterials.

Meanwhile, nanostructure catalyzed aerosol CL was developed and used as detector after HPLC or capillary electrophoresis (CE) separation. [61, 62] This aerosol CL-based detector, in which HPLC or CE effluent was converted to aerosol and then generated CL emission on the surface of porous alumina, was composed of three main processes: ebuliztion of HPLC or CE effluent, CL emission on surface of porous alumina material, and optical detection. The CL emission could be generated due to the catalyzing oxidization of aim analytes, like saccharides, poly (ethylene glycol)s, amino acids, and steroid pharmaceuticals, on the surface of porous alumina. It could be an important supplement of HPLC and CE detectors for UV lacking compounds. Zhang et al. also found that the nanostructures catalyzed CTL can be quenched when introducing Ho^{3+}, Co^{2+} and Cu^{2+} into the nanosized catalyst, while new intensive CTL peaks appear when the catalyst was doped with Eu^{3+} or Tb^{3+}. [63]

More recently, Zhang et al. developed an optical sensor array based on chemiluminescent images from spots of nanomaterials, which was employed to recognize odorous samples. The images obtained from the array permit identification of a wide range of analytes, even homologous compounds. The sensors with each nanomaterial (porous alumina, ZrO_2, ZrO_2:Tb^{3+}, ZrO_2:Eu^{3+}, $SrCO_3$, Y_2O_3, Fe_2O_3, MgO, and WO_3) were made simply by spotting the nanomaterials onto a piece of ceramic chip, individually. Hydrogen sulfide, methanol, ethanol, n-propanol and n-butanol can be sensitively fingerprinted with the sensors array (Figure 5).

[64] Moreover, a new optical strategy to screen the catalytic activities of gold catalysts was also developed based on the chemiluminescence during the oxidation of CO. [65]

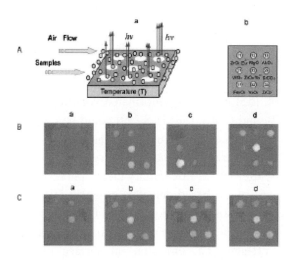

Figure 5. Schematic diagrams of the CL sensor array and the images recorded upon exposure to various samples. (A) Schematic diagrams of the CL sensor array (a) and the arrangement of nanomaterial spots (b). The sensor elements are arranged as follows: ZrO_2/Eu^{3+} (1,1), MgO (1,2), Al_2O_3(1,3), WO_3 (2,1), ZrO_2/Tb^{3+} (2,2), $SrCO_3$ (2,3), Fe_2O_3 (3,1), Y_2O_3 (3,2),and ZrO_2 (3,3). (B) Images obtained by the sensor array after exposure toair for 1 min without any sample (a), with ethanol vapor (b), hydrogen sulfide (c), and TMA vapor (d). (C) Images obtained by the sensor array upon exposure to four alcohol vapors: (a) methanol, (b) ethanol, (c) n-propanol, and (d) n-butanol. The integrated CL intensities were recorded. Reprinted with permission from [Ref. 64], N. Na, *et al.* J. Am. Chem. Soc., 128, 14420,(2006). Copyright @ American Chemical Society.

Zhu et al. synthesized $LaSrCuO_4$ nanowires using carbon nanotubes (CNTs) in the presence of citrate; $LaSrCuO_4$ nanoparticles were also prepared by a conventional citrate route in order to compare with the nanowires. The catalytic and CL properties for CO oxidation over $LaSrCuO_4$ catalysts with different morphologies were investigated further. The results revealed that CNTs and citrate played the key roles in controlling the morphology, crystallization and phase compositions of

LaSrCuO₄ catalyst; a high thermal stability of LaSrCuO₄ nanowires against calcination was observed, and a high activity for CO oxidation was maintained. And on the spot, nanostructures catalytic chemiluminescence has been used as an effective tool for characterizing the catalysis activity of novel nanomaterials. [70]

2.2. Nanostructure as luminophor in chemiluminescence

2.2.1. QDs chemiluminescence

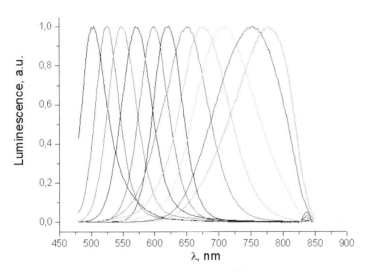

Figure 6. Photoluminescence emission spectra of CdTe quantum dots of different sizes.

Quantum dots (QDs), crystals composed of periodic groups of II-VI, III-V, or IV-VI materials, also referred as luminescent semiconductor nanocrystals, are semiconductor nanostructure that confines the motion of conduction band electrons, valence band holes, or excitons (bound pairs of conduction band electrons and valence band holes) in all three spatial directions. Small quantum dots, such as colloidal semiconductor nanocrystals, can be as small as 2 to 10 nanometers, corresponding to 10 to 50 atoms in diameter and a total of 100 to 100,000 atoms within the quantum dot volume. Self-assembled quantum dots are typically between 10 and 50 nm in size. Like in atoms, the energy levels of small quantum dots can be probed by optical spectroscopy techniques. In quantum dots that confine electrons and holes, the interband absorption edge is blue shifted due to the confinement compared to the bulk material of the host semiconductor material. As a consequence, quantum dots of the same material, but with different sizes, can emit light of different colors (Figure 6). [71-73]

The superior emitting properties of QDs attract the growing attention to applications of these materials in LED and display devices [74, 75] as well as in biological luminescent labels. [76] In quantum dots, atomic-like electronic energy levels are formed due to the charge carrier confinement. The spacing of the highest occupied and lowest unoccupied quantum confined orbitals is a pronounced function of nanocrystal size, providing the advantage of continuous band gap tunability over a wide range simply by changing the size of the nanocrystal (the quantum size effect). [77] High quality nanocrystal samples exhibit bright luminescence that is superior to conventional organic fluorescence dyes with respect to color purity (narrow emission band) and photostability. [74, 76, 78]

The various types of luminescence differ from the source of energy to obtain an excited state that can relax into a ground state with the emission of light. In today's applications of quantum dots, the required excitation energy is supplied either by absorption of a quantum of light given rise to photoluminescence (PL), by electrical injection of an electron-hole pair (electroluminescence, EL), or by electron impact resulting in cathodoluminescence. [79] Moreover, the other types of luminescence, chemiluminescence, which has developed to an important and powerful tool in biological and medical investigations, have also been observed for the novel class of quantum dot luminophores. [78]

Talapin et al reported the first observation of band gap CL of semiconductor quantum dots in solution and in nanoparticulate layers. The spectral position of the band gap CL of CdSe/CdS core shell and InP nanocrystals depends on their particle size, allowing thus an efficient tuning of the emission color with superior color purity inherent for monodisperse samples (Figure 7). The efficiency of nanocrystal CL can be dramatically enhanced by applying a cathodic potential to the nanoparticulate layers made from CdSe or CdSe/CdS core-shell nanocrystals. In this case electrochemical n-doping of the particles via electron injection provides a "quantum confined cathodic protection" against nanocrystal oxidative corrosion upon hole injection and allows the achievement of efficient and stable electrogenerated chemiluminescence. [78]

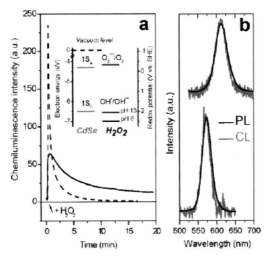

Figure 7. (a) Temporal evolution of the integral chemiluminescence intensity of a film of CdSe/CdS nanocrystals induced by the addition of H_2O_2 in 0.1 M KOH (solid line) and in 0.1 M Na_2SO_4 (dashed line) solutions. Inset shows proposed energy level diagram of CdSe/CdS nanocrystals in contact with an aqueous H_2O_2 solution. (b) CL spectra measured from the films of CdSe/CdS nanocrystals with different size of CdSe core (red lines) and PL spectra of the corresponding films (black lines). Reprinted with permission from [Ref. 78], S. K. Poznyak *et al.* Nano Lett. 4, 4693 (2004). Copyright @ American Chemical Society.

Li's group then synthesized CdTe nanocrystals (NCs) capped with thioglycolic acid (TGA) via a microwave-assisted method. The CL of CdTe NCs induced by directly chemical oxidation and its size-depended and surfactant-sensitized effect in aqueous solution were then investigated. It was found that oxidants, especially hydrogen peroxide and potassium permanganate, could directly oxidize CdTe NCs to produce strong CL emission in basic conditions. The oxidized CL of

CdTe NCs displayed size-dependent effect and its intensity increased along with increasing the sizes of the NCs. Moreover, the CL intensity could, if surfactants CTAB or β-cyclodextrin were added to the above CL system, be sensitized to some degree. The sensitized CL induced by CTAB and β-cyclodextrin was mainly contributing to the formation of aggregate nanostructure and the micellar micronano-environment, respectively. [80]

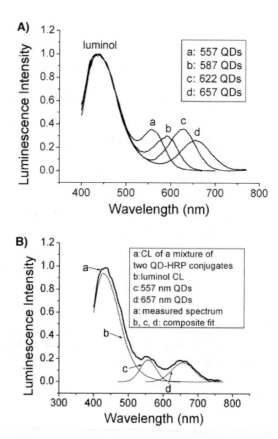

Figure 8. A) CL spectra of the QD-HRP-luminol system using luminol as donor and different sized QDs as accepters. B) CL spectra of the QD-HRP-luminol system using two QD accepters simultaneously. All spectra are normalized. Reprinted with permission from [Ref. 82], X. Huang, *et al.* Angew. Chem. Int. Ed. 45, 5140(2006). Copyright @ Wiley-VCH.

Similar to Li's work, Wang et al. also investigated the CL behaviors of mercaptoacetic acid (MA) capped water-soluble CdS NCs in aqueous solution. [81] They found that hydrogen peroxide directly oxidized the MA-capped CdS NCs and produced strong CL emission in basic conditions. And the CL of CdS NCs was size-dependent, the CL intensity increased with increasing CdS NCs size. UV-visible spectra, CL spectra, PL spectra, and TEM were used to investigate the CL reaction mechanism. Moreover, some biological molecules and metal ions were observed to inhibit the CL signal of the H_2O_2-CdS NCs system, which made it applicable for the detection of such species.

The research work of Ren et al. further utilized QDs as energy acceptor for chemiluminescence resonance energy transfer (CRET) assay. [82] In the study, different sized water-soluble CdTe QDs were synthesized in the aqueous phase using the reaction between Cd^{2+} and NaHTe solution in the presence of mercaptopropyl acid (MPA) as a stabilizer [83]. The MPA-coated CdTe QDs were conveniently conjugated to certain proteins (such as HRP and BSA) using EDC (1-ethyl-3-(3-dimethylaminopropyl) carbodiimide) as a coupling reagent. The mixtures were purified using ultrafiltration membrane. In the system, they chose the luminol/hydrogen peroxide CL reaction catalyzed by horseradish peroxidase (HRP) because this is one of the most sensitive CL reactions. In capillary electrophoresis with CL detection, the detection limit of HRP was below 10^{-19} mol in the presence of para-iodophenol (p-IP) as an enhancer. [84] And more importantly, the CL spectrum (425–435 nm) of luminol overlaps well with the absorption of the QDs (Figure 8). The principle of CRET is illustrated in Scheme 1. In Figure 9A, the CL donor, luminol, is not directly linked with the QDs, and the catalyst, HRP, is conjugated to the QDs. HRP can continuously catalyze the luminol/hydrogen peroxide CL reaction. In this system, the QD–HRP conjugates can be used as probes in cell and tissue imaging similar to BRET. [85] In Figure 9B, QDs are linked with bovine serum albumin (BSA), and HRP is conjugated with the BSA antibody (anti-BSA). When the anti-BSA–HRP binds to the BSA–QDs, CRET can occur. This system has potential to be used in immunoassay in non-competition and competition modes.

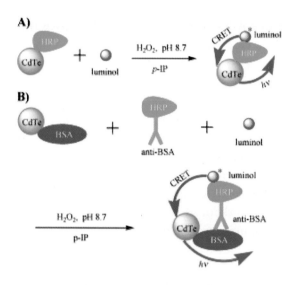

Figure 9. A) Schematic illustration of CRET based on luminol donors and HRP-labeled CdTe QD accepters, B) Schematic illustration of CRET for luminol donors and QD accepters based on the immuno-reaction of QD-BSA and anti-BSA-HRP. Reprinted with permission from [Ref. 82], X. Huang, *et al.* Angew. Chem. Int. Ed. 45, 5140(2006). Copyright @ Wiley-VCH.

2.2.2. Nanogold chemiluminescence

Catalytic property of nanogold has been investigated extensively in various fields. However, whether or not nanogold can be used as luminophor or energy acceptor to radiate light with certain wavelength, there is very few information dealing with it.

Recently, the research work from Cui's group seems changed the state. They observed the light emission at ~415 nm for gold particles with diameters of 2.6-6.0 nm dispersed in a solution containing bis(2,4,6-trichlorophenyl) oxalate and hydrogen peroxide. It was found that the light intensity was independent of the protecting reagents of the gold nanoparticles with similar size, the light intensity with gold nanoparticles of 5.0 and 6.0 nm in diameter was stronger than that with gold nanoparticles of 2.6 and 2.8 nm in diameter, and the light intensity increased linearly with the concentration of the gold nanoparticles using 6.0-nm gold nanoparticles. The gold nanoparticles were identified as emitting species, and the quantum yield was determined to be (2.8 ± 0.3) × 10^{-5} using 6.0-nm gold nanoparticles. The light emission was suggested to involve a sequence of steps: the oxidation reaction of bis(2,4,6-trichlorophenyl) oxalate with hydrogen peroxide yielding an energy-rich intermediate 1,2-dioxetanedione, the energy transfer from this intermediate to gold nanoparticles, and the radiative relaxation of the as-formed exited-state gold nanoparticles. The observed luminescence is expected to find applications in the field of bioanalysis owing to the excellent biocompatibility and relatively high stability of gold nanoparticles. [86]

3. Nanostructure presented electrochemiluminescence

3.1. QDs electrochemiluminescence

QDs have unique electronic properties depending on size and composition that can be probed by spectroscopic and electrochemical measurements. The properties can also be very sensitive to the surface structure because of the large surface-to-volume ratio of QDs compared to the bulk materials. [87-94] The electrochemistry of semiconductor NPs can sometimes reveal quantized electronic behavior as well as decomposition reactions upon reduction and oxidation. The special surface structure and wide band gap of QDs are then the vital factors to make QDs be used for ECL reaction.

In a bulk semiconductor, electrons and holes move freely throughout the crystal. However, in a nanocrystal, confinement of the electrons and holes leads to a variety of optical and electronic consequences, including size dependent molecular-like optical properties, greater electron/hole overlap for enhanced PL efficiencies, and discrete single-electron/hole charging. Because of their enormous surface area-to-volume ratios, nanocrystals (NCs) are highly susceptible to heterogeneous redox chemistry with the surrounding environment. Depending on the semiconductor and the surface chemistry, this chemical reactivity can lead to either fatal chemical degradation or new useful properties, such as reversible photocatalytic and electrochromic properties and redox reactivity.

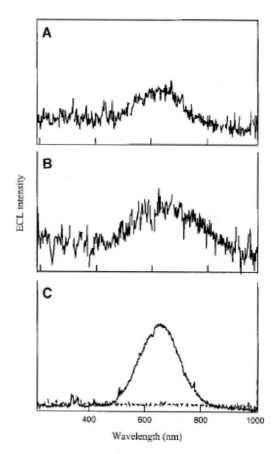

Figure 10. ECL spectra for (A) annihilation of cation and anion radicals generated by stepping the potential between 2.7 and -2.1 V at 10 Hz with an integration time of 30 min; (B) an oxalate coreactant system, stepping the potential between 0.1 and 3 V at 10 Hz, integration time 40 min; and (C) a persulfate coreactant system, stepping the potential between -0.5 and -2.5V at 10 Hz, integration time 10 min. The dotted curve (C) is the ECL spectrum for the blank solution. Reprinted with permission from [Ref. 96], Z. Ding, *et al.* Science, 296, 1293(2002). Copyright @American Association for the Advancement of Science.

In 1998, Kelly et al. reported the reduction mechanism of oxidizing agents at silicon and porous silicon electrodes in relation to light emission from the porous semiconductor in aqueous electrolyte. Hole injection at the open-circuit potential from certain oxidizing agents at porous silicon electrodes resulted in visible luminescence with characteristics similar to those of the emission observed during anodic oxidation in indifferent electrolyte. [95]

Subsequently, Bard's group systematically investigated the ECL of QDs or QDs assemblies in organic solvent and aqueous solution. The research work of Bard et al. revealed that electron transfer reactions between positively and negatively charged Si NCs (or between charged nanocrystals and molecular redox-active coreactants) occurred and led to electron and hole annihilation, producing visible light. The ECL spectra exhibited a peak maximum at 640 nanometers, a significant red shift from the PL maximum (420 nm) of the same silicon NC solution. These results demonstrated that the chemical stability of silicon NCs could enable their use as redox-active macromolecular species with the combined optical and charging properties of semiconductor quantum dots. [96]

In the ECL experiments, electron-transfer annihilation of electrogenerated anion and cation radicals results in the production of excited states (Figure 10A) [97, 98]

$$R^{\cdot-} + R^{+\cdot} \rightarrow R^* + R \tag{1}$$

$$R^* \rightarrow R + h\nu \tag{2}$$

In this case, R^- and R^+ refer to negatively and positively charged Si NCs electrogenerated at the Pt electrode, which then react in solution to give the excited state R^*.

Meanwhile, higher intensity light emission from the Si NC solution was observed when coreactants were added, which help overcome either the limited potential window of a solvent or poor radical anion or cation stability. [97] For example, by adding excess $C_2O_4^{2-}$ to the NC solution, ECL only requires hole injection and can be obtained by simply oxidizing the NCs. In this case, the oxidation of oxalate produces a strong reducing agent, $CO_2^{\cdot-}$, which can inject an electron into the LUMO of an oxidized Si NC to produce an excited state that then emits light (Figure 10B). [97, 99]

$$R^{+\bullet} + CO_2^{-\bullet} \rightarrow R^* + CO_2 \tag{3}$$

Similarly, ECL was observed in the potential region for NC reduction through the addition of excess $S_2O_8^{2-}$ to solution. Reduction of persulfate produces a strong oxidant, SO_4^-, which can then react with the negatively charged NCs by injecting a hole into the HOMO, producing an excited state (Figure 10C) [100, 101]

$$R^{-\bullet} + SO_4^{-\bullet} \rightarrow R^* + SO_4^{2-} \tag{4}$$

After the ECL experiments, the solutions showed the same PL as the original solution, so no bulk degradation of the Si NCs occurred.

The ECL spectra (Figure 10) in the above three cases all show a maximum wavelength of 640 nm, which is substantially red-shifted from that in the PL spectra. The orange ECL emission was not sensitive to NC size or the capping agent used. On the other hand, the Si NC PL is size-dependent. [102]

Bard et al. observed ECL from TOPO-capped CdSe nanocrystals dissolved in CH_2Cl_2 containing 0.1 M TBAP. Cyclic voltammetry and differential pulse voltammetry of this solution displayed no distinctive features, but light emission was observed through the annihilation of oxidized and reduced forms electrogenerated during cyclic potential scans or steps. The oxidized species was somewhat more stable than the reduced form. The ECL spectrum was substantially red shifted by 200 nm from the PL spectrum, suggesting that surface states play an important role in the emission process. [103] The ECL spectrum of CdSe/ZnSe NCs dispersed in a CH_2Cl_2 solution containing 0.1 M TBAP was obtained by stepping the potential between +2.3 and -2.3 V. Unlike the spectra from the ECL of Si or CdSe NCs, where the emission occurred in a single peak that was significantly red-shifted from the PL peak, ECL from CdSe/ZnSe produced a spectrum containing two peaks: a sharp peak whose position was almost identical to that in the PL spectrum and another broader peak with a red shift of 200 nm compared to that in the PL. This suggested emission from both surface states on the NCs and from the bulk in NCs where the surface states have been passivated. [104] Differential pulse voltammetry (DPV) of TOPO-capped CdTe NPs in dichloromethane and a mixture of benzene and acetonitrile showed two anodic and one cathodic peaks of the NPs themselves and an additional anodic peak resulting from the oxidation of reduced NPs. The electrochemical band gap (2.1 eV) between the first anodic and cathodic DPV peaks was close to the value (2 eV) obtained from the absorption spectrum. ECL of CdTe NPs was highly intense for scans into the negative potential region in dichloromethane. The fact that the ECL peak occurs at about the same wavelength as the band-edge PL peak indicates that, in contrast to CdSe NPs, the CdTe NPs as synthesized had no deep surface traps that can cause a substantially red shifted ECL. [105] ECL from the Ge NCs dispersed in DMF containing 0.1 M TBAP was observed during potential scans or pulses through the annihilation of electrogenerated oxidized and reduced species. The light emission intensity was higher during the oxidation scans and pulses, suggesting that the reduced forms were more stable than the oxidized ones. The ECL spectrum was obtained by potential stepping between the potentials for oxidation and reduction. The ECL spectrum was red-shifted from the PL spectrum by 200 nm, which implied, in agreement with previous studies, ECL emission predominantly via surface states and the importance of the surface passivation on ECL. [106, 107]

The most research works of Zhu et al. focused on the ECL of QDs and QDs assemblies in aqueous solution. Recently, they investigated the ECL of CdS NCs in aqueous solution and found its application in bioassay. Mercaptoacetic acid (RSH)-capped CdS NCs was demonstrated to be electrochemically reduced during potential scan and react with the coreactant $S_2O_8^{2-}$ to generate strong ECL in aqueous solution. Based on the ECL of CdS NCs, a novel label free ECL biosensor for the detection of low-density lipoprotein (LDL) has been developed by using self assembly and gold nanoparticle amplification techniques. [108] They synthesized CdS nanotubes using a double-template method and found the CdS nanotubes composed of compact nanocrystals exhibit strong ECL. [109] Then, its sensing application was studied by entrapping the CdS nanotubes in carbon paste electrode. Two ECL peaks were observed at -0.9 V (ECL-1) and -1.2 V (ECL-2), respectively, when the potentialwas cycled between 0 and -1.6 V. The electrochemically reduced nanocrystal species of CdS nanotubes could collide with the oxidized species in an annihilation process to produce the peak of ECL-1. The electron-transfer reaction between the reduced CdS nanocrystal species and oxidant coreactants such as $S_2O_8^{2-}$, H_2O_2, and reduced dissolved oxygen led to the appearance of the ECL-2 peak. Based on the enhancing effect of H_2O_2 on ECL-2 intensity, a novel CdS ECL sensor was developed for H_2O_2 detection. In addition, the ECL spectrum in aqueous solution also exhibited two peaks at 500 and 640 nm, respectively. [110] The strong ECL of CdS semiconductor NCs prepared by a solvothermal method was also observed by Chen et al. [111] Zhu et al. also synthesized the hollow spherical CdSe QD assemblies via a sonochemical approach that utilized β-cyclodextrin as a template reagent in aqueous solution. The as-prepared hollow nanospheres had an average diameter of 70 nm and were found to consist of an assembly of monodispersed 5 nm sized CdSe quantum dots. Following an electrochemical reaction with persulfateions, strong ECL was observed from the CdSe nanoassemblies suspended in an aqueous solution of pH ≤ 7.95. The study indicated that the morphology of the 70 nm nanoassembly played an important role in generating the stable ECL since individually dispersed quantum dots did not exhibit any significant ECL. [112] Furthermore, the same group synthesized Bi_2Te_3 hexagonal nanoflakes with controllable edge length via an ultrasonic-assisted disproportionation route, and observed the ECL of as-prepared Bi_2Te_3 for the first time. [113]

Ju et al. elucidated the detailed ECL process of the thioglycolic acid-capped CdSe QDs film/peroxide in aqueous system. The QDs were first electrochemically reduced to form electrons injected QDs -1.1 V, which then reduced hydrogen peroxide to produce $OH^{·-}$ radical. The intermediate $OH^{·-}$ radical was a key species for producing holes-injected QDs. The ECL emission with a peak at -1.114 V was demonstrated to come from the $1Se - 1Sh$ transition emission. Using thiol compounds as the model molecules to annihilate the $OH^{·-}$ radical, their quenching effects on ECL emission were studied. This effect led to a novel strategy for ECL sensing of the scavengers of hydroxyl radical. The detection results of thiol compounds showed high sensitivity, good precision, and acceptable accuracy, suggesting the promising application of the proposed method for quick detection of both scavengers and generators of hydroxyl radical in different fields. [114] They have yet proposed a simple strategy for the fabrication of the first biosensor based on the intrinsic ECL of QDs coupled with an enzymatic reaction with glucose oxidase as a model, which could be applied in more bioanalytical systems for oxidase substrates. [115]

Zhang et al. observed a band gap-ECL of ZnS nanoparticles (NPs) in alkaline aqueous solution at a platinum electrode during the potential applied between -2.0 V (versus Ag/AgCl, saturated KCl) and +0.86 V. The ECL peak of ZnS NPs in 0.10 M sodium hydroxide solution appeared at +0.86 V, and the ECL peak wavelength of the ZnS NPs was ~460 nm. The ECL scheme of the ZnS NPs in alkaline aqueous solution was proposed, indicating that the surface passivation effect and the core/shell structure of $ZnS/Zn(OH)_2$ played a significant role in the ECL process and that the similarity of the ECL and PL spectra of semiconductor NPs was dependent on the extent of the surface passivation. The ECL intensity of ZnS NPs in alkaline aqueous solution was greatly enhanced by an addition of $K_2S_2O_8$. [116]

3.2. Nanostructure assisted electrochemiluminescence

Nanostructures, including Si nanoparticles, SiO_2 nanoparticles, carbon nanotube, and clay nanoparticles, due to their electric charge and absorption properties, were also utilized as immobilized substrate to bind with tris(2,2'-bipyridyl)ruthenium (II) (($Ru(bpy)_3^{2+}$) or luminol for ECL sensor preparation.

Dong et al. proposed serials of ECL sensor based on the immobilization of $Ru(bpy)_3^{2+}$ with nanoparticles. They proposed a simple method of preparing $\{SiO_2/Ru(bpy)_3^{2+}\}$n multilayer films. Positively charged $Ru(bpy)_3^{2+}$ and negatively charged SiO_2 nanoparticles were assembled on ITO electrodes by a layer-by-layer method. The multilayer films containing $Ru(bpy)_3^{2+}$ was used for ECL determination of TPA, and the sensitivity was more than 1 order of magnitude higher than that observed for previous reported immobilization methods for the determination of TPA. The multilayer films also showed better stability for one month at least. The high sensitivity and stability mainly resulted from the high surface area and special structure of the silica nanoparticles. [117] The same group also developed a novel ECL sensor based on $Ru(bpy)_3^{2+}$-doped silica (RuDS) nanoparticles conjugated with a biopolymer chitosan membrane. These uniform RuDS nanoparticles

(~40 nm) were prepared by a water-in-oil microemulsion method and were characterized by electrochemical and transmission electron microscopy technology. The $Ru(bpy)_3^{2+}$-doped interior maintained its high ECL efficiency, while the exterior nanosilica prevented the luminophor from leaching out into the aqueous solution due to the electrostatic interaction. This sensor shows a detection limit of 2.8 nM for tripropylamine, which is 3 orders of magnitude lower than that observed at a Nafion-based ECL sensor. Furthermore, the present ECL sensor displays outstanding long-term stability. [118] They either immobilized $Ru(bpy)_3^{2+}$ in $\{clay/Ru(bpy)_3^{2+}\}_n$ multilayer films and ion-exchanged $Ru(bpy)_3^{2+}$ in carbon nanotube (CNT) /Nafion composite films to construct more effective ECL sensors. [119, 120]

Zheng et al. synthesized the core-shell luminol-doped SiO_2 nanoparticles and immobilized it on the surface of chitosan film coating graphite electrode by the self-assembled technique. Then, a ECL sensor for pyrogallol was developed based on its ECL enhancing effect for the core-shell luminol-doped silica nanoparticles. The ECL analytical performances and the sensing mechanism of this ECL sensor for pyrogallol were investigated in detail. The corresponding results showed that: compared with the conventional ECL reaction procedures by luminol ECL reaction system, the electrochemical (EC) reaction of pyrogallol and its subsequent CL reaction occurred in the different spatial region whilst offering a high efficiency to couple the EC with the CL reaction to form the ECL procedures. In this case, this new sensing scheme offered more potential to improve the analytical performances of the ECL reaction. Under the optimum experimental conditions, this ECL sensor showed less than 5% decrease in continuums over 100 times ECL measurements, the detection limit was 1.0×1.0^{-9} mol/L for pyrogallol. The linear range extended from 3.0×10^{-9} mol/L to 2.0×10^{-5} mol/L for pyrogallol. [121]

$Ru(bpy)_3^{2+}$-doped silica nanoparticles ($Ru(bpy)_3^{2+}$-doped SNPs) have been used as DNA tags for sensitive ECL detection of DNA hybridization. In this protocol, ($Ru(bpy)_3^{2+}$-doped SNPs) was used for DNA labeling with trimethoxysilylpropydiethylenetriamine (DETA) and glutaraldehyde as linking agents. The ($Ru(bpy)_3^{2+}$-doped SNPs labeled DNA probe was hybridized with target DNA immobilized on the surface of polypyrrole (PPy) modified Pt electrode. The hybridization events were evaluated by ECL measurements and only the complementary sequence could form a double-stranded DNA (dsDNA) with DNA probe and give strong ECL signals. A three-base mismatch sequence and a non-complementary sequence had almost negligible responses. Due to the large number of $Ru(bpy)_3^{2+}$ molecules inside SNPs, the assay allows detection at levels as low as 1.0×10^{-13} mol l^{-1} of the target DNA. The intensity of ECL was linearly related to the concentration of the complementary sequence in the range of 2.0×10^{-13} to 2.0×10^{-9} mol l^{-1}. [122]

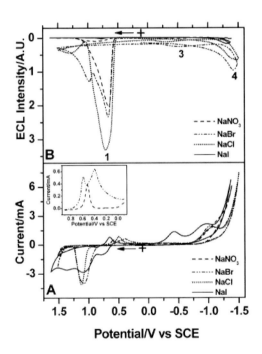

Figure 11. Effect of electrolytes on CV (A) and IECL-E (B) curves of luminol. Electrolytes: 0.1 mol/L NaNO$_3$ (– – –), 0.1 mol/L NaBr (- ·· -), 0.1 mol/L NaCl (- - -), 1×10^{-3} mol/L NaI (–). Inset shows the enlarged CVs from 0.90 to -0.10 V. Reprinted with permission from [Ref. 123], H. Cui, *et al.* Anal. Chem. 76, 4002(2004). Copyright @ American Chemical Society.

3.3. Nanogold presented electrochemiluminescence

The catalysis of gold nanoparticles was still considered and often utilized to amplify the ECL signals in ECL reactions.

Cui et al. examined the ECL behavior of luminol on a gold nanoparticle self-assembled electrode in neutral and alkaline conditions under conventional cyclic voltammetry (CV). The gold nanoparticle self-assembled electrode exhibitied excellent electrocatalytic property and redox reactivity to the luminol ECL system. In neutral solution, four ECL peaks were observed at 0.69, 1.03, -0.45 (Figure 11), and -1.22 V (vs SCE) on the curve of ECL intensity versus potential. Compared with a bulk gold electrode, two anodic and one cathodic ECL peaks were greatly enhanced, and one new cathodic ECL peak appeared. In alkaline solution, two anodic ECL peaks were obtained at 0.69 and 1.03 V, which were much stronger than those on a bulk gold electrode. These ECL peaks were found to depend on gold nanoparticles on the surface of the electrode, potential scan direction and range, the presence of O_2 or N_2, the pH and concentration of luminol solution, NaBr concentration, and scan rate. The emitter of all ECL peaks was identified as 3-aminophthalate by analyzing the ECL spectra. [123] Similarly, the ECL of lucigenin on a gold nanoparticle self-assembled gold electrode in neutral and alkaline solutions was also studied. The emitter of all ECL peaks was identified as N-methylacridone (NMA). [124] They further studied ECL of luminol on a gold-nanorod-modified gold electrode and obtained five ECL peaks under conventional CV in both neutral and alkaline solutions. Their results indicated that a gold-nanorod-modified gold electrode had a catalytic effect on luminol ECL different from that of a gold-nanosphere-modified gold electrode, revealing that the shape of the metal nanoparticles had an important effect on the luminol ECL behavior. [125] Moreover, the same group compared the ECL behavior of luminol on various electrodes modified with gold nanoparticles of different size in neutral solution by CV. The results demonstrated that the gold nanoparticle modified electrodes could generate strong luminol ECL in neutral pH conditions. The catalytic performance of gold nanoparticle modified electrodes on luminol ECL depended not only on the gold nanoparticles but also on the substrate. Gold electrode and glassy carbon electrode were the most suitable substrates for the self-assembly of gold nanoparticles. Moreover, the gold nanoparticle modified gold and glassy carbon electrode had satisfying stability and reproducibility and did not need tedious pretreatment of electrode surface before each measurement. It was also found that luminol ECL behavior depended on the size of gold nanoparticles. The most intense ECL signals were obtained on a 16-nm-diameter gold nanoparticle modified electrode. [126]

Wang et al presented a gold nanoparticle amplification approach for ECL determination of a biological substance (bovine serum albumin (BSA) and immunoglobulin G (IgG) using 4-(Dimethylamino)butyric acid (DMBA) as label. With gold nanoparticle amplification, the ECL peak intensity was proportional to the concentration over the range 1-80 and 5-100 μg/mL for BSA and IgG consuming 50 μL of sample, respectively. A 10- and 6-fold sensitivity enhancement was obtained for BSA and IgG over their direct immobilization on an electrode using DMBA labeling. The relative standard deviations of five replicate determinations of 10 μg/mL BSA and 20 μg/mL IgG were 8.4 and 10.2%, respectively. [127]

Zhang et al. proposed gold nanoparticles carrying multiple probes for the ECL detection of DNA hybridization. 2,2'-bipyridine-4,4'-dicarboxylicacid-N-hydroxysuccinimide ester (Ru(bpy)$_2$(dcbpy)NHS) was used as a ECL label and gold nanoparticle as a carrier. Probe single strand DNA (ss-DNA) was self-assembled at the 3'-terminal with a thiol group to the surface of gold nanoparticle and covalently labeled at the Y-terminal of aphosphate group with Ru(bpy)$_2$(dcbpy)NHS and the resulting conjugate, (Ru(bpy)$_2$(dcbpy)NHS)-ss-DNA-Au, was taken as a ECL probe. When target analyte ss-DNA was immobilized on a gold electrode by self-assembled monolayer technique and then hybridized with the ECL probe to form a double-stranded DNA (ds-DNA), a strong ECL response was electrochemically generated. The ECL signal generated from many reporters of ECL probe prepared is greatly amplified, compared to the convention scheme which is based on one reporter per hybridization event. [128]

3.4. Other nanostructure presented electrochemiluminescence

Other nanostructures, like Pt nanoparticles, magnetic nanoparticles, polycrystalline diamond and Bi_2Te_3 hexagonal nanoflakes, were also used in ECL analysis.

Dong et al. treated colloidal Pt solution with $Ru(bpy)_3^{2+}$ and caused the assembly of Pt nanoparticles into aggregates. Most importantly, directly placing such aggregates on bare solid electrode surfaces can produce very stable films exhibiting excellent ECL behaviors. [129] Wang et al. proposed a facile synthesis of the novel platinum nanoparticles/EastmanAQ55D/ruthenium (II) tris (bipyridine) ($PtNPs/AQ/Ru(bpy)_3^{2+}$) colloidal material for ultrasensitive ECL solid-state sensors. The cation ion-exchanger AQ was used not only to immobilize ECL active species $Ru(bpy)_3^{2+}$ but also as the dispersant of Pt NPs. The electronic conductivity and electroactivity of Pt NPs in composite film made the sensor exhibit faster electron transfer, higher ECL intensity of $Ru(bpy)_3^{2+}$, and a shorter equilibration time than $Ru(bpy)_3^{2+}$ immobilized in pure AQ film. [130] Cui et al. studied the ECL behavior of lucigenin at a glassy carbon electrode in the presence of platinum nanoparticles dispersed in alkaline aqueous solutions. Two ECL peaks were observed at -0.65 and -2.0 V, respectively. ECL-1 was a conventional ECL peak of lucigenin also observed in the absence of platinum nanoparticles. ECL-2 was a new ECL peak appearing in the hydrogen-evolution potential region. It was found that ECL-1 decreased and ECL-2 increased with an increase in the concentration of platinum nanoparticles. [131]

Nafion-stabilized magnetic nanoparticles ($Nafion/Fe_3O_4$) formed on a platinum electrode surface by means of an external magnet have been fabricated for highly sensitive and stable $Ru(bpy)_3^{2+}$ ECL sensor. [132] ECL was used to image the spatial variations in electrochemical activity at the heavily doped polycrystalline diamond surface. ECL was generated by the reaction of $Ru(bpy)_3^{2+}$ and tripropylamine. Images of the CL patterns at the polycrystalline diamond surface were recorded photographically after magnification with optical microscopy to show the location and size of individual active regions. The spatial distribution for ECL intensity indicated that the electrochemical reactivity at polycrystalline diamond electrodes was microscopically heterogeneous. [133]

4. Conclusions and Outlook

We have reviewed the recent applications of nanomaterials and nanostructures to CL and ECL. QDs CL and ECL, metal nanostructures catalyzed CL, ECL and amplification technique, and the other nanostructures presented CL and ECL open new horizons for CL and ECL analysis, and they certainly may be developed as detection techniques in biomedical, environmental and related analysis. This not only takes CL and ECL analysis into nanoscale substrates from molecules and atom level, but would made possible for the application of novel nanotechnologies utilizing the CL and ECL characteristics of nanostructures.

Acknowledgements

This work was partly supported from "973" National Key Basic Research Program of China (Grant No. 2007CB310500) and National Natural Science Foundation of China (No. 20435010 and No. 20675044).

References

1. E. Katz, I. Willner, *Angew. Chem. Inter. Ed.*, 43, 6042(2004).
2. K. Y. Chumbimuni-Torres, Z. Dai, N. Rubinova, Y. Xiang, E. Pretsch, J. Wang, E. Bakker, *J. Am. Chem. Soc.*, 128, 13676(2006).
3. N. T. K. Thanh, Z. Rosenzweig, *Anal. Chem.*, 74, 1624(2002).
4. M. Bruchez, M. Moronne, P. Gin, S. Weiss, A. P. Alivisatos, *Science*, 281, 2013(1998).
5. W. C. W. Chan, S. M. Nie, *Science*, 281, 2016(1998).
6. Y. Li, G. Hong. *Luminescence,* 124, 297(2007).
7. J. E. Wampler, Instrumentation: *Seeing the Light and Measuring It, in Chemi- and Bioluminescence*, J.G. Burr, ed., Marcel Dekker, New York, 1-44 (1985).
8. A. K. Campbell, *Detection and Quantification of Chemiluminescence, in Chemiluminescence Principles and Applications in Biology and Medicine,* Ellis Horwood, Chichester, 68-126 (1988).

9. F. Berthold, *Instrumentation for Chemiluminescence Immunoassays, in Luminescence Immunoassay and Molecular Applications,* K. Van Dyke and R. Van Dyke, eds., CRC Press, Boca Raton, 11-25 (1990).

10. T. Nieman, *Chemiluminescence: Theory and Instrumentation, Overview,* in Encyclopedia of Analytical Science, Academic Press, Orlando, 608 (1995).

11. A. M. Garc-Campa, W. R. G. Baeyens, *Chemiluminescence in analytical chemistry,* Marcel Dekker, New York . (2001).

12. A. M. Powe, K. A. Fletcher, N. N. St. Luce, M. Lowry, S. Neal, M. E. McCarroll, P. B. Oldham, L. B. McGown, I. M. Warner, *Anal. Chem.,* 76, 4614(2004).

13. J. Yakovleva, R. Davidsson, A. Lobanova, M. Bengtsson, S. Eremin, T. Laurell, J. Emneus, *Anal. Chem.,* 74, 2994(2004).

14. F. M. Li, C. H. Zhang, X. J. Guo, W. Y. Feng, *Biomed. Chromatogr.,* 17, 96(2003).

15. J. G. Lv, Z. J. Zhang, J. D. Li, L.R. Luo, *Forensic Sci. Int.* 148, 15(2005).

16. C. Dodeigne, L. Thunus, R. Lejeune, *Talanta,* 51, 415(2000).

17. J. H. Lin, Ju, H. X. *Biosens. Bioelectron.,* 20, 1461 (2005).

18. A. Dapkevicius, T. A. van Beek, H. A.G. Niederlander, A. de Groot, *Anal. Chem.,* 71, 736(1999).

19. J. Wang, W. Huang, Y. Liu, J. Cheng, J. Yang, *Anal. Chem.,* 76, 5393(2004).

20. J. K. Leland, M. J. Powell, *J. Electroanal. Chem.,* 318, 91(1991).

21. A. J. Bard, J. D. Debad, J. K. Leland, G. B. Sigal, J. L. Wilbur, J. N. Wohlstadter, Encyclopedia of Analytical Chemistry; R. A. Meyers, Ed.; Wiley: Chichester, U.K., 9842 (2000).

22. G. F. Blackburn, H. P. Shah, J. H. Kenten, J. Leland, R. A. Kamin, J. Link, J. Peterman, M. J. Powell, A. Shah, D. B. Talley, *Clin. Chem.,* 37, 1534(1991).

23. K. A. Fahnrich, M. Pravda, G. C. Guilbault, *Talanta,* 54, 531(2001).

24. R. D. Gerardi, N. W. Barnett, S. W. Lewis, *Anal. Chim. Acta,* 378, 1(1999).

25. A. W. Knight. *Trends Anal. Chem.,* 18, 47(1999).

26. W. Y. Lee, *Microchim. Acta,* 127, 19(1997).

27. A. W. Knight, G. M. Greenway, *Analyst,* 119, 879(1994).

28. J. G. Velasco. *Bull Electrochem.,* 10, 29(1994).

29. N. N. J. Rozhitskii, *Anal. Chem.* 47, 1288(1992).

30. J. G. Velasco, *Electroanalysis,* 3, 261(1991).

31. H. Cui, Z.-F. Zhang, M.-J. Shi, *J. Phys. Chem. B,* 109, 3099(2005).

32. Z.-F. Zhang, H. Cui, C.-Z. Lai, L.-J. Liu, *Anal. Chem.,* 77, 3324 (2005).

33. Z.-F. Zhang, H. Cui, M.-J. Shi, P*hys. Chem. Chem. Phys.,* 8, 1017(2006).

34. C. Duan, H. Cui, Z. Zhang, B. Liu, J. Guo, W. Wang, *J. Phys. Chem. C,* 111, 4561(2007).

35. Z. P. Wang, J. Q. Hu, Y. Jin, X. Yao, J.H. Li, *Clin. Chem.* 52, 1958(2006).

36. L. Wang, P. Yang, Y. Li, *Talanta,* 72, 1066 (2007).

37. S.L. Xu, H. Cui, *Luminescence,* 22, 77(2007).

38. T. Niazov, B. Shlyahovsky, I. Willner, *J. Am. Chem. Soc.,* 129, 6374(2007).

39. J. Z. Guo, H. Cui, *J. Phys. Chem. C,* 111, 12254(2007).

40. B. A. Gorman, P. S. Francis, D. E. Dunstanb, N. W. Barnett, *Chem. Commun.*, 395(2007).

41. A. Fan, C. Lau, J. Lu, *Anal. Chem.*, 77, 3238(2005).

42. Z.-P. Li, Y.-C. Wang, C.-H. Liu, Y.-K. Li, *Anal. Chim. Acta*, 551, 85 (2005).

43. C.-H. Liu, Z.-P. Li, B.-A. Du, X.-R. Duan, Y.-C. Wang, *Anal. Chem.*, 78, 3738(2006).

44. L. R. Luo, Z. J. Zhang, L. Y. Hou, *Anal. Chim. Acta*, 584, 106(2007).

45. M. Breysse, B. Claudel, L. Faure, M. Guenin, R.J. Williams, *J. Catal.*, 45, 137(1976).

46. B. Claudel, M. Breysse, L. Faure, M. Guenin, *Rev. Chem. Int.*, 2, 75(1978).

47. M. Nakagawa, N. Fujiwara, Y. Matsuura, T. Tomiyama, I. Yamamoto, K. Utsunomiya, T. Wada, N. Yamashita, Y. Yamashita, *Bunseki Kagaku*, 39, 797 (1990).

48. K. Utsunomiya, M. Nakagawa, T. Tomiyama, I. Yamamoto, Y. Matsuura, S.M. Chikamori, T. Wada, N. Yamashita, Y. Yamashita, *Sensors Actuators B*, 13–14, 627(1993).

49. M. Nakagawa, *Sensors Actuators B*, 29, 94(1995).

50. M. Nakagawa, T. Okabayashi, T. Fujimoto, K. Utsunomiya, I. Yamamoto, T. Wada, Y. Yamashita, N. Yamashita, *Sensors Actuators B*, 51, 159(1998).

51. T. Okabayashi, T. Fujimoto, I. Yamamoto, K. Utsunomiya, T. Wada, Y. Yamashita, M. Nakagawa, *Sensors Actuators B*, 64, 54(2000).

52. T. Okabayashi, T. Toda, I. Yamamoto, K. Utsunomiya, N. Yamashita, M. Nakagawa, *Sensors Actuators B*, 74, 152(2001).

53. P. McCord, S.L. Yau, A.J. Bard, *Science*, 257, 68(1992).

54. Y.F. Zhu, J.J. Shi, Z.Y. Zhang, C. Zhang, X.R. Zhang, *Anal. Chem.*, 74, 120 (2002).

55. J.J. Shi, J.J. Li, Y.F. Zhu, F. Wei, X.R. Zhang, *Anal. Chim. Acta*, 466, 69(2002).

56. Z.Y. Zhang, C. Zhang, X.R. Zhang, *Analyst*, 127(2002) 792.

57. J.J. Shi, R.X. Yan, Y.F. Zhu, X.R. Zhang, *Talanta*, 61, 157(2003).

58. G.H. Liu, Y.F. Zhu, X.R. Zhang, B.Q. Xu, *Anal. Chem.*, 74, 6279(2002).

59. Z. Miao, Y. Wu, X. Zhang, *J. Mater. Chem.*, 17, 1791(2007).

60. J. Shi, Y. Zhu, X. Zhang, W. R.G. Baeyens, A. M. García-Campaña, *Trends Anal. Chem.* 23, 351(2004).

61. Y. Lv, S. Zhang, G. Liu, M. Huang, X. Zhang, *Anal. Chem.*, 77, 1518(2005).

62. G. Huang, Y. Lv, S. Zhang, C. Yang, X. Zhang, *Anal. Chem.*, 77, 7356(2005).

63. Z. Y. Zhang, K. Xu,; W. R. G. Baeyens, X. R. Zhang, *Anal. Chim. Acta*, 535, 145(2005).

64. N. Na, S. Zhang, S. Wang, X. Zhang, *J. Am. Chem. Soc.*, 128, 14420(2006).

65. X. Wang, N. Na, S. Zhang, Y. Wu, X. Zhang, *J. Am. Chem. Soc.*, 129, 6062(2007).

66. M. Breysse, B. Claudel, L. Faure, M. Guenin, R. J. Williams, *J. Catal.* 45, 137(1976).

67. K. Nakao, S. Ito, K. Tomishige, K. Kunimori, *J. Phys. Chem. B*, 109, 17553(2005).

68. K. Nakao, S. Ito, K. Tomishige, K. Kunimori, *J. Phys. Chem. B*, 109, 24002(2005).

69. H. Liu, A.I. Kozlov, A. P. Kozlova, T. Shido, K. Asakura, Y. Ivasawa, *J. Catal.,* 185, 252(1999).

70. F. Teng, B. Gaugeu, S. Liang, Y. Zhu, *Appl. Catal. A*, 294, 158(2007).

71. W. Chen, J. Z. Zhang, A. G. Joly, *J. Nanosci. Nanotech.,* 4, 919(2004).

72. S. W. Chen, L. A. Truax, J. M. Sommers, *Chem. Mater.,* 12, 3864(2000).

73. L. H. Qu, X. G. Peng, *J. Am. Chem. Soc.*, 124, 2049(2002).

74. S. Coe, W. K. Woo, M. Bawendi, V. Bulovic, *Nature*, 420, 800(2002).

75. N. Tessler, V. Medvedev, M. Kazes, S. H. Kan, U. Banin, *Science*, 295, 1506(2002).

76. M. P. Bruchez, M. Moronne, P. Gin, S. Weiss, A. P. Alivisatos, *Science*, 281, 2013(1998).

77. S. V. Gaponenko, *Optical Properties of Semiconductor Nanocrystals*, Cambridge University Press: Cambridge (1998).

78. S. K. Poznyak, D.V. Talapin, E. V. Shevchenko, H. Weller, *Nano Lett.,* 4, 4693(2004).

79. J. Rodriguez-Viejo, K. F. Jensen, H. Mattoussi, J. Michel, B. O. Dabbousi, M. G. Bawendi, *Appl. Phys. Lett.*, 70, 2132(1997).

80. Z. P. Wang, J. Li, B. Liu, J. Q. Hu, X. Yao, J. H. Li, *J. Phys. Chem. B*, 109, 23304(2005).

81. Y. X. Li, P. Yang, P. Wang, X. Huang, L. Wang, *Nanotechnology,* 18, 25602-1(2007).

82. X. Huang, L. Li, H. Qian, C. Dong, J. Ren, *Angew. Chem. Int. Ed.,* 45, 5140(2006).

83. L. Li, H. F. Qian, J. C. Ren, *J. Lumin.*, 116, 59(2006).

84. J. N. Wang, J. C. Ren, *Electrophoresis,* 26, 2402(2005).

85. M. K. So, C. Xu, A. M. Loening, S. S. Gambhir, J. Rao, *Nat. Biotechnol.*, 24, 339(2006).

86. H.Cui, Z.-F.Zhang, M.-J. Shi, Y. Xu, Y.-L. Wu, *Anal. Chem.,* 77, 6402(2005).

87. A. Henglein, *Chem. Rev.*, 89, 1861(1989).

88. M. L. Steigerwald, L. E. Brus, *Acc. Chem. Res.*, 23, 183(1990).

89. M. G. Bawendi, M. L. Steigerwald, L. E. Brus, *Annu. Rev. Phys. Chem.*, 41, 477(1990).

90. Y. Wang, Y. Herron, *J. Phys. Chem.*, 95, 525(1991).

91. H. Weller, *Adv. Mater.*, 5, 88(1993).

92. H. Weller, *Angew. Chem., Int. Ed.*, 32, 41(1993).

93. A. Hagfeldt, M. Grätzel, *Chem. Rev.*, 95, 49(1995).

94. A. P. Alivisatos, *J. Phys. Chem.*, 100, 13226(1996).

95. E. S. Kooij, K. Butter, J. J. Kelly, *J. Electrochem. Soc.*, 145, 1232(1998).

96. Z. Ding, B. M. Quinn, S. K. Haram, L. E. Pell, B. A. Korgel, A. J. Bard, *Science*, 296, 1293(2002).

97. A. J. Bard, L. R. Faulkner, Electrochemical Methods, Fundamentals and Applications, Wiley, New York (2001).

98. L. R. Faulkner, A. J. Bard, in Electroanalytical Chemistry, A. J. Bard, Ed., Dekker, New York, 20, 1-95(1977).

99. T. C. Richards, A. J. Bard, *Anal. Chem.*, 67, 3140 (1995).

100. H. S. White, A. J. Bard, *J. Am. Chem. Soc.,* 104, 6891(1982).

101. A. W. Knight, G. M. Greenway, *Analyst,* 119, 879(1994).

102. J. D. Holmes, *J. Am. Chem. Soc.,* 123, 3743(2001).

103. N. Myung, Z. Ding, A. J. Bard, *Nano Lett.,*2, 315(2002).

104. N. Myung, Y. Bae, A. J. Bard, *Nano Lett.*, 3, 1053(2003).

105. Y. Bae, N. Myung, A. J. Bard, *Nano Lett.*, 4, 1153(2004).

106. N. Myung, X. Lu, K. P. Johnston, A. J. Bard, *Nano Lett.*, 4, 183(2004).

107. Y. Bae, D. C. Lee, E. V. Rhogojina, D. C. Jurbergs, B. A. Korgel, A. J. Bard, *Nanotechnology*, 17, 3791(2006).

108. G. Jie, B. Liu, H. Pan, J. Zhu, H. Chen, *Anal. Chem.*, 79, 5574(2007).

109. J.-J. Miao, T. Ren, L. Dong, J.-J. Zhu, H.-Y. Chen, *Small*, 1, 802(2005).

110. G. F. Jie, B. Liu, J. J. Miao, J. J. Zhu, *Talanta*, 71, 1476(2007).

111. M. Chen, L. J. Pan, Z. Q. Huang, J. M. Cao, Y. D. Zheng, H. Q. Zhan, *Mat. Chem. Phys.*, 101, 317(2007).

112. B. Liu, T. Ren, J. R. Zhang, H. Y. Chen, J. J. Zhu, C. Burda, *Electrochem. Comm.*, 9, 551(2007).

113. B. Zhou, B. Liu, L. P. Jiang, J.J. Zhu, *Ultrason. Sonochem.*, 14, 229(2007).

114. H. Jiang, H. Ju, *Anal. Chem.*, 2007, 79, 6690(2007).

115. H. Jiang, H. X.Ju, *Chem. Comm.*, 404(2007).

116. L. Shen, X. Cui, H. Qi, C. Zhang, *J. Phys. Chem. C*, 111, 8172(2007).

117. Z. Guo, Y. Shen, M. Wang, F.G. Zhao, S. Dong, *Anal. Chem.*, 76, 184(2004).

118. L. Zhang, S. Dong, *Anal. Chem.*, 78, 5119(2006).

119. Z. Guo, Y. Shen, M. Wang, S. Dong, *Analyst*, 129, 657(2004).

120. Z. Guo, S. Dong. *Anal. Chem.*, 76, 2683(2004).

121. L. Zhang, X. Zheng, *Anal. Chim. Acta*, 570, 207 (2006).

122. Z. Chang, J. Zhou, K. Zhao, N. Zhu, P. He, Y. Fang, *Electrochim. Acta*, 52, 575(2006).

123. H. Cui, Y. Xu, Z.-F. Zhang, *Anal. Chem,.* 76, 4002(2004).

124. H. Cui, Y. P..Dong, J. *Electroanal. Chem.*, 595, 37(2006).

125. Y.-P. Dong, H.Cui, C.-M Wang, *J. Phys. Chem. B,* 110, 18408(2006).

126. Y.-P. Dong, H. Cui, Y. Xu, *Langmuir*, 23, 523(2007).

127. X.-B. Yin, B. Qi, X. Sun, X. Yang, E. Wang, *Anal. Chem.,* 77, 3525(2005).

128. H. Wang, C. X. Zhang, Y. Li, H. L. Qi, *Anal. Chim. Acta*, 575, 205(2006).

129. X. P. Sun, Y. Du, L. X. Zhang, S. J. Dong, E. K Wang, *Anal. Chem.,* 78, 6674(2006).

130. Y. Du, B. Qi, X. R. Yang, E. K. Wang, *J. Phys. Chem. B*, 110, 21662(2006).

131. J.-Z. Guo, H. Cui, S.-L. Xu, Y.-P. Dong, *J. Phys. Chem. C*, 111, 606(2007).

132. D.-J. Kim, Y.-K. Lyu, H. N. Choi; I.-H. Min, W.-Y. Lee, *Chem. Comm.*, 2966(2005).

133. K. Honda, T. Noda, A. Yoshimura, K. Nakagawa, A. Fujishima, *J. Phys. Chem. B*, 108, 16117(2004).

CHAPTER 3

EXCITONS IN NANOSCALE SYSTEMS: FUNDAMENTALS AND APPLICATIONS

Gregory D. Scholes and Garry Rumbles

Department of Chemistry 80 St. George Street, Institute for Optical Sciences, and Centre for Quantum Information and Quantum Control, University of Toronto, Toronto, Ontario M5S 3H6 Canada
E-mail: gscholes@chem.utoronto.ca
National Renewable Energy Laboratory, Chemical and Biosciences Center, MS3216, 1617 Cole Boulevard, Golden, Colorado 80401-3393 U.S.A.
E-mail: garry_rumbles@nrel.gov

The focus if this review is to ask: What is a nanoscale exciton and how should we think about its photo-induced dissociation? An overview of the field provides a perspective and identifies the questions presently being examined. Specific examples are discussed, including conjugated polymers, carbon nanotubes, semiconductor nanocrystals (quantum dots), as well as hybrid systems. On one hand it is shown why stable excitons are the primary photo-excitations in nanoscale systems. On the other hand, we discuss why and how these states can be dissociated. In that context we relate nanoscale excitons to potential applications in photovoltaics and light-emitting devices.

1. Introduction

Nanosystems represent the most compact kind of device or machinery, much like their biological counterparts: proteins and enzymes. Nanoscale systems are forecast to be a means of integrating desirable attributes of molecular and bulk regimes into easily processed materials. Notable examples include plastic light-emitting devices and organic solar cells, the operation of which hinge on the formation and control of electronic excited states, nanoscale excitons. The spectroscopy of nanoscale materials reveals details of their collective excited states, characterized by atoms or molecules working together to capture and redistribute excitation. In this review we build and expand upon our recent article [1]. Our goal is to establish the nature and special properties of excitons in nanometer-sized materials. Furthermore, we extend the scope of our previous discussion by exploring the concept and observations of exciton dissociation; that is, the formation of free carriers. Does free carrier formation occur directly upon photoexcitation, or does the predominant pathway involve exciton dissociation in the presence of intrinsic defects or at interfaces?

We concluded previously that the new aspect of light absorption that is prevalent—or even that defines—nanoscale systems is that the physical size and shape of the material significantly influences the properties of electronic levels and excited states [1]. That is of interest in the field because: (a) Optical properties can be engineered in a material by the arrangement of building blocks; (b) The spatial compression of the exciton accentuates many of its interesting physical properties, exposing them for examination; (c) Our understanding is challenged: Nanoscale materials provide both a test-bed and an inspiration for new approaches leading to elucidation of the electronic properties of large systems. An exciting aspect of nanoscience, therefore, is that relationships between structure and electronic properties are being revealed through a combination of synthesis, structural characterization, chemical physics, and theory.

The study of small molecules in the gas phase has revealed the importance of just a few nuclear coordinates (e.g. bond stretching directions), quantum effects can be important, and reactions are described in terms of potential energy surfaces and conical intersections [2-6]. These systems have served as test beds for increasing the precision and sophistication of quantum chemical calculations [7,8]. The examination of large molecules, such as dyes and other chromophores that absorb in the ultraviolet and visible region of the spectrum, has shown how the nuclear degrees of freedom of the molecule and its surroundings are so numerous that they are best grouped into just a few effective nuclear coordinates (often called a reaction coordinate) [9-12].

Photophysical processes are typically described in terms of free energy curves as a function of these reaction coordinates, highlighting the statistical nature of measurements and the diminished importance of quantum modes in determining the primary processes being studied. Precise chemical design and assembly—supramolecular chemistry [13]—led to demonstrations of how structural arrangements of molecular building blocks have a significant impact on excited state properties and dynamics [14,15].

As we venture into the investigation of nanoscale systems, new challenges as well as new areas of focus are emerging. For example, are the nuclear degrees of freedom more weakly coupled to nanoscale excitons than large molecules in the condensed phase? Energetic disorder is potentially much larger in nanoscale systems owing to size distributions of the exciton. What are the implications of that? Furthermore, how do nanoscale excitons evolve in complex systems such as polymer blends, polymer–nanocrystals blends, and so on? If excitons are the stable electronic excited states of nanoscale systems, what promotes photo-induced free carrier formation?

2. What is an Exciton in a Nanoscale System?

Nanoscale excitons are simply electronic excitations. These excitations are characteristically large in terms of their spatial extent, and it can be convenient to think of this size to evolve through interactions among subunits that make up the structure. Those subunits may be atoms, such as carbon atoms in a single-wall carbon nanotube (SWNT), or they can be molecules or molecular subunits, as is the case in aggregates, crystals, and macromolecules. A prototypical nanoscale exciton is therefore described as a delocalized excitation, perhaps involving charge transfer—or sharing of electron density—among constituent subunits of the system. Each exciton state is a ladder of levels, converging to a continuum (at zero temperature) as one or more dimensions of the system approaches infinite size. The size of a nanoscale exciton can help excitation and charges to disperse rapidly over long length scales. It is therefore not surprising that nanoscale excitons have great potential in applications such as light emitting or photovoltaic devices [16-18].

2.1 Limiting Cases of Excitons

Excitons are typically discussed in two limiting cases [19,20]. In the first limiting case, it is assumed that the electronic interaction between the subunits is large. Then molecular orbitals that are delocalized over entire system are a good starting point for describing electronic states. Photo-excitation introduces an electron into the conduction orbitals, leaving a "hole" in the valence orbitals. The essentially free motion of the resulting electron and hole leads to formation of a Wannier-Mott type of exciton, characterized by a weak mutual attraction of the electron and hole, which are on average separated by several subunits. The strength of the electron-hole attraction determines the "binding" of the lower energy, optically allowed, states compared to the dense manifold of charge transfer (CT) exciton states, as shown in a recent paper examining these concepts [21]. In a common assumption the electron and hole move under their mutual attraction in a dielectric continuum, and then the exciton energy levels are found as a series analogous to the Rydberg series. Such a model cannot capture details of bonding and structure. This first limit naturally converges to the free carrier limit where the electron-hole attraction is negligible compared to thermal energies.

In the second limiting case, electronic excitation is delocalized over the subunits (usually molecules), but the electron and hole are together localized on individual subunits. That situation arises when the subunits are separated from each other by ~5 Ångstrom or more, in which case sharing of electron density among subunits is negligibly small in magnitude. In other words, the orbital overlap between molecules is small owing to the exponential decay of wavefunction tails. This limit is known as the Frenkel exciton limit and has been usefully applied to numerous systems, including assemblies of molecules in crystals [22], photosynthetic light-harvesting proteins [23,24], and molecular aggregates [25,26].

2.2 A General Picture of Nanoscale Excitons

What are the distinguishing features of nanoscale excitons and their properties? Size tuning of the electronic excitation energies has attracted much attention in the field, and examples are shown in Figure 1. Indeed, this phenomenon is not unique to nanoscale systems and has been of interest for many years [27]. The predominant factor controlling this size-tuning is the band gap, and therefore we can conclude that size-tuning of absorption and emission spectra is not a distinguishing feature of excitons, and is therefore not an incisive probe of the properties of excitons [1]. That idea has motivated us to ask what other properties are size-tunable and why? However, since the concept of an exciton can be quite confusing given the various limiting models and associated languages that prevail in the literature, first we aim to establish an intuitive picture for describing the electronic properties of excitons.

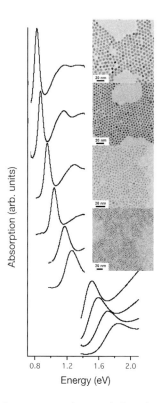

Figure 1. Electronic absorption spectra and transmission electron micrographs of a series of colloidal PbS nanocrystals showing the size-tunable optical properties [28].

It was recently shown that a critical parameter in deciding the nature of the exciton states in a nanoscale system is the distance-dependence of the electron-hole interaction, and the resulting series of energy states was likened to a Rydberg series [21]. However, these configurations in the series mix, the result being a partitioning of the series of states into two manifolds: the bound excitons and the charge transfer exciton (CTX) states (confined free carriers). This is summarized in Figure 2.

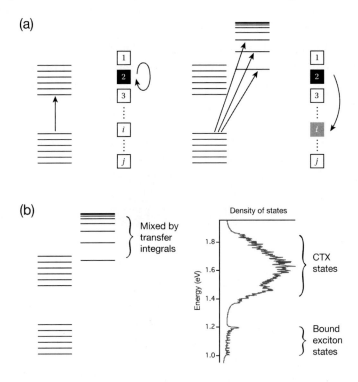

Figure 2. (a) Bound exciton states derive predominantly from excitation localized on or near constituent subunits of the material, whereas the nanoscale free carriers (CTX states) involve excitation from one subunit to another, relatively distant, subunit. (b) A characteristic of exciton states is that closely-spaced levels in the progression of CT configurations are mixed by the transfer integrals that promote hopping of electrons and holes from one subunit to another. The resulting density of states is plotted, showing the ladder of lower levels, the bound exciton states that are associated with light absorption and emission, and the dense manifold of upper levels called the charge transfer exciton (CTX) states wherein the electron and hole behave as independent particles.

The mixing among the series of configurations is instigated by transfer integrals that promote hopping of electrons and holes from one site to another (their magnitude is determined largely by center-to-center separations of the subunits). The relative magnitude of the transfer integrals compared to spacing between states in the ladder of CT configurations decides the extent of mixing. It was found that the closely separated electron-hole pair configurations dominated the lower energy state composition because of the significance of electron-hole attraction in these configurations, and these states persisted as a ladder of bound exciton levels after mixing. On the other hand, configurations containing extended electron-hole pairs were found to be closely spaced in energy and were therefore strongly mixed by the transfer integrals, thus leading to formation of a dense manifold of CTX eigenstates.

This picture further led to the definition of free carriers confined to nanoscale materials. When the transfer integrals are greater than the separation between configuration energies, the electron and hole have a propensity to hop from one site to another rather than remain bound by the electron-hole attraction. Thus the electron and hole act as independent particles in the manifold of CTX states, providing the nanoscale analog to free carriers.

2.3 Exciton Binding Energy

The lowest energy set of states, dominantly comprised of closely-bound electron-hole pair configurations which make up the exciton states, are found to be clearly distinguished from the abrupt onset of the vast number of CTX (nanoscale free carrier) states. It is surprising that the exciton binding energy is seen so clearly (in the absence or disorder or line broadening). We can think of the exciton binding energy as the energy required to ionize an exciton. Notably this ionization energy is significantly reduced compared to small, molecular systems because the many different ways that the electron and hole can be separated are coupled by a quantum mechanical interaction known as the transfer integral (the matrix element for electron or hole transfer).

In high dielectric constant *bulk* semiconductor materials the exciton binding energy is typically small: 27 meV for CdS, 15 meV for CdSe,

5.1 meV for InP, and 4.9 meV for GaAs. Excitons are therefore not a distinctive feature in the spectroscopy of such materials at room temperature, making those materials well suited for photovoltaic applications. On the other hand, in molecular materials, the electron-hole Coulomb interaction is substantial—usually a few eV. In nanoscale materials we find a middle ground where exciton binding energies are significant in magnitude—that is, excitons are important.

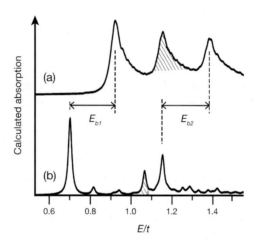

Figure 3. Calculation electronic transitions for an (8,0) SWNT using the Pariser-Parr-Pople (PPP) Hamiltonian. The shaded peaks are transverse excitations, the other peaks are longitundinal. Adapted from Ref. [29] with permission. (a) Transitions calculated at the Hartree-Fock level of theory. Here electrons and holes are independent so the excitons are unbound. (b) Transitions calculated using the single configuration interaction (SCI) method, which accommodates electron-hole binding and exchange. The significant exciton binding energies for the two dominant transitions are indicated (E_{b1} and E_{b2}).

Specific examples of nanoscale systems will be mentioned below. Here we highlight one example that illustrates the prediction of binding of excitons in SWNTs based on a semiempirical quantum chemical Hamiltonian. Zhao and Mazumdar calculated the absorption spectrum of various SWNTs in two ways [29]. Their results are reproduced in Figure 3. First the spectrum is calculated for the assumption that free carriers only are formed upon electronic excitation. To do that the energy difference between orbitals calculated at the Hartree-Fock level of theory is determined. In other words, this is the simple model for spectroscopy, often introduced in undergraduate teaching, where one electron is promoted from an occupied to unoccupied orbital by the absorption of a photon. Note that these orbitals include electron correlation, a concept often mentioned in the literature, which is the idea that electron motion has some correlated aspects because the electrons tend to avoid each other [30].

However, in real systems electronic excitations are not so simple as predicted by the orbital energy difference (Koopman's Theorem), and a more sophisticated quantum chemical treatment is essential. That is shown in the calculation that models the excited state as a linear combination of the various ways that the electron can be promoted from one orbital to another. The single configuration interaction (SCI) model, and closely related variants such as the Tamm-Dancoff and random phase approximations [31], are known to be the minimal predictive methods for calculating electronic excited states of molecules [32]. The SCI method introduces interactions between the electrons in these various orbitals, which captures the electron-hole attraction and thus leads to a significant stabilization of the excited states relative to those estimated with the more primitive model of orbital energy differences. That stabilization energy is the exciton binding energy, predicted here for SWNTs to be ~0.3–0.5 eV, which was later confirmed by experiment [33,34]. Yaron, et al. make an interesting comparison between calculations of exciton binding in conjugated polymers and inorganic semiconductors [35].

2.4 Singlet-Triplet Splitting and the Exchange Interaction

On the basis of the picture described above for electronic excitations, it can be imagined that there are two ways of promoting an electron from an occupied to an unoccupied orbital: either the spin of that electron remains unchanged, or it is flipped. Assuming that the ground state is a singlet (usually the case) and spin orbit coupling is negligible (only to simplify this explanation), it becomes clear that the SCI model predicts four similar looking excited states. These are the $S = 0$, $M_s = 0$ singlet excited state and three triplet states, $S = 1$, $M_s = -1$, 0, +1. What is not immediately obvious is that the singlet and triplet manifolds are split by an energy equal to twice the exchange interaction [36], the precise form of which can be found elsewhere [37,38]. The exchange interaction typically ranges in magnitude from hundreds of meV upwards for organic systems and is a few meV upwards in quantum dots [1,36,39-41].

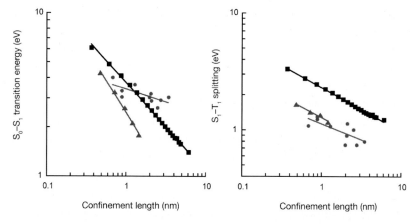

Figure 4. Calculations and experimental data showing on log–log plots (a) the lowest energy dipole-allowed electronic transition energy, and (b) the corresponding singlet-triplet splitting (i.e. twice the exchange interaction) for some quasi-1D exciton systems. The squares represent calculated data for linear polyene chains, the circles are for various conjugated oligomers, and the triangles are linear polyacenes. Confinement length is taken to be the length of the long axis of the π-electron system. Note how singlet-triplet splitting allows for direct comparison of exciton size and shape among different systems.

By surveying a range of molecules and nanoscale systems it was concluded that the singlet-triplet splitting (or exchange interaction) can provide insights into the size of the exciton [1]. The concept is that exchange interaction is obtained by subtracting the singlet excited state transition energy from that of the triplet state, with account of spin-orbit coupling when appropriate. Owing to the common origin of these two electronic states within the SCI framework, all information about the band gap is subtracted away, which is desirable because these mainly one-electron integral terms do not contain information specific to excitons. On the other hand, the two-electron integral that remains, the exchange integral, scales with the size of the nanoscale exciton, and appears to allow distinctly different kinds of materials to be compared directly [1].

A comparison is made between two quasi-one dimensional (quasi-1D) systems in Figure 4 (data are taken from refs [27,36,42-46]). Examination of the singlet-triplet splitting reveals a subtle effect: These quasi-1D excitons actually have a shape. For example, at each confinement length the polyene singlet-triplet splitting is about twice that of the polyacenes and conjugated oligomers. That diminished exchange interaction is a direct result of the expansion of the π–electron system from a linear conjugation in the case of the polyenes to a ring conjugation in the case of the other systems. The average electron-electron repulsion is reduced, providing a clear indicator for the exciton shape.

2.5 Exciton-Vibration Coupling

In the present review we focus largely on purely "electronic" aspects of excitons. In real systems, however, spectroscopy as well as the basic picture of the exciton can be significantly modified through the influence of nuclear motions. The coupling between electronic and nuclear degrees of freedom is manifest in spectroscopy as line broadening, the Stokes shift, and vibronic structure in absorption and photoluminescence spectra. The aims of this section are to obtain some insights into this problem and provide some specific examples of the consequences of disorder and exciton-bath coupling.

To understand the primary exciton-bath coupling effects we will start by recalling that the principal components of the lowest energy exciton are the site-localized energies of tightly bound electron-hole pairs and the coupling between these sites. Assuming that the electron and hole are strongly associated, then their coupling to the environment is correlated and we need to consider only the total site energy. To simplify this qualitative discussion we ignore the explicit interaction between sites.

Coupling between the site energies and the nuclear configurations of the system causes the site energies to fluctuate, or to exhibit disorder when we think of the ensemble of site energies (these two viewpoints are equivalent if the system satisfies ergodicity). These fluctuations (disorder) are caused by random motions in the positions of atoms, their electronic polarization, and the interactions among the nanoscale system and these atoms [47-49]. The atoms comprising the bath include the atoms of the nanoscale system itself as well as those of the surroundings (for example, the solvent). The interaction between excitons and nuclear motions can introduce time-dependent confinement effects, known as exciton self-trapping, where the exciton becomes trapped in a lattice deformation [50,51]. Those nuclear motions are contributed by intramolecular vibrations and the environment [52]. Exciton self-trapping occurs through a local collective structural change, connected to random nuclear fluctuations by the fluctuation-dissipation theorem. The resultant exciton is also called a polaron-exciton. Hence the size and electronic make-up of an organic exciton can change markedly on short time-scales after photoexcitation (tens of femtoseconds). That is understood as the tendency of molecules to change their equilibrium geometry in the excited state compared to the ground state, which is observed, together with solvation, as spectral diffusion [53-55]. The associated reorganization energy is equal to half the Stokes shift.

The coupling between the exciton and the bath has quite complicated consequences. For example, consider the spectral density of frequencies that couple to an exciton. Focusing on the low frequency motions (compared to thermal energies kT), it is found that when the fluctuations at each site in our nanoscale system are completely uncorrelated, then the site energies fluctuate randomly with time, and as a consequence the eigenstate composition fluctuates with time. When the root mean square amplitude of the fluctuations is large compared to the electronic coupling between sites, then exciton-bath coupling localizes the exciton onto one or just a few sites. On the other hand, completely correlated fluctuations preserve the eigenstate composition because at each sampling time interval they simply add an equal random energy offset to each site. Correlated fluctuations thus preserve the exciton coherence and do not contribute to self-trapping. The possible significance of these kinds of bath motions for modifying excitonic coherence and the efficiency of electronic energy transfer has been recently postulated [56,57].

The time scale of the nuclear fluctuations is important in delimiting the effects as static (on the time scale of the measurement), or fluctuating. The former gives rise to inhomogeneous line broadening, the latter to homogeneous line broadening. Inhomogeneous line broadening can be revealed in the frequency domain using fluorescence line narrowing, hole burning, or single molecule fluorescence experiments [58,59]. It has been discovered that the local structure around a molecule at low temperature differs from one site to another in an ensemble. Thus the differential solvation energy between ground and excited states varies throughout an ensemble. A seminal demonstration of how the transition energy of a single molecule also changes over a long time scale (seconds) was reported by Ambrose and Moerner [60]. It is clear that those energy modifications are significantly greater than the homogenous line width, and they reflect significant fluctuations of the structure in the immediate surroundings of the molecule. Nanoscale systems typically exhibit another important contribution to inhomogeneous line broadening that arises from the polydispersity of samples and the prevalence of size tunable optical properties in combination. In many cases this can be the dominant line broadening mechanism in spectroscopy [61,62].

3. Single-Wall Carbon Nanotubes

The photophysics of semiconducting single wall carbon nanotubes (SWNTs) have been studied intensively in recent years [1,63-72]. It has emerged that the lowest energy excited states are strongly bound excitons, with transition energies determined by the SWNT diameter [29,33,34,73-75]. The challenges to elucidating details of the excited state dynamics characteristic of SWNTs are twofold. Firstly, sample preparation is complicated and even the highest quality samples are difficult to study owing to inhomogeneous distributions of SWNTs in the ensemble. Secondly, models for describing the photophysics often sit at the convergence of solid-state physics and molecular spectroscopy. A major driver for the theory described in section 2 of this chapter is one attempt at obtaining some intuition regarding this frequently-encountered topic. An important aspect of the primary inhomogeneous line broadening in SWNT spectra is that, in contrast to the continuous distributions of transition energies in conjugated polymer or nanocrystal spectra, the SWNT spectrum is a sum of certain tube types. Therefore a simple photoluminescence excitation-emission map permits one to deconvolve the spectrum, Figure 5.

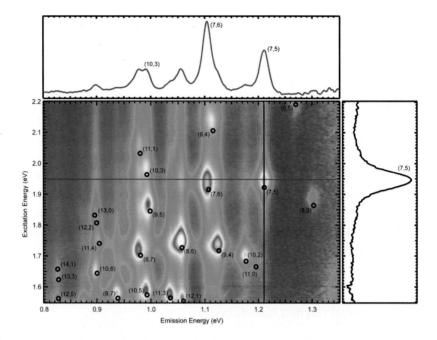

Figure 5. A Photoluminescence contour map of a sample of HiPco SWNTs in aqueous solution solubilized with SDBS [76]. The two axes correspond to excitation into E_2, while the other axes represent emission from E_1. The two spectra to the top and right signify a PL spectrum and a PLE spectrum indicated by the lines. Note the PLE spectrum only covers energies within the E_2 spectral region. The discrepancy between the measured and predicted [77] values is due to the latter being determined for an SDS surfactant, indicating the subtle, but distinct influence of the excitonic transitions on the local environment.

The most direct measurement of the exciton binding energy for SWNTs takes full advantage of the relative allowedness of the one- and two-photon transitions that occur at the optical band-edge of each individual nanotube species. As can be seen in Figure 5, SWNT samples contain many isolated nanotubes that can only be identified by scanning both the excitation and emission wavelengths within a 2-dimensional plot of the photoluminescence. Interestingly, had only the original studies of this phenomenon contained fewer nanotube species, accurate identification would have been far more difficult [64]. Closer examination of the two axes reveals that excitation is predominantly in to E_2, while emission emanates from the region of E_1, or at least the two exciton states described above that reside within E_1. By repeating this experiment using not one, but two photon excitation, a higher-energy exciton state that lies close to the CTX states could be identified by observing the emission from E_1. This experiment therefore allowed the exciton binding energy to be identified almost unequivocally, with even evidence for the absorption to the underlying CTX states.

The optically-allowed (bright) E_1 and E_2 states have been a focus of investigation to this point. Quantum chemical calculations suggest that the lowest lying excited state for all tube types is an optically-forbidden (dark) exciton [29,78,79], and that conclusion has been used to rationalize the very low estimated fluorescence quantum yield of the material [80] ($\Phi_f \sim 10^{-3}-10^{-4}$). However, a detailed model also explaining the temperature-dependence of Φ_f and the fluorescence decay kinetics has not yet been elucidated. In other words, open questions include: What are the important non-radiative decay routes for the excitons and is intersystem crossing important? A key component in current theories is the assumption of a Boltzmann distribution of population among the singlet states. That supposition is similar to Kasha's rule for molecular photophysics [81,82], which states that only the lowest energy singlet state of a molecule is fluorescent. This rule is followed when the rates of internal conversion (IC) from higher excited states (e.g. S_n) to the lowest excited state (e.g. S_1) of any spin multiplicity and intramolecular vibrational relaxation (IVR) within each electronic state are much greater than the rate of deactivation of the lowest excited state (e.g. the fluorescence rate). A recent report on the detailed analysis of dual fluorescence features [83] from surfactant-isolated SWNTs proposed that there are at least two low-lying excited states and that the relative population of each is controlled by the kinetics of radiationless transitions between the states in competition with radiative decay channels. In other words, among the manifold of electronic excited states close to the E_1 energy, Kasha's rule is not obeyed [84]; a result that has significant consequences for the photophysics.

The main focus of present SWNT research is on the preparation of samples that contain only one SWNT species described by a single (n.m) index. While progress is slow, there have been a few advances, with the separation of metal and semiconducting nanotubes [85], and selective solvation using lessons learned from nature by deploying single-stranded DNA [86]. The use of 2D-PL maps, such as that shown in Figure 5, are useful in aiding this process, but caution must be exercised, as these maps show only those SWNTs that are isolated and can thus emit. Non-emitting metallic SWNTs and any remaining bundled SWNTs could still reside within these solutions making the more advanced transient absorption spectroscopy or other sophisticated techniques difficult to interpret.

4. Conjugated Polymers

Conjugated polymers are highly conjugated linear macromolecules that display semiconductor-type properties [87-95]. Many chemical types are known, each typically containing 100 to 300 repeat units. A few examples are alkyl polyfluorenes, substituted polythiophenes, and functionalized poly(phenylenevinylene) (PPV) such as poly[2-methoxy-5-(2´-ethyl)hexyloxy-1,4-phenylene vinylene] (MEH-PPV). The primary interest in these materials is stimulated by their ease of processing relative to crystalline materials. Because excitons can be formed directly upon photoexcitation or through charge combination after electrical carrier injection, potential applications include displays, lighting, lasers, sensors, and solar cells [16,96-101].

4.1 The Basic Picture of Conjugated Polymer Excitons

Electronic structure calculations have played an important role in driving our understanding of excitons in conjugated polymers and oligomers, as reviewed elsewhere [70,89,102-106]. Early models based on a tight-binding Hamiltonian, for example the Su-Schreiffer-Heeger (SSH) theory yielded the first insights. Later it was found that more sophisticated treatments were crucial for the accurate prediction of electronic properties [35,102,107-111]. In particular, a realistic treatment of electron-electron interactions has been found to be crucial. Electron correlation effects, such as double excitations from the ground state reference determinant, are particularly important for the symmetric (A_g) excited states of conjugated polymers, as can be seen in calculations for the 10-ring oligomer of poly(paraphenylenevinylene) shown in Figure 6, which has been adapted from the work of Beljonne, et al. [110].

The ground and excited state spectroscopy of many conjugated polymers have been investigated, showing that the basic picture is similar for most of these π-conjugated systems. A number of absorption bands are evident in the spectra, which complicates analysis and assignment of the excitons; it is here that input from theory has been important. The alternating antisymmetric (B_u) and symmetric (A_g) singlet electronic states provide distinctive excited state absorption ($B_u \rightarrow A_g$) features [70], which also play a significant role in determining the nonlinear optical properties of these materials [104,112,113].

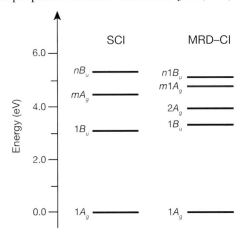

Figure 6. Energy diagram for the essential excited states relevant for third-order nonlinear response in the single-chain model for poly(phenylenevinylene) (PPV). Vertical excitation energies extrapolated to a 10-ring oligomer were calculated using the INDO Hamiltonian. Left: results calculated at the SCI level. Right: results calculated at the multireference double configuration interaction (MRD-CI) level, which effectively includes up to fourth-order excitations from the Hartree-Fock reference determinant. Note how the level of theory has a significant impact on the ordering of states, and particularly influences symmetric (dark) states (the A_g states). Adapted from Ref. [110].

Electroabsorption measurements allow the charge transfer character of excited states to be measured [114]. Such experiments have been used to establish where the charge transfer character becomes important for conjugated polymers [115,116], thus providing a link with the model for nanoscale excitons sketched in Figure 2. Molecular orbital calculations also show that as the excitation energy increases, the charge transfer character of the excitations increases [117].

4.2 Conformational Subunits and Their Interactions

The optical properties and excited state dynamics of conjugated polymers are strongly influenced by polymer chain conformation [118-124]. Recent work has demonstrated how that chain conformation can be controlled using anisotropic solvation [125]. Other work has demonstrated a novel "nanoparticle" polymer morphology [126,127]. In typical solutions, however, most conjugated polymers undergo surprisingly facile rotational motions about bonds along the chain, which thereby disrupt the rigid structure and concomitantly disturb the π-electron conjugation [128,129]. For example, single molecule studies and simulations of the polymer chain established that MEH-PPV adopts a defect cylinder chain conformation, implying that there is a combination of many minor and relatively few substantial chain kinks in a typical polymer chain [130]. These structure-imposed chain twists break the polymer into a series of chromophores known as conformational subunits.

It is thought that the resultant "conformational disorder" impacts the spectroscopy by introducing a kind of inhomogeneous line broadening. According to site-selective fluorescence spectroscopy and low-temperature single molecule fluorescence work, the excitation energy absorbed by a polymer chain can take a range of energies, depending on the absorbing conformational subunit. After absorption, excitation is subsequently funneled to low energy chromophores along the chain [119,120]. The primary mechanism for that kind of energy funneling is energy migration via Förster-type electronic energy transfer (EET) [131,132]. Evidence for such EET is implied by measurements of polarization anisotropy decay, which lead to the conclusion that the energy migration is a complicated dynamical process that proceeds mainly on a 1–50 ps time scale [133-136]. Interchain EET is significantly more efficient than intrachain EET [137], which has been explained by quantum chemical calculations as being due to the larger electronic coupling between cofacial conformational subunits compared to those in a linear arrangement [103,138].

In addition, observations that the initial anisotropy of these data is lower than 0.4 suggests that an extremely rapid ultrafast process changes the exciton transition dipole moment direction after excitation and prior to EET [139]. Given the time scale of this process, it is likely to be a combination of exciton relaxation (from higher to lower energy quasi-delocalized states) in competition with self-trapping to form localized excitations on the conformational subunits.

A further consequence of conformational disorder and chain conformation is that we can imagine two ways that conformational subunits can interact with each other. These scenarios lead to the idea that there are two basic types of exciton: intrachain and interchain [103,138]. The former are formed by the extended π-conjugation along sections of the polymer backbone that may encompass more than a single conformational subunit. Interchain excitons form when two intrachain excitons couple through-space, either because the chains are nearby to each other in a solid film, or because the chain is folded back on itself. The electronic coupling between conformational subunits in the latter case have been found to be the most significant, meaning that interchain species should be lower energy than intrachain species.

There are two main pools of evidence in support of the existence and importance of two distinct kinds of exciton in conjugated polymers. Yu and Barbara discovered that the distribution of emitting chromophores in the relatively flexible MEH-PPV polymer is bimodal [140,141]. Two distinct kinds of emissive states were observed. A red emitting species was found in about one third of single chain spectra and was attributed to a complex formed by interchain contact. The other spectra were blue emitters. Similarly, a pair of emitting species was identified by Rothberg and co-workers based on dilute solution studies in mixed solvents [121,142]. It was shown how the conformation of MEH-PPV could be controlled by solvent quality and the change in steady state emission spectra could be modeled as linear combinations of a blue and a red spectral form. The red spectral form was assigned to a species formed by aggregation of conformational subunits of a single chain.

In a recent report, Clark, et al. propose that the emission from regioregular poly(3-hexylthiophene) is dominated by aggregate emission. Indeed, it is well known that a unique aspect of this polymer in films is that it forms ordered, lamella domains. These morphological structures are thought to be important in determining the photophysics as well as desirable electrical properties [143-145].

Recent work has aimed to understand better how to think about conformational subunits and their role in absorption and ultrafast dynamics. Even simple models for electronic coupling (e.g. dipole-dipole coupling) suggest that adjacent conformational subunits should be electronically coupled. Careful quantum chemical investigations furthermore reveal that it is difficult to define conformational subunits with respect to torsional angles—there is not clear point when the strong interactions that depend on π–π overlap "switch off" and, even when weak, those interactions are important for defining the nature of the excited states and the delocalization length [146]. Beenken and Pullerits conclude that conformational subunits arise concomitantly with dynamic localization of the excitation (exciton self-trapping) [147]. That suggests an interplay and connection between conformational disorder, torsional motions, and self-trapping, which we address in the follow section.

It is clear that the size of a conformational subunit plays a role in determining its optical properties—longer conjugation length suggests a more red-shifted absorption/emission energy and a stronger transition dipole strength. The size-scaling of the linear as well as nonlinear optical properties is of interest and has been studied through experiment as well as theory [148-151]. A notable conclusion is that the scaling saturates at conjugation lengths of about 50 double bonds, meaning that typical, real π-conjugated molecules cannot be thought of as infinite 1D exciton systems.

4.3 Exciton-Phonon Coupling and the Role of Torsional Modes

The characteristic non-mirror image absorption-emission spectra found for conjugated polymers is mostly attributable to the effects of torsional modes. Evidence in support of that conclusion is that conjugated oligomers show similar spectra at room temperature to that of polymers, thus ruling out a significant effect of conformational disorder [152]. Quantum chemical studies of these systems reveal that there is a substantially different torsional potential in the ground state chromophore compared to the excited state [153,154]. A broad distribution of torsional angles in the ground state at room temperature results in a large frequency distribution for absorption to the steeper excited state potential, while rapid relaxation and equilibration in the excited state renders the distribution of emission frequencies narrow in comparison, Figure 7. At low temperature, or in ladder-type polymers, the conformational distribution in the ground state potential is narrowed sufficiently so that the absorption and emission spectra are mirror images.

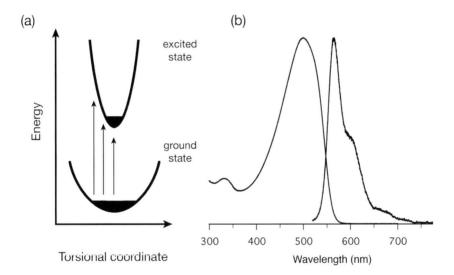

Figure 7. (a) Schematic representation of a torsional vibration potential in the ground state and excited state of a conjugated polymer (or oligomer). These modes exhibit a marked frequency change, but no displacement between ground and excited electronic states. The black shading portrays the thermal population of each potential, which leads to a broader spread of energies in absorption (upward arrows) than emission. (b) Absorption and emission spectra of a dilute solution (chlorobenzene, 293 K) of MEH-PPV. Note the absence of mirror image symmetry and the large apparent Stokes shift.

In conjugated polymers the scenario is more complicated, and the many torsional modes couple to excitons, as suggested by Berg, Yaron and co-workers [155,156]. These torsional motions are also thought to play a role in exciton self-trapping [51,157]. Bittner, et al. recently suggested that the aggregation state of a conformational subunit can also affect the exciton-vibrational coupling [158].

4.4 Ultrafast Dynamics of Excitons

There are several processes that occur on a hierarchy of time scales after photo-excitation of conjugated polymers (either isolated chains in solution or films). The fastest dynamics are typically an entanglement of processes that are difficult to elucidate. Such processes include relaxation through somewhat delocalized exciton states, self-trapping of the exciton, vibrational cooling, and finally energy migration and trapping. Exciton photodissociation occurs in parallel to the above scheme and will be considered in a subsequent section. Conjugated polymers provide models for disordered π-electron systems and therefore it is of general interest to learn about the pathways and processes that dictate how the exciton evolves after optical preparation. In particular, the transition from relaxation/trapping dynamics to EET "hopping" transport along and between chains is of fundamental interest. Furthermore, there are clear relationships between chain morphology, conformation, and EET rates that are gradually being clarified.

The fastest time snap-shot of the evolution of a conjugated polymer exciton has been obtained using the three-pulse photon echo peak shift (3PEPS) method for examining MEH-PPV in dilute solution [159,160]. The 3PEPS data allow an estimation of spectral diffusion within the frequency window defined by the laser pulse spectrum [53,63,161-163]. Using appropriate, though complicated, analysis a correlation function for frequency fluctuations (homogenous line broadening) can be extracted, even though such information is obscured in condensed phase spectra. In the case of MEH-PPV, however, the bath fluctuations were found to be overwhelmed in significance by a relaxation process that enabled the excitation (of the ensemble) to explore rapidly many frequencies. It was proposed that a sequence of events on different time scales characterize the excitons. It was proposed that absorption occurs from the ground state into a delocalized exciton manifold. After photoexcitation, a spectral diffusion process, associated with relaxation to lower energy exciton states with concomitant localization of excitation, occurs on a time scale of ~25 fs [159]. It is possible that the localization is driven by the large reorganization energy associated with planarization of the torsional coordinates since these play such a significant role in the absorption line shapes.

Electronic energy transfer (EET) occurs in the next stage of the dynamics [132,133,135,136,139,164-167]. As one example, in recent work both *intermolecular* and *intramolecular* energy transfer processes were assessed experimentally and theoretically for a covalently linked donor-acceptor system wherein poly(indenofluorene) chains acting as donors for EET were end-capped with red-emitting perylene derivatives [132]. Contributions of *interchain* and *intrachain* processes to the overall EET dynamics were determined by recording time-resolved photoluminescence and absorption spectra of the system under investigation in both solution and thin film. In solution, the EET process was found to occur on a 500 ps time scale and it therefore competes with both radiative and non-radiative decay of the excitations. EET was found to be much more efficient (a few tens of ps) in the solid state, leading to a complete quenching of the polyindenofluorene luminescence. This difference in dynamical behavior is considered to be related to the emergence of additional channels for the excitation migration in films as a result of the presence of close contacts between adjacent chains. To rationalize the different dynamics observed in solution and in the solid state, the intrachain and interchain EET processes were extensively examined using a series of models with increasing level of sophistication.

Many studies of the ultrafast dynamics of conjugated polymer excitons have examined charge separation (carrier formation). Such work has been reviewed recently [92].

5. Quasi-One-Dimensional Systems

Quasi-1D systems are characterized by strong quantum confinement in two dimensions and an essentially infinite third dimension. This means that in ideal systems (at zero temperature) one expects a band-structure for exciton states rather than discrete densities of states.

SWNTs are a good example of a quasi-1D system, and they have recently been compared and contrasted with quasi-1D conjugated polymers [39,70,168]. The realization that useful comparisons can be made has particularly benefited advances in understanding SWNTs photophysics and electronic structure because the foundation has been established during many years of research and controversy regarding models for the electronic structure of conjugated polymers.

A characteristic of quasi-1D nanoscale systems is that, not only the nonradiative rates, but also radiative rate constants vary with temperature. That is in contrast to molecules where all the temperature dependence can be attributed to the nonradiative processes. The radiative rate of a quasi-1D exciton is temperature-dependent because the exciton band in fact consists of a ladder of states, where the lowest state can radiate, but the upper states are typically dark. For example, in a SWNT of infinite length, those upper states are dark owing to the requirement of momentum conservation for exciton recombination [79,84]. Thermal population of this ladder of states therefore lengthens the effective radiative rate of the exciton because the observed radiative rate is controlled by population of the lowest level in the band. At high temperatures population is transferred to dark states in the band, whereas at low temperature the population resides mainly in the lowest, bright, exciton state. This complication in the photophysics of SWNTs—that the radiative decay rate depends on temperature—is common to quasi-1D molecular aggregates, such as J-aggregates [169-171].

An interesting contrast to the conjugated polymers discussed above is polydiacetylene. Highly ordered single chains, resembling organic semiconductor quantum wires, have been investigated extensively by Schott and co-workers [172-174]. Key properties that differentiate these chains from many other conjugated polymers are their structural order on all length scales, their rigid, crystalline environment, and weak exciton-phonon coupling. Strikingly, it has been observed that the exciton state is spatially coherent over a length scale of tens of microns [175].

6. Semiconductor Nanocrystals

Inorganic semiconductors are well known as model systems for investigating excitonic phenomena. Indeed, much of our current language and models originated from this field. A huge amount of important work established the concept of quantum confinement, whereby reduction of one or more physical dimensions of a bulk semiconductor leads to confinement of wavefunctions in the dimension, with a consequent impact on electronic and optical properties. In the early 1980s it was predicted that confinement in all three spatial dimensions would replace the band structure with discrete levels, forming a so-called "artificial atom". That is, these quantum dots (QDs) possess discrete electronic energy levels and excited state transitions, more reminiscent of molecules than bulk semiconductors [1,176-191]. The resulting field has largely bifurcated according to fabrication techniques: Self-assembly of QD "islands" in semiconductor substrates and colloidal growth of nanocrystalline QDs (nanocrystals). Here we focus on colloidal QDs.

The electronic spectroscopy of nanocrystalline semiconductor QDs has attracted much attention in recent years [1]. The widespread interest in the electronic properties of these materials encompasses investigations of laser media [17,192-201], electroluminescence [202-205], photovoltaics [206-208], multiple exciton generation for solar cells [209-211], magnetic doping [212], electrical control of excitons [213], electrochromism [214], fine structure dynamics [215-217], biological labeling [218-220], and quantum information [221,222].

Elucidation of the properties of QD excitons has followed advances in sample preparation and characterization. For example, the primary excitonic features in the absorption spectra of QDs were identified through a combination of theory and experiment only once polydispersity was reduced enough to see clear excitonic peaks and aspects of the photophysics such as identification of surface states and trap emission were clarified [179]. Key advances involved the optimization of nucleation and control of growth, largely inspired by the discovery of the organometallic route for synthesis of cadmium chalcogenides [223]. Recent work has paved the way toward the study of shape-tunable properties by finding that nanocrystalline semiconductors can be grown in a variety of morphologies [224-233], Figure 8. That work has stemmed from the discovery that certain surfactants adsorb selectively to specific crystal faces, hence slowing growth in those directions relative to other, more active crystal faces [234].

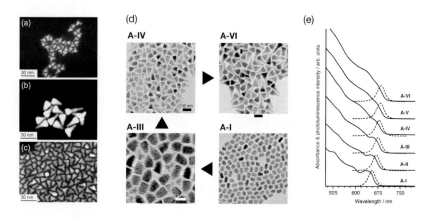

Figure 8. (a) CdSe multipods, (b) CdSe trigonal pyramids, and (c) CdSe multipod pyramids [235]. The 3D topology of these nanocrystals is clearly seen in these dark-field TEM images owing to the high contrast ratio and lack of Bragg scattering artifacts. The bright spots represent local NC structure that bulges towards the viewer. (d) TEM images showing the sequence of steps a multiple injection strategy for evolutionary shape control [233]. CdSe rods are transformed to bullet-shaped rods and finally to pyramids by control of reagent and ligand addition to the growth solution. (e) Absorption spectra (solid lines) and photoluminescence spectra (dashed lines) showing the evolution of spectroscopic properties during growth of the CdSe pyramids. The systematic red shift is due to the particle growth.

6.1 Size-Tunable Properties

An important characteristic of QDs is their size-dependent properties, especially prominent in spectroscopic measurements. The size-dependence of QD spectroscopy derives from confinement of the exciton by the crystal to a size much smaller than its Bohr radius [236-238]. If the electron and hole wavefunctions are considered independently in the effective mass approximation and the finite size of the QD is considered to impose an infinite potential outside the QD, then the electron and hole wavefunctions are each located in a ladder of discrete states split off from the band continua. These one-electron levels have been recently seen using scanning tunneling microscopy [239,240].

The quantum confinement effect for excitons comes into play when the physical size of the QD is comparable to, or smaller than, the electron and/or hole Bohr radii, or more simply, the exciton Bohr radius. As a consequence, the exciton transitions, which derive at zeroth-order from band-to-band excitations, are size-dependent. For example, the exciton Bohr radius of PbS is ~20 nm and its bulk band-gap is 0.41 eV. Absorption spectra for PbS QDs of radii ranging from ~1.3 to ~3.5 nm are shown in Figure 1, revealing QD exciton energies in the range 0.7 to 1.5 eV. Our current understanding of size-tunable spectroscopy in QDs has evolved from the simple picture where an electron is promoted from valence to conduction orbital to one where it is clear that interactions between these singly excited configurations are important in deciding the electronic states, known as the exciton fine structure (see below) [41,241]. In other words, in spectroscopy we probe electronic states where the electron and hole are correlated.

6.2 Surface Passivation

A further challenge that particularly hinders the use of photoluminescence to examine QD excitons quantitatively is that QD exciton properties and dynamics are greatly influenced by surface effects [242-244]. For example, bright exciton states are quenched by charge separation to surface traps, but subsequent slow recombination processes can yield long time tails to photoluminescence decays; all of which combine to make QD photoluminescence decays highly non-exponential [245].

Colloidal QDs are normally synthesized in a solvent comprised of surfactant-like organic molecules that play an import role in the nucleation and growth processes to ensure formation of narrow size dispersions. In addition, they act as ligands to confer colloidal stability upon the particles and passivate the surface. It is well known that trap states lying energetically within the optical gap quench photoluminescence, and these traps are formed when the QD surface is ineffectively passivated with ligands. For example, the elucidation of that picture is beautifully illustrated by the seminal work of Henglein and coworkers [246]. At the surface of a QD, bonds are disrupted because the crystal unit cell is not infinitely repeating. Non-coordinated, dangling bonds at the QD surface are thought to be capable of trapping electrons or holes, hence diminishing the photoluminescence yield. Passivation is the process whereby molecules bond or coordinate to these dangling bonds on the surface. That is traditionally achieved with ligands like alkyl phosphines, long-chain amines, or thiols. For example, CdSe particles are normally prepared in the presence of a mixture of trioctylphosphine (TOP) and trioctylphosphine oxide (TOPO). A detailed study of ligands, their adsorption and desorption, and photoluminescence of CdSe QDs has been reported by Bullen and Mulvaney [247].

In recent work oligomers and polymers have been demonstrated to be effective ligands for QDs. These kinds of ligands confer exceptional colloidal stability to the colloidal particles because of the multidentate binding effect [248,249].

6.3 Shells and Heterostructures

Altering the optical properties of colloidal NCs by grafting different semiconductors together is called exciton engineering. Exciton engineering opens possibilities for improving surface passivation, changing excited state dynamics of the exciton, or manipulating wavefunctions. Fusing one type of nanocrystal over another enables further control of material properties that are dependent on the relative alignments of their energy levels. The first example of a QD heterostructure was the (CdSe)ZnS core shell QD [250,251]. The ZnS shell has energy levels arranged to confine the exciton to the core, thus it serves as a surface passivating layer, thereby substantially increasing the photoluminescence quantum yield. In analogy to solid state physics terminology, that is a type I heterostructure. In general these materials are based on the core-shell concept. A recent paper reviews the synthetic aspects of core-shell nanocrystals [252].

Since the initial work, a great diversity of QD heterostructures has been reported. Quite complex onion type layered QDs can be prepared, such as the CdS/HgS/CdS "quantum dot quantum well" developed Mews and co-workers [253]. A systematic study of spatially engineered electron and hole wavefunctions in (ZnSe)CdSe inverted core-shell QDs has been recently reported by Balet, *et al.* [254]. Lifshitz and co-workers have examined how excitons are tuned through the composition of lead salt core-shell structures [255].

Type II QD heterostructures based on cadmium chalcogenides show qualitatively different behaviour because the electronic levels of the two semiconductor components are aligned such that the lowest energy state involves charge transfer between core and shell [256]. As a result of the charge transfer, these heterostructures exhibit distinct absorption and emission features in the near-infrared spectral region that are absent in either the individual core or shell counterparts. For (CdSe)CdTe, photo-excitation leads to a state where the hole is mostly confined to the CdTe shell while the electron is confined within the CdSe core [256-259]. In solid-state language, radiative recombination of the electron-hole pair occurs across the core-shell interface. Type II core-shell heterostructures can thus be considered to have effective band gaps that are then determined by band offsets. In addition, the thickness of the shell and core size also play a role in determining the effective band gap through quantum confinement effects. Thus by changing the core size and shell thickness, the emission behavior of these type II quantum dot heterostructures can be tuned. Shieh *et al.* presented a novel route to prepare nanorod heterostructures where a shape transition from a heterostructure nanorod to a spherical CdSeTe alloyed nanocrystal was observed to occur via a dumbbell-shaped morphology [260].

On a molecular scale, synthesis of supramolecular compounds has inspired advancement in theories for photo-induced charge transfer. Heterostructured nanocrystals potentially provide a nanoscale analog of such systems. Recently a method for preparing heterostructured nanocrystalline rods has been reported, and these systems showed photo-induced charge separation vectorially along the rod axis [259], Figure 9. It was found that the energy and lifetime of charge transfer photoluminescence band could be tuned by changing the relative

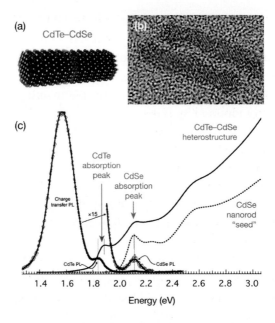

Figure 9. (a) Schematic illustration of a CdSe–CdTe nanorod heterotructure. (b) High resolution TEM image of CdSe–CdTe nanorod heterotructures (~6 × 25 nm) (c) Typical absorption and photoluminescence spectra of a CdSe–CdTe quantum rod heterostructure sample and that of the seed CdSe rods from which they were grown. See Ref. [259] for details.

alignment of band edges in CdSe/CdTe heterostructures nanorods. Slow charge recombination is an important attribute of materials for solar photoconversion, and and it was found that the long-lived charge transfer states (~4 μs) in these type II semiconductors may make them attractive for photovoltaic applications. The photophysics of these linear CdSe–CdTe nanorod heterostructures were examined in detail [261]. An important step was the identification of charge transfer emission and absorption bands. Analysis of those bands revealed the factors governing photoinduced electron transfer from CdTe to CdSe and it was thereby shown how quantum confinement effects decide the thermodynamic parameters of Marcus-Hush theory. An important finding was the very small reorganization energy associated with the nuclear degrees of freedom (~20 meV in toluene), which seems to be a characteristic of these nanoscale donor-acceptor systems and differentiates them from most analogous molecular systems. Therefore Marcus "inverted region" behavior was found to be typical for these systems.

6.4 Line Broadening and Fine Structure

An obvious challenge to uncovering detailed information regarding the QD exciton absorption and optical properties is that spectral features are hidden by spectral inhomogeneity that arises mainly from sample size dispersions. QD materials can now be prepared with a Gaussian size distribution of standard deviation ~5%. Nevertheless, sample size dispersion has obfuscated even simple observables such as the Stokes shift. The reason for that is that an important manifold of electronic states is hidden by the inhomogeneous line broadening. These states are known as the exciton fine structure, and are roughly analogous to the singlet and triplet manifolds of molecules. Detailed descriptions of these states, their origin, and some implications can be found elsewhere [41,216,262,263].

The basic picture of the fine structure states for CdSe QDs is that there are 8 states spanning an energy range of only a few to a few tens of meV (depending on the QD size). The lowest energy pair of degenerate states (labeled by their total angular momentum $F = \pm 2$) are essentially triplet states, and are therefore "dark", that is, they do not contribute to the absorption spectrum. However, these states are the origin of low temperature, long lifetime, photoluminescence and are the emitting states seen in fluorescence line narrowing experiments [264]. The next highest states are the lower $F = \pm 1$ exciton states, which are the primary luminescent states at room temperature. Above that lies a dark $F = 0$ state, then the upper $F = \pm 1$ states, which carry most of the dipole strength for optical absorption in CdSe spherical QDs. Above those states is the bright $F = 0$ state. Recent calculations have suggested how this fine structure manifold depends on shape for CdSe QDs [265], and how the fine structure is considerable more complex in lead salt semiconductors like PbSe [262].

The photoluminescence anisotropy of spherical QDs is depolarized owing to the circularly polarized selection rules for the $F = \pm 1$ exciton recombination. In contrast, it was discovered that emission from single rod-shaped CdSe QDs is substantially linearly polarized [266]. It has been shown recently that there are two factors that together explain this property of nanorods [267]. Firstly, the fine structure splitting is predicted to be different for nanorods than QDs. The lowest state, 0_d, is optically dark, the next states are the allowed $F = \pm 1$ states, and the upper state is the bright 0_b state which has a linearly polarized transition moment. Secondly, the 0_b state has a significantly larger dipole strength than the $F = \pm 1$ states, related to the elongation of the nanorod. It is thermal population of the 0_b state that contributes to the linearly polarized photoluminescence.

As a result of this fine structure and the inhomogeneous line broadening, a "non-resonant" Stokes shift is defined analogously to the usual Stokes shift as the energy difference between the exciton absorption band maximum and that of the photoluminescence peak. It is measured by exciting far to the high-energy side of the absorption spectrum. That Stokes shift is strongly affected by the size-distribution of QDs in the ensemble because the principal absorbing states are the upper $F = \pm 1$ states (in the case of CdSe QDs), whereas the emitting states are the lower $F = \pm 1$ states [41]. A "resonant" Stokes shift is measured though fluorescence line-narrowing spectroscopy, where a narrow distribution of QDs are photo-selected from the band-edge of a sample at low-temperature using laser excitation. That measurement determines the bright-dark exciton splitting. These principals have been reviewed recently [191].

Key differences between QDs and organic materials, such as the nature of nuclear motions that couple to excitons, stem from the rigid, crystalline structure of QDs—a particularly notable contrast to the flexible structures of organic materials. There are two characteristic vibrations in a QD: the LO-phonon modes, typically found at ~200 cm^{-1}, and acoustic phonon modes, in the range 5 to 40 cm^{-1}, depending on the QD size. The acoustic phonons of QDs are discrete torsional and spheroidal motions that have been modeled according to in such modes of an elastic sphere. A great deal of work has gone into studying interactions between QD excitons and the environment—principally the acoustic phonon modes of the QD—since those interactions dictate spectral line shapes and dephasing. The reader is referred to Ref. [61] for a summary of this literature.

6.5 Multiple Exciton Generation

It is well known that several important factors, in concert, determine the energy conversion efficiency (ECE) of solar cells. One particular limiting factor, common to all single junction devices, is that absorbed photon energy that is in excess of the bandgap is dissipated and hence not converted to electrical energy. That observation imposes the thermodynamic limit for solar cell efficiency (somewhat over 33%, depending on the bandgap [268]). There have been several approaches proposed for overcoming this limit, and such ideas found the next generation of photovoltaics [269]. Organic and/or nanocrystalline semiconductor based solar cells are an attractive alternative to traditional solar cells because the production cost can be dramatically reduced owing to ease of processing [270]. In this arena, Nozik proposed an intriguing possibility for stepping beyond the thermodynamic limit for ECE whereby it might be possible to convert incident single photons with energies greater than twice the bandgap into multiple excitons, and perhaps ultimately multiple free carriers [209].

An important property of QDs is that they can support multiple exciton populations: biexcitons, triexcitons, and so on [197]. Multiexciton states are very short lived in organic materials because excitons resident on proximate, but spatially distinct parts of a macromolecule or aggregate, annihilate efficaciously by a resonance energy transfer mechanism to form a higher electronic state that relaxes rapidly through a radiationless transition [271,272]. On the other hand, two excitons in a nanostructure can occupy almost orthogonal states, precluding annihilation. Thus biexcitons in CdSe QDs have recombination times of the order of tens of picoseconds and those in PbS and PbSe live for hundreds of picoseconds. Nozik proposed that these special properties of multiexcitons in QDs could be harnessed to increase the energy conversion efficiency of solar cells [209]. That process is now known as multiple exciton generation (MEG) or carrier multiplication [210,211].

There is mounting recent evidence that this phenomenon does indeed occur in a fairly wide variety of nanocrystalline QDs as well as in QD films [210,273-277]. However, some important questions remain to be addressed in future work. In particular, there is still discussion regarding the mechanism of MEG [273,278-280]. The challenge is two-fold. Firstly a calculation of electronic excited states must be carried out for a system containing an extraordinary number of electrons ($>10^5$ for a small PbS nanocrystal). Secondly, the higher excited states need to be calculated rather than the lowest few excited states. That is a daunting proposition given that the density of electronic excited states increases dramatically with energy.

To make an impact on the ECE recorded under solar illumination, the bandgap (E_g) of the active layer needs to be ~1.3 eV and the MEG onset needs to begin abruptly at $2E_g$ [281]. In most systems examined so far the onset is gradual and happens at $>2E_g$ (typically closer to $3E_g$). How may these observation be explained, and based on that new understanding, how can systems be designed that possess the key requirement of an abrupt onset of MEG at $2E_g$? To address such questions we need to elucidate more clearly the mechanism of MEG, but it is also important to identify the relative energies of high energy exciton states composed primarily of single electron-hole pair configurations compared to the multiexciton states. For example, to ensure a prompt MEG onset at $2E_g$, the biexciton state must lie slightly lower in energy that the nearest one-photon allowed single exciton state.

As a concluding thought, MEG is fascinating from a fundamental viewpoint, and is certainly one of the important breakthroughs of recent years. As the next step we need to ask: Can we dissociate these excitons fast enough to capture multiple charge carriers? Demonstrating multiple carriers actually being harnessed subsequent to single photon absorption is the crucial next step connecting MEG to its observation in a device.

7. Nanoscale Charge Separation

There are a number of applications that take advantage of the light-emitting properties of nanoscale systems, with diodes and lasers, single-photon sources, and bio-labels being three notable examples to highlight.

However, a growing and very important area of application for these nanoscale systems is in the field of renewable energy and the harvesting of solar photons. The preceding sections have identified one of the key attributes of excitons confined to nanoscale systems - that of tuning the optical absorption spectrum using size, shape and composition. Figures 1 and 8 have already provided two elegant examples of how this might be achieved using colloidal nanocrystals as the active medium. The absorption spectra of SWNTs extend from the visible well in to the infra-red. The shifting of the absorption spectrum of conjugated polymers from the visible to the near-ir has been accomplished using some very elegant molecular synthesis but has yet to prove as successful as would be desired.

However, the preceding sections also serve to identify one of the important issues associated with nanoscale excitons: the electron and hole are bound and therefore in order to extract the energy of the original photon, these two species must be separated in order that they can be used to do useful work. Photosynthetic reaction centers have proven particularly effective in this capacity, and these systems provide an excellent role model. Similarly, the conventional photographic process demonstrates that excitons confined to molecular J-aggregates can be used to sensitize silver halide crystals by the injection of electrons [282]. The dye-sensitized solar cell (DSSC) [283] takes this concept through to a photovoltaic device, where an excited dye molecule injects an electron into a TiO_2 nanoparticle. This electron and the hole that remains on the dye molecule are then transported away to external electrodes where the energy can be utilized. Dissociating the exciton using acceptor species has proven successful in a number of other, similar nanostructured systems. A blend of PCBM (Phenyl-C_{61}-Butyric-Acid-Methyl Ester), a soluble derivative of C_{60}, in a conjugated polymer such as MEH-PPV, or more commonly P3HT (poly(3-hexylthiophene)) have also demonstrated very good photovoltaic performances. Better-known as the bulk heterojunction, the excitons are photo-induced in the polymer and are dissociated at the polymer PCBM interface [284]. Using a 50:50 blend of the two components, the resulting electron and hole can then migrate to external electrodes and be available to useful work. Similar functionality has been found when using colloidal nanocrystals [285-291] instead of

PCBM, and even attempts to use SWNTs have been reported [292], but these have met with far less success. Collectively, these types of devices, even the DSSC, are referred to as excitonic solar cells; an apt name as it identifies the primary excitation as a bound exciton, and not an unbound electron and hole pair.

An interesting issue that arises in all these cases, and one that is pertinent to this chapter is not why the exciton dissociates at the interface, but why the electron and hole do not immediately recombine. Intrinsic to the stabilization of an exciton, on a simple level, is a low dielectric constant that results in a large exciton binding energy [21]. The cases of conjugated polymers and SWNTs relative to the colloidal nanocrystals are a good demonstration of this phenomenon. A low dielectric constant reduces screening and stabilizes the exciton. Therefore it is somewhat surprising that the electron and hole at the dissociation interface do not immediately recombine, and the reason why this does not occur readily remains a topic of discussion and debate [293-295]. It should be noted that in the DSSC, the high dielectric constant of TiO_2 is often used as one of the reasons to explain why the injected electron does not immediately recombine with the oxidized dye molecule.

While the use of acceptors to promote exciton dissociation is successful, it does introduce some interesting problems that continue to plague the field. First, the localization of an exciton on a conjugated polymer chain, or between chains, or on a nanoscale structure is only part of the problem. The excitons must now be transported to the dissociating interface, and then the same system must be used to transport the carriers away. Thus, good coupling between the nanoscale systems is key to good performance. If the coupling is too strong, quantum confinement can be lost, too weak and the transport process is inhibited. This is particularly noticeable for colloidal nanocrystals, where good transport of the carriers (often electrons) only occurs when the capping group is removed completely [285], or made extremely small [296]. As discussed in the previous section on surface passivation, complete removal of the surface ligand has a detrimental impact on the electronic properties of the nanocrystals, introducing electron (and hole) traps at the surface.

Finally, the exciton model proposed for nanoscale systems and depicted in Figure 1, distinguishes between the bound exciton states and the CTX states or in some cases, such as the SWNTs, free carriers. Based on this model, excitation into the bound states should reveal evidence for no 'free carriers', until the energy of the exciting photon exceeds the lowest CTX state. Indeed, this is perhaps the incisive experiment that verifies the existence of the bound excitonic states. As simple as this sounds, results are far from conclusive and this has resulted in many experiments that have questioned the exciton model [87]. This is especially true for conjugated polymers, and the same uncertainties need to be considered when studying SWNTs.

The difficulty arises when free (or separated) electrons and holes are observed when exciting directly into the bound exciton states. Does this undermine the exciton model? Or is there something else that is taking place that has not been considered? A simple, and perhaps incorrect, interpretation is that the nanoscale system contains an intrinsic dissociation site, such as a chemical defect, a surface state, or in the case of 1D systems, an active end-termination. Extrinsic factors like residual catalysts, oxygen [297] or other trace chemical impurities might also be responsible. Distinguishing among this collection of possibilities continues to drive many research projects, but underlying this plethora of options is the possibility that there is an intrinsic mechanism that has not been captured by the exciton model and that may provide a novel mechanism for separating the charge carriers. There are models [298] that propose a branching ratio that divides the absorbed photon between the bound exciton state and free carriers. This idea is depicted in Figure 10, which identifies three distinct pathways for an absorbed photon.

The following section identifies some key experiments that address this important topic, and how these have been used to identify the presence of unbound electrons and holes. The application of these techniques to the study of conjugated polymers, quantum-confined structures and SWNTs, are then explored. This is far from being an exhaustive study, but is used to highlight this important topic. A point to note is the distinction between experiments conducted on isolated species, and those conducted on condensed assemblies such as polymer films, quantum dot arrays and SWNT bundles. This distinction is important as the production of an unbound electron and hole in a single nanoscale species can be strongly influenced by the surroundings, where exciton dissociation is promoted by the surroundings and is not intrinsic to the isolated species itself.

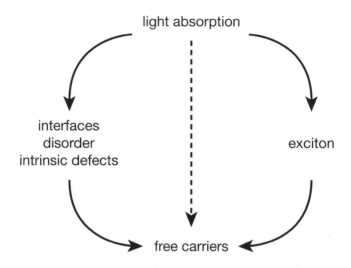

Figure 10. Three possible pathways for the primary photoexcitation in a nanoscale system. Photo-excitation predominantly forms nanoscale excitons owing to their large absorption cross-sections and binding energies. Subsequently, excitons may dissociate to form free carriers, and this typically requires an interface or defect to overcome the exciton binding energy. Direct photo-excitation to form free carriers is possible if the CTX states garner oscillator strength, perhaps by coupling to higher energy exciton resonances. It is possible that intrinsic defects, disorder, or interfaces provide more effective means of providing direct routes for free carrier formation.

7.1 Techniques and Studies

Of the techniques available for such investigations, transient absorption spectroscopy offers the most flexibility, providing a capability of following the fate of photo-induced species from femtoseconds to milli-seconds [94,142,299]. However, the species that are optically detected can be both the neutral exciton, the unbound electron and hole, often identified as a negative and positive polaron, as well as other species such as bound electron-hole pairs and bipolarons [300]. Reverting to a molecular language adds charge transfer states, excimers and exciplexes. All of which must be identified within the transient absorption spectrum. Techniques that are only sensitive to electrons and holes offer a far more incisive probe of the existence of the electrons and holes, but are still subject to uncertainties and questions. Optically detected magnetic resonance and electron spin resonance provide means of studying directly both electrons and holes via their 1/2 spin character and have proven to be excellent tools, especially for conjugated polymers and colloidal quantum dots [301,302]. Stark spectroscopy and electro-absorption take advantage of the influence of low frequency electric fields to probe the charged species. While transient microwave conductivity [303-307] is also excellent in this capacity, pushing the frequency of the electric field to the GHz region of the spectrum. A new technique that has emerged recently extends this frequency to the terahertz (THz) spectral region [308], offering not only the ability to probe the existence of mobile carriers, but to do so on a femtosecond time-scale [309].

Attempts to cover equally conjugated polymers, SWNTs, and QDs is difficult as there is vastly more information on the polymers; an interesting observation that results almost certainly from their application in light-emitting devices. However, it should be noted that regardless of the nanoscale system under investigation, the issues are the same and are therefore relevant.

Of the three nanoscale systems, it is the colloidal nanocrystals that are the most difficult to study. While excitonic states in these systems have been described [310], studies to provide experimental evidence is scarce [311]. Not only is the binding energy extremely small, but the unbound carriers are difficult to probe. Neither of the high frequency (GHz and THz) techniques offer the sensitivity that would be ideal, as "mobility" of the carriers, if this is the correct word, is extremely low due to the confinement of the carrier with the nanocrystal walls acting as reflecting boundaries for the charge density [312]. A similar effect is found for charges confined to isolated polymer chains [313]. Transient EPR can observe carriers but only after they are trapped [302].

The study of excitons in single-wall carbon nanotubes is still a field in its infancy. Like theory, there is a great deal about these systems that can be learned from the work on conjugated polymers. The task is not easy, however, as the availability of samples of isolated SWNTs of a single (*n,m*) index are not yet available. Enhanced photoconductivity of samples that contain SWNTs has been observed [314], but these contain not only a mixture of both semiconducting and metal (semimetal) tubes, but the nanotubes themselves are highly bundled. The studies do demonstrate the availability of photoinduced carriers, but the actual source of these carriers cannot be identified uniquely.

For conjugated polymers, there are numerous excellent papers that focus on the issue with some very good reviews that summarize the current level of understanding. Many of the original studies focused on the important question: is the primary excitation excitonic, or free carrier? The same question is re-visited here, but extended over a wider range of nanoscale systems. The observation of photoconductivity at the onset of the absorption spectrum [315] seems to suggest the latter, although this is not the most commonly-accepted answer, with the balance of opinion appearing to favour excitons as the primary photoexcitation, as promoted earlier in this chapter. The formation of electrons and holes in conjugated polymers, often described as polaronic species as the charges are dressed in phonons, can be readily observed in transient absorption spectra appearing almost instantaneously with the excitation light pulse [316]. However, it is the quantum yield of this process that must be determined in order to assess whether it is the primary process. Using transient THz spectroscopy on films of MEH-PPV, the quantum yield for free carrier production has been shown to be < 1% [298]. A more recent study using transient photomodulation spectroscopy [317] reveals the important role that film morphology plays, with carrier yields as high as 30% for ordered films of regioregular poly(3-hexylthiophene). With excitation at 3.2 eV, which is high into the polymer absorption band, a correlation was found between the film order, PL quantum yield and carrier quantum yield; high order correlating with more carriers and less PL. Isolated chains in solution also revealed significant carrier yields, specifically for MEH-PPV. This high molecular weight polymer can coil in solution and even aggregate, increasing the probability of chains interacting and hence promoting polaron formation via exciton dissociation at the contact point. This interaction has been proposed [300] as a mechanism to promote a polaron pair, where the oppositely charged species reside on adjacent chains at the position of interaction. Alternatively, these carriers may arise from exciton dissociation on a single chain, with a defect, such as a carbonyl group or bound oxygen [297] molecule being the site of dissociation.

For conjugated polymers, the branching ratio between excitons and free carriers upon photo-excitation is promoted by the interactions found in films, but is not eliminated by isolating the chains. The small number of carriers produced in these situations can be attributed to unwanted defects, but might still be an intrinsic mechanism simply associated with electronic disorder along the polymer chain. A comparison of the quality of available polymer samples with those currently available for SWNTs gives an insight into the difficulties ahead for answering the same question of the branching ratio in these relatively new 1D nanoscale systems.

An issue that plagues all excitons resident in a nanoscale system, and one that is of direct relevance to solar energy conversion, is the interaction between an exciton and a charged carrier. There is growing evidence of efficient annihilation of excitons by polarons in conjugated polymers [318-321], leading to an enhancement of the non-radiative decay mechanism. Similarly, in colloidal nanocrystals, the mechanism proposed to explain blinking is the quenching of an exciton by a hole that results from the ejection of an electron to the surroundings that was created by a previous exciton. This annihilation process must be considered seriously if these nanoscale systems are to be implemented in solar harvesting applications.

8. Outlook

The past 20 years has seen the emergence of a number of new nanoscale systems to be added to the existing list of J-aggregates, photosynthetic reaction centers, and antenna proteins. Unique to these systems is a primary photoexcitation that is excitonic in nature. Understanding the particular properties of these species is key, not only to initiating and advancing technological applications, but also to ascertaining an understanding of the fundamental properties. In this respect, the nanoscale is the ideal starting point for developing new theories, providing a link between molecular and bulk, and acting as the keystone between the two disciplines. Improvements in the theories used to understand the electronic structure of these nanoscale systems combined with the spectroscopic techniques that are the window into this structure are essential if the full potential is to be realized.

Our present understanding of condensed aromatic hydrocarbons stems from the availability of pure samples [322]. The difference between benzene, pyrene, tetracene, chrysene, triphenylene is analogous to the difference sizes and shapes of the colloidal nanocrystals. Pivotal to the success of this understanding, however, was the availability of samples of high purity and the same need is required for the nanoscale studies. SWNTs of a single (n,m) index, defect-free, capped, all of the same length and unbundled is perhaps a tall order, but is a goal worth striving for. Monodispersed samples of colloidal nanocrystals with perfect passivation, and conjugated polymers of perfect regioregularity are equally challenging.

Access to such nanoscale systems will not only serve to extend our fundamental knowledge, but will prove to be invaluable in applications, specifically in solar energy research. Our current understanding of how to treat excitations in these systems is still in its infancy, but is advancing. Learning how to extract the energy stored in the excitonic states efficiently is a key stepping stone that should motivate us.

Acknowledgements

This work was supported by Natural Sciences and Engineering Research Council of Canada and the U.S. Department of Energy's Solar Photochemistry Program within the Office of Science, Office of Basic Energy Sciences, Division of Chemical Sciences, Geosciences, and Biosciences. GDS acknowledges the support of an E. W. R. Steacie Memorial Fellowship.

References

1.   GD Scholes, G Rumbles: Nature Mater. 5 (2006) 683.
2.   PA Schultz, AS Sudbo, DJ Krajnovich, HS Kwok, YR Shen, YT Lee: Annu. Rev. Phys. Chem. 30 (1979) 379.

3. YT Lee: Science 236 (1987) 793.
4. AH Zewail: J. Phys. Chem. A 104 (2000) 5660.
5. J Michl, V Bonacic-Koutecky: Electronic Aspects of Organic Photochemistry, Wiley-Interscience, New York, 1990.
6. CG Elles, FF Crim: Annu. Rev. Phys. Chem. 57 (2006) 273.
7. H Koppel, W Domcke, LS Cederbaum: Adv. Chem. Phys. 57 (1984) 59.
8. F Bernardi, M Olivucci, MA Robb: Acc. Chem. Res. 23 (1990) 405.
9. RA Marcus: Rev. Mod. Phys. 65 (1993) 599.
10. RA Marcus, N Sutin: Biochim. Biophys. Acta 811 (1985) 265.
11. JN Onuchic, DN Beratan, JJ Hopfield: J. Phys. Chem. 90 (1986) 3707.
12. B Bagchi, FG R., DW Oxtoby: J. Chem. Phys. 78 (1983) 7375.
13. JM Lehn: Angew. Chem. Int. Ed. Engl. 27 (1988) 89.
14. MN Paddon-Row: Aust. J. Chem. 56 (2003) 729.
15. MR Wasielewski: Chem. Rev. 92 (1992) 435.
16. RH Friend, RW Gymer, AB Holmes, JH Burroughes, RN Marks, C Taliani, DDC Bradley, DA Dos Santos, JL Bredas, M Logdlund, WR Salaneck: Nature 397 (1999) 121.
17. VI Klimov, AA Mikhailovsky, S Xu, A Malko, JA Hollingsworth, CA Leatherdale, HJ Eisler, MG Bawendi: Science 290 (2000) 314.
18. BA Gregg: J. Phys. Chem. B 107 (2003) 4688.
19. RJ Elliott, in C.G. Kuper, G.D. Whitfield (Eds.), Polarons and Excitons. Plenum Press, New York, 1962, p. 269.
20. RS Knox, in B.D. Bartolo (Ed.), Collective Excitations in Solids. Plenum Press, New York and London, 1981, p. 183.
21. GD Scholes: (2007) submitted.
22. F Spano: Annu. Rev. Phys. Chem. 57 (2006) 217.
23. R Jimenez, SN Dikshit, SE Bradforth, GR Fleming: J. Phys. Chem. 100 (1996) 6825.
24. GD Scholes, GR Fleming: Adv. Chem. Phys. 132 (2005) 57.
25. H Fidder, J Knoester, DA Wiersma: J. Chem. Phys. 95 (1991) 7880.
26. D Möbius: Adv. Mater. 7 (1995) 437.
27. HB Klevens, JR Platt: J. Chem. Phys. 17 (1949) 470.
28. MA Hines, GD Scholes: Adv. Mater. 15 (2003) 1844.
29. H Zhao, S Mazumdar: Phys. Rev. Lett. 93 (2004) 157402.
30. A Szabo, NS Ostlund: Modern Quantum Chemistry: Introduction to Advanced Electronic Structure Theory, McGraw-Hill, New York, 1989.
31. AL Fetter, JD Walecka: Quantum Theory of Many-Particle Systems, McGraw-Hill, New York, 1971.

32. A Dreuw, M Head-Gordon: Chem. Rev. 105 (2005) 4009.
33. F Wang, G Dukovic, LE Brus, TF Heinz: Science 308 (2005) 838.
34. Y-Z Ma, L Valkunas, SM Bachilo, GR Fleming: J. Phys. Chem. B 109 (2005) 15671.
35. D Yaron, EE Moore, Z Shuai, JL Brédas: J. Chem. Phys. 108 (1998) 7451.
36. A Köhler, D Beljonne: Adv. Funct. Mater. 14 (2004) 11.
37. RD Harcourt, GD Scholes, KP Ghiggino: J. Chem. Phys. 101 (1994) 10521.
38. R McWeeny: Methods of Molecular Quantum Mechanics, Academic Press, London, 1992.
39. S Tretiak: Nano Lett. 7 (2007) 2201.
40. A Fanceschetti, A Zunger: Phys. Rev. Lett. 78 (1997) 915.
41. AL Efros, M Rosen, M Kuno, M Nirmal, DJ Norris, MG Bawendi: Phys. Rev. B 54 (1996) 4843.
42. SP McGlynn, T Azumi, M Kinoshita: Molecular Spectroscopy of the Triplet State, Prentice-Hall, Englewood Cliffs, New Jersey, 1969.
43. J Catalán, JLG de Paz: J. Chem. Phys. 120 (2004) 1864.
44. D Hertel, S Setayesh, H-G Nothofer, U Scherf, K Müllen, H Bässler: Adv. Mater. 13 (2001) 65.
45. D Wasserberg, P Marsal, SCJ Meskers, RAJ Janssen, D Beljonne: J. Phys. Chem. B 109 (2005) 4410.
46. CY Chi, C Im, G Wegner: J. Chem. Phys. 124 (2006) 24907.
47. GR Fleming, M Cho: Annu. Rev. Phys. Chem. 47 (1996) 109.
48. R Kubo, K Tomita: J. Phys. Soc. Jpn. 9 (1954) 888.
49. RA Marcus: J. Chem. Phys. 43 (1965) 1261.
50. S Nakajima: The physics of elementary excitations, Springer-Verlag, New York, 1980.
51. I Franco, S Tretiak: J. Am. Chem. Soc. 126 (2004) 12130.
52. V Chernyak, S Mukamel: J. Chem. Phys. 105 (1996) 4565.
53. GR Fleming, SA Passino, Y Nagasawa: Phil. Trans. Royal Soc. Lond. Series A 356 (1998) 389.
54. M Maroncelli, GR Fleming: J. Chem. Phys. 86 (1987) 6221.
55. RM Stratt, M Maroncelli: J. Phys. Chem. 100 (1996) 12981.
56. GS Engel, TR Calhoun, EL Read, T-K Ahn, T Mancal, Y-C Cheng, RE Blankenship, FG R.: Nature 446 (2007) 782.
57. H Lee, Y-C Cheng, FG R.: Science 316 (2007) 1462.
58. M Orrit, J Bernard, RI Personov: J. Phys. Chem. 97 (1993) 10256.
59. WE Moerner: J. Phys. Chem. B 106 (2002) 910.
60. QP Ambrose, WE Moerner: Nature 349 (1991) 225.

61. MR Salvador, MW Graham, GD Scholes: J. Chem. Phys. 125 (2006) 184709.

62. MR Salvador, MA Hines, GD Scholes: J. Chem. Phys. 118 (2001) 9380.

63. T Ando: J. Phys. Soc. Japan 66 (1997) 1066.

64. SM Bachilo, MS Strano, C Kittrell, RH Hauge, RE Smalley, RB Weisman: Science 298 (2002) 2361.

65. ML Cohen: Materials Science & Engineering C-Biomimetic and Supramolecular Systems 15 (2001) 1.

66. S Reich, C Thomsen, J Maultzsch: Carbon nanotubes: Basic concepts and physical properties, Wiley, New York, 2004.

67. MJ O'Connell, SM Bachilo, CB Huffman, VC Moore, MS Strano, EH Haroz, KL Rialon, PJ Boul, WH Noon, C Kittrell, J Ma, RH Hauge, RB Weisman, RE Smalley: Science 297 (2002) 593.

68. MS Dresselhaus, G Dresselhaus, R Saito, A Jorio: Physics Reports-Review Section of Physics Letters 409 (2005) 47.

69. MY Sfeir, T Beetz, F Wang, LM Huang, XMH Huang, MY Huang, J Hone, S O'Brien, JA Misewich, TF Heinz, LJ Wu, YM Zhu, LE Brus: Science 312 (2006) 554.

70. H Zhao, S Mazumdar, CX Sheng, M Tong, ZV Vardeny: Phys. Rev. B 73 (2006) 075403.

71. H Htoon, MJ O'Connell, SK Doorn, VI Klimov: Physical Review Letters 94 (2005).

72. A Hartschuh, HN Pedrosa, L Novotny, TD Krauss: Science 301 (2003) 1354.

73. CL Kane, EJ Mele: Physical Review Letters 93 (2004).

74. E Chang, G Bussi, A Ruini, E Molinari: Phys. Rev. Lett. 92 (2004) 196401.

75. CD Spataru, S Ismail-Beigi, LX Benedict, SG Louie: Phys. Rev. Lett. 92 (2004) 77402.

76. TJ McDonald, M Jones, C Engtrakul, RJ Ellingson, G Rumbles, MJ Heben: Review of Scientific Instruments 77 (2006).

77. RB Weisman, SM Bachilo: Nano Letters 3 (2003) 1235.

78. V Perebeinos, J Tersoff, P Avouris: Nanolett. 5 (2005) 2495.

79. CD Spataru, S Ismail-Beigi, RB Capaz, SG Louie: Phys. Rev. Lett. 95 (2005) 247402.

80. M Jones, C Engtrakul, WK Matzger, RJ Ellingson, AJ Nozik, MJ Heben, G Rumbles: Phys. Rev. B 71 (2005) 115426.

81. M Kasha: Discussions of the Faraday Society 14 (1950).

82. M Kasha, SP McGlynn: Annual review of Physical Chemistry (1956).

83. WK Metzger, TJ McDonald, C Engtrakul, JL Blackburn, GD Scholes, G Rumbles, MJ Heben: Journal of Physical Chemistry C 111 (2007) 3601.

84. GD Scholes, S Tretiak, TJ McDonald, WK Metzger, C Engtrakul, G Rumbles, MJ Heben: J. Phys. Chem. C 111 (2007) 11139.

85. MS Arnold, AA Green, JF Hulvat, SI Stupp, MC Hersam: Nature Nanotechnology 1 (2006) 60.

86. M Zheng, A Jagota, MS Strano, AP Santos, P Barone, SG Chou, BA Diner, MS Dresselhaus, RS McLean, GB Onoa, GG Samsonidze, ED Semke, M Usrey, DJ Walls: Science 302 (2003) 1545.

87. NS Sariciftci: Primary Photoexcitations in Conjugated Polymers: Molecular Exciton versus Semiconductor Band Model, World Scientific, Singapore, 1997.

88. JH Burroughs, DDC Bradley, AR Brown, RN Marks, K Mackay, RH Friend, PL Burns, AB Holmes: Nature 347 (1990) 539.

89. JL Brédas, D Beljonne, V Coropceanu, J Cornil: Chem. Rev. 104 (2004) 4971.

90. MD Watson, A Fechtenkötter, K Müllen: Chem. Rev. 101 (2001) 1267.

91. G Hadziioannou, PF van Hutten (Ed.)^(Eds.), Semiconducting Polymers: Chemistry, Physics and Engineering. Wiley-VCH, Weinheim, 2000.

92. IG Scheblykin, A Yartsev, T Pullerits, V Gulbinas, V Sundström: J. Phys. Chem. B 111 (2007) 6303.

93. FC Grozema, LDA Siebbeles, GH Gelinck, JM Warman: Top. Curr. Chem. 257 (2005) 135.

94. LJ Rothberg, M Yan, F Papadimitrakopoulos, ME Galvin, EW Kwock, TM Miller: Synth. Met. 80 (1996) 41.

95. U Scherf, EJW List: Adv. Mater. 14 (2002) 477.

96. IDW Samuel, GA Turnbull: Chem. Rev. 107 (2007) 1272.

97. S Gunes, H Neugebauer, NS Sariciftci: Chem. Rev. 107 (2007) 1324.

98. SW Thomas, GD Joly, TM Swager: Chem. Rev. 107 (2007) 1339.

99. CD Müller, A Falcou, N Reckefuss, M Rojahn, V Wiederhirn, P Rudati, H Frohne, O Nuyken, H Becker, K Meerholz: Nature 421 (2003) 829.

100. PE Burrows, SR Forrest, SP Sibley, ME Thompson: Appl. Phys. Lett. 69 (1996) 2959.

101. SR Forrest: Nature 428 (2004) 911.

102. J-L Brédas, J Cornil, D Beljonne, DA dos Santos, Z Shuai: Acc. Chem. Res. 32 (1999) 267.

103. D Beljonne, G Pourtois, C Silva, E Hennebicq, LM Herz, RH Friend, GD Scholes, S Setayesh, K Müllen, JL Brédas: Proc. Natl. Acad. Sci. 99 (2002) 10982.

104. S Tretiak, S Mukamel: Chem. Rev. 102 (2002) 3171.

105. J Gierschner, J Cornil, H-J Engelhaaf: Adv. Mater. 19 (2007) 173.

106. JL Brédas: Adv. Mater. 7 (1995) 263.

107. M Chandross, S Mazumdar, S Jeglinski, X Wei, ZV Vardeny, EW Kwock, TM Miller: Phys. Rev. B 50 (1994) 14702.

108. M Chandross, S Mazumdar, M Liess, PA Lane, ZV Vardeny, M Hamaguchi, K Yoshino: Phys. Rev. B 55 (1997) 1486.

109. A Shukla, H Ghosh, S Mazumdar: Phys. Rev. B 67 (2003) 245203.

110. D Beljonne, Z Shuai, J Cornil, DA Dos Santos, JL Brédas: J. Chem. Phys. 111 (1999) 2829.

111. D Beljonne, J Cornil, RH Friend, RAJ Janssen, JL Brédas: J. Am. Chem. Soc. 118 (1996) 6453.

112. S Mukamel, A Takashashi, HW Wang, G Chen: Science 265 (1994) 250.

113. ZG Shuai, D Beljonne, JL Brédas: J. Chem. Phys. 97 (1992) 1132.

114. LM Blinov, SP Palto, G Ruani, C Taliani, AA Tevosov, SG Yudin, R Zamboni: Chem. Phys. Lett. 232 (1995) 401.

115. JM Leng, S Jeglinski, X Wei, RE Benner, ZV Vardeny, F Guo, S Mazumdar: Phys. Rev. Lett. 72 (1994) 156.

116. M Liess, S Jeglinski, ZV Vardeny, M Ozaki, K Yoshino, Y Ding, T Barton: Phys. Rev. B 56 (1997) 15712.

117. A Köhler, DA dos Santos, D Beljonne, Z Shuai, J-L Brédas, AB Holmes, A Kraus, K Müllen, RH Friend: Nature 392 (1998) 903.

118. BJ Schwartz: Annu. Rev. Phys. Chem. 54 (2003) 141.

119. H Bässler, B Schweitzer: Acc. Chem. Res. 32 (1999) 173.

120. S Heun, RF Mahrt, A Greiner, U Lemmer, H Bässler, DA Halliday, DDC Bradley, PL Burn, AB Holmes: J. Phys.: Condens. Matter 5 (1993) 247.

121. CJ Collison, LJ Rothberg, V Treemaneekarn, Y Li: Macromol. 34 (2001) 2346.

122. SV Chasteen, SA Carter, G Rumbles: J. Chem. Phys. 124 (2006) 214704.

123. KF Wong, MS Skaf, C-Y Yang, PJ Rossky, B Bagchi, D Hu, J Yu, PF Barbara: J. Phys. Chem. B 105 (2001) 6103.

124. R Chang, J-H Hsu, WS Fann, KK Liang, CH Chang, M Hayashi, J Yu, SH Lin, EC Chang, KR Chuang, SA Chen: Chem. Phys. Lett. 317 (2000) 142.

125. PF Barbara, WS Chang, S Link, GD Scholes, A Yethiraj: Annu. Rev. Phys. Chem. 58 (2007) 565.

126. CF Wu, C Szymanski, J McNeill: Langmuir 22 (2006) 2956.

127. C Szymanski, CF Wu, J Hooper, MA Salazar, A Perdomo, A Dukes, J McNeill: J. Phys. Chem. B 109 (2005) 8543.

128. S Yaliraki, RJ Silbey: J. Chem. Phys. 104 (1996) 1245.

129. G Rossi, RR Chance, RJ Silbey: J. Chem. Phys. 90 (1989) 7594.

130. D Hu, J Yu, K Wong, B Bagchi, PJ Rossky, PF Barbara: Nature 405 (2000) 1030.

131. GD Scholes: Annu. Rev. Phys. Chem. 54 (2003) 57.

132. E Hennebicq, G Pourtois, C Silva, S Setayash, AC Grimsdale, K Müllen, JL Brédas, D Beljonne: J. Am. Chem. Soc. 127 (2005) 4744.

133. MM-L Grage, PW Wood, A Ruseckas, T Pullerits, W Mitchell, PL Burn, IDW Samuel, V Sundström: J. Chem. Phys. 118 (2003) 7644.

134. SCJ Meskers, J Hübner, M Oestreich, H Bässler: Chem. Phys. Lett. 339 (2001) 223.

135. R Chang, M Hayashi, SH Lin, J-H Hsu, WS Fann: J. Chem. Phys. 115 (2001) 4339.

136. KM Gaab, CJ Bardeen: J. Phys. Chem. A 108 (2004) 10801.

137. T-Q Nguyen, J Wu, SH Tolbert, BJ Schwartz: Adv. Mater. 13 (2001) 609.

138. J Cornil, DA Dos Santos, X Crispin, R Silbey, JL Brédas: J. Am. Chem. Soc. 120 (1998) 1289.

139. MH Chang, MJ Framptin, HL Anderson, LM Herz: Phys. Rev. Lett. 98 (2007) 027402.

140. ZH Yu, PF Barbara: J. Phys. Chem. B 108 (2004) 11321.

141. PF Barbara, AJ Gesquiere, S-J Park, YJ Lee: Acc. Chem. Res. 38 (2005) 602.

142. LJ Rothberg, in G. Hadziioannou, G.G. Malliaras (Eds.), Semiconduncting Polymers: Chemistry, Physics and Engineering. Wiley-VCH, Weinheim, 2006, p. 179.

143. R Österbacka, CP An, XM Jiang, ZV Vardeny: Science 287 (2000) 839.

144. T-A Chen, X Wu, RD Rieke: J. Am. Chem. Soc. 117 (1995) 233.

145. H Sirringhaus, PJ Brown, RH Friend, MM Nielsen, K Bechgaard, BMW Langeveld-Voss, A Spiering, JRA J., EW Meijer, P Herwig, DM de Leeuw: Nature 401 (1999) 685.

146. E Hennebicq, C Deleener, JL Brédas, GD Scholes, D Beljonne: J. Chem. Phys. 125 (2006) 054901.

147. WJD Beenken, T Pullerits: J. Phys. Chem. B 108 (2004) 6164.

148. IDW Samuel, I Ledoux, C Dhenaut, J Zyss, HH Fox, RR Schrock, RJ Silbey: Science 265 (1994) 1070.

149. J Grimme, M Kreyenschmidt, F Uckert, K Müllen, U Scherf: Adv. Mater. 7 (1995) 292.

150. S Tretiak, K Igumenshchev, V Chernyak: Phys. Rev. B 71 (2005) 033201.

151. S Tretiak, V Chernyak, S Mukamel: Phys. Rev. Lett. 77 (1996) 4656.

152. J Cornil, D Beljonne, CM Heller, IH Campbell, BK Layrich, DL Smith, DDC Bradley, K Müllen, JL Brédas: Chem. Phys. Lett. 278 (1997) 139.

153. J Gierschner, H-G Mack, L Lüer, D Oelkrug: J. Chem. Phys. 116 (2002) 8596.

154. G Heimel, M Daghofer, J Gierschner, EJW List, AC Grimsdale, K Müllen, D Beljonne, JL Brédas, E Zojer: J. Chem. Phys. 122 (2005) 054501.

155. LT Liu, D Yaron, MI Sluch, MA Berg: J. Phys. Chem. B 110 (2006) 18844.

156. LT Liu, D Yaron, MA Berg: J. Phys. Chem. C 111 (2007) 5770.

157. S Tretiak, A Saxena, RL Martin, AR Bishop: Phys. Rev. Lett. 89 (2002) 97402.

158. ER Bittner, S Karabunarliev, LM Herz: J. Chem. Phys. 126 (2007) 191102.

159. X Wang, TE Dykstra, GD Scholes: Phys. Rev. B 71 (2005) 45203.

160. TE Dykstra, V Kovalevskij, GD Scholes: Chem. Phys. 318 (2005) 21.

161. M Cho, J Yu, T Joo, Y Nagasawa, S Passino, G Fleming: J. Phys. Chem. 100 (1996) 11944.

162. MH Cho, GR Fleming: J. Chem. Phys. 123 (2005) 114506.

163. W deBoeij, M Pshenichnikov, D Wiersma: J. Phys. Chem. 100 (1996) 11806.

164. A Ruseckas, P Wood, IDW Samuel, GR Webster, WJ Mitchell, PL Burn, V Sundström: Phys. Rev. B 72 (2005) 115214.

165. LJ Rozanski, CW Cone, DP Ostrowski, DA Vanden Bout: Macromol. 40 (2007) 4524.

166. S Westenhoff, C Daniel, RH Friend, C Silva, V Sundström, A Yartsev: J. Chem. Phys. 122 (2005) 094903.

167. C Silva, DM Russell, AS Dhoot, LM Herz, C Daniel, NC Greenham, AC Arias, S Setayesh, K Müllen, RH Friend: J. Phys.: Condens. Matter 14 (2002) 9803.

168. D Prezzi, E Molinari: phys. stat. solidi (a) 203 (2006) 3602.

169. H Fidder, J Terpstra, DA Wiersma: J. Chem. Phys. 94 (1991) 6895.

170. VF Kamalov, IA Struganova, K Yoshihara: J. Phys. Chem. 100 (1996) 8640.

171. M Bednarz, VA Malyshev, J Knoester: Phys. Rev. Lett. 91 (2003) 217401.

172. F Dubin, J Berréhar, R Grousson, T Guillet, C Lapersonne-Meyer, M Schott, V Voliotis: Phys. Rev. B 66 (2002) 113202.

173. R Lécuiller, J Berréhar, JD Ganière, C Lapersonne-Meyer, P Lavallard, M Schott: Phys. Rev. B 66 (2002) 125205.

174. T Guillet, J Berréhar, R Grousson, J Kovensky, C Lapersonne-Meyer, M Schott, V Voliotis: Phys. Rev. Lett. 87 (2001) 087401.

175. F Dubin, R Melet, T Barisien, R Grousson, L Legrand, M Schott, V Voliotis: Nature Physics 2 (2006) 32.

176. SV Gaponenko: Optical Properties of Semiconductor Nanocrystals, Cambridge University Press, Cambridge, 1998.

177. A Henglein: Chem. Rev. 89 (1989) 1861.

178. MG Bawendi, ML Steigerwald, LE Brus: Annu. Rev. Phys. Chem. 41 (1990) 477.

179. H Weller: Adv. Mater. 5 (1993) 88.

180. H Weller: Angew. Chem. Int. Ed. Engl. 32 (1993) 41.

181. AP Alivisatos: Science 271 (1996) 933.

182. AP Alivisatos: J. Phys. Chem. 100 (1996) 13226.

183. M Nirmal, LE Brus: Acc. Chem. Res. 32 (1999) 407.

184. A Eychmüller: J. Phys. Chem. B 104 (2000) 6514.

185. AL Efros, M Rosen: Annu. Rev. Mater. Sci. 30 (2000) 475.

186. AD Yoffe: Adv. Phys. 50 (2001) 1.

187. D Bimberg, M Grundman, NN Ledentsov: Quantum Dot Heterostructures, Wiley, Chichester, 1999.

188. C Burda, XB Chen, R Narayanan, MA El-Sayed: Chem. Rev. 105 (2005) 1025.

189. EH Sargent: Adv. Mater. 17 (2005) 515.

190. AL Rogach, A Eychmüller, SG Hickey, SV Kershaw: Small 3 (2007) 536.

191. VI Klimov: Annu. Rev. Phys. Chem. 58 (2007) 635.

192. M Brumer, M Sirota, A Kigel, A Sashchiuk, E Galun, Z Burshtein, E Lifshitz: Appl. Optics 45 (2006) 7488.

193. DV Vezenov, BT Mayers, RS Conroy, GM Whitesides, PT Snee, Y Chan, DG Nocera, MG Bawendi: J. Am. Chem. Soc. 127 (2005) 8952.

194. V Sukkovatkin, S Musikhin, I Gorelikov, S Cauchi, L Bakueva, E Kumacheva, EH Sargent: Opt. Lett. 30 (2005) 171.

195. G Oohata, Y Kagotani, K Miyajima, M Ashida, S Saito, K Edamatsu, T Itoh: Physica E 26 (2005) 347.

196. AV Malko, AA Mikhailovsky, MA Petruska, JA Hollingsworth, VI Klimov: J. Phys. Chem. B 108 (2004) 5250.

197. VI Klimov: J. Phys. Chem. B 110 (2006) 16827.

198. VI Klimov, SA Ivanov, J Nanda, M Achermann, I Bezel, JA McGuire, A Piryatinski: Nature 447 (2007) 441.

199. SA Ivanov, J Nanda, A Piryatinski, M Achermann, LP Balet, IV Bezel, PO Anikeeva, S Tretiak, VI Klimov: J. Phys. Chem. B 108 (2004) 10625.

200. B Fisher, J-M Caruge, Y-T Chan, J Halpert, MG Bawendi: Chem. Phys. 318 (2005) 71.

201. Q Darugar, W Qian, MA El-Sayed: Appl. Phys. Lett. 88 (2006) 261108.

202. N Tessler, V Medvedev, M Kazes, SH Kan, U Banin: Science 295 (2002) 1506.

203. L Bakueva, S Musikhin, MA Hines, T-WF Chang, M Tzolov, GD Scholes, EH Sargent: Appl. Phys. Lett. 82 (2003) 2895.

204. BO Dabbousi, MG Bawendi, O Onitsuka, MF Rubner: Appl. Phys. Lett. 66 (1995) 1316.

205. JS Steckel, P Snee, S Coe-Sullivan, JR Zimmer, JE Halpert, P Anikeeva, LA Kim, V Bulovic, MG Bawendi: Angew. Chem. Int. Ed. Engl. 45 (2006) 5796.

206. WU Huynh, JJ Dittmer, AP Alivisatos: Science 295 (2002) 2425.

207. SA McDonald, G Konstantatos, S Zhang, PW Cyr, EJD Klem, L Levina, EH Sargent: Nature Mater. 4 (2005) 138.

208. S Kumar, GD Scholes: Microchim. Acta (2007) DOI: 10.1007/s00604.

209. AJ Nozik: Physica E 14 (2002) 115.

210. RD Schaller, VI Klimov: Phys. Rev. Lett. 92 (2004) 186601.

211. RJ Ellingson, MC Beard, JC Johnson, P Yu, OI Micic, AJ Nozik, A Shabaev, AL Efros: Nanolett. 5 (2005) 865.

212. SC Erwin, LJ Zu, MI Haftel, AL Efros, TA Kennedy, DJ Norris: Nature 436 (2005) 91.

213. K Becker, JM Lupton, J Müller, AL Rogach, DV Talapin, H Weller, J Feldmann: Nature Mater. 5 (2006) 777.

214. CJ Wang, M Shim, P Guyot-Sionnest: Science 291 (2001) 2390.

215. VM Huxter, V Kovalevskij, GD Scholes: J. Phys. Chem. B 109 (2005) 20060.

216. GD Scholes, K J., CY Wong: Phys. Rev. B 73 (2006) 195325.

217. J Kim, CY Wong, S Nair, KP Fritz, S Kumar, GD Scholes: J. Phys. Chem. B 110 (2006) 25371.

218. M Bruchez, M Moronne, P Gin, S Weiss, AP Alivisatos: Science 281 (1998) 2013.

219. IL Medintz, HT Uyeda, ER Goldman, H Mattoussi: Nature Mater. 4 (2005) 435.

220. JM Klostranec, WC Chan: Adv. Mater. 18 (2006) 1953.

221. JM Smith, PA Dalgarno, RJ Warburton, AO Govorov, K Karrai, BD Gerardot, PM Petroff: Phys. Rev. Lett. 94 (2005) 197402.

222. DD Awschalom, N Samarth, D Loss (Ed.)^(Eds.), Semiconductor Spintronics and Quantum Computation. Springer-Verlag, Berlin, 2002.

223. CB Murray, DJ Norris, MG Bawendi: J. Am. Chem. Soc. 115 (1993) 8706.

224. Y Yin, AP Alivisatos: Nature 437 (2005) 664.

225. Y Cheng, Y Wang, F Bao, D Chen: J. Phys. Chem. B 110 (2006) 9448.

226. Y-W Jun, MF Casula, J-H Sim, SY Kim, J Cheon, AP Alivisatos: J. Am. Chem. Soc. 125 (2003) 15981.

227. S Kumar, T Nann: Small 2 (2006) 316.

228. S-M Lee, S-N Cho, J Cheon: Adv. Mater. 15 (2003) 441.

229. L Manna, DJ Milliron, A Meisel, EC Scher, AP Alivisatos: Nature Mater. 2 (2003) 382.

230. L Manna, E Scher, AP Alivisatos: J. Am. Chem. Soc. 122 (2000) 12700.

231. DJ Milliron, SM Hughes, Y Cui, L Manna, J Li, J-W Wang, AP Alivisatos: Nature 430 (2004) 190.

232. X Peng, L Manna, W Yang, J Wickham, E Scher, A Kadavanich, AP Alivisatos: Nature 404 (2000) 59.

233. PS Nair, KP Fritz, GD Scholes: Small 3 (2007) 481.

234. T Svedberg: Colloid Chemistry, Chemical Catalog Co., New York, 1924.

235. J Kim, PS Nair, CY Wong, GD Scholes: Nano Lett. (2007) in press.

236. AI Ekimov, F Hache, MC Schanne-Klein, D Ricard, C Flytzanis, IA Kudryavtsev, TV Yazeva, AV Rodina, AL Efros: J. Opt. Soc. Am. B 10 (1993) 100.

237. LE Brus: J. Chem. Phys. 80 (1984) 4403.

238. AI Ekimov, AA Onushchenko, AG Plyukhin, AL Efros: Sov. Phys. JETP 61 (1985) 891.

239. U Banin, O Millo: Annu. Rev. Phys. Chem. 54 (2003) 465.

240. U Banin, Y Cao, D Katz, O Millo: Nature 400 (1999) 542.

241. DJ Norris, MG Bawendi: Phys. Rev. B 53 (1996) 16338.

242. MG Bawendi, PJ Carroll, WL Wilson, LE Brus: J. Chem. Phys. 96 (1991) 946.

243. M Kuno, JK Lee, BO Dabbousi, FV Mikulec, MG Bawendi: J. Chem. Phys. 106 (1997) 9869.

244. AM Kapitonov, AP Stupak, SV Gaponenko, EP Petrov, AL Rogach, A Eychmüller: J. Phys. Chem. B 103 (1999) 10109.

245. M Jones, J Nedeljkkovic, RJ Ellingson, AJ Nozik, G Rumbles: J. Phys. Chem. B 107 (2003) 11346.

246. L Spanhel, M Haase, H Weller, A Henglein: J. Am. Chem. Soc. 109 (1987) 5649.

247. C Bullen, P Mulvaney: Langmuir 22 (2006) 3007.

248. M Wang, JK Oh, TE Dykstra, X Lou, GD Scholes, MA Winnik: Macromol. 39 (2006) 3664.

249. M Wang, TE Dykstra, X Lou, MR Salvador, GD Scholes, MA Winnik: Angew. Chem. Int. Ed. Engl. 45 (2006) 2221.

250. MA Hines, P Guyot-Sionnest: J. Phys. Chem. 100 (1996) 468.

251. BO Dabbousi, J RodriguezViejo, FV Mikulec, JR Heine, H Mattoussi, R Ober, KF Jensen, MG Bawendi: J. Phys. Chem. B 101 (1997) 9463.

252. J van Embden, J Jasieniak, DE Gómez, P Mulvaney, M Giersig: Aust. J. Chem. 60 (2007) 457.

253. A Mews, A Eychmüller, M Giersig, D Schss, H Weller: J. Phys. Chem. 98 (1994) 934.

254. LP Balet, SA Ivanov, A Piratinski, M Achermann, VI Klimov: Nanolett. 4 (2004) 1485.

255. M Brumer, A Kigel, L Amirav, A Sashchiuk, O Solomesch, N Tessler, E Lifshitz: Adv. Funct. Mater. 15 (2005) 1111.

256. S Kim, B Fisher, HJ Eisler, MG Bawendi: J. Am. Chem. Soc. 125 (2003) 11466.

257. K Yu, B Zaman, S Romanova, DS Wang, JA Ripmeester: Small 1 (2005) 332.

258. CY Chen, CT Cheng, CW Lai, YH Hu, PT Chou, YH Chou, HT Chiu: Small 1 (2005) 1215.

259. S Kumar, M Jones, S Lo, GD Scholes: Small 3 (2007) 1633.

260. F Shieh, AE Saunders, BA Korgel: J. Phys. Chem. B 109 (2005) 8538.

261. GD Scholes, M Jones, S Kumar: J. Phys. Chem. C 111 (2007) 13777.

262. JM An, A Franceschetti, A Zunger: Nano Lett. 7 (2007) 2129.

263. A Franceschetti, H Fu, LW Wang, A Zunger: Phys. Rev. B 60 (1999) 1819.

264. M Nirmal, DJ Norris, M Kuno, MG Bawendi, AL Efros, M Rosen: Phys. Rev. Lett. 75 (1995) 3728.

265. Q Zhao, PA Graf, WB Jones, A Franceschetti, J Li, L-W Wang, K Kim: Nano Lett. (2007) (in press).

266. J Hu, L-S Li, W Yang, L Manna, L-W Wang, AP Alivisatos: Science 292 (2001) 2060.

267. A Shabaev, AL Efros: Nanolett. 4 (2004) 1821.

268. W Shockley, HJ Queisser: J. Appl. Phys. 32 (1961) 510.

269. MA Green: Third Generation Photovoltaics: Advanced Solar Energy Conversion, Springer, New York, 2006.

270. G Dennler, C Lungenschmied, H Neugebauer, NS Sariciftci, A Labouret: J. Mater. Res. 20 (2005) 3224.

271. FC De Schryver, T Vosch, M Cotlet, M Van der Auweraer, K Müllen, J Hofkens: Acc. Chem. Res. 38 (2005) 514.

272. H Sternlicht, GC Nieman, GW Robinson: J. Chem. Phys. 38 (1963) 1326.

273. RD Schaller, VM Agranovich, VI Klimov: Nature Physics 1 (2005) 189.

274. RD Schaller, M Sykora, JM Pietryga, VI Klimov: Nano Lett. 6 (2006) 424.

275. MC Beard, KP Knutsen, Y P., JM Luther, Q Song, WK Metzger, RJ Ellingson, AJ Nozik: Nano Lett. 7 (2007) 2506.

276. JM Luther, MC Beard, Q Song, M Law, RJ Ellingson, AJ Nozik: Nano Lett. 7 (2007) 1779.

277. JJH Pijpers, E Hendry, MTW Milder, R Fancuilli, J Savolainen, JL Herek, D Vanmaekelbergh, S Ruhman, D Mocatta, D Oron, A Aharoni, U Banin, M Bonn: J. Phys. Chem. C 2007 (2007) 4146.

278. A Franceschetti, JM An, A Zunger: Nano Lett. 6 (2006) 2191.

279. M Califano, A Zunger, A Franceschetti: Nano Lett. 4 (2004) 525.

280. A Shabaev, AL Efros, AJ Nozik: Nano Lett. 6 (2006) 2856.

281. MC Hanna, AJ Nozik: J. Appl. Phys. 100 (2006) 074510.

282. CE Kenneth, TH James: The Theory of the Photographic Process, MacMillan, New York, 1977.

283. B Oregan, M Gratzel: Nature 353 (1991) 737.

284. S Gunes, H Neugebauer, NS Sariciftci: Chemical Reviews 107 (2007) 1324.

285. NC Greenham, XG Peng, AP Alivisatos: Physical Review B 54 (1996) 17628.

286. WU Huynh, JJ Dittmer, AP Alivisatos: Science 295 (2002) 2425.

287. WU Huynh, JJ Dittmer, WC Libby, GL Whiting, AP Alivisatos: Advanced Functional Materials 13 (2003) 73.

288. WU Huynh, JJ Dittmer, N Teclemariam, DJ Milliron, AP Alivisatos, KWJ Barnham: Physical Review B 67 (2003).

289. BQ Sun, HJ Snaith, AS Dhoot, S Westenhoff, NC Greenham: Journal of Applied Physics 97 (2005).

290. BQ Sun, NC Greenham: Physical Chemistry Chemical Physics 8 (2006) 3557.

291. P Wang, A Abrusci, HMP Wong, M Svensson, MR Andersson, NC Greenham: Nano Letters 6 (2006) 1789.

292. E Kymakis, GAJ Amaratunga: Reviews on Advanced Materials Science 10 (2005) 300.

293. Y Wang, A Suna: Journal of Physical Chemistry B 101 (1997) 5627.

294. M Koehler, MC Santos, MGE da Luz: Journal of Applied Physics 99 (2006).

295. VI Arkhipov, P Heremans, H Bassler: Applied Physics Letters 82 (2003) 4605.

296. DV Talapin, CB Murray: Science 310 (2005) 86.

297. J Yu, DH Hu, PF Barbara: Science 289 (2000) 1327.

298. E Hendry, JM Schins, LP Candeias, LDA Siebbeles, M Bonn: Physical Review Letters 92 (2004).

299. B Kraabel, VI Klimov, R Kohlman, S Xu, HL Wang, DW McBranch: Physical Review B 61 (2000) 8501.

300. E Conwell, Primary Photoexcitations in Conjugated Polymers: Molecular Exciton versus Semiconductor Band Model. World Scientific, Singapore, 1997.

301. L Lanzani: Photophysics of Molecular Materials: from Single Molecules to Single Crystals, Wiley-VCH, Weinheim, 2006.

302. OI Micic, AJ Nozik, E Lifshitz, T Rajh, OG Poluektov, MC Thurnauer: Journal of Physical Chemistry B 106 (2002) 4390.

303. JE Kroeze, TJ Savenije, JM Warman: Comptes Rendus Chimie 9 (2006) 667.

304. G Dicker, MP de Haas, LDA Siebbeles, JM Warman: Physical Review B 70 (2004).

305. G Dicker, MP de Haas, JM Warman, DM de Leeuw, LDA Siebbeles: Journal of Physical Chemistry B 108 (2004) 17818.

306. TJ Savenije, JE Kroeze, MM Wienk, JM Kroon, JM Warman: Physical Review B 69 (2004).

307. JM Warman, MP de Haas, G Dicker, FC Grozema, J Piris, MG Debije: Chemistry of Materials 16 (2004) 4600.

308. MC Beard, GM Turner, CA Schmuttenmaer: Journal of Physical Chemistry B 106 (2002) 7146.

309. X Ai, MC Beard, KP Knutsen, SE Shaheen, G Rumbles, RJ Ellingson: Journal of Physical Chemistry B 110 (2006) 25462.

310. A Franceschetti, H Fu, LW Wang, A Zunger: Physical Review B 60 (1999) 1819.

311. RJ Ellingson, JL Blackburn, J Nedeljkovic, G Rumbles, M Jones, HX Fu, AJ Nozik: Physical Review B 67 (2003).

312. MC Beard, GM Turner, JE Murphy, OI Micic, MC Hanna, AJ Nozik, CA Schmuttenmaer: Nano Letters 3 (2003) 1695.

313. P Prins, FC Grozema, JM Schins, S Patil, U Scherf, LDA Siebbeles: Physical Review Letters 96 (2006).

314. E Kymakis, GAJ Amaratunga: Journal of Applied Physics 99 (2006).

315. D Moses, H Okumoto, CH Lee, AJ Heeger, T Ohnishi, T Noguchi: Physical Review B 54 (1996) 4748.

316. M Wohlgenannt, XM Jiang, ZV Vardeny: Physical Review B 69 (2004).

317. CX Sheng, M Tong, S Singh, ZV Vardeny: Physical Review B 75 (2007).

318. EJW List, CH Kim, W Graupner, G Leising, J Shinar: Materials Science and Engineering B-Solid State Materials for Advanced Technology 85 (2001) 218.

319. EJW List, CH Kim, W Graupner, G Leising, J Shinar: Synthetic Metals 119 (2001) 511.

320. EJW List, CH Kim, AK Naik, U Scherf, G Leising, W Graupner, J Shinar: Physical Review B 64 (2001).

321. S Moller, G Weiser, C Lapersonne-Meyer: Synthetic Metals 116 (2001) 23.

322. JB Birks: Photophysics of Aromatic Molecules, Wiley-Interscience, London, 1970.

CHAPTER 4

SILICON NANOCRYSTAL ASSEMBLIES: UNIVERSAL SPIN-FLIP ACTIVATORS?

Dmitri Kovalev

Department of Physics, University of Bath, Bath BA2 7AY, United Kingdom,
Tel.: +4401225383113, E-mail: d.kovalev@bath.ac.uk

Minoru Fujii

Department of Electrical and Electronic Engineering, Graduate School of
Engineering, Kobe University, Rokkodai, Nada, Kobe 657-8501, Japan

Oxygen molecules are one of the substances playing an important role in many biological and chemical reactions. Its excited singlet states are extraordinary chemically reactive because they are energy-rich and oxidation reactions of many organic molecules become spin-allowed. The reduction in size often modifies many properties of materials that undergo significant changes in a certain size range. For example, different forms of silicon nanocrystal assemblies have common entirely new physical properties due to morphological and quantum size effects. Most of those are governed by a large accessible surface area of hydrogen-terminated silicon nanocrystals and size-tuneable energies of excitons that have specific spin structures. These features result in new emerging functionality of nanosilicon: it is a very efficient spin-flip activator of oxygen and different organic molecules. In this review article, we describe activities towards understanding the fundamental details of the electronic interaction between photoexcited silicon nanocrystals and adsorbed molecules (in particular oxygen molecules). Furthermore, we demonstrate that silicon nanostructures have the extraordinary property of acting as facilitators for their photoexcitation. We argue that the whole effect is based on the energy transfer from long-lived electronic excitations, confined in Si nanocrystals, to the surrounding oxygen molecules via an exchange of electrons having mutually opposite spins. We further demonstrate that an identically efficient energy transfer processes mediated by silicon nanocrystals can be possible for a large variety of organic molecules having a ground singlet and a first excited triplet state what makes nanosilicon a chemically- and biologically-active material. Finally, we discuss implications of these findings for physics, chemistry and medicine.

1. Introduction

One of the strategic objectives of nanotechnology is the development of new materials having nanometer sizes which have entirely new physical properties with respect to bulk systems and, therefore, new functionalities. The main scientific question which can be asked regarding "nano" concept is: what new properties or behavior we can expect from nanomaterials which they do not have in a larger size scale. There are many examples of nanomaterials which indeed demonstrate unusual and frequently unexpected properties: metal nanoparticles, carbon nanostructures, semiconductor quantum dots or nanocrystals etc. At first sight, one might expect the interaction between silicon and oxygen to be no more than a simple oxidation process resulting in formation of SiO_2. However, we discovered that, at the nanoscale, the interaction becomes much more subtle, interesting and controllable. In this review we would like to concentrate on the property of Si nanostructures to act as facilitators for indirect photoexcitation of adsorbed molecules via energy transfer from electronic excitations confined in Si nanocrystals (excitons) to the surrounding molecules. The photoexcitation mechanism is likely to be universally applicable to a wide range of other inorganic and organic molecules.

1.1 Nanosilicon, current status

Silicon (Si) is an elemental semiconductor and its functionality in bulk form is limited. To prepare nanostructured materials two simple approaches can be used. Nanostructures can be prepared from atomic or molecular precursors in gas or liquid phase. Another approach relies on reducing the dimensions of bulk materials. This can be done using standard methods: photolithography, electron beam lithography, chemical or electrochemical etching, etc. Historically, nanostructured Si was first produced by Ingeborg and Arthur Uhlir at Bell Labs in the 1950's [1]. They were studying electropolishing of Si surfaces using aqueous solutions of HF and found that, at low current densities, electrochemical etching results in a sponge-like structure. In the 1970's, researches revealed the porous nature of the material, but intended to use it only as a precursor for making low dielectric constant layers [2]. Despite the observation of photoluminescence from porous silicon (PSi) at cryogenic temperatures in 1984 [3], wide attention to optical properties of PSi and other nanosilicon-containing systems was drawn only after reports on visible light emission at room temperature by Canham [4] and on blue shift of the absorption by Lehmann and Gösele [5].

An ideal functional device should combine electronic and optoelectronic components in the same chip. However, bulk Si, being a main material for the semiconductor industry, due to its indirect band-gap electronic structure, is a very inefficient light emitter. This is why in recent years most research efforts have been directed towards developing different approaches to improve the efficiency of light emission from nanosilicon-based structures. The key idea is that the reduction of the size of Si nanocrystals results, due to quantum size effects, in a widely tunable confinement energy of excitons and a partial breaking of the indirect band-gap nature of bulk Si [6, 7].

Si nanostructures can be produced according to different technological procedures. Structural investigations have confirmed that they all consist of Si nanocrystals of different size (typically a few nm) and shape that retain the diamond lattice structure of bulk Si. The most widely discussed system is PSi [4–8] prepared via anodization of bulk Si wafers in HF-based solutions. Depending on the type of dopants (p- or n-types) and the doping level of the wafer the sizes of pores and remaining Si crystals can be varied from micrometers to nanometers [6, 8]. This preparation procedure has attracted much interest due to its simplicity, unlike costly lithographic or epitaxial techniques that were at the time the conventional approaches to realise nanosized semiconductor structures. Other examples include Si nanocrystal assemblies prepared via ion implantation in a SiO_2 matrix [9], by reactive Si deposition onto quartz

[10], by plasma enhanced chemical vapor deposition [11] and by magnetron sputtering [12] (for further details see comprehensive review [13]). Recently, a laser pyrolysis technique has been introduced to achieve relatively narrow size distribution of Si nanocrystals [14]. Probably, the simplest preparation procedure of nanosilicon proposed so far is stain etching of bulk crystalline or polycrystalline Si wafers and films [15]. Due to successive oxidation of Si and removal of the grown oxide in $HF:HNO_3:H_2O$ solutions luminescing nanosilicon layers can be formed. For industrial applications of nanostructured materials, it is important that manufacturing costs are not unduly high and large scale production is viable. For this simple wet chemical synthesis route, the cost and complexity are orders of magnitude lower than for any other synthesis methods.

The interest in luminescence properties of Si nanocrystal assemblies was caused not only by the demonstration in 1990 that PSi can emit visible photoluminescence very efficiently at room temperature [4] but also by the confidence that Si-based efficient light-emitting device operating in the visible range can be realized. Since this time, much progress has been made. All features of the structural, optical and electronic properties of the material have been subjected to in-depth scrutiny [6, 7]. Later, in addition to light emission, PSi has been investigated for many other applications. The remarkable structural properties and morphology of this material [6, 8] in conjunction with the ability to tailor its surface chemistry [16] have led to interesting new applications, such as chemical and biological sensing [17], fuel cells [18], photosynthesis [19], drug delivery [20], explosives [21], adsorbers etc.

1.2 Singlet oxygen: physics, chemistry and applications

The interest that oxygen molecules (O_2) have attracted in various scientific fields, (e.g. molecular physics and photochemistry), stems from its particular electron spin configuration. Faraday in 1847 was probably the first to notice that oxygen molecules have an intrinsic magnetic moment. He simply found that oxygen-filled soap bubbles were driven into the region of a strong magnetic field [22]. Now it is well known that the ground state of O_2 has triplet nature ($^3\Sigma$) [23,24]. It has also been realised that spin states of energetically or chemically interacting substances can, in a large extent, control the interaction process. One of the most important examples is the interaction of organic molecules with O_2. The chemical reactions between singlet organic molecules and triplet O_2 forming new singlet organic molecules are forbidden by the spin selection rule. Thus, the triplet multiplicity of O_2 is the reason why most reactions between oxygen and organic substances at room temperature are very inefficient. A consideration of spin conservation restrictions answers a fundamental question: why is organic life so stable in the oxygen ambient? Surprisingly, it appears that, in fact, organic life is a spin-dependent phenomenon.

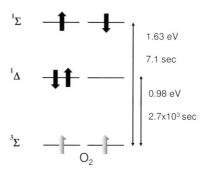

Figure 1. Electronic spin configurations and spectroscopic labelling of molecular oxygen. Superscript denotes the spin multiplicity.

The two lowest excited states of O_2, $^1\Delta$ and $^1\Sigma$, are singlets [23,24]. The corresponding electron spin configurations of O_2, and its energy levels, are indicated in Figure 1. Kautsky and DeBruijn [25] first demonstrated the existence of these species, singlet oxygen molecules ($^1O_2$) experimentally. Because of its singlet multiplicity, no spin-restriction exists for reactions of $^1O_2$ with singlet organic molecules. This, combined with their excitation energies of 0.98 eV and 1.63 eV, make singlet oxygen molecules extremely chemically reactive. The singlet states mediate fundamental processes in chemistry and biology [26-29]. They react with many organic compounds including aromatics, steroids, vitamins, amino acids, proteins, etc. They also find applications in bleaching and disinfection reactions and are involved in the modification of biological structures [27,29,30]. An important example of the medical application of $^1O_2$ is the photodynamic therapy of cancer [29].

The transition from the $^3\Sigma$ ground state to one of the excited $^1O_2$ states, and vice versa, requires a change of the electron spin state (spin-flip process). However, a direct conversion of spin states via absorption/emission of photons is spin-forbidden in the first approximation. This causes extremely long radiative lifetimes of $^1\Delta$ and $^1\Sigma$ states of isolated oxygen molecule being 2.7×10^3 sec and 7.1 sec, respectively [23,24]. Therefore other $^1O_2$ generation methods involving indirect O_2 excitation using light have been developed. For instance $^1O_2$ can be produced via gaseous discharge or chemical reactions [23,26]. However, a large variety of practical applications, especially in medicine, requires its generation in organic solvents or human tissues in a controlled manner. Therefore, the most common $^1O_2$ generation procedure involves photosensitizers [23,24] which mediate energy transfer to O_2.

1.3 Energy transfer processes

Photosensitization is an important process employed for the excitation of molecules exhibiting optically forbidden electronic transitions. Direct electronic excitation of atoms, molecules and nano-objects by light can be spin-, total momentum- or parity-forbidden. There are many molecules and nano-objects in which the ground and excited states have different spin multiplicities, e.g., the ground state might be a triplet and the excited state a singlet (or vice versa). The direct optical excitation is then forbidden by spin selection rule and other mechanisms of excitation should be involved. One particularly important excitation method is through energy transfer from a suitable sensitizing medium in close proximity. The energy transfer can proceed according to selection rules that are different from those for direct absorption of light, thus enabling triplet to singlet or singlet to triplet transitions to occur.

Historically, strongly light-absorbing organic dye molecules have been used as the photosensitizer (or energy "donor") [23,24]. After light absorption, such donors are usually efficiently excited in the long-lived triplet states. Provided there is an energy match, the donors can return to their ground states by energy transfer to the energy-accepting species ("acceptors"). We use the terms "donor" and "acceptor" in the sense of energy donation/acceptance, rather than in the more common semiconductor notion of electron donation/acceptance). In this way, the spin selection rules that prevent direct photoexcitation of the acceptor can be overcome. Conceptually this process is represented in Figure 2.

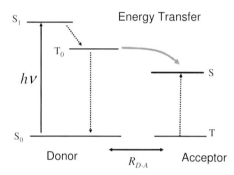

Figure 2. Schematic illustration of the energy transfer process between light-absorbing donor and acceptor. S denotes a singlet state and T a triplet state. R_{D-A} denotes spacing between interacting species.

The ideal donor material should have the following characteristics:

- Photoexcitation of the donor material should be easy; therefore excitation in its excited singlet state is required (allowed by dipole approximation).

- The photoexcitations should have a long lifetime (in the absence of energy acceptors) since energy transfer is a competitive process with respect to radiative relaxation of the donor.

- The photoexcitations in the donor material should have energies that match those which can be transferred to the acceptor. If they don't match energy conservation is maintained by the vibrational and rotational modes of donors and acceptors being recipients of the excess energy. This makes the process less efficient.

- It should be possible to bring large numbers of donors and acceptors into close proximity since energy exchange is a short range process.

- The spin states of the photoexcitations in the donor material and the energy transfer mechanism should be such as to overcome the selection rules for the excitation of the acceptor molecules.

Other desirable features include readily available light sources for the selective excitation of the sensitizer.

Energy transfer from an excited donor to an acceptor proceeds via dipole-dipole or Coulombic, Förster [31] process, or direct electron exchange interaction, Dexter process [32]. Both interactions are short-range: the efficiency of the Förster process scales as R_{D-A}^{-6} (in the far field approximation) while the Dexter process as $e^{-4\alpha R_{A-D}}$, where R_{D-A} is the space separation between donor and acceptor and α being a typical decay coefficient of the electronic wavefunctions. The Förster process relies on induction of a dipole oscillation in an acceptor induced by a dipole which represents an electronically excited donor. Since it depends on the oscillator strength of electronic transitions in the donor and the acceptor, this mechanism is mainly applicable for dipole-allowed transitions. The Dexter process is based on an electron exchange between donors and acceptors and is important mostly in the case of triplet states. In both scenarios all key requirements formulated above have to be fulfilled to assure the high efficiency of the energy transfer [23,24].

1.4 Singlet oxygen photosensitizers

The interaction of the excited triplet state of a dye molecule with the triplet ground state of O_2 results in the relaxation of the dye molecule to the ground state while the O_2 is activated via a spin flip process. This mechanism is frequently referred as triplet-triplet annihilation. In the last decades, a large number of different substances with the ability to generate singlet oxygen molecules and many photochemical reactions involving oxygen molecules have been studied. Efficiencies of the sensitized generation of $^1O_2$ have been determined for hundreds of photosensitizers, because of the importance of $^1O_2$ as a chemical and a biological reagent [23,24,29]. For instance in photodynamic therapy

(PDT), light, oxygen and photosensitizers are combined to produce a selective therapeutic effect in localized tumors [29]. Typical photosensitizers used in PDT are members of a dye molecules family known as porphyrins [29]. These dye molecules have properties that are essential for the PDT: they are soluble in water, stable under illumination, non-toxic and efficiently absorb visible light. Due to the small diffusion length of singlet oxygen in water, their action is localized in the area where the photosensitizer is located or where it is excited by light. The main limitation of the photodynamic therapy is a small light penetration depth into the tissue. Mainly, due to strong light scattering in human tissue, only light with a wavelength longer than 650 nm has the ability to penetrate relatively deeply in the human body. The current generation of photosensitizers efficiently absorbs light between 630 and 700 nm, which penetrates only a few millimetres into tissue. Therefore it is necessary to synthesize new compounds which will be able to absorb light in the 800 nm wavelength range; at this wavelength the penetration depth of light is a few centimetres.

Recently, the potential of nanomaterials as photosensitizers or carriers for singlet oxygen photosensitizers and their potential applications in PDT have been exploited [33-38]. In particular, large progress has been achieved in the use of nanoparticles including semiconductor nanocrystals/quantum dots as photosensitizers and polymer-based nanocomposites as photosensitizer carrier. At normal conditions, most semiconductor nanocrystals have direct band gaps and, therefore, the radiative recombination of excitons is very fast if the optical transition is dipole-allowed. Theory, however, predicts that the lowest exciton state of direct band gap CdSe nanocrystals is the triplet state [39], frequently referred as a "dark state". This result is potentially important because the triplet structure of excitons is required to undergo energy transfer from excitons to O_2 in a similar manner to dye molecules. Unfortunately, in direct band-gap semiconductors, the exciton lifetime at room temperature is controlled by thermally excited optically allowed singlet "bright" states and is in the submicrosecond time domain [40]. Since the time of the energy transfer to acceptors always competes with the lifetime of electronic excitations of the donor, the fast relaxation of excitons would drastically reduce the photosensitizing ability of direct band gap nanocrystal assemblies. Photosensitizing properties of carbon nanostructures, more specifically fullerenes and their derivatives, have been also studied in details [33,38]. However, despite of a very high efficiency of $^1O_2$ generation, C_{60} and C_{70} fullerenes do not absorb photons effectively in the visible and near-infrared spectral ranges. Therefore their potential use as *in vivo* sensitizers can be considered only if additional structures acting as a light-harvesting antenna are appended to their skeleton [38].

These different systems have certain advantages and shortcomings as far as their application for photodynamic therapy is concerned (for details see Ref. 33). For instance, TiO_2, ZnO are wide band-gap materials and can be excited only by ultraviolet light. For this spectral range the penetration depth of light in the tissue is negligible. On the other hand the application of CdS and CdSe nanocrystals in PDT is highly questionable because these materials have high toxicity.

Since 1990 [4,5], the efficiency of light emission from nanosilicon-based structures has been significantly improved. Recently it has been recognized that despite a tunable photoluminescence (PL) energy and a high quantum yield (up to 50% [41]), a long exciton lifetime is an inherent limitation for light emitting applications of Si nanocrystal assemblies [6,7]. Indeed at room temperature the operation frequency of nanosilicon-based devices should be of the order of inverse exciton lifetime which is in the range of 10 kHz.

However, a very long exciton lifetime is certainly a great advantage for photosensitizing applications since it implies very efficient energy storage and, therefore, a large probability of energy transfer. We found recently that this combination of new physical properties of Si at nanoscale is exceptionally favourable for the transfer of the energy from photoexcited Si nanocrystals to O_2 followed by the generation of $^1O_2$ and the efficiency of this process is approximately 100 % at low temperatures and 90 % at room temperature [19,42].

1.5 Content of this article

In the following chapters of this article we will review experimental work that has been performed towards detailed understanding of the energy transfer from photoexcited Si nanocrystals to O_2. The paper is structured as follows: Chapter 2 will cover the issues related to the morphological and optical properties of Si nanocrystals. Optical properties of Si nanocrystal assemblies will be highlighted with emphasis on the physics of excitons confined in Si nanocrystals. Chapter 3 is the main chapter of this review. We will describe in detail the mechanism of energy transfer from photoexcited Si nanocrystals to oxygen molecules. We will demonstrate that the unique optical properties of Si nanocrystals allow the investigation of the mechanism of energy transfer in much more detail than for ordinary photosensitizers. Finally, with a few examples we will argue that identical efficient energy transfer processes mediated by Si nanocrystals should be possible for a large variety of substances having a ground singlet and a first excited triplet states.

2. Physics of silicon nanocrystals

2.1 Morphological properties of Si nanocrystal assemblies

Transmission electron microscope (TEM) technique has been intensively used to provide some of the most detailed information on the internal structure of Si nanocrystal assemblies. Direct images of the structural elements can be obtained with resolutions down to the atomic scale. In Figure 3 we demonstrate high resolution TEM (HRTEM) images of different types of Si nanocrystal assemblies. Electrochemical or chemical etching of bulk Si wafers or powders results in a sponge-like structure containing interconnected undulating Si nanowires and pores with diameters down to 5 nm (see Figure 3a). The material is completely crystalline, as confirmed by HRTEM and diffraction measurements [43].

Since the etching is performed in HF-based solutions, almost all surface Si bonds are passivated by hydrogen and as-prepared PSi contains essentially no oxygen. FTIR measurements have confirmed that all possible surface atomic configurations are present: Si-H, S-H_2 and Si-H_3 [6]. The overall structure of PSi layers depends very strongly on the anodization conditions and the resistivity and doping type of the bulk Si wafer. Pore diameters and spacing can vary over a wide range from the nanometre scale up to the micrometre scale [7]. The natural ageing in a time scale of months or thermal annealing of PSi in air at temperatures below ~ 300 °C result in the incorporation of a monolayer of oxygen atoms back-bonded to the surface of nanocrystals while hydrogen atoms still remain at the surface. The effusion of hydrogen from the surface of PSi starts at 300 °C and at annealing temperatures above 700 °C the surface of nanocrystals can be completely oxidized.

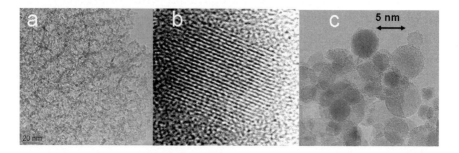

Figure 3. HRTEM images of different types of Si nanocrystals assemblies. a) Porous Si powder. b) Si nanocrystals in SiO$_2$ matrix prepared according to the procedure described in Ref. 12. c) Si nanospheres synthesized from the gas phase.

In Figure 3b we demonstrate a HRTEM image of an individual Si nanocrystal imbedded in a SiO$_2$ matrix. These nanosilicon structures are usually prepared by annealing non-stoichiometric SiO$_2$ (SiO$_x$) at temperatures between 900 and 1200 °C. The lattice fringes in the HRTEM image of Figure 3b correspond to the (111) planes of Si nanocrystals, thus Si nanocrystals retain the diamond crystalline structure of bulk Si. Recently, another promising technique based on synthesis of Si nanocrystals having a crystalline core from the gas phase (silane) has been developed. It results in spherical Si nanoparticles having nanometer size (Figure 3c) [41, 44].

The surface of Si nanocrystals has a key influence on their light emission properties. Surface Si dangling bonds are nonradiative mid-gap states [6,7]. In Si nanocrystals nonradiative processes efficiently compete with a slow radiative recombination. Since incomplete hydrogen passivation of the surface or the poor electronic quality of the Si/SiO$_2$

interface results in a very low PL quantum yield different strategies for an improvement of the surface passivation have been developed.

2.2 Luminescence properties of Si nanocrystal assemblies

As it has been mentioned before, bulk Si, due to its indirect band-gap structure and nonpolar lattice is a spectacularly poor light emitter. Since the radiative time of the indirect transitions is extremely long and the transport of excitons is efficient, the main decay channel for free excitons or electron-hole (*e-h*) pairs is their capture in bound exciton states or nonradiative recombination. This results in a very low quantum yield of light emission. Even at liquid He temperatures it is of the order of 10^{-4}-10^{-6}. The spatial confinement of three carriers in the vicinity of a charged impurity center leads to a very high effective *e-h* concentration of ~10^{18}-10^{19} cm^{-3}. Therefore, the quantum efficiency for bound exciton transitions is determined by the ratio of the indirect optical transition probability to the nonradiative Auger process probability and is extremely small as well [45].

In a nanocrystal the situation changes. Firstly, spatial confinement has to shift both absorbing and luminescing states to higher energies and results in atomic-like electronic states due to a rising of the minimum kinetic energy and quantization. Secondly, according to the uncertainty principle, the geometrical confinement leads to a delocalization of carriers in the crystal quasimomentum space thus allowing zero phonon optical transitions and significantly enhancing the oscillator strength of the zero phonon transitions in small Si nanocrystals [46]. Thirdly, due to the better overlap of electron and hole envelope wavefunctions one can expect a strong enhancement of the *e-h* exchange interaction inducing a splitting of the exciton levels [47]. Finally, since photoexcited carriers are strongly geometrically localized in nanocrystals they were created, recombination has the geminate character. Therefore the recombination statistics is quite different from that used for bulk crystals. Measured PL lifetimes can be considered to a large extent to be radiative. This can be understood by regarding Si nanocrystal assemblies as a granular-like materials consisting of luminescing (internal quantum yield is equal to 1) and dark nanocrystalls. The first type of crystallites belong to those which do not contain nonradiative centres while the second ones have at least one surface nonradiative defect. Under this assumption the observed PL decay time is the time of radiative recombination [48].

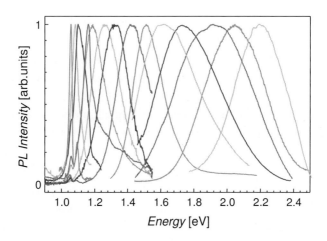

Figure 4. Tunability of PSi PL band. To achieve size variation of Si nanocrystals different levels of bulk Si substrate doping and various etching conditions (current density and etching solution concentration) have been used. E_{ex}=2.54 eV. Reproduced with permission from Ref. 49, D. Kovalev et al, Adv. Mater., **17**, 2531 (2005), Copyright @ Wiley-VCH Verlag GmbH & Co. KGaA.

Clear evidence that emitting states are driven to higher energies by the confinement is coming from the PL measured under high energy of optical excitation. Under these conditions all crystallites in the distribution having different sizes and, therefore, confinement energies are excited. The PL, depending on the mean size of Si nanocrystals and their size distribution, can be continuously tuned with small increments over a very wide spectral range from the Si band-gap to almost the green region. We demonstrate this in Figure 4 using PL spectra of PSi samples but very similar tunability has been convincingly demonstrated for other types of Si nanocrystal assemblies [13]. Large spectral width of these PL spectra (full width at half of maximum up to 500 meV) is governed by the variation of size and shape of Si nanocrystals. Thus, the confinement energy can be more than 1.5 eV, larger than the fundamental Si band-gap itself (1.12 eV). This observation is essential for the energy transfer processes: the energy of excitons can be adjusted to any desirable value from 1.12 eV to 2.5 eV simply by a proper choice of Si nanocrystal sizes.

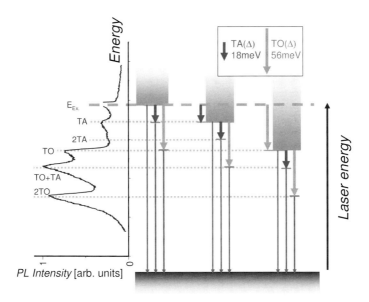

Figure 5. Left side: Resonant PL spectra of naturally oxidized PSi. Each peak in the PL spectrum corresponds to the additional no-phonon and phonon-assisted processes in the absorption/emission cycle which are allowed at this particular energy. Dashed lines show the energy position of Si TA and TO momentum-conserving phonons with respect to the exciton ground state. Right side: Sketch representing 3 groups of Si nanocrystals having different band-gaps involved into the absorption/emission cycle under resonant optical excitation. Arrows indicate momentum-conserving phonons participating in this cycle. Green arrows: emission of TO momentum-conserving phonons, blue arrows: emission of TA momentum-conserving phonons. Energy of the laser is indicated by the horizontal green dashed line. Red arrows demonstrate different recombination channels of Si nanocrystals. Reproduced with permission from Ref. 49, D. Kovalev et al, Adv. Mater., **17**, 2531 (2005), Copyright @ Wiley-VCH Verlag GmbH & Co. KGaA.

We would like to note that no distinct emission features allowing a determination of the nature of the luminescing centres are observed. This spectroscopic information is obscured by the residual nanocrystal size and shape distributions. This has probably stimulated a wide variety of models explaining the emission from Si nanocrystals. The standard method used to lift the inhomogeneous PL line broadening is the resonant PL excitation experiment. The essence of this approach is the selectivity of the optical excitation. When the exciting laser light is chosen to fall inside the PL band, only the subset of emitters having the excitation threshold below the laser energy is probed. This type of experiment on PSi was first performed by Calcott et al. [50] and the authors found that under resonant excitation conditions distinct steps or peaks in PL spectra can be observed. The energy of the onsets is identical to that of the TO and the TA momentum conserving phonons of bulk Si crystals (c-Si) (56 meV and 18 meV, respectively, as an example of resonant PL spectrum see left side of Figure 5). Since two strong structures are only observed (for each type of phonon) the only possible transitions in the absorption-emission cycle involve zero, one or two momentum-conserving phonons. Taking into account that absorption and emission are mirror-like events, these processes correspond to no-phonon transitions in both emission and absorption, to a one phonon-assisted process in either the absorption or the emission, and to phonon-assisted transitions in both the emission and the absorption (see sketch on the right side of Figure 5). It has been stated by the authors that these PL signatures provide evidence that "the luminescing material has the electronic and vibrational band structures of c-Si" (for detailed discussion see Ref. 50). Later a number of works [51-56] have confirmed the general observations reported in this paper for different systems containing Si nanocrystals. This type of experiments is characterized by a high energy- and, therefore, nanocrystals size-selectivity but the spectral width of the PL features (5-10 meV) is still much larger than that measured for individual nanocrystals or quantum dots [57,58]. Quantum confinement theory predicts that the density of electronic states for the "particles in a box" should be atomic-like. This also implies that the PL spectrum of individual nanocrystals should be very narrow. An ultimate PL experiment with a single Si nanocrystal has been performed by Ilya Sychugov et al. [57]. This group demonstrated that the linewidth of the PL from individual Si nanocrystals is ~ 2 meV at $T=35$ K. This value, clearly below the thermal broadening at this temperature, proves the atom like emission from Si quantum dots subject to quantum confinement [57].

When the size of the crystallite is so small that it contains only a few unit cells, any selection rules that derived from the translational symmetry of the bulk material crystalline lattice should be strongly broken. The presence of TO- and TA momentum-conserving phonon replicas in emission spectra of Si nanocrystals under resonant excitation, however, evidences that, despite of an efficient breakdown of the quasi-momentum conservation rule, Si nanocrystals still partially retain their indirect band-gap nature. Therefore even nanometer-size Si crystallites do not become a direct band-gap semiconductor. Si nanocrystals are simply not sufficiently small to assure direct band-gap optical transitions. Unfortunately, despite the high PL yield, they still behave to a large extent as an indirect band gap semiconductor.

In bulk Si the radiative lifetime of electron-hole excitations is extremely long and can not be measured directly since the non-radiative processes dominate the recombination statistics. The contribution of non-radiative processes results in an extremely low optical emission quantum yield. Already first measurements of the temporal evolution of the PL emitted by PSi demonstrated that the exciton lifetime, depending on the temperature, is in the microsecond – millisecond time domain. For comparison, in low-dimensional direct band gap semiconductors radiative exciton lifetimes are in the nanosecond time domain [59]. The geometrical confinement of the exciton results not only in a blue shift of the optical transitions but also in a modification of the oscillator strengths of no-phonon and phonon-assisted processes. In smaller nanocrystals (having larger confinement energies) both zero-phonon transition (due to an efficient breakdown of crystal quasi-momentum conservation rule) and phonon-assisted transition probabilities (due to a better overlap of *e-h* wavefunctions) should be significantly increased. This can be seen experimentally as a shortening of the exciton radiative lifetime. We illustrate this effect in Figure 6 which demonstrates the spectral dependence of the PL decay time measured at 200 K when the PL quantum yield has a maximum value. Lifetimes measured for other systems containing Si nanocrystals have very similar values. In this temperature range the PL decay time to a large extent can be considered as radiative [7, 60]. The PL decay time varies from almost a millisecond in the vicinity of the bulk Si band gap down to a microsecond for the green spectral range. The PL decay times measured for other systems containing Si nanocrystals have very similar values. A very strong spectral dispersion of the exciton recombination time is a direct consequence of quantum confinement effects. In smaller nanocrystals the emission energy and the oscillator strength of radiative transitions are larger in accordance with predictions of quantum confinement theory.

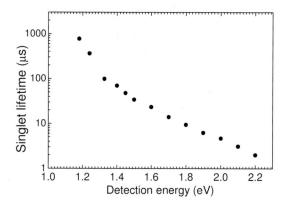

Figure 6. Spectral dispersion of singlet exciton lifetime. T=200 K. $E_{ex.}$=3.67 eV. Reproduced with permission from Ref. 49, D. Kovalev et al, Adv. Mater., **17**, 2531 (2005), Copyright @ Wiley-VCH Verlag GmbH & Co. KGaA.

The study of the temporal PL behaviour is very important since it allows the prediction of possible applications for which a particular luminescent material may be appropriate. Long lifetime of excitons confined in Si nanocrystals, in the range of 10^{-3} s -10^{-6} s, implies long-term storage of the energy of electronic excitations. According to criteria listed above, this can make energy or charge transfer to other substances possible. Therefore Si nanocrystals are promising candidates for energy/charge donors. The lifetime of excitons varies from ~ 10 μs at room temperature to a few milliseconds at He temperatures while the PL quantum yield increases insignificantly. Therefore, the increase of *radiative* exciton lifetime should be governed by different selection rules at room and cryogenic temperatures [6,7,50].

2.3 Spin structure of excitons confined in Si nanocrystals

The exchange interaction between electrons and holes having certain mutual spin orientation is a very weak perturbation. For instance, for exciton in bulk Si the exciton exchange splitting energy is about 150 μeV and does not play a role in the optical transitions [61]. Its value is strongly dependent on the spatial overlap of electron and hole wavefunctions which is different for parallel and antiparallel spins. When the size of the crystallite approaches the bulk exciton Bohr radius a drastic enhancement of the effect can be measured. Recently it has been demonstrated that the electron-hole exchange interaction plays an important role in the description of the emission properties of nanocrystal assemblies and quantum dots [50,62-66]. In most of these systems the ground state of the exciton has triplet nature. The first experimental evidence for the importance of this type of interaction in Si nanocrystals was provided by Calcott et al. [50] and later was confirmed by a number of groups [7, 65]. The upper and lower exciton states are assumed to be an optically active spin-singlet ($S=0$) and an optically passive spin-triplet ($S=1$), respectively. The exchange interaction splits these two states and triplet state has a lower energy. Although the spin-orbit interaction in Si is weak, it has been shown to play an important role in Si nanocrystals [67]. Due to this type of interaction there is admixture of singlet character to the triplet transitions and they become weakly allowed.

According to the authors of Ref. 50, the optical absorption-emission cycle of Si nanocrystals is characterized by different spin structures of the absorbing and luminescing states: absorption takes place via the optically allowed transition to the singlet exciton state. At low temperatures, after a fast spin-flip process the exciton relaxes to the dipole-forbidden triplet   state with a following slow electron-hole annihilation (see sketch of Figure 7). In Figure 7 the resonant PL (a) and photoluminescence excitation spectra (b) measured very near to the excitation energy are shown. The spectral gap $\Delta_{exch.}$ of the order of a few meV between the excitation line and the onset of the emission is clearly

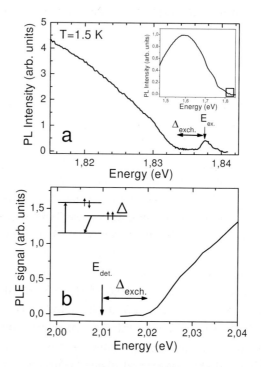

Figure 7. Resonant a) PL and b) PLE spectra measured very near to the excitation energy. Detection energies are shown by vertical arrows; values of the spectral gap $\Delta_{exch.}$ are indicated by horizontal arrows. Inset of Figure 7a: resonant PL spectrum, the square is the area which has been magnified. Inset of Figure 7b: sketch of spin structure of exciton: photons excite exciton in a singlet state (antiparallel spins) while after spin-flip process exciton emits photon from its ground triplet state (parallel spins). Reproduced with permission from Ref. 7, D. Kovalev et al, Phys. Stat. Sol. (b) **215**, 871 (1999), Copyright @ Wiley-VCH Verlag GmbH & Co. KGaA.

seen. This gap can be detected in all resonant PL spectra and its value is strongly excitation energy-dependent. This very small energy splitting results in a strong temperature dependence of the exciton lifetimes. At low temperatures when $k_B T \ll \Delta_{exch.}$ only the lowest triplet state is occupied and the decay time is very long, about several milliseconds. In the other limit $k_B T \gg \Delta_{exch.}$ both states are equally occupied and the transition takes place mainly from the faster singlet state. However the lifetime of the indirect gap singlet state is long as well, from several microseconds to several hundreds of microseconds, depending on the size of Si nanocrystals. Thus, contrary to all other semiconductor nanocrystals, for Si nanocrystals the exciton lifetime is extremely long over the entire temperature range. Since the exciton lifetime is 4-5 orders of magnitude longer than that in other systems containing direct band-gap semiconductor nanocrystals or quantum dots, Si nanocrystal assemblies seem to be favourable candidates for energy or charge transfer interactions.

3. Silicon nanocrystals as a singlet oxygen photosensitizer

We already formulated above the key requirements for high efficiency of the energy transfer process. Si nanocrystals accomplish all of them and seem to be an almost ideal candidate for donors, more specifically, for an efficient energy transfer to O_2 and other molecules having ground singlet and excited triplet states:

- Photoexcitation of Si nanocrystals is easy; it is excited in a singlet state.

- Excitons have very long radiative time (from microseconds to milliseconds).

- By adjusting the size of Si nanocrystals the energies of confined excitons can be tuned over a wide range from 1.1 eV to almost 2.5 eV. Therefore, the energy of excitons can always match those which can be transferred to the acceptor (1.1 eV-2.5 eV).

- Si nanocrystal assemblies have a huge internal area (up to 500 m^2/cm^3) being accessible for O_2 or other molecules which can be physisorbed on extended nanosilicon surfaces [6,8]. Therefore large number of donors and acceptors can be brought into close proximity.

- The first excited state of excitons is a triplet state and the excitation of an acceptor from a ground triplet/singlet state to an excited singlet/ triplet state is a spin-allowed process.

3.1 Main observations. Low temperatures

The first indication of an interaction between photoexcited Si nanocrystals and O_2 came soon after the discovery of efficient photoluminescence from PSi. Authors of Ref. 68 found that the photodegradation of PSi is very fast in oxygen ambient and is particularly a result of the photooxidation of the PSi surface which introduces an additional non-radiative recombination channel: Si dangling bonds. However, no distinct microscopic mechanism of PSi photodegradation and photooxidation has been proposed. Later, Harper and Sailor reported on the quenching of PSi emission in O_2 ambient at room temperature and ascribed this effect to "a mechanism involving transient non-radiative electron transfer from the luminescent chromophore in PSi to a weakly chemisorbed O_2 molecule" [69]. However, attempts to observe the characteristic $^1O_2$ emission line were not successful. Additional experiments with chemical traps of $^1O_2$ also gave negative results.

Recently, we have shown that, owing to the overlap of the energy levels of Si nanocrystal assemblies and O_2, PSi can be successfully employed for the photosensitized singlet oxygen generation [21,42]. We will start with a first demonstration of the interaction of excitons confined in Si nanocrystals with oxygen molecules at cryogenic temperatures. These studies give a general evidence of this interaction and allow monitoring details of the energy-transfer process that are obscured at elevated temperatures as a result of thermal broadening effects. The broad photoluminescence spectrum of the nanocrystal assembly probes energy transfer from excitons to oxygen molecules.

Figure 8 demonstrates a strong interaction between photoexcited Si nanocrystals and oxygen molecules. The low-temperature PL spectrum of PSi measured in vacuum (Figure 8, dashed line) is characterized by a broad, featureless emission band located in the visible and near-infrared spectral range. This reflects the wide band-gap distribution in the Si nanocrystal assemblies. This emission spectrum is drastically modified by the physisorption of oxygen molecules, even performed at very low pressures. (Figure 8, dotted and solid lines). The PL band is quenched and low temperature spectra exhibit fine structure. A complete PL suppression is observed at energies above 1.615 eV (indicated by a vertical dotted line), which almost coincides with the $^1\Sigma$ state excitation energy for isolated O_2. The desorption of oxygen molecules leads to a complete recovery of the initial emission properties of PSi, which indicates the reversibility of the quenching mechanism. Direct proof for

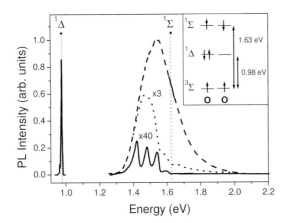

Figure 8. PL spectra of as-prepared PSi layer. $E_{ex.}$=2.41 eV. Dashed curve: T =50 K; layer is in vacuum. Doted curve: T =50 K; O_2 pressure is 10^{-2} mbar. Solid curve: T =5 K; O_2 pressure is 10^{-4} mbar. The narrow emission line at 0.98 eV is due to the $^1\Delta$-$^3\Sigma$ transition. The broad weak background of the Si dangling bonds PL band is subtracted for clarity. Energies of $^1\Delta$ and $^1\Sigma$ states are shown by vertical dotted lines. PL scaling factors are shown. Inset: sketch of the energy levels of oxygen molecules depending on the electron spin configuration. Labelling and energies of the transitions are indicated. Reproduced with permission from Ref. 19, D. Kovalev et al, Phys. Rev. Lett. **89**, 137401 (2002), Copyright @ American Physical Society.
http://link.aps.org/abstract/PRL/v89/e137401

the generation of $^1O_2$ is the detection of the light emission during its relaxation to the $^3\Sigma$ ground state of O_2. Fast relaxation of the $^1\Sigma$ state prevents the experimental observation of the $^1\Sigma$-$^3\Sigma$ transition [23,24]. We found that the quenching of the PL at low temperature is always accompanied by the appearance of a narrow PL line at 0.98 eV (Figure 8a, narrow atomic-like PL peak) resulting from the $^1\Delta$–$^3\Sigma$ transition of O_2, i.e., obviously there is energy transfer from annihilated excitons to O_2. This indicates that these two characteristic energies are entirely relevant to the interacting systems, the Si nanocrystal assembly and O_2. We would like to mention here that the $^1\Delta$–$^3\Sigma$ transition is most probably the most improbable in nature because it is simultaneously spin-, orbital angular momentum-, and parity-forbidden [24]. Its clear observation for photoexcited micrometer-thick PSi layers evidences the extremely high efficiency of $^1O_2$ generation.

In Figure 8 distinct spectroscopic features can be seen in the presence of oxygen molecules on the surface of Si nanocrystals. We would like to mention that resonant features governed by the involvement of momentum-conserving phonons have been previously seen at resonant optical excitation [50-56]. However at nonresonant excitation conditions these features obviously appears due to energy transfer from excitons to O_2: the spectral positions of the features do not depend on the excitation energy. Therefore local spectral minima in Figure 8 correspond to the most efficient energy transfer while maxima to inefficient energy transfer. To investigate the nature of the spectroscopic fine structure, the tunable emission properties of Si nanocrystal assemblies have been employed. Different sample preparation procedures were used to vary the size distribution of the Si nanocrystal assemblies and to shift the luminescence energies from the band-gap of bulk Si up to 2.2 eV. Specifically, two samples having different nanosilicon size distributions have been prepared. The fist sample has the PL maximum at 1.3 eV and can efficiently couple only to $^1\Delta$ state while the second sample having PL maximum at 1.9 eV interacts mainly with $^1\Sigma$ state of O_2. To resolve the spectral features above the $^1\Sigma$ state that is strongly coupled to excitons confined in Si nanocrystals, a weak PL suppression has been realized by a low concentration of adsorbed oxygen molecules. For both samples the quenched emission spectra reveal a fine structure which is present in the entire probed spectral range and the features have identical spectral positions for samples having different PL bands (see Figure 9).

As follows from Figure 9, the exact shape of the quenched PL spectra is defined by the convolution of the "envelope function," i.e., the Si nanocrystal size distribution and the spectral dependence of the coupling strength between excitons and oxygen molecules. To eliminate the influence of the size distribution on the shape of the quenched PL spectrum we define the strength of quenching as the ratio of the PL intensity measured in vacuum to that measured in quenched condition. Figure 10 demonstrates the results of this procedure for the spectra shown in Figure 9.

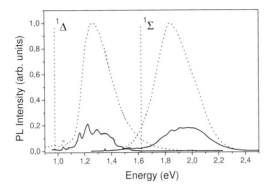

Figure 9. PL spectra of PSi in vacuum (dotted lines) and with adsorbed oxygen molecules (solid lines). $T=55$ K. Energies of $^3\Sigma$-$^1\Delta$ and $^3\Sigma$-$^1\Sigma$ transitions are indicated by dotted lines. Reproduced with permission from Ref. 42 , E. Gross et al, Phys. Rev. B **68**, 115405 (2003), Copyright @ American Physical Society.
http://link.aps.org/abstract/PRB/v68/e115405

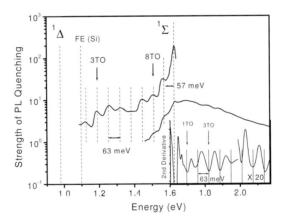

Figure 10. Spectral dependence of the PL quenching strength of PSi ($T=55$ K). Spectroscopic features, related to multiple TO-phonon emission, are representatively labelled at two spectral positions (3 TO and 8 TO). Vertical dotted lines are guide for the eye. Free exciton emission energy of bulk Si (FE(Si)) and energies of $^1O_2$ molecules ($^1\Delta$ and $^1\Sigma$) are also indicated by dashed lines. Inset: Second derivative of the quenching strength curve shown above. For convenient presentation, the data have been partially scaled by the indicated multiplication factor. Reproduced with permission from Ref. 42 , E. Gross et al, Phys. Rev. B **68**, 115405 (2003)), Copyright @ American Physical Society.
http://link.aps.org/abstract/PRB/v68/e115405

To resolve the weak spectral modulation of the curve covering the energy region above the $^1\Sigma$ state energy its second derivative is used (partial scaling is used for convenient presentation). The spectral dependence of the PL suppression strength allows a detailed description of the energy-transfer mechanism. Quenching is the strongest for nanocrystals having band-gap energies that coincide with the $^3\Sigma$-$^1\Sigma$ transition of O_2. Since Si nanocrystals also luminesce 57 meV below their band-gap energy due to the emission of a momentum-conserving TO phonon [6,7], they do not contribute to the PL while transferring the excitation. Therefore, an additional maximum in the quenching strength is observed 57 meV below the $^1\Sigma$ state energy. It is evident from Figure 10 that nanocrystals whose band-gaps do not match resonantly the excitation energies of O_2 singlet states participate in the energy transfer as well. The excess of the exciton energy with respect to the energies of the $^1\Delta$ and $^1\Sigma$ states is released by the emission of phonons. The probability of phonon emission scales with the phonon density of states which for nanocrystals is much higher than for molecules having local vibration modes.

In Figure 11 the mechanism of energy transfer from excitons to O_2 is sketched. Since real electronic states below the nanocrystal band gap are absent, energy dissipation should be governed by multiphonon emission rather than a phonon cascade. This process is most probable for phonons having the highest density of states which in bulk Si are transversal optical phonons being almost at the centre of the Brillouin zone with an energy of 63 meV [70] . This can be seen as multiple local maxima in the PL suppression curve (see Figure 10,11): energy transfer process is most efficient under these conditions. If the band-gap energy of Si nanocrystals does not coincide with the excitation energy of the O_2 singlet state plus an integer number of the energy of those phonons, the additional emission of acoustical phonons is required to conserve the energy. This process has a smaller probability and the efficiency of energy transfer is reduced what can be seen as local minima in the PL suppression spectra (Figure 10). Consequently, equidistant maxima and minima in the spectral dependence of the quenching strength appear which experimentally evidences the phonon-assisted energy transfer. In Si the exciton-phonon coupling is weak, but surprisingly, the simultaneous emission of up to eight phonons during the energy transfer to the $^1\Delta$ and $^1\Sigma$ states is detected with comparable probability. This process has to be similar for excitons coupling to $^1\Delta$ and $^1\Sigma$ states of O_2 and spectroscopic features relevant to both transitions can be clearly seen in Figure 10.

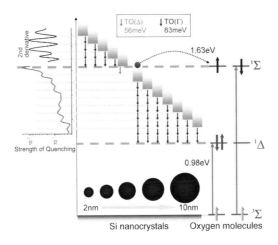

Figure 11. Sketch of the energy-level diagram of O_2 and different groups of Si nanocrystals participating in the electron-exchange process most efficiently. Principal steps occurring in the energy-transfer process are shown. Energy exchange occurs when a photoexcited electron (indicated by the red sphere), initially belonging to a nanocrystal, is exchanged with a non-excited electron initially belonging to O_2 (indicated by the grey doted arrow). This process results in the formation of singlet O_2 states and compensation of the holes confined in Si nanocrystals. This process can be viewed as triplet-triplet annihilation. Participation of a number of TO phonons required to conserve energy is indicated by blue vertical arrows. The momentum-conserving TO phonon is marked by a light-blue arrow. Blue spheres are inserted for clarity. They indicate how the energy of Si nanocrystals depends on nanocrystal size. Reproduced with permission from Ref. 49, D. Kovalev et al, Adv. Mater., **17**, 2531 (2005), Copyright @ Wiley-VCH Verlag GmbH & Co. KGaA.

At intermediate temperatures (T=110-250 K) a second quenching band in the spectral region of 1.75–1.95 eV can be seen, which becomes better pronounced with increasing oxygen concentration (see Figure 12). The energy of this new PL feature coincides with the double energy of the $^3\Sigma$-$^1\Delta$ transition of O_2. Therefore we attribute this PL suppression channel to the energy transfer from excitons confined in Si nanocrystals to the oxygen dimer $(O_2)_2$. The O_2 dimer is known as a complex of ground-state oxygen molecules induced by a weak van der Waals interaction [71,72]. The discrete electronic transition in $(O_2)_2$ corresponding to the 2 ($^3\Sigma$-$^1\Delta$) transition occurs at 1.95 eV and is thermally and collisionally broadened in the gas phase. The 2 ($^1O_2$)-related quenching band in the spectral dependence of the PL quenching strength has an energy below 1.95 eV. We believe that the difference in energy is governed by the physisorption energy of oxygen molecules. Excitation of the bound complex involves spin-conserving electron exchange among the two $^3\Sigma$ states having mutually opposite spins (see inset of Figure 12), whereas the exciton provides the energy to activate the process. Consequently, the PL of PSi is quenched in the considered spectral range and energy transfer to the dimer state is enhanced at higher pressures due to an increased probability of oxygen dimer formation. For temperatures higher than 250 K the energy transfer to $2(O_2)$ cannot be resolved spectroscopically due to weaker physisorption of O_2 molecules and reduced probability of dimer formation.

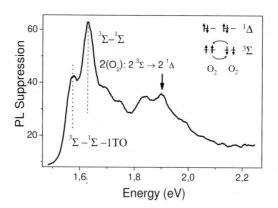

Figure 12. Spectral dependence of the quenching strength of PSi emission in the ambient of oxygen gas. T=110 K, O_2 pressure is 100 mbar. The energy of the $^3\Sigma$–$^1\Sigma$ transition of O_2 and of the $2(^3\Sigma$–$^1\Delta)$ simultaneous transitions (O_2 dimer) mediated by excitons confined in Si nanocrystals are indicated. Upper right side: sketch of electron exchange between two neighbouring O_2 molecules.

The efficiency of the singlet $^1O_2$ generation is usually defined as the ratio of the number of incident photons to the number of generated singlet oxygen molecules. Direct excitation of O_2 dimers implies that one photon, in general, can create two $^1O_2$ molecules. Therefore, theoretically, singlet O_2 generation efficiency in our system can exceed 100 %.

3.2 Microscopical mechanism of energy transfer from Si nanocrystals to oxygen molecules

While all basic observations evidence efficient energy transfer from excitons confined in Si nanocrystals to oxygen molecules, the mechanism of energy transfer has to be examined experimentally in detail. The dipole-dipole interaction [31] and the direct electron exchange [32] are possible candidates for energy transfer from excitons confined in Si nanocrystals to oxygen molecules. Förster showed that the dipole-dipole interaction can dominate the energy transfer mechanism only when the donor and the acceptor are characterized by dipole-allowed transitions [31]. However, in our system both radiative recombination of excitons and activation of O_2 are spin-forbidden processes. Dexter demonstrated that the spin states of donor and acceptor can be changed simultaneously if the energy transfer is governed by direct electron exchange [32]. Therefore, the simultaneous transfer of a photoexcited electron to oxygen molecule and the compensation of a hole in Si nanocrystal by an electron from oxygen molecule, i.e., triplet exciton annihilation and spin-flip excitation of an oxygen molecule (triplet-triplet annihilation), are allowed processes. O_2 physisorbed on the surface of nanocrystals should play a role similar to that of mid-gap deep centres or surface states in a semiconductor. Trapping of carriers on those states is usually accompanied by phonon emission cascade [73] that is consistent with our observations.

To clarify the energy exchange mechanism we performed similar experiments on naturally and thermally oxidized PSi layers having monolayers of oxygen atoms backbonded to the surface Si atoms or SO_2 shell. The surfaces of nanocrystals play a key role in virtually all of their properties, from light emission to solubility of nanocrystals in water. For

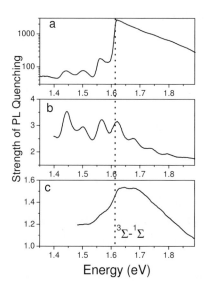

Figure 13. The spectral dependence of the PL quenching strength. a) As-prepared H-terminated PSi. b) Partially oxidized PSi having approximately one monolayer of back-bonded oxygen. c) Si/SO$_2$ core-shell nanostructures (heavily oxidized PSi). Doted line indicates energy of $^3\Sigma-^1\Sigma$ transition of O$_2$. T=5 K. The same concentration of O$_2$ ambient (10^{-4} mbar) has been used. Reproduced with permission from Ref. 19, D. Kovalev et al, Phys. Rev. Lett. **89**, 137401 (2002), Copyright @ American Physical Society. http://link.aps.org/abstract/PRL/v89/e137401

H-terminated Si nanocrystals spacing between nanocrystals core and adsorbed O$_2$ molecule is ~1.5 Å. For Si nanocrystals having a monolayer of back-bonded oxygen the increased spacing between confined excitons and adsorbed oxygen molecules is on the order of 3 Å (double the length of the Si-O bond) [74]. For Si/SiO$_2$ core-shell structures this spacing is in the range of 10 Å. This critically affects the efficiency of the electron exchange interaction. Contrary to a strong coupling for hydrogen-terminated nanocrystals (Figure 13a), while all spectral features relevant to the $^1\Delta$ and $^1\Sigma$ states of O$_2$ are still present, the PL quenching efficiency (and electron exchange rate) is reduced by orders of magnitude if a thin oxide barrier is present (Figure 13b). For Si/SiO$_2$ structures the energy transfer process is almost absent (see Figure 13c). Because the transition from H-terminated to O-terminated surfaces can be done smoothly via successive nanocrystals surface oxidation, the photosensitizing efficiency of Si nanocrystal assemblies, contrary to many other systems, can be accurately controlled.

We would like to note that if dipole-dipole interaction would be responsible for the process, the variation of spacing between the nanocrystal core and O_2 on a subnanometer scale should not affect significantly the efficiency of energy transfer. Indeed, the radius of exciton confined in Si nanocrystals is significantly larger and, therefore, this would be relatively long-range interaction. Dexter interaction is essentially a short-range interaction: a monolayer of incorporated oxygen implies an additional potential barrier for the mutual tunnelling of electrons and the efficiency of this process depends exponentially on the spacing between interacting species. This evidences that spin-flip activation of oxygen molecule is governed by the direct exchange of electrons between photoexcited Si nanocrystals and O_2.

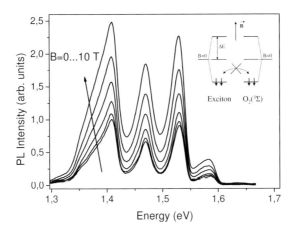

Figure 14. PL spectrum of PSi in the presence of physisorbed O_2 at various strength of magnetic field. T=10 K. The magnetic field increment is equal to 2 T. Inset: Zeeman splitting of a triplet exciton and $^3\Sigma$ ground state of oxygen molecules. The spin orientations for electrons in the lowest-lying levels are indicated by arrows. Energy transfer from exciton to ground-state O_2 via electron exchange among these states is prohibited by conservation of the total magnetic quantum number (in magnetic field at low temperatures mainly spin-down states of excitons and O_2 are occupied). Reproduced with permission from Ref. 42 , E. Gross et al, Phys. Rev. B **68**, 115405 (2003), Copyright @ American Physical Society. http://link.aps.org/abstract/PRB/v68/e115405

The direct electron exchange process is a spin-dependent phenomenon. Therefore, the control of the energy exchange efficiency can be properly achieved via the manipulation of electron spin. While the involved transitions are spin forbidden in the first approximation for isolated Si nanocrystals (recombination of triplet exciton) and oxygen molecules (their spin-flip activation) they become allowed through exchange interaction. For the energy transfer to occur the exchanged electrons must have opposite mutual spin orientation to conserve the total magnetic quantum number of the coupled system. To demonstrate the influence of spin statistics on the energy-transfer rate we measured the magnetic-field dependence of the PL quenching efficiency (see Figure 14). If no magnetic field is present the energy levels belonging to different spin orientations of triplet excitons and the triplet ground state of O_2 are threefold degenerated and populated with equal probability. Thus, the spin requirements are fulfilled for all excitons and all oxygen molecules, and energy transfer occurs most efficient. A magnetic field introduces a common quantization axis for the spins and the degeneracy is lifted (see inset of Figure 14). In general, the number of possible states participating in the electron exchange is reduced and the decreased energy-transfer rate results in a weaker PL quenching. Raising the magnetic field increases the Zeeman splitting and the occupation number of the thermally populated higher-lying states decreases. At low temperatures a magnetic field results in preferential occupation of ''spin-down'' states for both O_2 and excitons, while to proceed with energy exchange ''spin-up states'' are required. For magnetic fields of 10 T and a temperature of 10 K the relevant energies $k_B T$=0.8 meV and $E_{Zeeman} \sim$ 1.1 meV [75] are comparable, and a significant reduction of the PL quenching is observed. In the limit of zero temperature energy exchange process becomes completely prohibited in the presence of external magnetic field.

These experiments evidence the importance of spin states of the interacting species for energy transfer process and, remarkably, a very small magnetic energy ($\sim$1 meV) can efficiently control energy exchange processes at the scale of eV by aligning the spins of the interacting species.

3.3 Dynamics of energy transfer

The fine structure of PL spectra in the presence of O_2 on the surface of Si nanocrystals implies that at certain energies energy transfer is most efficient (energy of singlet-triplet splitting of O_2 plus integer number of TO phonon energies, 63 meV). As it was mentioned before, energy transfer competes with the radiative recombination of excitons. The energy-transfer time from one Si nanocrystal to a single oxygen molecule is not accessible experimentally, since a large number of nanocrystals contribute to the PL at certain emission energy. Though the concentration of physisorbed O_2 can be varied, it is subjected to statistical fluctuations and the absolute number of artificially introduced nanocrystal mid-gap nonradiative states cannot be determined exactly. However measurement of the PL lifetime in the presence of O_2 molecules allows estimation of the energy transfer time. Figure 15 demonstrates the spectral dispersion of the PL decay time which is almost identical to the spectral shape of the PL band. The PL decay time at the energies of PL minima is about 300 μs and almost twice shorter than that measured at PL maxima energies. It is ~15 times shorter than the triplet exciton recombination time [6,7] measured in vacuum (~5 ms) and, therefore, it gives the time of the energy transfer from excitons confined in Si nanocrystals to the $^1\Delta$ state of O_2 (at this particular PL quenching level). These experiments show that the energy transfer rate is maximal when simultaneous emission of only TO phonons is required to fulfil the energy conservation law. At energies above 1.63 eV the decay of extremely weak remnant PL is very fast and the time of the energy transfer to the $^1\Sigma$ state is faster than our experimental time resolution (1 μs).

For PSi containing physisorbed O_2, under continuous optical excitation the strong coupling to the $^1\Sigma$ state results in an almost complete suppression of PL band above the energy of the $^1\Sigma$ state. Therefore ordinary continuous wave (CW) spectroscopic investigations do not give details of the energy transfer mechanism. Figure 16a demonstrates different strength of coupling of excitons to different $^1O_2$ states: coupling to the $^1\Sigma$ state is almost 3 orders of magnitude stronger. During the energy transfer process forming the $^1\Sigma$ state, orbital angular

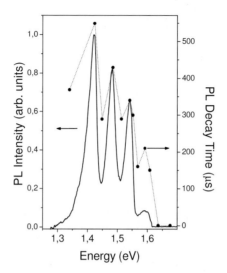

Figure 15. The PL spectrum from as-prepared PSi (solid curve) and the spectral dispersion of the PL decay time (circles) at $T = 5$ K; O_2 pressure is 10^{-4} mbar. Measurements have been performed at a weak level of PL quenching to keep the signal-to-noise ratio on a reliable level for time-resolved measurements. Reproduced with permission from Ref. 19, D. Kovalev et al, Phys. Rev. Lett. **89**, 137401 (2002), Copyright @ American Physical Society. http://link.aps.org/abstract/PRL/v89/e137401

momentum conservation is fulfilled, while for the $^1\Delta$ state, a change of angular momentum of O_2 ($\Delta L=2$) is required. This restriction results in relatively weak coupling between excitons and O_2 forming the $^1\Delta$ state, while their interaction followed by generation of the $^1\Sigma$ state is very efficient.

To estimate the energy transfer time to this state we performed detailed time-resolved PL measurements [76]. Figure 16b shows time-resolved PL spectra of a PSi layer in vacuum and that containing physisorbed O_2 on the surface of Si nanocrystals. The gate width is kept constant 100 ns while the delay time with respect to the excitation pulse is varied from 80 ns to 3.08 µs. In vacuum, the PL spectral shape and its intensity do not change significantly within the time scale investigated (lifetime of triplet exciton is in millisecond range). On the other hand, under presence of O_2, already at 80 ns delay time with respect to PL

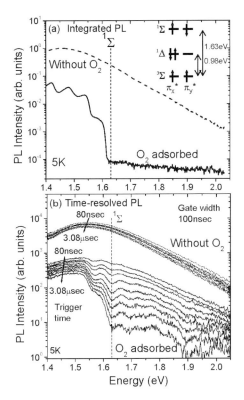

Figure 16. a) CW PL spectra of a PSi layer in vacuum (dashed line) and a layer having physisorbed oxygen molecules on the surface (solid line) at T=5 K. Electron-spin configurations and spectroscopic labelling of O_2 states are shown in the inset. b) Time-resolved PL spectra at 5 K in vacuum (dashed lines) and with physisorbed oxygen molecules (solid lines). The measurement gate width is 100 ns. PL spectra measured at different delay time with respect to the excitation pulse: 0.08, 0.18, 0.28, 0.48, 0.78, 1.08, 1.48, 2.08, 3.08 μs are shown. Reproduced with permission from Ref. 76, M. Fujii et al, Phys. Rev. B **72**, 165321 (2005), Copyright @ American Physical Society. http://link.aps.org/abstract/PRB/v72/e165321

excitation pulse strong almost energy-independent PL quenching can be resolved and a very strong variation of the PL spectral shape appears as the delay time of the PL measurements is increased (note the semilogarithmic scale of Figure 16). The spectra shown in Figure 16b, apparently, can be divided into two spectral ranges. In the low energy range (below 1.55 eV) the PL lifetime is relatively long and no distinct

PL features are observed. In the high energy range the PL lifetime is much shorter and a number of PL signatures related to the energy transfer process can be clearly distinguished. This observation allows estimating the time of energy transfer to O_2 molecules (for $^3\Sigma$-$^1\Sigma$ transition). The energy transfer time is shortest at 1.63 eV, and is about 320 ns. The energy transfer time becomes longer with an increasing number of TO energy-conserving phonons emitted during the energy exchange. It is already about 450 ns when one TO-phonon is involved in the process.

It should be noted that already at 80 ns delay time the PL from PSi containing physisorbed O_2 is ~ 10 times weaker than that from PSi in vacuum. This implies that another very fast and efficient non-radiative recombination process is additionally introduced by adsorbed O_2. A large fraction of Si nanocrystals does not contribute to the emission. The fact that the quenching at 80 ns delay time occurs almost uniformly in the entire spectral range, even at very low energy, where excitons can only inefficiently couple to the $^1\Delta$ state, indicates that the PL quenching is not related to the energy transfer.

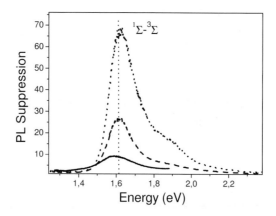

Figure 17. PL suppression spectra of PSi in oxygen ambient at 1 bar. Doted line: T=110 K. Dashed line: T=220 K. Solid line: T=295 K. The energy of $^3\Sigma$-$^1\Sigma$ transition is indicated by vertical line.

Another process which is known to be responsible for the fast PSi PL quenching is charging of Si nanocrystals. It has been demonstrated that charged or doped Si nanocrystals do not contribute to the PL due to a very efficient non-radiative Auger process [77,78]. Adsorption of NO_2 molecules on the surface of Si nanocrystals results also in very efficient structureless PL quenching. NO_2 molecules are very efficient electron acceptors and the hole remaining in Si nanocrystals quenches completely the PL [79]. One of the possible reasons for the structureless PL quenching is the formation of superoxide, $^-O_2$ state mediated by excitons. O_2 is a very efficient electron acceptor and electrons can be donated by photoexcited excitons. In this scenario nanocrystals become positively charged and cannot contribute to the PL due to a highly efficient Auger process.

3.4 Energy transfer at elevated temperatures

Spectroscopic experiments at cryogenic temperatures clarify the details of the energy transfer mechanism. However, the generation of singlet O_2 at room temperatures in gas and liquid environments is much more important due to its involvement in photochemical reactions and in biological or medical systems [26-30]. Figure 17 shows the strength of the PL quenching obtained by dividing the intensities of PL spectra taken in vacuum by those in oxygen gas ambient at 1 bar at 120 K, 220 K and 295 K. Contrary to cryogenic temperatures, the conditions for the optimal exciton-O_2 interaction are not fulfilled. A small spatial separation is realized only during the short time of collisions between oxygen molecules and the nanocrystal surface. Additionally, the exciton lifetime and the occupation number of the spin-triplet state of the exciton decreases with raising temperature [6,7]. Therefore, a weaker PL suppression that scales with the collision rate, i.e. the gas pressure, occurs and the energy transfer to the $^1\Sigma$ state is seen as a relatively broad spectral resonance (still around 1.63 eV) which becomes broader towards higher temperatures due to thermal broadening effects. This PL suppression in the presence of O_2 implies that each exciton which has been lost from the emission necessarily creates a singlet oxygen molecule. Since the PL suppression level at room temperature is about 9 the quantum yield of energy transfer process is ~ 90% (calculated with respect to *luminescing* Si nanocrystals). This allows us to estimate the generation rate of singlet O_2 within the pores of PSi. For ambient O_2 pressure and 1 W/cm^2 illumination intensity this value is ~ 5×10^{20} singlet oxygen molecules/cm^3 s.

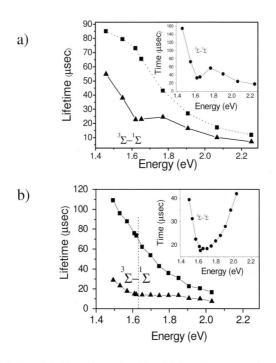

Figure 18. a). Spectral dispersion of exciton lifetime in vacuum (squares) and in oxygen ambient (triangles). Inset: spectral dispersion of the energy transfer time. b). Spectral dispersion of exciton lifetime in degassed water (squares) and in oxygen-saturated water (triangles). Inset: spectral dispersion of the energy transfer time.

For biological and medical applications of singlet oxygen, its generation in oxygen-containing aqueous solutions is crucial. Unfortunately, fast non-radiative vibrational relaxation processes of singlet oxygen in H_2O makes the detection of singlet oxygen dissolved in water extremely difficult. However, even if the $^1O_2$ emission line is not detected, the energy transfer can be indirectly probed by monitoring the shape of the PL band and the lifetimes of excitons in the presence of dissolved O_2. To obtain the spectral dependence of the energy transfer efficiency for oxygen-saturated water we use the procedure identical to that employed for gaseous O_2 ambient. Although the PL suppression level is weaker, its spectral shape is almost identical to that measured in gaseous O_2. This evidences the formation of singlet oxygen in water and demonstrates that the energy exchange mechanism should be identical. From the PL suppression level it follows that the efficiency of energy transfer in O_2-saturated water is equal to 75%.

Energy transfer from Si nanocrystals to O_2 would imply the activation of a non-radiative decay channel which can be experimentally seen as an energy-selective PL quenching and shortening of the exciton lifetime. Under the simple assumption that an exciton can recombine either radiatively or non-radiatively via an energy transfer process, the measured PL lifetime ($\tau_{meas.}$) for PSi samples containing O_2 molecules is a combination of the radiative exciton lifetime ($\tau_{rad.}$) and the time of the energy transfer ($\tau_{tr.}$): $1/\tau_{meas.} = 1/\tau_{rad.} + 1/\tau_{tr.}$. Slightly simplified assumption that $\tau_{meas.} = \tau_{rad.}$ for samples measured in vacuum allows us to measure the energy transfer time for gaseous O_2 ambient and O_2-saturated water. Figure 18a demonstrates results of these measurements for gaseous O_2 ambient and Figure 18b for O_2-saturated water. The exciton lifetimes for Si nanocrystals immersed in degassed water are identical to those measured in vacuum. Oxygen ambient or oxygen dissolved in water causes its significant shortening over all the spectral range investigated. The spectral dependence of the energy transfer time in oxygen-saturated water is very similar to that observed for gaseous oxygen ambient (see inset of Figs. 18). The energy transfer time of ~ 20 µs is still shorter than the exciton lifetime what is in good agreement with the observed small difference in the PL suppression levels. This observation clearly demonstrates the importance of long exciton lifetimes for the efficiency of the energy transfer process. If exciton lifetimes were in nanosecond range, as it is in direct band-gap semiconductors, the energy transfer process would be inefficient.

For a practical application of Si nanocrystal assemblies as singlet oxygen generators in photochemistry, biology and medicine, their photosensitizing efficiency at room temperature is crucial. In particular, in most of applications, the generation of singlet oxygen in organic and aqueous solutions is required. Up to now we presented only indirect evidences for energy transfer in solutions and there still remains a room to dispute whether singlet oxygen is really generated by energy transfer from Si nanocrystals in organic or aqueous solvents. Therefore, the formation at room temperature must be proved without doubts by detecting the near infrared emission from singlet oxygen. Although the intrinsic radiative lifetime of $^1O_2$ in the lowest excited state $^1\Delta$ is extremely long, intermolecular interactions lead to an enhancement of the transition rate. The radiative transition rate is three to four orders of magnitude larger in solution than in diluted gas phase [80]. However, in most solvents, the deactivation of $^1O_2$ is radiationless by collisional electronic to vibrational energy transfer from $^1O_2$ to a solvent and oxygen molecules. The most probable energy-accepting oscillator of a solvent molecule is its terminal atom pairs with the highest vibrational energy (e.g., O-H, C-H) [81,82]. Molecules composed of low energy oscillators such as C-F and C-Cl act as poor quenchers whereas those with high energy oscillators such as O-H and C-H are strong quenchers. In fact, the lifetime of $^1O_2$, varies over a wide range, from 4 µs to 100 ms, depending on the kind of solution [82]. The lifetime of $^1O_2$ in H_2O is very short ~3.1 µs because it contains high frequency O-H bond [24]. The radiative lifetime of $^1\Delta$ state of O_2 is extremely long and its PL quantum yield scales with its non-radiative lifetime in solutions. Therefore, to obtain reliable luminescence data, solvents consisting of poor quenchers should be chosen. The second important requirement on the solvent is that it should not quench the PL from PSi.

As a solvent which satisfies these requirements, we employed hexafluorobenzene C_6F_6 [83] and D_2O [84]. The singlet oxygen lifetime in C_6F_6 is about 25 msec [81], which is about three orders of magnitude longer than that in benzene C_6H_6. Because of the lower frequency of D–O bond oscillations the singlet oxygen lifetime in D_2O, 68 µs, is much longer than that for water [24], and the quantum yield of the singlet oxygen PL is higher. Therefore, in the PL measurements aiming the detection of singlet oxygen generation, we employ D_2O instead of H_2O. Since the solubility of O_2 in D_2O is comparable to that in H_2O [85] the singlet oxygen generation efficiency in D_2O is considered to be comparable to that in H_2O.

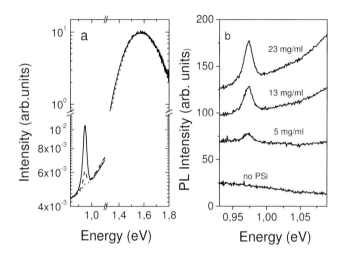

Figure 19. a) PL spectrum of porous Si dispersed in C_6F_6 solution at room temperature. The peak at around 0.975 eV corresponds to the emission from singlet oxygen ($^1\Delta-^3\Sigma$). Logarithmic scale is used for the vertical axis. Doted line: after substitution of dissolved oxygen by nitrogen. Dashed line: after partial substitution. Solid line: oxygen-saturated solution. b). PL spectra from PSi powder dissolved in oxygen-saturated D_2O. Concentrations of powder are indicated. Reproduced with permission from Ref. 83, M. Fujii et al, Phys. Rev. B **70**, 085311 (2004), Copyright @ American Physical Society (Figure19a) and with permission from Ref. 84, M. Fujii et al, J. Appl. Phys. **100**, 124302 (2006), Copyright @ American Institute of Physics (Figure 19b). http://link.aps.org/abstract/PRB/v70/e085311

To demonstrate singlet oxygen generation at room temperature in oxygen-saturated solutions we study PL from PSi powder dispersed in C_6F_6 and in D_2O. Figure 19a demonstrates emission from PSi dissolved in C_6F_6. If oxygen, naturally dissolved in solution kept in ambient conditions, is substituted by nitrogen via its bubbling no emission around 0.98 eV can be detected. However if nitrogen is afterwards partially substituted by oxygen a distinct PL line at 0.98 eV appears which has maximum amplitude when nitrogen is completely substituted by oxygen. This PL line is a fingerprint of singlet oxygen what evidences its generation in organic solution. Figure 19b shows PL spectra of PSi dispersed in oxygen-saturated D_2O. Again the same characteristic PL line can be seen and its amplitude scales with the concentration of PSi in the solution. These results demonstrate that PSi and other forms of H-terminated Si nanocrystals can be used as a singlet oxygen photosensitizer in organic and aqueous solutions and, therefore different forms of nanosilicon have potential for photochemistry, biology and medicine.

4. Photochemical activity of singlet oxygen generated by Si nanocrystals

Singlet oxygen has a very high chemical reactivity and is involved in the chemical transformation of photosensitizers as well. The common notation for this process is "photobleaching". Each photosensitizer has certain stability against photogenerated $^1O_2$ but the duration of its photosensitizing activity is limited. We have demonstrated that surface oxidation of Si nanocrystals results in the reduction of the singlet oxygen generation rate. Since singlet oxygen is generated at the surface of Si nanocrystals it can also oxidize them. This process has two consequences. First, due to surface oxidation, additional non-radiative centres (dangling Si bonds) are created and the PL quantum yield is reduced. Because only luminescing Si nanocrystals can transfer energy to O_2, the efficiency of $^1O_2$ generation is reduced. Second, as it was demonstrated above, oxidation of Si nanocrystal surfaces further reduces the efficiency of the energy transfer process. Since generation rate of $^1O_2$ scales with the concentration of excitons, photooxidation of Si nanocrystals should be more efficient for samples having higher PL quantum yield. To confirm this conjecture we performed photooxidation experiments with samples having different PL quantum yield and, therefore, different singlet oxygen generation rate. We demonstrate this effect in Figure 20. As-prepared PSi samples have H-terminated surfaces and their photosensitizing efficiency is very high. While all H-terminated samples illuminated in vacuum possess a completely stable PL, in oxygen ambient the PL intensity exhibits a fast photodegradation (in timescale from minutes to hours, depending on the illumination intensity) [86].

Figure 20 shows the IR absorption spectra of PSi layers in the spectral range of Si-H$_x$, x=1,2,3 and the O-Si-O vibration bands [87].While as-prepared layers exhibit only Si-H$_x$ surface bonds, the illumination in the oxygen ambient results in the oxidation of the nanocrystal surfaces. The oxidation is always more efficient for samples with a high quantum yield. A complete rearrangement of the Si-H$_x$ stretching bond configuration shows that oxygen atoms backbonded to surface Si atoms are introduced during the illumination (Figure 20, dashed line).

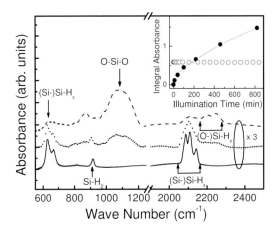

Figure 20. IR absorbance of porous Si. Solid curve: H-terminated layer. Dotted curve: weakly luminescing sample after illumination in oxygen ambient at 1 bar. T=300 K. I_{ex}=100 mW/cm^2. E_{ex}=2.54 eV; irradiation time is 1 h. Dashed curve: strongly luminescing sample after irradiation at the same conditions. Arrows label surface vibration bonds. Inset: Integral absorbance due to O–Si–O bonds (solid circles) and due to Si-Si-H$_x$ and O-Si-H$_x$ stretching bonds (between 2050 and 2300 cm^{-1}, open circles) vibrations. Reproduced from Ref. 86, D. Kovalev et al, Appl. Phys. Lett. **85**, 3590 (2004), Copyright @ American Institute of Physics.

The additional manifestation of the oxidation can be seen as an efficient suppression of the S-H$_x$ wagging mode at 650 cm^{-1} as well as the Si-H$_2$ scissor mode at 912 cm^{-1} [87]. This effect is almost absent for the samples with a small quantum yield and the only noticeable signature of oxidation appears due to strongly absorbing O-Si-O stretching bonds (Figure 20, dotted line).

The time evolution of the Si-O bond concentration during the illumination in oxygen ambient at 1 bar is shown in the inset of Figure 20 (closed circles). On the initial stage of the oxidation the decrease in the number of sites available for oxidation can be neglected. Instead, the sublinear rate of oxidation is governed by a reduced singlet oxygen generation rate due to photodegradation: again there is a clear anticorrelation between the oxidation rate and the PL quantum yield. The integral absorbance governed by the surface O-Si-H_x and Si-Si-H_x bond vibrations [87] does not change during the photooxidation and evidences a constant number of surface H atoms (inset of Figure 20, open circles). Thus the efficient photodegradation of H-terminated Si nanocrystals is mediated by photosensitized singlet oxygen molecules. This effect is very similar to that known for light-emitting polymers or dye molecules. This process is an inherent limitation for photosensitizing activity of Si nanocrystals which reduces significantly in a time scale from minutes to days depending on the illumination intensity [84].

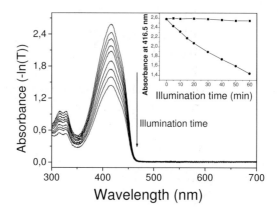

Figure 21. Absorption spectra of DPBF-dissolved in benzene as a function of the irradiation time (each curve represents additional 10 minutes of illumination.). The wavelength and power of the irradiation light are 514.5 nm and 80 mW/cm², respectively. Inset: absorbance of DPBF at 416 nm versus illumination time. Squares: without PSi present in the solution. Circles: with PSi present in the solution. Reproduced from Ref. 88, M. Fujii et al, J. Appl. Phys. **95**, 3689 (2004), Copyright @ American Institute of Physics.

Due to its extraordinary high chemical reactivity $^1O_2$ reacts with a large variety of substances. Up to date hundreds of photochemical reactions involving oxygen molecules have been studied. For many applications in photochemical and biological fields, formation of singlet oxygen in solution is required, because singlet oxygen mediated reactions proceed usually in solutions. The standard method to realize and detect photochemical reaction mediated by $^1O_2$ is to use biochemical traps (singlet oxygen acceptors) and to analyze a specific reaction product or monitor the decrease in the amount of acceptor materials. Typical biochemical traps are cholesterol, 1,3-diphenylisobenzofuran (DPBF), *p*-nitrosodimethylalanine, sodium azide, etc. [88].

To demonstrate the photochemical activity of Si nanocrystals we employed DPBF as a singlet oxygen acceptor [88]. DPBF readily undergoes a 1,4-cycloaddition reaction with singlet oxygen forming endoperoxides, which in turn decompose to yield irreversible product (1,2-dibenzoylbenzene). This process can be monitored by the decrease in the intensity of the absorption band of DPBF centered at 416 nm [89, 90].

Figure 21 shows absorption spectra of DPBF-dissolved in benzene containing H-terminated PSi powder. A strong absorption band centred at 416 nm is due to DPBF. The absorption of porous Si does not appear in the spectra in Figure21 because of its small chosen concentration. With an increase of the illumination time the absorbance decreases, implying that DPBF is decomposed. The change in the absorbance was observed only when PSi powder is added to the solution. Inset of Figure21 demonstrates the degree of the absorbance change versus irradiation. Without irradiation, PSi powder does not exert any effects on DPBF what demonstrates that decomposition of DPBF is not due to a chemical reaction with PSi. This observation shows the potential of application of Si nanocrystal assemblies in photochemistry. Already at very low concentrations (for experiments described above it was ~ 1 mg/ml) it can mediate chemical reactions through generation of singlet oxygen.

5. Biomedical applications

In the photodynamic cancer therapy photosensitizing agents and oxygen are combined to produce a selective therapeutic effect under light illumination [29]. The selectivity is based on the concentration of photosensitizer in or between the cancer cells and choice of the illuminated area. Singlet oxygen is considered as the main reason for photocytotoxicity in PDT because it causes oxidation and degradation of cancer cells [91]. Typical photosensitizers used in PDT are members of a family of dye molecules known as porphyrins [29]. However recently the question "do quantum dots possess a potential to be photosensitizers?" (for PDT) has been raised [34]. We demonstrated before that Si nanocrystals assemblies have very high efficiency of singlet oxygen generation. Therefore it is tempting to examine whether they can act in a similar manner to conventional photosensitizers.

The authors of Ref. 92 used photoexcited Si nanocrystals to suppress the division of cancer cells. In biophysical experiments they used 3T3 NIH-line cancer cells (modified mouse fibroblasts) grown using the standard in vitro subcultivation procedure [93] in Petri dishes and special plates consisting of 96 wells. Before the addition of the Si nanocrystal powder, the growth medium was changed to the fresh one. The cells under investigation were divided into three groups. The aqueous suspension containing a certain amount of Si nanocrystals was added to the first and second groups, whereas the third group was a reference group (Si nanocrystals were not added to this group). The cells of the first and second groups were exposed for one hour to the light of a mercury lamp whose radiation was passed through a distilled water filter (in order to suppress the thermal component of the spectrum) and a glass filter with the transmission band of 350–600 nm. The light intensity on the sample was equal to ~1 mW/cm^2. The cells of the second group were not irradiated. After the experiment, the growth medium in all groups of cells was changed to the fresh one, cells were placed for 20 hours for cultivation, and then their number and composition were determined. At all stages of irradiation and cultivation of cells, a temperature of $T=37\ ^0$C and a medium acidity of pH = 7.2 were maintained. Several standard methods were used to count the number of cells and to analyze their composition.

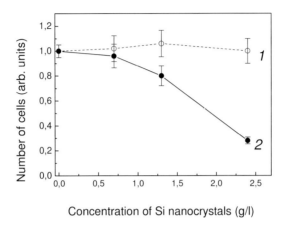

Concentration of Si nanocrystals (g/l)

Figure 22. Number of cancer cells normalized to the values in the reference group, where Si nanocrystals were not added, versus the concentration of silicon nanocrystals. 1. In darkness. 2. After illumination. Number of living cells was determined from the change in the optical density of the cells. Reproduced from Ref. 92, V. Yu. Timoshenko, JETP Letts. **83**, 423 (2006), with kind permission of Springer Science and Business Media.

In the first method, cells were stained in a 0.2 % solution of a crystalline violet dye in methyl alcohol and the optical density of the content of the wells was determined using the absorption of light at 540 nm. In the second method, cells in the Petri dishes were washed off by trypsin–EDTA and their number was counted in a hemocytometer (under a microscope). In addition, the DNA of cells was analyzed in a flow cytofluometer (for details of the methods, see Ref. 94, 95).

Figure 22 shows the numbers of living mouse fibroblast cells in the first and second groups after the termination of the cultivation as a function of the concentration of Si nanocrystals. These numbers are normalized to the value in the reference group. It is seen that the number of living cells after the irradiation in the growth medium with a Si nanocrystal concentration of ~0.5 g/l or higher decreases strongly as compared to the reference group. The death of 80 % of cells was detected for a concentration of 2.5 g/l. At the same time, the effect of Si nanocrystals was almost absent over the entire concentration range in darkness. Therefore, the suppression of the cancer cell proliferation can be attributed to the action of active oxygen produced by the photoexcitation of Si nanocrystals. Analysis of the DNA of cells shows that they die through the apoptosis (programmed cell death) mechanism [94] after the irradiation in the presence of Si nanocrystals with a concentration of more than 0.1–0.5 g/l and the concentration dependence of the number of living cells is close to that plotted in Figure 22. The death of cancer cells likely occurs due to the action of photosensitized active oxygen, in particular, due to the oxidation of cell substance by singlet oxygen. In addition, the effect of other active forms of oxygen is also possible, e.g., so-called superoxide radicals. To reveal the particular mechanisms of the effect of photoexcited Si nanocrystals on biological objects, additional extensive investigations are required. However, these first results show that employment of Si nanocrystals for the suppression of the division of cancer cells is viable.

6. Photosensitization of other materials

Molecules having a ground triplet electronic state are very rare in nature: most of them are in the singlet state. As mentioned, spin-flip triplet to singlet and singlet to triplet electronic transitions are mirror-like processes. Therefore, an identical, efficient energy transfer processes mediated by Si nanocrystals should be possible for a large variety of substances having a ground singlet and a first excited triplet state. To verify this conjecture we have chosen a family of organic molecules - anthracene and one of its derivatives, dimethylanthracene, β-carotene and naphthalene to study details of the energy transfer process according to the predicted scenario. All these molecules have a ground singlet and a first excited triplet state. The spin splitting energies of the first two molecules fall inside the PL band of Si nanocrystals E_{S-T}=1.84 eV and E_{S-T}=1.73 eV, for isolated molecules [96], while for the last two their energies are smaller (E_{S-T}=0.9 eV) or higher (E_{S-T}=2.63 eV), respectively [97]. The size of these molecules is still smaller than the typical pore diameter of PSi, therefore, they can be incorporated in the pores using their organic solutions and a small spatial donor-acceptor separation can be achieved.

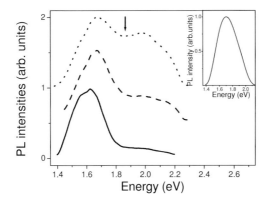

Figure 23. PL spectra of PSi samples having pores impregnated with anthracene molecules measured at different temperatures. Doted line: T=300 K, dashed line: T=200 K, solid line T=100 K. Spectra have been shifted vertically for clarity. Inset: PL spectrum of a reference PSi sample containing no anthracene. Energy of singlet-triplet splitting of an isolated anthracene molecule is indicated by the arrow. E_{ex}=2.54 eV. Reproduced with permission from Ref. 97, B. Goller et al, Phys. Rev. B **75**, 073403 (2007), Copyright @ American Physical Society. http://link.aps.org/abstract/PRB/v75/e073403

The inset of Figure 23 shows the PL spectrum of a H-terminated reference PSi sample. The large variety of shapes and sizes of Si nanocrystals causes a broad featureless PL band. The incorporation of anthracene molecules in the pores of PSi samples results at room temperature in a selective PL quenching at around 1.85 eV. With a gradual decrease of the temperature the PL quenching becomes more prominent and at T=100 K almost all emission above ~1.8 eV disappears (see Figure 23). This energy selective PL quenching indicates a spectrally-dependent energy transfer from excitons confined in Si nanocrystals (energy donors) to anthracene molecules (energy acceptors). Since the electronic states of an acceptor molecule are discrete, the most efficient energy transfer (and the fastest rate of the energy transfer) is expected to be realized when the exciton energy coincides with the energy of a certain electronic level of the acceptor molecule (resonant energy transfer). At lower temperatures the exciton lifetime becomes longer [6,7] due to the suppression of non-radiative channels and nonresonant energy transfer from excitons having larger energies than the singlet-triplet splitting energy of anthracene becomes efficient as well. In fact, measurement of the threshold energy is itself a useful way to determine the singlet-triplet structure of the adsorbed molecules.

To estimate the time of energy transfer we performed time-resolved PL measurements at different delay times after the pulsed excitation in different time domains. The resonant PL quenching at around 1.9 eV can be already seen at 100 ns delay time after a pulsed excitation. For larger delay times PL quenching becomes more pronounced at larger detection energies and finally the overall spectral shape of the PL band becomes almost independent from the delay time. Energy transfer from Si nanocrystals would imply the activation of a non-radiative decay channel which can be experimentally seen as an energy-selective PL quenching and shortening of the exciton lifetime. Therefore, using a simple relation derived previously for the interaction of Si nanocrystals with O_2, the energy transfer time can be deduced directly. Its shortest value is again measured at 1.9 eV, i.e. for a resonant energy transfer. Below this energy it rapidly increases and at ~ 1.7 eV its value goes to infinity (no energy transfer is present). The shortest measured energy transfer time is relatively long ~20 µs but it is still significantly shorter than the lifetime of excitons confined in Si nanocrystals and the efficiency of energy transfer is again very high. We would like to mention that a spectral cut-off associated with a resonant energy transfer is much broader than one can expect from thermal broadening effects. We believe that it is governed by the difference of the singlet-triplet splitting energies of isolated anthracene molecules and molecular crystals. Since in PSi the statistic short range order of anthracene molecules is a matter of the pore filling, the singlet-triplet splitting energy has no single discrete value.

Basically, again dipole-dipole [31] or direct electron exchange [32] interactions can account for the energy transfer from excitons to organic molecules. Since long-range multipole interaction is based on optically allowed transitions of a donor and an acceptor, it can not be applied for dipole-forbidden (spin-flip) transitions in Si nanocrystals and organic molecules. For the electron exchange mechanism these spin restrictions are lifted and the triplet exciton annihilation accompanied by the spin-flip excitation of an organic molecule is an allowed process. The energy transfer rate is defined by the spatial overlap of the electronic

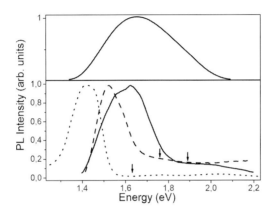

Figure 24. Upper part: the PL spectrum of a H-terminated reference PSi sample. Lower part: PL spectra of PSi containing anthracene molecules (solid line), dimethylanthracene molecules (dashed line) and O_2 molecules (dotted lines) confined in the pores. $T=$ 80 K. The energy of the singlet-triplet splitting of the corresponding molecules is indicated by arrows. Reproduced with permission from Ref. 97, B. Goller et al, Phys. Rev. B **75**, 073403 (2007), Copyright @ American Physical Society. http://link.aps.org/abstract/PRB/v75/e073403

wavefunctions of the interacting species and depends exponentially on the donor-acceptor distance [32]. As it has been mentioned before, the advantage of our system is that a controlled variation of the donor-acceptor separation is possible via modification of the nanocrystal surfaces. As-prepared PSi exhibits a hydrogen-terminated surface while after the annealing a monolayer of oxygen atoms back-bonded to the surface of Si nanocrystals is formed. The increase of the spacing between the core of a nanocrystal and a foreign molecule critically affects the efficiency of the electron exchange interaction. Identical experiments performed with oxidized PSi samples show only a very weak modification of the emission properties of Si nanocrystals by all incorporated organic molecules.

Finally, we would like to demonstrate that Si nanocrystal assemblies seem to be almost ideal spin-flip activators for a variety of substances whose electronic states split by an exchange interaction to singlet and triplet states. Figure 24 shows PL spectrum from PSi samples containing oxygen, anthracene and dimethylanthracene molecules in the pores (their singlet-triplet transition energies are indicated by arrows). Despite a different order of the singlet and triplet states, PL quenching and therefore energy transfer from nanocrystals having energies above singlet-triplet splitting energies of molecules is always efficient. Identical experiments performed with naphthalene show no PL quenching effect while for β-carotene a strong and energy-dependent PL quenching is observed. The PL quenching level continuously rises towards lower energies; at 600 nm its value is about 3 while at 900 nm it is already 7 times larger. Clearly, since the singlet-triplet splitting of naphthalene (E_{S-T}=2.56 eV) is larger than exciton energies no energy transfer is possible while for β-carotene (E_{S-T}=0.8 eV) all excitons can participate in the energy exchange process. We would like to mention that the energy transfer time measured for all organic molecules is very similar and is almost two orders of magnitude longer than that for oxygen molecules [42]. Since the electron exchange rate critically depends on the overlap of the wavefunctions of the donor and the acceptor, the energy transfer time should be much longer for larger molecules.

7. Conclusion

We have demonstrated that quantum confinement and morphological effects result in a very high photosensitizing activity of Si nanocrystal assemblies. Most of the presented experimental data are based on measurements performed with PSi. However other forms of hydrogen-terminated Si nanocrystals assemblies (for instance, synthesized from gaseous phase) have almost identical optical and photosensitizing properties. Microscopically the effect is governed by a spin-dependent electron exchange, in accordance with the basic laws of quantum mechanics. This energy transfer process does not depend on the order in energy of the acceptor triplet and singlet states and seems to be universal. Si nanocrystal assemblies can therefore be viewed as almost ideal candidates to study energy transfer processes and to probe the electronic structure of aromatic molecules and clusters (e.g., C_{60}, C_{70}, carbon nanotubes). A very broad absorption band of Si nanocrystal assemblies covering all the visible range implies that nanosilicon can utilize light in all visible range for singlet oxygen formation. Controllable chemical functionalization of Si nanocrystal surfaces suggests that the photosensitizing properties of Si nanocrystals can be accurately tailored. Finally, Si nanocrystals sensitizers can be produced in large amounts chemically in completely controllable manner from a fine Si powder. We believe that these findings are relevant for different disciplines: physical chemistry, material science and condensed matter physics. Additional emerging fields of their application are biology and medicine. In particular, a large amount of effort has been directed toward making nanosilicon a biologically relevant material [98]. Si nanocrystals can be considered as chemical reagents which may be dissolved in fluids containing organic molecules or biological objects. From the point of view of practical applications, specifically in medicine, Si nanocrystals in colloidal form should act in a similar way to conventional dye molecules. Therefore, we believe that this research represents a step towards functionalization of the most commercially used semiconductor for chemical, biological and medical applications.

Acknowledgements

We would like to thank the many colleagues and co-workers who have contributed ideas, theoretical and experimental work and are not formally authors of this review. In particular, Bernhard Goller, Egon Gross, Nicolai Künzner and Viktor Timoshenko deserve mention here. A part of this work was supported by the Industrial Technology Research Grant Program from the New Energy and Industrial Technology Development Organization (NEDO), Japan, by Commission of the European Communities, 6th Framework Program (STRP 013875) and by EPSRC grant (EP/F012659/1).

References

1. A. Uhlir, Bell Syst. Tech. **35**, 333 (1956)
2. Y. Watanabe and T. Sakai, Rev. Electr. Commun. Lab. **19**, 899 (1971); Y. Arita, F. Kato and T. Sudo, IEEE Trans. Electron Devices **24**, 756 (1977)
3. C. Pickering, M. I. J. Beale, D. J. Pearson and R. G. Greef, Physica C **17**, 6535 (1984)
4. L. T. Canham, Appl. Phys. Lett. **57**, 1046 (1990)
5. V. Lehmann and U. Goesele, Appl. Phys. Lett. **58**, 856 (1991)
6. A. G. Cullis, L. T. Canham and P. D. J. Calcott, J. Appl. Phys. **82**, 909 (1997)
7. D. Kovalev, H. Heckler, G. Polisski and F. Koch, Phys. Stat. Sol. (a) **215**, 871 (1999)
8. O. Bisi, S. Ossicini and L. Pavesi, Surf. Sci. Rep. **38**, 1 (2000)
9. M.L. Brongersma, P.G. Kik, A. Polman, K.S. Min and H.A. Atwater, Appl. Phys. Lett. **76**, 351 (2000)
10. L. Khriachtchev, M. Räsänen, S. Novikov and J. Sinkkonen, Appl. Phys. Lett. **79**, 1249 (2001)
11. L. Del Negro, M. Cazzanelli, L. Pavesi, S. Ossicini, D. Pacifici, G. Franzo and F. Priolo, Appl. Phys. Lett. **82**, 4636 (2003)
12. M. Fujii, M. Yoshida, Y. Kanzawa, S. Hayashi and K. Yamamoto, Appl. Phys. Lett. **71**, 1198 (1997); J. Ruan, P. M. Fauchet, L. Del Negro, M. Cazzanelli and L. Pavesi, Appl. Phys. Lett. **83**, 5479 (2003)
13. J. Heitmann, F. Müller, M. Zacharias and U. Gösele, Adv. Mater. **17**, 795 (2005)
14. F. Huisken, G. Ledoux, O. Guillois and C. Reynaud, Adv. Mater. **14**, 1861 (2002)
15. A. J. Steckl, J. Xu and H. C. Mogul, Appl. Phys. Lett. **62**, 2111 (1992); G. Anaple, R. Burrows, Y. Wu and P. Boolchand, J. Appl. Phys. **78**, 4273 (1995)
16. J. M. Buriak, Chem. Rev. **102,** 1271 (2002)
17. M. J. Sailor, Sensor Applications of Porous Silicon, in: Properties of Porous Silicon, edited by L. T. Canham, Short Run Press Ltd., London, (1997); S. Content, W. C. Trogler and M. J. Sailor, Chem. Eur. J. **6**, 2205 (2000); M. P. Stewart and J. M. Buriak, Adv. Mater. **12**, 859 (2000)
18. S. Gold, K. L. Chu, C. Lu, M. A. Shannon and R. I. Masel, J. Power Sources **135**, 198 (2004)
19. D. Kovalev, E. Gross, N. Künzner, F. Koch, V. Yu. Timoshenko and M. Fujii, Phys. Rev. Lett. **89**, 137401 (2002)
20. L. T. Canham and R. Aston, Phys. World **27** (2001)
21. D. Kovalev, V. Yu. Timoshenko, N. Künzner, E. Gross and F. Koch, Phys. Rev. Lett. **87**, 68301 (2001); F. V. Mikulec, J. D. Kirtland and M. J. Sailor, Adv. Mater. **14**, 38 (2002)
22. M. Faraday, Philosophical Magazine **XXXI**, 401 (1847)
23. N. J. Turro, Modern Molecular Photochemistry, University Science Books, Sausalito, CA, (1991)
24. C. Schweitzer and R. Schmidt, Chem. Rev. **103**, 1685 (2003) and references therein
25. H. Kautsky and H. DeBruijn, Naturwissenschaften, **19**, 1043 (1931)

26. Methods in Enzymology, Singlet Oxygen, UV-A, and Ozone, Eds: L. Packer, H. Sies, Academic Press, London (2000)

27. D. L. Gilbert and C. A. Colton, Reactive Oxygen Species in Biological System, Plenum Pub. Corp. (1999)

28. M. C. DeRosa and R. J. Crutchley, Coord. Chem. Rev. **351**, 233 (2002)

29. J. G. Moser, Photodynamic Tumor Therapy: 2nd and 3rd Generation Photosensitizers, G & B Science Pub (1998)

30. D. B. Min and J. M. Boff, Chemistry and Reactions of Singlet Oxygen in Foods, Vol. 1, Comprehensive Reviews in Food Science and Food Safety, http://www.ift.org/cms/ (2002)

31. T. Förster, Ann. Phys. (N.Y.) **2**, 55 (1948)

32. D. L. Dexter, J. Chem. Phys. **21**, 836 (1953)

33. S. Wang, R.Gao, F. Zhou and M. Selke, J. Mat. Chem. **14**, 487 (2004)

34. R. Bakalova, H. Ohba, Z. Zhelev, T. Nagaze, R. Jose, M. Ishikawa and Y. Baba, Nanoletters, **4**, 1567 (2004); R. Bakalova, H. Ohba, Z. Zhelev, M. Ishikawa and Y.Baba, Nature Biotechnology **22**, 1360 (2004)

35. M. K. Nissen, S. M. Wilson and M. L. W. Thewalt, Phys. Rev. Lett. **69**, 2423 (1992)

36. J. W. Arbogast, A. P. Darmanyan, C. S. Foote, Y. Rubin, F. Diederich, M. M. Alvarez, S. J. Anz and R. L. Whetten, J. Phys. Chem. **95**, 11 (1991)

37. N. Tagmatarchis, H. Kato and H. Shinohara, Phys. Chem. Chem. Phys., **3**, 3200 (2001)

38. P. Cheng, S. R. Wilson and D. I. Schuster, Chem. Commun. **1**, 89 (1999)

39. M. Nirmal, D. J. Norris, M. Kuno and M. G. Bawendi, Phys. Rev. Lett. **75**, 3728 (1995)

40. M. Nirnal and L. Brus, Acc. Chem. Res. **32**, 407 (1999)

41. D. Jurbergs, E. Rogojina, L. Mangolini and U. Kortshagen, Appl. Phys. Lett. **88**, 233116 (2006)

42. E. Gross, D. Kovalev, N. Künzner, J. Diener, F. Koch, V. Yu. Timoshenko and M. Fujii, Phys. Rev. B **68**, 115405 (2003)

43. A. G. Cullis and L. T. Canham, Nature **353**, 335 (1991)

44. J. Knipping, H. Wiggers, B. Rellinghaus, P. Roth, D. Konjhodzic and C. Meier, Journ. .of Nanoscience and Technology **4**, 1039 (2004); C. Meier C, A. Gondorf, S. Luttjohann, A. Lorke and H. Wiggers, J. Appl. Phys **101**, 103112 (2007)

45. W. Schmid, Phys. Stat. Sol. **84**, 529 (1977)

46. M. S. Hybertsen, Phys. Rev. Lett. **72**, 1514 (1994); M. S. Hybertsen in "Porous Silicon Science and Technology" ed. by J.-C.Vial and J. Derrien, Springer (1995)

47. Al. L. Efros, M. Rosen, M. Kuno, M. Nirmal, D. J. Norris and M. Bawendi, Phys. Rev. B **54**, 4843 (1996)

48. J. Diener, D. Kovalev, G. Polisski, H. Heckler and F. Koch, Phys. Stat. Sol. (b) **214**, R13 (1999)

49. D. Kovalev and M. Fujii, Adv. Mater. **17**, 2531 (2005)

50. P. D. J. Calcott, K. J. Nash, L. T. Canham, M. J. Kane and D. Brumhead, J. Phys. Condens. Matter **5**, L91 (1993)

51. T. Suemoto, K. Tanaka, A. Nakajima and T. Itakura, Phys. Rev. Lett. **70**, 3659 (1993)

52. M. Rosenbauer, S. Finkbeiner, E. Bustarret, J. Weber and M. Stutzmann, Phys. Rev. B **51**, 10539 (1995)

53. L. E. Brus, P. F. Szajowski, W. L. Wilson, T. D. Harris, S. Schuppler and P. H. Citrin, J. Am. Chem. Soc. **117**, 2915 (1995)

54. D. Kovalev, M. Ben-Chorin, J. Diener, B. Averboukh, G. Polisski and F. Koch, Phys. Rev. Lett. **79**, 119 (1997)

55. Y. Kanemitsu and S. Okamoto, Phys. Rev. B **58**, 9652 (1998)

56. G. F. Grom, D. J. Lockwood, J. P. Mccaffrey, H. J. Labbe, P. M. Fauchet, B. White, J. Diener, D. Kovalev, F. Koch, and L. Tsybeskov, Nature **407**, 358 (2000)

57. I. Sychugov, R. Juhasz, J. Valenta and J. Linnros, Phys. Rev. Lett. **94**, 087405 (2005); I. Sychugov, R. Juhasz, J. Valenta, A. Zhang, P. Pirouz and J. Linnros, Appl. Surf. Science **252**, 5249 (2006)

58. S. A. Empedocles, D. J. Norris and M. G. Bawendi, Physical Review Letters **77**, 3873 (1996).

59. Spectroscopy of Isolated and Assembled Semiconductor Nanocrystals, ed. by L.E. Brus, Al.L.Efros, T. Itoh, Journal of Luminescence **70**, (1996)

60. R. J. Walters, J. Kalkman, A. Polman, H. A. Atwater and M. J. A. de Dood, Phys. Rev. B **73**, 132302 (2006)

61. J. C. Merle, M. Capizzi, P. Fiorini and A. Frova, Phys. Rev. B **17**, 4821 (1978)

62. M. Nirmal, D. J. Norris, M. Kuno, M. G. Bawendi, Al. Efros and M. Rosen, Phys. Rev. Lett. **75**, 3728 (1995)

63. D. Gammon, E. S. Snow, B. V. Shanabrook, D. S. Katzer and D. Park, Phys. Rev. Lett. **76**, 3005 (1996)

64. T. Takagahara, Phys. Rev. B **47**, 4569 (1993); T. Takagahara, Phys. Rev. B **53**, R4205 (1996)

65. E. Martin, C. Delerue, G. Allan and M. Lannoo, Phys. Rev. B **50**, 18258 (1994)

66. A. Franceschetti and A. Zunger, Phys. Rev. Lett. **78**, 915 (1997)

67. K. J. Nash, P. D. J. Calcott, L. T. Canham and R. J. Needs, Phys. Rev. B **51**, 17698 (1995)

68. M. A. Tischler, R. T. Collins and J. H. Stathis and J. C. Tsang, Appl. Phys. Lett. **60**, 639 (1992)

69. J. Harper and M. J. Sailor, Longmuir **13**, 4652 (1997)

70. W. Weber, Phys. Rev. B **15**, 4789, (1977)

71. A. Campargue, L. Biennier, A. Kachanov, R. Jost, B. Bussery-Honvault, V. Veyret, S. Chussary and R. Bacis, Chem. Phys. Lett. **288**, 734 (1998)

72. J. Goodman and L.E. Bruce, J. Chem. Phys. **67**, 4408 (1977)

73. V. N. Abakumov, V. I. Perel and I. N. Yassievich, in Nonradiative Recombination in Semiconductors, edited by V.M. Agranovich and A. A. Maradudin, Modern Problems in Condensed Matter Sciences Vol. 33, North-Holland, Amsterdam (1991)

74. CRC Handbook of Chemistry and Physics, 78th ed. ed: D. R. Linde, CRC Press, New York (1997-1998)
75. H. Heckler, D. Kovalev, G. Polisski, N.N. Zinov'ev and F. Koch, Phys. Rev. B **60**, 7718 (1999)
76. M. Fujii, D. Kovalev, B. Goller, S. Minobe, S. Hayashi and V. Yu. Timoshenko, Phys. Rev. B **72**, 165321 (2005)
77. D. Kovalev, H. Heckler, B. Averboukh, M. Ben-Chorin, M. Schwarzkopff and F. Koch, Phys. Rev. B **57**, 3741 (1998)
78. M. Fujii, Y. Yamaguchi, Y. Takase, K. Ninomiya and S. Hayashi, Appl. Phys. Lett. **85**, 1158 (2004)
79. E. A. Konstantinova, L. A. Osminkina, K. S. Sharov, E. V. Kurepina, P. K. Kashkarov and V. Y. Timoshenko, JETP **99**, 741, (2004)
80. M. Hild and R. Schmidt, J. Phys. Chem. A **103**, 6091 (1999)
81. R. Schmidt, J. Am. Chem. Soc. **111**, 6983 (1989)
82. R. Schmidt, F. Shafii and M. Hild, J. Phys. Chem. **103**, 2599 (1999)
83. M. Fujii, S. Minobe, M. Usui, S. Hayashi, E. Gross, J. Diener and D. Kovalev, Phys. Rev. B **70**, 085311 (2004)
84. M. Fujii, N. Nishimura, H. Fumon, S. Hayashi, D. Kovalev, B. Goller and J. Diener, J. Appl. Phys. **100**, 124302 (2006)
85. B. A. Cosgrove and J. Walkley, J. Chromatogr. **216**, 161 (1981)
86. D. Kovalev, E. Gross, V. Yu. Timoshenko and M. Fujii, Appl. Phys. Lett. **85**, 3590 (2004)
87. W. Theiss, Surf. Sci. Rep. **29**, 91 (1997)
88. M. Fujii, M. Usui, S. Hayashi, E. Gross, D. Kovalev, N. Künzner and J. Diener, J. Appl. Phys. **95**, 3689 (2004)
89. R. H. Young, K. Wehrly and R. L. Martin, J. Am. Chem. Soc. **93**, 5774 (1971)
90. M. Nowakowska, M. Kepczyn'ski and K. Szczubiatka, Macromol. Chem. Phys. **196**, 2073 (1995)
91. J. M. Fernandes, M. D. Bilgin, L. I. Grossweiner and J. Photochem. Photobiology B **37**, 131 (1999)
92. V. Yu. Timoshenko, A. A. Kudryavtsev, L. A. Osminkina, A. S. Vorontsov, Yu. V. Ryabchikov, I. A. Belogorokhov, D. Kovalev and P. K. Kashkarov, JETP Letters **83**, 423 (2006)
93. W. Kueng, E. Silber and U. Eppenberger, Anal. Biochem. **182**, 16 (1989)
94. H. Steller, Science **267**, 1445 (1995)
95. O. E. Grichenko, V. V. Shaposhnikova, A. A. Kudryavtsev and Yu. N. Korystov, Izv. Ross. Akad. Nauk, Ser. Biol. **3**, 274 (2004)
96. S. Murov, I. Carmichael and G. Hug, Handbook of Photochemistry, 2nd ed. Marcel Dekker, New York, (1993)
97. B. Goller, S. Polisski and D. Kovalev, Phys. Rev. B **75**, 073403 (2007)
98. J. M. Buriak, Chemical Reviews, **102**, 1271 (2002)

CHAPTER 5

DNA-TEMPLATED NANOWIRES: CONTEXT, FABRICATION, PROPERTIES AND APPLICATIONS

Qun Gu[1] and Donald T. Haynie[2]*

[1]*Pacific Nanotechnology, Inc., 3350 Scott Blvd, #29, Santa Clara, CA 95054*
[2]*Artificial Cell Technologies, Inc., 5 Science Park at Yale, Third Floor, New Haven, CT 06511 National Dendrimer and Nanotechnology Center, Department of Chemistry, Central Michigan University, Mt Pleasant, MI 48859*
*E-mail: haynie@jhu.edu

Nanowires will be important components in future nanoscale devices and systems. Great effort is therefore being invested in the development of novel strategies for reliable realization of these tiny one-dimensional structures. The size and structure of the deoxyribonucleic acid (DNA) molecule make it promising for "bottom-up" DNA-templated nanowire fabrication. A key advantage of DNA is its "natural" ability to localize by molecular recognition. In earlier work we reviewed methods of DNA nanowire fabrication [1]. Here, we discuss DNA-templated nanowires in a broader context, covering recent progress on DNA nanowire characterization and prospective applications, in addition to developments on fabrication. DNA-templated nanowires are compared with other kinds of nanowires, fabrication methods are classified, and physical properties are summarized.

1. Introduction

Nanoscience can be defined as the study of the structure and behavior of matter, and the manipulation of matter, at the level of atoms, molecules, and clusters of atoms and molecules [2]. Sub-topics include nanoparticles, nanopores, nanoribbons, nanoropes, nanocells, nanotubes, nanomanipulators, nanostructured materials, nanomechanics, and nanoelectronics. Related topics are properties which can be discriminated by electron microscopy, solvent evaporation and lubrication at the molecular scale, the miniaturization of computational devices, storage devices, switching devices, transistors, and the inspiration that can be derived from molecular biosystems. The list is hardly exhaustive, and some of the named topics are perhaps better called nanotechnology than nanoscience. *Nanomanufacturing* is a broad term denoting the repetitive fabrication of objects at the nanoscale. The meaning of *nanomanufacturing* will seem clear enough to nanotechnology researchers, but it is nonetheless interesting that, taken literally, the word is nonsense: the unaided hand (Latin, *manus*) is totally incapable of making (Latin, *facere*) things 100 nanometers or smaller!

The present review concerns nanometer-scale wires, and more especially nanowires made of DNA. These objects are a type of one-dimensional nanoscale material or device. How are DNA nanowires prepared? What are their physical properties? Are DNA nanowires mere objects of scientific and engineering curiosity? Or could they one day be useful elements of commercial products? We discuss the topic in the broader context of different approaches to nanowire fabrication. We also discuss envisioned applications of the technology. Electrically conductive and semi-conductive nanowires receive particular emphasis, the reason being that these areas have been the object of most of the reported research activity. What has motivated the development of electrically conductive and semi-conductive nanowires?

1.1. Moore's Law

The rapid growth of the semiconductor industry in the past several decades has been fueled by the continual push for innovative development of reliable "top-down" fabrication of structures of ever-decreasing dimensions. The increased scaling of conventional complementary metal-oxide-semiconductor (CMOS) device technology has translated into ever higher circuit densities and increased device performance. The well-known prognostication of Intel co-founder Gordon Moore, popularly termed "Moore's Law," says that the number of transistors on a chip will double about every other year. Indeed, the minimum feature size of commercial devices has steadily shrunk from 2 μm in 1980 to below 100 nm today. The 2005 International Technology Roadmap for Semiconductors Industry shows the scaling trend reaching performance limits in 2012 [3].

There are hard barriers, however to the reduction of device size. Quantum mechanical limitations are inevitably encountered as CMOS device size approaches the level of largish clusters of atoms and still smaller particles. Conventional photolithography is limited by the wave nature of light for features smaller than 100 nm. The cost of building, operating and maintaining fabrication facilities, especially lithography, increases significantly with each generation of CMOS technology. Lithographic methods such as X-ray lithography and extreme UV lithography, though effective, are too expensive for commodity-type nanoscale device fabrication. Electron beam lithography provides a means of patterning polycrystalline metal nanowires as small as 20 nm in diameter, but again the cost is so high that implementation has been limited to research.

Various electronic devices that operate by quantum mechanical principles are nevertheless of interest [4]. Some may even show superior functionality to their conventional counterparts or novel useful properties. These "quantum devices" or "nanoelectronic devices" are typically smaller than 100 nm. Examples include resonant tunneling diodes, single-electron transistors, and quantum cellular automata. What will be the best ways to prepare such structures? Can cues be taken from nature? Protein-based light-harvesting complexes in plants, for instance, have dimensions on the order of 10 nanometers and carry out a complicated series of exquisitely efficient electron transfer reactions. These "nanomachines" play a vital role in transducing the energy of the Sun into plant material and animal material for food. It is believe that the analysis of the structure and function of such complexes, which self-organize by molecular recognition, will advance biomimetic fabrication methods and the realization of novel nanometer-scale devices and systems.

1.2. Self-Assembly and Molecular Recognition

It is not generally known, but investigations in biology and medicine have played a key role in the discovery and development of several remarkably basic ideas in physics.   Notable examples from late eighteenth and nineteenth century Europe include conduction of electricity (Galvani, Volta), conservation of energy (Mayer, Helmholtz), and diffusion of matter (Brown, Pfeffer).   In the twentieth century the Hungarian-American mathematician John von Neumann adopted the vertebrate central nervous system as a model of electronic signal transmission, computer memory and storage capacity [5].   Here we mention how the basic concepts of self-assembly and molecular recognition are illustrated by biology.   These concepts are then used to analyze DNA nanowire fabrication, characterization, and application.

Self-assembly refers to the spontaneous generation of organized matter on all length scales, from clusters of atoms or molecules ("nanoparticles" or "nanoparticulates") to clusters of entire galaxies.   In some cases the self-assembly process is reversible; in all cases pre-existing components come together to form larger structures or patterns. Examples abound.   The most profound ones are in biology.   The four subunits of the protein hemoglobin, for instance, spontaneously organize into the multimeric molecule which plays a vital role in vertebrate respiration.   A second example is bacteriophage M13, a type of filamentous bacterial virus, the components of which assemble spontaneously (a virus is not alive).   A third example is provided by the two "strands" of double-helical DNA, the repository of genetic information in all known living organisms and many viruses.   DNA strands, when sufficiently long, will overcome the dissipative tendency due to thermal energy and bind to each other entirely spontaneously, specifically, and persistently.   Others examples of self-organizing systems are planets, weather patterns, and geological features.   Self-assembly is "static" when a process results in an ordered state at equilibrium which does not dissipate energy and "dynamic" otherwise. DNA nanowire fabrication involves static self-assembly.

Molecular recognition is a term used to describe the specific interaction between two or more molecules by non-covalent bonding. Various kinds of non-covalent bonds can play a role in molecular recognition: hydrogen bonds, metal coordination bonds, hydrophobic interactions, van der Waals interactions, π-π interactions, coulombic interactions. Some non-covalent interactions are more significant than others in specific cases. The molecules or particles involved in molecular recognition generally exhibit both shape complementarity (geometrical fit) and charge complementarity (e.g. positive surface binding to negative, hydrophobic to hydrophobic, hydrogen bond donor to acceptor).

Molecular recognition is a basic feature of the macromolecular constituents of the living cell. Many examples could be given: ligand-receptor interactions, protein-protein interactions, protein-DNA interactions, DNA-DNA interactions, and so on. The protein hemoglobin, for instance, binds diatomic oxygen with high specificity at several distinct sites. Each site is formed in part by a non-covalently bound molecule called porphyrin, a rather complicated ring structure that is synthesized by the endogenous molecular machinery of all known living organisms. In DNA, one molecular "strand" of the double helix "recognizes" its complement, forming a large number of hydrogen bonds between structurally complementary chemical moieties. Certain small molecules show considerably higher specificity for the ribosomes of bacteria than the ribosomes of vertebrates, making these small molecules useful as antimicrobial agents. Chemists have designed artificial molecular systems to display a type of molecular-scale recognition. Crown ethers, for example, can be selective for specific cations. All such nanoscale structures imitate biology in some respects. These nanoscale structures are fascinating and should be studied. Much of the world or what can be done with the atoms the world is made of is not yet known! Will it be possible to turn novel nanoscale structures into devices of practical value? Will it be possible to develop practical methods of producing these structures at a scale that will support the commercialization of the devices?

1.3. Nanoscience, Nanotechnology and Nanosystems

We have noted that "top-down" fabrication approaches become prohibitively expensive as device dimensions approach the nanometer range. Academic scientists and engineers and industry visionaries have therefore been keen to prospect for novel strategies for submicron fabrication methods, in the hope that it might be possible to transform the more reliable strategies from a laboratory-scale approach into a mainstream process. "Bottom-up" approaches to device or materials fabrication, which typically involve self-assembly or molecular recognition, have been a subject of intense interest since about 2000, when state sponsorship of research and development in nanotechnology rose sharply. It is generally agreed that bottom-up approaches can indeed provide viable alternate pathways for the fabrication of nanoscale devices, and that such devices can display different or more desirable properties than counterparts made by conventional methods.

An example of a bottom-up approach that has been explored in recent years is a generic synthetic method of rational design of multiply-connected and hierarchically-branched nanostructures [6]. The structures are built inside nano-channels in anodic Al_2O_3 templates. The nano-channels enable controlled fabrication of branched nanostructures, including carbon nanotubes and metallic nanowires which display several hierarchical levels of multiple branching. The number of branches, the frequency of branching, and the overall dimensions and architecture of the nanostructures can be controlled precisely by way of pore design and templated assembly. The technique is useful for making nanostructures of considerable morphological complexity and therefore could have influence the design of future nanoscale systems.

Many researchers in nanoscience and nanotechnology have found interesting ideas molecular-scale biology and begun applying them in new ways. The biomacromolecules of the living cell, particularly proteins, are aptly called "nanobiomachines." Ordered peptide structure was first visualized at atomic resolution in the early 1950s, the structure of the DNA double helix appeared in 1953, and the first high-resolution protein structures became available at the end of the 1950s. It was not until relatively recently, however, that molecular-scale biology came to play a key role in inspiring nanosystem design [7, 8], despite several references to biology in Feynman's famous talk, "(Plenty of) Room at the Bottom" [9]. Some will find it counterintuitive, but there is a plausible sense in which nanotechnology was born of biology and medicine [10].

Biomolecular systems can be used to investigate the physics of nanostructures. From the *yin* point of view, so to speak, there is the use of nanosized metal clusters as a tool, for example, to manipulate biomolecular systems or to develop nanoscale sensors [11-14]. And from the *yang* perspective, there is using biomolecules as templates to build inorganic nanostructures [15-20]. Both approaches are reviewed below. Both general areas of nanobiotechnology have commercial potential. As we shall see, DNA is a promising candidate material for forming nanowires and nanostructured assemblies of wires from about 5 nm to up to several microns.

2. Nanowires

Interest in nanowires and related quasi-1D structures, often called nanotubes or nanorods, has risen sharply in recent years. The nanowire has come to be regarded as a key building block for bottom-up fabrication of different micro-/nanosystems. Unique material properties exhibited by nanowires have considerable potential for the development of functional devices, for example electronic devices such as diode logic gates [21], bipolar transistors [22], and field effect transistors (FETs) [23-25]. Cobalt, nickel and iron nanowires having giant magneto-resistance are potentially useful for high-density storage memory [26-28]. These can devices can in principle be integrated into systems of devices, as with their micro-scale counterparts.

There are different bottom-up approaches to one-dimensional nanostructure fabrication. The most successful methods of this type have been vapor-liquid-solid (VLS) growth, laser-assisted catalytic growth (LCG), and template-based methods (TBMs). Various other non-biomolecular nanowire assembly approaches are being investigated, for example the use of electric and magnetic fields, laminar flow in microfluidic channels, and Langmuir-Blodgett compression. In this section we consider nanowires in general, providing background information useful for assessing the advances and future prospects of results presented in the section on DNA-templated nanowires.

VLS was first used to grow single-crystal silicon whiskers by Wagner and Ellis in the mid-1960s [29]. The approach was developed by Givargizov in the following decade [30]. More recently, Yang and coworkers have described real-time growth of germanium nanowires by VLS [31]. To produce these nanowires, material from the vapor is incorporated into the growing nanowire via a liquid catalyst, commonly a low melting point eutectic alloy. There are three general stages in the process (Figure 1). In the first, gold nanocrystals and germanium vapor form an alloy that liquefies at high temperature. Liquid alloy condenses on a decrease of temperature and some of the germanium segregates as a crystal. In the second stage, germanium nuclei precipitate at the liquid-solid interface. Germanium vapor continues to dissolve into the alloy, furthering segregation. In the third stage, condensed germanium grows into solid single crystal nanowires.

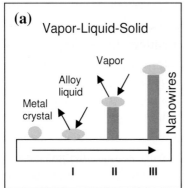 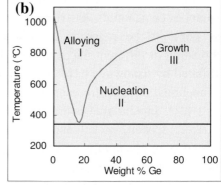

Figure 1. The mechanism of VLS. (a) The three-stages of growth: Alloying, nucleation, and axial growth. (B) Relationship of temperature and composition. Adapted from ref. 31. Copyright 2001 American Chemical Society.

LCG, developed by Lieber and colleagues, is an improvement on the VLS method for nanowire fabrication [32]. 3-6 nm Si and Ge nanoclusters are generated by a pulsed laser, initiating nanowire growth. Photochemical energy is used to condense vapor species into solid nanoscale crystals. Subsequent steps resemble VLS. LCG has been used to fabricate semiconductor nanowires made of compounds from the III-V groups or II-VI groups, including GaN [33], GaAS, GaP, InAs, InP, GaAs/P, InAs/P, ZnS, ZnSe, CdS, and CdSe [34]. The average diameter of these nanowires is 11-30 nm, and the orientation is mainly <111>. A wide range of different nanowires can be prepared by LCG because a laser can be used to generate nanoseeds of many different materials.

There are different TBM approaches. One is the negative template method. Nanowires are deposited into long cylindrical pores, which serve as arrays of negative templates. Pore structure, orientation and distribution on the template together determine width, length, and distribution of the nanowires. In track-etched substrates, for example, discrete cylindrical pores can be generated by ion bombardment followed by chemical etching. Possin *et al.* have fabricated metal nanowires with a diameter of ~40 nm by filling track-etched pores on mica [35]. Anodic alumina can also be used for this purpose. Well-ordered honeycomb-like nanopores can be realized by anodization of aluminum in an acidic electrolyte. The size of pores and distance between adjacent pores are mainly determined by anodization conditions. Other negative templates, for example polymethylmethacrylate, polystyrene, glass and silica, have also been used for nanowire fabrication [36].

Nanowire fabrication by electrochemical deposition, introduced by Martin and coworkers [37], is the most widely used template-based method. An advantage is that different kinds of nanowires can be prepared: the material for the nanowire can be a metal [38-47], a semiconductor [48, 49], or a conducting polymer [50-53]. The template is soaked in electrodeposition solution. A metal film deposited on the template then serves as an electrode for electrodeposition. Extraction of nanowires from the template post fabrication can be achieved by chemical dissolution of the template (if it is a thin membrane) or by application of an AC field. Nanowires thus fabricated can be highly conductive. Electron transfer during electrodeposition is rapid along a highly conductive pathway. Other negative template methods of nanowire fabrication are chemical vapor deposition, electroless plating, and sol-gel chemistry [36].

Table 1. Applications of Nanowires				
Potential applications		Nanowires	Fabrication method	Width (nm)	Ref.
Nanoelectronics	Diode logic gates	Si/GaN	VLS/LCG	10-30	21
	Bipolar transistor	Si	LCG	20-50	22
	Field effect transistor	InP/Si/CdS	LCG/CVD	5-75	23-25
	Storage memory	Fe/Co/Ni	TBM	~200	26-28
	Interconnection	GaN/InP/Si	LCG	NA	67
Optoelectronics	Laser	ZnO/GaN/CdS	VLS/LCG	10-200	54-58
	Waveguides	ZnO/SnO$_2$	VLS	40-350	58-59
	Photodetector/ Switch	ZnO/InP	VLS/LCG	20-60	61-63
	Light emitting diode	InP/Si/GaN	LCG/VLS	45	23, 60
NEMS	Sensors	Si/SnO$_2$	LCG/VLS	10-20	64-66

Prospective applications of nanowires encompass nanoelectronics, optoelectronics, and nanoelectromechanical systems (NEMS) (Table 1). Some of the fabrication methods enable single crystal growth of semiconductor nanowires of controlled length and diameter. These nanowires can display a high electrical conductance by *n*- or *p*-type doping. Elemental electronic devices, diodes, logic gates, bipolar transistors, and FETs, which are conventionally integrated on a planar silicon substrate, have been realized in 1-D nanowires [21-25]. P-N junctions, which are formed by combining P-type and N-type semiconductors in very close contact, display properties which are useful in modern electronics. A single nanowire in a P-N junction is potentially advantageous for various devices, for example lasers [54-58], waveguides [58, 59], light emitting diodes [60], photodetectors and optical switches [61-63]. Wide band-gap semiconductor materials, for example ZnO, GaN and InP, are useful as laser cavities and can emitting monochromic light from the UV to the IR. The first electrically driven nanowire-based laser was fabricated by Lieber and coworkers [56]. Yang and coworkers have shown that nanowires made of ZnO emit monochromic light at 385 nm after optical excitation [54]. GaN nanowires have been reported to function as a blue UV laser.

Nanowire-based sensors could be used to detect trace quantities of biomolecules and chemicals in NEMS devices [64-66]. Cui *et al.*, for example, have reported that the conductance of modified silicon nanowires scales linearly with the pH of solution [64]. These nanowires could be used to fabricate tiny pH sensors. Single nanowires could also be useful for array-based screening of chemicals and *in vivo* diagnostics. Hahm *et al.* have demonstrated than a silicon nanowire-based device can be used to detect nucleic acid with high fidelity and sensitivity [66]. Nanowire-based NEMS could possibly be used to detect proteins, viruses, and other pathogens with high sensitivity.

Besides metallic and semiconductor nanowires, Bi and Bi_2Te_3 nanowires have been shown to exhibit unique thermoelectric properties [68]. Resistivity decreases monotonically with temperature increase. These nanowires could be useful in a miniature cooling system.

Recent work has emphasized control over the fabrication and advanced technology applications of conductive nanowires [69-71], magnetic nanowires [72], semiconductive nanowires [73-88], and other nanowires [89-93]. Great gains in methods and understanding have been made. Nevertheless, further research and development are needed to understand the unique properties of nanowires made of different materials. Although much remains to be learned, it seems clear enough that some nanowire-based devices and systems are closer than others to becoming a commercial reality.

3. DNA-Templated Nanowires

3.1. Introduction

DNA-based nanotechnology is an active and growing field. DNA molecules have been put to use in many different ways. Nevertheless, all of these approaches have made use of the structural or specific molecular recognition properties of DNA. A fascinating illustration of the molecular recognition properties of DNA is the self-assembly of novel supramolecular DNA nanostructures [94]. See Figure 2. Another intriguing example is the use of DNA molecules in a novel type of computing [95].

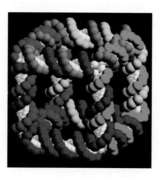

Figure 2. Cube formed from microscopic DNA "bricks." Each strand is shown in a different color. Copyright 1997 Connect: Information Technology at NYU. Reprinted with permission.

A new field of DNA-based nanotechnologies has developed since the late 1990s. These technologies exploit natural or synthetic DNA molecules as a templating material to assemble nanostructures made of other materials. In the work of Alivisatos *et al.* [96] and Mirkin *et al.* [97], for instance, gold nanoparticles were assembled into organized structures with nanometer precision by molecular recognition of DNA. Later, Loweth *et al.* [98] utilized the Watson-Crick base pairing to assemble two or three individual gold nanocrystals on specific sites of a single stranded DNA (ssDNA) molecule.

DNA-templated nanowires are a distinctive class of functional one-dimensional nanostructures. The nanowires prepared by the LCG and VLS methods, which have been used exclusively for the fabrication of semiconductor nanowires, can be "decorated" and thereby "functionalized" with macromolecules, including DNA. DNA-templated nanowires, by contrast, are "functionalized" with metal nanoparticles, semiconductor nanoparticles, or conducting polymers.

The range of possible properties of DNA-templated nanowires is very large. It has nevertheless been found that key aspects of nanowire fabrication can generally be described in terms of established principles of wet chemistry, notably electrostatic interaction, complexation, redox reactions, and nucleation and growth. No expensive reaction chamber is required for DNA-templated nanowire fabrication. Nor is there a need for high temperature. Modest infrastructure needs are attractive for low-cost manufacturing.

Nanowires prepared by DNA templating tend to have different structural properties from nanowires fabricated by LCG or VLS methods. This is because nanoparticle arrays on DNA molecules tend to lack crystallinity and uniformity. Nanowire fabrication by electrochemical deposition, by contrast, results in structures and properties that depend essentially on template preparation. DNA-templated nanowire fabrication is done on DNA molecules, uniform linear polyanions having a diameter of 2 nm. The length of DNA molecules is tunable from nanometers to microns. The exquisite molecular recognition capability of a single-stranded DNA (ssDNA) can be used for precise positioning of DNA nanowires on a 2D substrate.

Fabrication of conductive silver nanowires on bacteriophage λ DNA connecting two micron-spaced microelectrodes by Braun *et al.* [99] in 1998 marked the birth of DNA-templated nanowires. Since then, DNA nanowires with electrical properties have been the object of considerable attention. Silver [99-107, 121, 135], gold [100, 101, 106, 108-125], palladium [126-129], platinum [130-133], and copper [134-137] have been used to "metallize" DNA molecules in the preparation of nanowires. Electroless chemical plating on λ DNA templates, which are ~16 μm long, has been the main approach. Synthetic DNA molecules [102-104, 106] and polyribonucleic acid (RNA) molecules [124, 125] have been employed as nanowire templates. Magnetic [138-143], semiconductor [144-151], conducting polymer [152-156] nanowires have successfully been fabricated on nucleic acid templates. Advanced applications of DNA-templated nanowires or systems have been studied for the development of nanoscale devices [157-163], DNA detection systems [157], FETs [158], and quantum interference devices [160].

This section of the review provides background information on DNA structures, DNA template stretching and positioning, DNA nanowire fabrication methodologies, electrical DNA nanowires, magnetic DNA nanowires, semiconducting DNA nanowires, conducting polymer DNA nanowires, and applications of DNA nanowires. Collectively, the reviewed results provide insight on the feasibility of using DNA nanowire in the development of nanodevices.

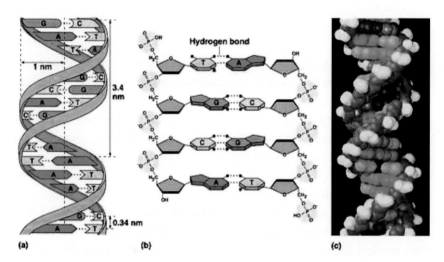

Figure 3. Schematic diagram of the structure of double-stranded DNA. (a) Double helix; (b) chemical structure; (c) molecular model. Each subunit consists of a phosphate group, a sugar and one of four bases: adenine (A), thymine (T), guanine (G) or cytosine (C). Molecular recognition in Watson-Crick base pairing means that A pairs with T only, and G pairs with C only. Two classes of binding site for DNA-templated metallic nanowire fabrication are shown: Negatively charged phosphate groups in the polymer backbone, and the N7 atom in bases G and A and the N3 atom in bases C and T. The distance between two adjacent bases is approximately 0.34 nm. The diameter of a duplex molecule is approximately 2 nm. Copyright Pearson Education, Inc.

3.2. DNA Structure

Linear double-stranded DNA has a crystallographic diameter of 2 nm and a length of 0.34 nm per subunit (Figure 3a). The overall length of the molecule can be orders of magnitude longer than its width. This large aspect ratio makes DNA especially well suited for the template-based bottom-up fabrication described here. Molecular length and copy number can be controlled by established methods of molecular biology, for example DNA ligation, enzymatic digestion, and polymerase chain reaction.

The utility of DNA for nanoassembly is directly related to its unique molecular recognition properties. These properties are due to the basic units of a DNA molecule, known as nucleosides or bases. Each consists of a 5'-carbon sugar (deoxyribose), a nitrogenous base, and a single phosphate group in the polymer backbone (Figure 3b). There are four different bases in DNA – adenine (A), thymine (T), guanine (G) and cytosine (C). A pairs only with T to form two hydrogen bonds in normal double-stranded DNA, and G pairs only with cytosine C to form three hydrogen bonds. The sequence of the subunits in a DNA strand can be prepared at will by modern synthetic methods and determined with great accuracy by modern sequencing methods.

Two polynucleotide chains with complementary sequences form a single double-stranded molecule in which the strands twist together to form the familiar right-handed helical spiral (Figure 3a). The bases are on the inside of the helix, stacked atop each other like the steps of a spiral staircase (Figure 3c). Bases in one strand match up spontaneously by molecular recognition with bases in a complementary strand. This base pairing has been adapted for the assembly and positioning of nanowires. Precise spatial arrangement of assembled nanostructures can be achieved on the length scale of a few nanometers. The molecular recognition properties of DNA also enable the spontaneous formation of complex nanostructures.

The backbone of each polymer strand comprises a sugar moiety and a negatively charged phosphate group. DNA thus has a "natural" affinity for ions, nanoparticles, organic molecules, and other "building blocks:" non-covalent electrostatic interactions with cations or coordination coupling with various metals. The polyanionic backbone of DNA

(Figure 3b) can bind positively charged nanoparticles. Metal ions such as Pt(II) and Pd(II) are coordinated by the N7 atoms of DNA bases. The first step in most approaches to DNA-templated nanowire fabrication is the self-assembly of nanowire precursors, mostly ions or nanoparticles, on DNA templates by electrostatic attraction.

3.3. DNA Stretching and Positioning

Fabricating well-defined DNA-templated nanowires requires aligned surface deposition. Single-stranded DNA is especially useful for specific positioning by molecular recognition. Migration of a DNA molecule for localization on a surface can be achieved by bulk fluid flow or relative motion of the air-water interface. Localization efficiency can be improved by surface modifications favoring DNA-substrate interaction. Precise positioning on a microstructured substrate is critical for the integration of nanoscale structures in devices. Parallel processes will have to be established to facilitate technologically relevant integration densities.

Methods of immobilization, stretching, and positioning of DNA molecules are tools for controlling template and nanowire structure, as well as spatial orientation and localization. A DNA molecule in aqueous solution will be a random coil due to thermal fluctuations. The persistence length of a DNA molecule will be governed by the charge of the sugar-phosphate background and the associated counterions. Entropy will shorten the end-to-end distance, often to a much smaller size than the contour length. DNA molecules must therefore be stretched to serve as templates for organized nanowire fabrication. Strand aggregation can occur, giving rise to an uneven distribution of cation binding sites and therefore irregular structures.

Various approaches have been developed to stretch and orient DNA molecules. These include molecular combing [164-174], electrophoretic stretching [175-185], hydrodynamic stretching [186-190], van der Waals interactions [191]. Molecular combing is straightforward: no chemical modification of DNA molecules or substrate is required. The process can nevertheless yield a well-dispersed parallel array of molecules on surfaces of different hydrophobicity or hydrophilicity. "Combed" DNA template molecules are not only oriented by combining but generally also become strongly bound to the substrate, which is favorable for subsequent metallization and nanowire characterization. Electrophoretic stretching with an AC field is useful for stretching and positioning nanowire templates directly between electrodes. The approach facilitates characterization of some properties of the resulting wires. However, no report on DNA-templated nanowire fabrication based on electrophoretic stretching has yet appeared. Spin stretching, a type of hydrodynamic stretching, does not require chemical modification of DNA. It has been used successfully in nanowire fabrication. These approaches are discussed in detail in the present section. The detailed molecular mechanics of these processes, though interesting, will not be covered here.

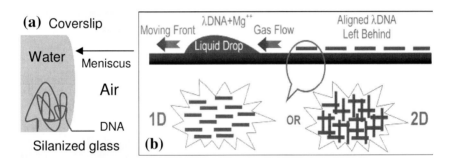

Figure 4. Molecular combing for DNA stretching. (a) Schematic diagram of the principle. Reprinted from ref. 1. Copyright 2006 IOP Publishing Ltd. (b) Schematic diagram of the "gas-flow" method of stretching. Reprinted from ref. 128 with permission. Copyright 2003 the American Chemical Society.

In molecular combing [164], a single DNA molecule or bundle of DNA molecules is stretched by a receding meniscus between a substrate and coverslip (Figure 4a). DNA is deposited on the substrate, for example silanzied glass, and the terminus of a molecule or bundle reacts with surface vinyl groups, anchoring the molecule or bundle to the substrate. Surface tension promotes the extension of DNA molecules during a dewetting step. The extension force is greater than the entropic force but smaller than the force needed to break covalent bonds.

Molecules thus become oriented parallel to the direction of meniscus movement.

The end-to-end distance of combed bacteriophage λ DNA molecules depends on the substrate surface [165]. In a recent study the length was 21–24 μm on a silanized hydrophobic surface and 16–18 μm on hydrophilic anti-dioxigenin. Other surfaces, for example polystyrene, polymethylmethacrylate (PMMA), and polylysine, are reportedly useful for molecular combing [166]. DNA combing efficiency was highest on a hydrophobic surface in pH 5.5 solution. Interactions with the hydrophobic surface allowed the molecule to be overstretched to 130–150% of the contour length. Gueroui *et al.* [167] have reported that a 1-dodecanol coating will reduce surface tension and decrease the end-to-end length of stretched DNA to close to its crystallographic length.

Various improvements on the basic concept of molecular combing have been developed. Examples include mechanical control of meniscus movement [168], movement of substrate rather than coverslip [169], and dipping of silanzied glass into DNA solution (rather than deposition of DNA on glass) followed by extraction of the glass at a constant rate [170]. The length distribution of DNA molecules can be narrowed by moving the meniscus at a steady speed. The constant extension force on the molecules is useful for template preparation for nanowire fabrication.

Li *et al.* have used a modification of molecular coming to align DNA molecules on mica [171]. Nitrogen gas flow was applied at a 45° angle to a DNA-coated surface. The DNA molecules extended to their full length as the liquid-air-mica interface was blown dry. A key advantage of this "gas-flow" method is ease of use. No modification of DNA or substrate is required. Deng *et al.* have used the gas-flow method to stretch DNA and fabricate 1D parallel and 2D crossed palladium nanowire arrays [128]. 2D DNA templates were prepared by applying gas-flow combing twice (Figure 4b). Compressed gas flow was applied to DNA solution on a mica substrate; the second DNA alignment was made by applying gas flow in the direction perpendicular to the parallel array of stretched DNA molecules. Kim *et al.* have reported a detailed comparative study on molecular combing, gas-flow combing, and spin stretching of DNA [172]. It was found that DNA molecules stretched on polystyrene or a polymethylmethacrylate substrate by the gas-flow were less extended than stretching by molecular combing or spin stretching.

A simple method has been developed for the reproducible creation of highly aligned DNA nanowires in the absence of surface modification or special equipment [173, 174]. Stretched DNA molecules on a polydimethylsiloxane (PDMS) surface were transferred to another surface by transfer printing from a relief pattern. Fluorescent microscopy and atomic force microscopy then revealed that many DNA molecules were highly aligned on the second surface. Two-dimensional assembly of DNA nanowires was realized by repeating the transfer process at 90° relative to the first transfer. The approach is potentially useful for the fabrication of DNA chips and functional electronic circuits of DNA-based 1D nanostructures.

Electrophoretic stretching involves stretching DNA molecules in an applied DC or AC field. In DC electrophoretic stretching, a uniform field results in the migration of polyanionic chains of DNA toward the anode, thus stretching the molecules [175-177]. This technique requires DNA immobilization in a gel matrix prior to stretching, making it less than ideal for the development of DNA-templated nanowires. In an AC field, by contrast, stretched DNA molecules are positioned between two electrodes and chemical modification of the molecules is not required.

Dielectrophoresis (DEP) is a non-uniform field effect which orients dipolar objects parallel to the direction of electric field [178-185]. When the field strength and frequency are high, DEP can be used to stretch DNA molecules to their full length. Washizu *et al.* [178-180] have studied DEP effects on DNA and found that large thermal fluctuation near the electrodes might be responsible for low efficiency stretching and positioning. Dielectrophoresis stretching has not yet been used for DNA nanowire fabrication. The floating electrode approach [180] could make dielectrophoresis for useful for this purpose.

Hydrodynamic stretching is another widely used method of stretching DNA molecules [186-190]. Dynamic flow is applied to DNA molecules which are tethered at one end. The drag force generated by the momentum gradient between the applied flow and the DNA chains align them in the direction of the flow. Extension of the molecules depends on the flow velocity and solution viscosity. A disadvantage of the method for nanowire fabrication is that the substrate or the DNA molecules must be chemically modified for anchoring.

Spin stretching is a simplified and effective hydrodynamic stretching method for aligning DNA molecules [189-190]. Droplets of DNA molecules are transferred onto silanized glass, which is then spun at 3-5 krpm. The rotational dynamic flow generates a viscous force, extending the DNA molecules. The average extension of DNA is greater at the periphery of the glass than near the spinning center, the higher linear speed at the periphery leading to a higher viscous force on the DNA molecules.

The development of straightforward and reliable methods of preparing the interface between a nanowire and a specific surface will be important for the development of nanoscale electronic devices. DNA-templated nanowires can be interconnected to electronic circuit elements by anchoring the DNA templates to a surface prior to metallization. In current DNA nanowire research, the major method for nanowire positioning is known as modified combing [102-103, 105, 111, 120, 122, 127-129, 152]. DNA molecules are deposited in droplets between electrodes and then stretched by drying. The process is simple and no modification of DNA is required, but the overall process tends to yield relatively poorly ordered nanowires.

DNA-based molecular recognition and gold-thiol coupling have been used to position single DNA molecules on gold microelectrodes [99]. In this approach "sticky ends" of double-stranded DNA, created by digestion with restriction enzymes, bond to oligonucleotides capped with a thiol group at their 3' ends by molecular recognition (Figure 5). A covalent bond is formed between the molecules by ligation, a process achieved with an enzyme called ligase. Dynamic flow perpendicular to the electrodes is then used to stretch the modified DNA molecules, which are bound to gold electrodes by gold-thiol coupling. The strength of this coupling prevents the DNA molecules from becoming unattached during subsequent template functionalization steps. Mbindyo *et al.* [192] have positioned oligonucleotide-modified gold nanowires on a 2D substrate coated with complementary oligonucleotides. After hybridization, only nanowires capped with oligonucleotides were positioned on the surface.

Highly efficient and precise positioning of DNA on a microstructured surface has been demonstrated recently. Maubach *et al.* [193] have used a positively charged surface to achieve positioning of exactly one DNA molecule between two electrodes. The same group [194] has utilized guiding microelectrodes to facilitate template alignment during combing, improving the yield of positioned DNA molecules.

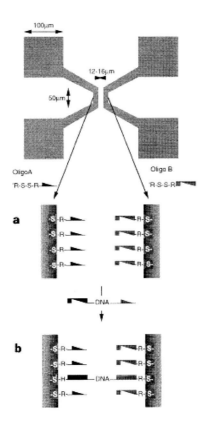

Figure 5. Attachment of DNA to gold electrodes. The top image shows the electrode pattern used in the experiments. (a) Oligonucleotides with two different sequences attached to the electrodes. (b) DNA bridge connecting the two electrodes. Reprinted with permission from ref. 99. Copyright 1998 Macmillan Publishers Ltd.

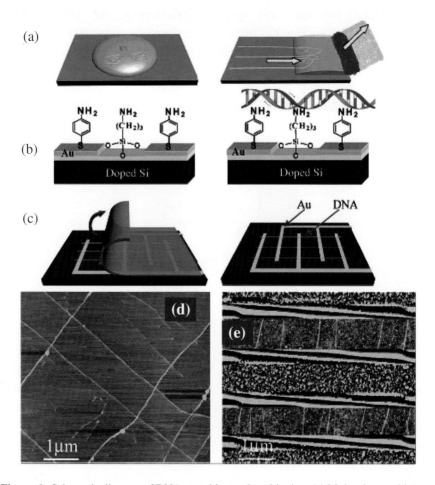

Figure 6. Schematic diagram of DNA stretching and positioning. (a) Molecular combing of DNA on a PDMS stamp; (b) silicon chip with patterned gold electrodes is modified to amino-terminated surface, which strongly binds DNA; (c) transfer aligned DNA from a PDMS stamp to a modified silicon chip by microcontact printing; (d) crossed DNA structures prepared by repetitive contact printing at right angles; (e) DNA strands positioned between gold electrodes. Reprinted from ref. 195 with permission. Copyright 2005 American Chemical Society.

Zhang *et al.* have adapted the methods described in refs 173-174 and demonstrated a multi-step approach for stretching and positioning DNA on gold electrodes that takes advantage of both molecular combing and microcontact printing [195]. Parallel DNA arrays were combed on a PDMS stamp (Figure 6a). Then, patterned gold electrodes on silicon were modified to yield an amino group-terminated surface (Figure 6b). Microcontact printing was then used to transfer aligned DNA molecules from the stamp to the silicon surface (Figure 6c). Figure 6d shows 2D crossed DNA structures stretched by repetitive contact printing. The second layer of aligned DNA was achieved by changing the direction of printing direction. Figure 6e shows stretched DNA molecules positioned between two gold electrodes. This method allows tuning the orientation of DNA strands relative to the electrodes by changing the printing angle. Moreover, control over the density, linearity, and extension of the stretched DNA molecules is greater than by other methods, in part because the PDMS surface is more uniformly hydrophobic than a silanized surface.

3.4. Nanowire Fabrication Methodologies

Functional DNA-templated nanowires are typically made of at least two types of material: DNA templates and nanoparticles having certain physical properties. In the case of electrical nanowires, for example, which are proposed to function as interconnecting elements in future nanocircuitry, the nanowire resulting from the fabrication process must be electrically conductive. The poor conductivity of unmodified DNA [196-200] prohibits its direct use in electronic circuits. Assembly of nanoparticles of good electrical conductors on linear template DNA, a process called metallization, enhances conductivity. Similarly, magnetic nanowires can be achieved by deposition of magnetic nanoparticles or nanoclusters on DNA templates.

Many different approaches have been developed to deposit target nanomaterials on DNA templates to achieve "functionalization." Figure 7 illustrates the three main methods used to fabricate DNA-templated nanowires. A common strategy is a two-step chemical process – nucleation and growth. The result is continuous nanowire in aqueous solution. In the first step, which is optional in some cases, DNA-nanoparticle complexes are prepared. These nanowire precursors then serve as active nuclei for autocatalytic growth on nanowires in an appropriate electroless plating bath.

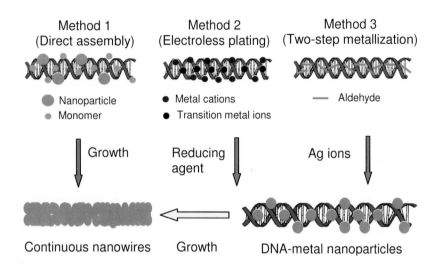

Figure 7. Schematic of the three main methods used to fabricate DNA-templated nanowires in a wet environment.

Method 1 in Figure 7 involves the direct assembly of nanoparticles on DNA. This approach is widely used to fabricate electrical, magnetic, semiconductor, and conducting polymer nanowires. Positively charged nanoparticles or polymer units bind to negatively charged DNA backbones (Figure 3b) by coulombic attraction. Positively charged gold nanoparticles have been used to make electrical nanowires [108-109, 111, 112]. Other kinds of positively charged nanoparticles, such as magnetic iron oxides [138-140], CdS [145], and CdSe [147-149], have been assembled into 1D nanostructures on DNA templates. Conducting polymer nanowires have been prepared by electrostatic interaction between charged moieties on non-DNA polymers [152-154] and the DNA backbone.

In different approaches, hydrogen bonding between complementary DNA bases, conventional gold-thiol coupling, and biotin-streptavidin binding have played an important role in work in nanowire studies

[114-117]. Gold nanoparticles, for example, have been attached to thiol-modified oligonucleotides which were then hybridized to single-stranded DNA (ssDNA) templates by molecular recognition. Streptavidin-coated gold nanoparticles have been used to bind biotinylated DNA, making use of the very strong association between streptavidin and biotin.

Method 2 involves multiple chemical reactions. In the first step, metal cations bind to DNA phosphate groups or chelating nitrogenous bases (Figure 3b) and thereby activate the DNA template. Next, metal ions bound to DNA are reduced to nanoparticles. Finally, nanoparticles or nanoclusters thus formed on DNA catalyze nanowire growth in an electroless plating baths. This general approach is also called electroless plating. The activation step is crucial for nanowire growth. It has been shown that nanoparticles or clusters do not deposit on non-activated DNA [126, 127, 130-133, 141, 142]. Metal cations such as Ag(I) [99], Cu(II) [134-137], Cd(II) [144, 151] have been used to bind to polyanionic DNA by electrostatic interaction.

Certain transition metal ions bind to the nitrogen atoms of the DNA bases and form metal-DNA complexes by coordination coupling involving to d orbitals. The N7 atoms of the bases guanine and adenine form strong complexes with Pt(II) and Pd(II) ions [201-203], and the N3 atoms of the bases thymine and cytosine strongly interact with Pd(II) ions [204], Pd [126-129] and Pt [130-133]. Such binding has enabled nanowire fabrication on the basis of coordination coupling. The advantage of this method is that certain metal nanoparticles can be used to catalyze growth of nanowires made of different materials ("heterogeneous growth"). Magnetic Co [141] and Ni [142] nanowires have been fabricated by catalysis of Pd nanoparticles on DNA templates.

In the first step of Method 3, known as "two-step metallization," a reducing agent substitutes for the nanoparticles or ions in Methods 1 and 2. The reducing agent glutaraldehyde, for example, attaches to DNA by a Michael-type addition. During incubation of DNA-aldehyde with Ag(I) solution, Ag(0) nanoclusters deposit uniformly on DNA by virtue of localization of the reducing agent. This method has been used successfully to prepare silver [100-105] and gold [100] nanowires. In this case gold nanowire growth was based on the binding of silver nanoparticles to DNA. Both double-helical λ DNA and synthetic DNA templates, such as 4 × 4 DNA tiles and triple-crossover DNA tiles, have been used in silver nanowire fabrication.

Most DNA-templated nanowires have been fabricated by the three methods illustrated in Figure 7. Some less often used methods have also been used with success. Quake *et al.* [119], for example, have developed a dry method to metallize DNA: sputtered gold was deposited directly on a DNA molecule stretched between two electrodes. Later, this method was used to fabricate superconducting $Mo_{21}Ge_{79}$ nanowires [160], which are potentially useful in the development of quantum interference devices. Patolsky *et al.* [120] have developed a UV-assisted approach to nanowire assembly. Berti *et al.* [121] have also used UV light to achieve highly efficient reduction of Ag(I) ions bound to λ DNA.

Polymer nanowire fabrication is discussed in Section G. Sections D-F discuss details of electrical, magnetic, and semiconductor nanowire fabrication and physical properties of the resulting structures.

3.5. Electrical Nanowires

A major potential application of DNA-templated nanowires is to provide interconnection between devices in future nanoelectronic circuits. These nanowires must be conductive. As noted above, unmodified DNA is as an electrical insulator. Study in the electrical DNA-templated nanowire field has focused on converting DNA templates in conductors. Silver [99-107, 121, 135], gold [100, 101, 106, 108-125], palladium [126-129], platinum [130-133], and copper [134-137] nanowires have been fabricated from DNA templates. The conductivities of these nanowires have been measured on model devices with the nanowire providing interconnection. The methods shown in Figure 7 are main ones used in the assembly of electrical nanowires.

3.5.1. Silver/Gold

100 nm-wide conductive nanowires of silver have been fabricated on molecular DNA templates stretched between micro-electrodes [99]. Figure 8 shows the microstructure and I-V properties of these nanowires. The metallization process used in this work (Method 2) has become a

model for DNA-templated nanowire fabrication in a wet environment. Two 12-mer oligonucleotides capped with a thiol group were hybridized and ligated to λ DNA templates. The templating DNA was then attached by gold-thiol coupling to gold electrodes separated by a 16 μm gap (Figure 5). Dynamic flow of solvent was used to orient and stretch the modified λ DNA molecules between the electrodes. During metallization, silver cations became bound to DNA by Ag/Na ion exchange. Hydroquinone was then used to reduce the Ag(I)-DNA complexes. The resulting Ag(0) nanoparticles nucleated further metallization of DNA, creating a nanowire. A significant decrease in resistance was observed. The nanowire shown in Figure 8a had a resistance of 30 MΩ (Figure 8c). A resistance of 7 MΩ was found following further deposition of silver (Figure 8d). I-V behavior was ohmic (dashed line in Figure 8d). The negative control shows that DNA was insulating prior to metallization (insets in Figure 8d). The calculated resistivity of these nanowires, 3.4×10^{-3} Ω-m, is higher than that of bulk silver. Drawbacks of the method include low shape uniformity and low electrical conductivity.

An advantage of nanowires made with DNA is that highly specific molecular recognition can be used for precise patterning of the interwiring components in the development of complex nanoelectronics. Keren *et al.* [100] have developed "sequence-specific molecular lithography" for selective metallization of molecular DNA templates. The approach is a novel tool for the self-assembly of molecule-sized devices. RecA bind to segments of DNA having a specific nucleotide sequence, forming nucleoprotein filaments (2027 bases, ~689 nm). See Figure 9a(i). Bound RecA filaments then functioned as photoresist in conventional lithography, the RecA protein preventing the binding of aldehyde to DNA (Figure 9a(iii)). Bound aldehyde then reduced Ag(I) to Ag(0) in regions of DNA not coated by RecA. Silver nanoparticles could not growth on the RecA-DNA segment during silver metallization (Method 3). See Figure 9a(iv). Panels c-f of Figure 9 show RecA-DNA, Ag(0)-DNA, and Au(0)-DNA structures prepared by this approach to molecular lithography. An insulating gap was present on the DNA nanowire at sites occupied by RecA after silver and gold metallization (Figure 9d-f). Single silver nanowires with multiple insulating gaps were fabricated by engineering DNA molecules [101]. DNA fragment A (2052 bp, ~700 nm, treated with aldehyde) and DNA fragment B (1106 bp, ~350 nm, not treated with aldehyde) were digested with restriction enzymes and ligated. This led to longer pieces of DNA which alternated between A and B. Figure 9b shows that molecular lithography with RecA yielded 350 nm insulating gaps on silver nanowires.

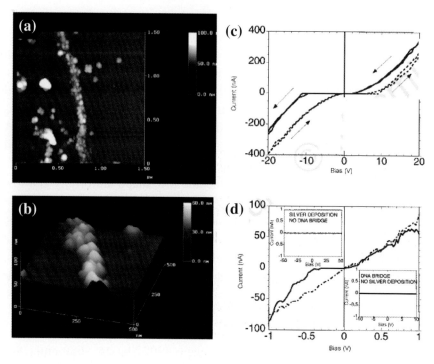

Figure 8. The first DNA-templated nanowires. (a) AFM micrograph of a 100-nm wide silver nanowire; (b) a three-dimensional magnified view of the same structure; (c-d) I-V characteristics of the nanowires. Arrows indicate the direction of voltage scans. Reprinted with permission from ref. 99. Copyright 1998 Macmillan Publishers Ltd.

Molecular lithography offers exciting new possibilities for precise patterning with nanowires at the molecular level. Fabrication by Method 3, however, though it allows confinement of grown metal nanoparticles on DNA templates, is limited if the nanowires are fabricated in aqueous solution. Metal ions are inevitably reduced in solution and deposit at random locations on the substrate. This leads to nanowires that display highly heterogeneous structures, for example irregular branching and necklace structures (Figure 8a). In Method 3, reduced Ag(0) clusters deposit predominantly on the template where the reducing agent is localized prior to metallization. As Figure 9f shows, it was possible to grow gold nanowires with low background by the catalysis of localized Ag(0) nuclei.

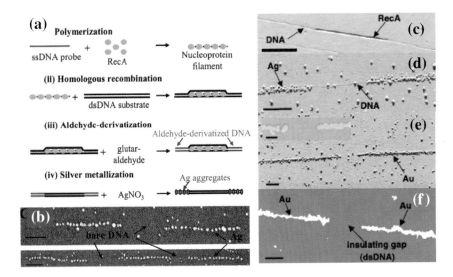

Figure 9. Patterned silver and gold nanowires by molecular lithography. (a) Schematic diagram of the process; (b) Ag(0) deposited on ligated DNA fragments. Upper panel: Two 700 nm silver clusters are separated by a 350 nm gap. Lower panel: Three 700 nm silver clusters are separated by two 350 nm gaps. (c) ~2 kb RecA nucleoprotein filament bound to λ DNA; (d) silver nanoparticles grown on naked DNA; (e-f) AFM and SEM images of the nanowires after gold metallization. Scale bars: 330 nm in (b); 500 nm in (c-f); 250 nm in inset of (e). (a-b) reprinted with permission from ref. 101. Copyright 2004 American Chemical Society; (c-f) reprinted from ref. 100. Copyright 2002 American Association for the Advancement of Science.

Double-helical λ DNA has been the most widely used source of DNA for nanowire templates. Synthetic DNA sequences have also proved interesting for nanotechnology [102-104]. Yan *et al.* [102], for example, have designed and assembled ~5 µm long DNA nanoribbons with a uniform width of ~60 nm. This novel DNA structure has crystal-like lattices with periodic square cavities. The basic unit is 4 × 4 tile, a four-arm junction made of nine oligonucleotides (Figure 10a-b). The two-step silver metallization process (Method 3) was used to produce highly conductive nanowires out of the DNA ribbons. Figures 10c and 10d show DNA nanoribbons before and after metallization, respectively. The micron-long silver nanowire has the average height of 35 nm and the width of 43 nm (Figure 10d). Relative to bare DNA, the metallized nanowires are taller and wider than the non-metallized structures; metallized 4 × 4 nanoribbons do not life flat on the surface. I-V measurement showed linear ohmic behavior for the silver nanowires in Figure 10e. The measured resistance of 200 Ω is similar to the bulk electrical resistivity of silver.

DNA tiles, well-organized DNA nanotubes [103], and DNA filaments [104] have shown promise as nanowire templates. The structures were assembled from synthetic oligonucleotides having programmed sequences to control base pairing. Two types of triple-crossover tiles and 1D triple-helix bundle (3HB) tiles are shown in Figures 11a and 11e, respectively. These unit tiles have been assembled into micron-long, lattice-like nanowire templates. "Nanotubes" with three coplanar double-helical domains, anti-parallel strand exchange points (the strands change direction on crossing over), and an odd number of helical half-turns between crossover points (TAO nanotubes) have been prepared. These structures have a uniform diameter of ~25 nm (Figure 11c-d). 3HB filaments have a diameter of ~4 nm, corresponding to the width of single DNA tile (Figure 11f). These synthetic DNA-based structures have been metallized with silver by a two-step metallization process (Method 3). The resulting metallized TAO nanotubes were ~35 nm high, ~40 nm wide, and up to ~5 μm long. I-V measurement revealed a resistivity of 1.4-3.2×10^{-5} Ω-m, significantly higher than that of bulk polycrystalline silver (1.6×10^{-8} Ω-m). The silver nanowires formed from 1D-3HB DNA filaments had an average diameter of ~ 30 nm and length of ~2 μm. The diameter ranged from ~20 nm to ~ 50 nm. I-V measurement has been done on electrodes with different gap spaces. These nanowires displayed ohmic behavior and had a resistivity of 2.25-2.57×10^{-6} Ω-m, much higher than the bulk resistivity of silver. The morphology of the silver nanowires assembled from DNA nanotubes and filaments was very similar to the nanowires in Figure 11d-e. Nanowires built on synthetic DNA tiles are generally more uniform and less grainy and than nanowires fabricated from B-form ds-DNA templates. A plausible explanation is that the wider DNA tile

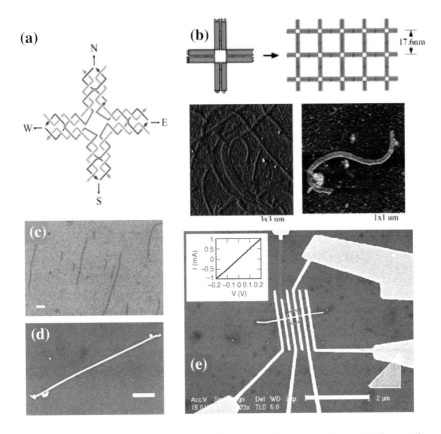

Figure 10. Silver nanowires fabricated on DNA nanoribbons. (a) The 4 ×4 tile contains nine oligonucleotides which form a four-arm junction; (b) schematic and AFM images of the self-assembled nanoribbon structures; (c) SEM image of the 4 ×4 nanoribbons prior to metallization; (d) SEM image of metallized DNA nanoribbons; (e) a silver nanowire deposited on electrodes and its I-V characterization. Scale bars: 500 nm in (c-d). Reprinted from ref. 102. Copyright 2003 American Association for the Advancement of Science.

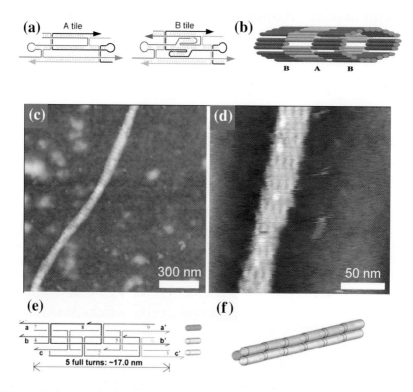

Figure 11. DNA nanotubes and filaments for nanowire templates. (a) Two types of triple-crossover DNA tiles. Each tile is made by four oligonucleotides. (b) 25 nm-wide TAO nanotube with eight tiles each layer; (c-d) AFM images of TAO nanotubes; (e) detailed structure of a one-dimensional three-helix-bundle DNA tile, composed of nine oligonucleotides; (f) a three-helix-bundle DNA filament nanowire template. (a-d) reprinted with permission from ref. 103. Copyright 2004, National Academy of Sciences, USA. (e-h) reprinted from ref. 104. Copyright 2005 American Chemical Society.

assemblies have a greater capacity to bind aldehyde, leading to greater deposition of silver nuclei in the first step of metallization.

The two-step metallization process (Method 3) on B-DNA or synthetic DNA localizes aldehyde on the template and thereby limits metal deposition off the template. Park *et al.* have optimized this method on λ DNA and on synthetic ds-DNA [105]. The diameter and length of silver nanowires have been found strongly dependent on DNA-glutaraldehyde complexation time, dialysis time, and incubation time of

DNA with Ag(I). Nanowires fabricated under optimal conditions were 3 μm long and 35 nm wide. Thinner, 15 nm wires have also been fabricated. Two-terminal I-V measurement indicated linear ohmic behavior of the nanowires at 300 K and 77 K. Fischler *et al.* [106] have modified the two-step metallization process with synthetic DNA (900 bps) containing alkyne-modified cytosines. Click chemistry was used to convert the alkyne groups on DNA to sugar triazole derivatives, which then reacted by a Tollens reaction to produce tiny silver metal deposits. Uniform bimetallic silver/gold nanowires having a diameter below 10 nm were prepared by this approach.

Direct assembly of nanoparticles on DNA (Method 1) has also been successful for gold nanowire fabrication. Four mechanisms have been used in the metallization process: electrostatic interaction between positively charged gold particles and polyanionic DNA, molecular recognition of DNA, gold-thiol coupling, and biotin-streptavidin interaction. Kumar *et al.* [108] and Sastry *et al.* [109], for example, have reported that linear arrays of lysine-coated colloidal gold (~4 nm) can be assembled on DNA by electrostatic attraction between the polyanionic DNA backbone and the positively charged gold nanoparticles. Harnack *et al.* [110] have used 1-2 nm trisphosphine-labeled gold particles to grow continuous nanowires. The gold nanoparticles bound to DNA with high density, providing numerous seeds for catalyzing nanowire growth. The resulting nanowires were 30-40 nm wide and showed ohmic behavior with a conductivity of 3×10^{-5} Ω-m. Ongaro *et al.* [111] have fabricated 20-40 nm gold nanowires in the catalysis of 4-(dimethylamino) pyridine-stabilized gold nanoparticles. The positively charged gold nanoparticles bound to calf thymus DNA by electrostatic attraction prior to nanowire growth. A method of forming highly ordered assemblies of gold nanoparticles along DNA molecules on substrates has been developed by Nakao *et al.* [112]. Well-stretched DNA templates were used to achieve well-aligned assemblies with long-range order. Also, oxidized aniline-capped gold nanoparticles became strongly attached to DNA through electrostatic interaction. Two different assembly methods were carried out, resulting in continuous deposition and necklace-like deposition of gold nanoparticles along DNA molecules.

Pre-programmed conductive nanowires are needed for the rational fabrication of nanoscale electronics based on molecular assembly. DNA is useful for the purpose. Nishinaka *et al.* [113] have described a method which relies on the DNA-RecA nucleoprotein filament template. DNA-functionalized gold nanoparticles were reacted with the thiol groups of cysteine-modified RecA derivatives (Method 1). The gold nanoparticles were then enhanced by chemical deposition to create uniformly metallized nanowires. The DNA-RecA protein templates have several advantages over DNA alone: higher stability, greater stiffness, increased homologous paring for site-selective metallization and junction formation. Site-selective gaps and three-way junctions were successfully demonstrated.

High uniformity and high conductivity important aims of electrical nanowire fabrication. ssDNA templates of repetitive sequence might be useful for achieving these aims by promoting the binding of nanoparticles to DNA with uniform inter-particle distance. This will in turn enable the fabrication of uniform nanowires by catalysis by highly ordered metal nuclei. Weizmann *et al.* [114] have used the ribonucleoprotein telomerase to synthesize ssDNA having a repetitive sequence. 1.4 nm gold nanoparticle-functionalized oligonucleotides were hybridized to the complementary sequences of the telomeric repeat unit (Method 1). Uniform gold nanowires (50-80 nm wide) were fabricated by catalytic gold enhancement on the template. In a second approach to metallization, amine groups were added to the template by telomerization in the presence of a modified nucleotide, namely, aminoallyl-functionalized dUTP. Active ester-modified gold nanoparticles then became covalently bound to amine groups in the template.

Beyer *et al.* [115] have synthesized long ssDNA nanotemplates by rolling circle amplification (RCA). In this efficient biological method, DNA polymerase continues copying a 74-nucleotide long circular DNA to generate a single strand with a length of up to several microns. The sequence is repetitive (Figure 12a). Biotinylated complementary oligonucleotides were hybridized to the ssDNA template (Figure 12b). ~5 nm streptavidin-tagged gold nanoparticles were attached to the ssDNA by binding to biotin (Figure 12c). 1D arrays of gold nanoparticles were thus assembled on the DNA template with an interparticle spacing of 30-50 nm.

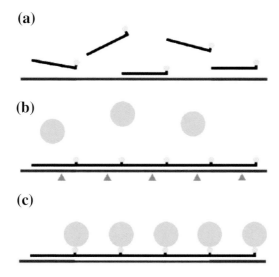

Figure 12. (a) ssDNA template with repetitive sequence (red); (b) biotinylated oligonucleotides (black) bind to the ssDNA template; (c) streptavidin-coated gold nanoparticles bind to the template by biotin-streptavidin interactions. Reprinted from ref. 115. Copyright 2005 American Chemical Society.

In a similar study, Deng *et al.* [116] used RCA to fabricate micron-long repetitive ssDNA templates for gold nanoparticle assembly. Thiolated 53-base oligonucleotides were attached to 5 nm gold nanoparticles by gold-thiol coupling. Templates having repeat units complementary to the gold-capped oligonucleotides were synthesized by RCA to be several microns long. Linear gold nanoparticle arrays with an inter-particle spacing of ~18.5 nm (corresponding to the length of the oligonucleotides) were formed by spontaneous base pairing by molecular recognition.

Li *et al.* [117] have utilized DNA triple crossover molecules to assemble highly ordered streptavidin and gold nanoparticle arrays. Such DNA molecules consist of three double-helix strands (in contrast to one double helix in B-DNA) and form three layers of helices in a single plane. The top and bottom layers were biotinylated. Two layers of streptavidin (~4 nm) in a linear array were assembled on the DNA template by interaction with biotin. In addition, streptavidin-conjugated gold nanoparticles were assembled into 1D strucutres. This assembly strategy could be useful for the development of a programmable logic in molecular-scale electronic circuits.

A drawback of DNA-templated nanowires is the necklace-like inter-particle spacing, which contributes to low electrical conductivity. Use of ssDNA prepared by the RCA method outlined above is one way of addressing the matter. Other approaches have been developed. The nanoconjunction method proposed by Lee *et al.* [118], for example, could be useful for "soldering" nanoscale spaces between adjacent particles. Oligonucleotide-modified gold colloids (8 nm) were linked to 12-, 21- and 42-mer oligonucleotides. Hybridization resulted in gold nanoparticles linked to duplex DNA molecules. Silver ions bound to the DNA were reduced to crystalline silver. For the 12- and 21-mer oligonucleotides, silver metal deposited on both the gold surface and on the DNA, fully capping the original gold nanoparticles and soldering the spaces between two adjacent gold nanoparticles. For the 42-mers, the spaces were not well soldered because silver preferred to deposit on gold. This method of silver nanoconjunction could be useful for repairing nanowire defects and achieving higher nanowire conductivity.

Some methods other than those illustrated in Figure 7 have yielded promising results. Quake *et al.* [119] have reported that gold evaporation on a single DNA template can be used to form thin nanowires which display quantum mechanical properties. DNA molecules were first stretched across a 60 nm-wide trench created by EBL. Evaporation of gold at an oblique angle led to 5-8 nm-wide nanowires. Interestingly, "diving-board" nano-resonators (i.e. broken nanowires) were detected over the trench. The authors considered it possible to fabricate high frequency (GHz) mechanical resonators out of these over-hanging nanowires. In any event, the approach represents yet another means of metallizing DNA: evaporation or sputtering a target metal on DNA molecules stretched across a trench. The expected advantage of the approach is that nanowires thus prepared can be very thin and highly uniform.

Patolsky *et al.* [120] have developed an approach of binding nanoparticles directly to DNA bases. ~1.4 nm gold particles labeled with amino psoralen become intercalated into a poly-A/poly-T DNA duplex by a photochemical reaction. UV irradiation then catalyzes covalent binding of amino psoralen to bases of DNA. Berti *et al.* [121] have used light at 254 nm to photoreduce Ag(I) ions which are bound to λ DNA by electrostatic attraction. The high reduction efficiency is associated with the DNA bases, which act as light sensing promoters. Irradiation at 388 nm wavelength, by contrast, does not result in the formation of silver nanoparticles on DNA. Typical nanoparticles thus formed had a diameter of 1.5-3 nm, independent of Ag(I) concentration and irradiation time. Changes in experimental conditions did influence homogeneity of coverage of DNA templates.

3.5.2. Palladium/Platinum/Copper

These DNA nanowire systems are formed by a selectively heterogeneous, template-controlled mechanism. Multi-step chemistry (Method 2) is the most common approach. DNA "platination," for example, refers to the direct bonding of Pt(II) ions to electron pair "donor" atoms (e.g. N and O) within the DNA molecule. The N7 atoms of purines (G and A) are favorable targets [201-203]. Such reactions are known by their relevance to the mechanism of certain anti-cancer drugs. Complex formation requires a labile ligand as a leaving group within the ligand coordination sphere. The other transition metal ion, Pd(II), binds to the N3 atoms of the bases thymine and cytosine by coordination coupling [204]. The complexation of Pt(II) or Pd (II) with DNA is a form of "activation" of the DNA template. Activation allows control over the fabrication process. Metal nanoparticles become deposited primarily on activated DNA during the electroless plating step, leading to nanowires with low background and high homogeneity.

Richter *et al.* [126] have studied the formation of nanoscale palladium clusters on a DNA template. A distinction is made between chains of separated clusters and a continuous coating to give a metal nanowire. Cluster deposition was used to metallize λ DNA (Method 2). This was accomplished by activating the DNA with palladium ions and then immersing the complex in a reducing bath. The result was the formation of ~50 nm Pd nanoclusters on the DNA. Increasing the duration of the reduction process led to well-separated palladium clusters on the DNA which became quasi-continuous with a grain-like structure.

Single double-stranded DNA molecules have been used to fabricate metallic palladium nanowires [127]. The DNA molecules were positioned between macroscopic gold electrodes prior to metallization. Low-resistance electrical interfacing was obtained by pinning the nanowires to the electrodes with electron-beam-induced carbon lines. The wires had an average diameter of 50 nm and an estimated conductivity of 2×10^4 S cm^{-1} (Figure 13). Transport behavior was ohmic at room temperature. Specific conductivity was only one order of magnitude below that of bulk Pd.

Deng and Mao have reported a simple method for fabricating 1-D parallel and 2-D crossed metallic nanowire arrays [128]. Molecular combing was used to form these ordered nanostructures by stretching and aligning linear DNA molecules into parallel or crossed patterns. The organized DNA molecules were subsequently metallized by electroless deposition of palladium.

The electrical conductivity of DNA-templated palladium nanowires has been measured at different temperatures [129]. Nanowires with a diameter of ~60 nm after fabrication by Method 2 were pinned between two electrodes with a space of 5 μm. These nanowires exhibited ohmic transport behavior at room temperature, and resistance decreased linearly with decreasing temperature. Below 30 K, a logarithmic increase of resistance was found with decreasing temperature. This quantum effects in a disordered metallic film. Annealing of palladium nanowires was found to improve conductivity by increasing structural order. A 9-10 fold decrease in resistance was found after heating at 200 °C.

The synthesis and preliminary characterization of platinum nanoparticles on DNA has been achieved by chemical reduction of platinated DNA (Method 2) [130-133]. Activation of DNA is a critical step in this approach to DNA nanowire assembly. The character of metal ion binding to DNA determines the coverage of metal deposits on the templates and the heterogeneity of the resulting nanowires. Ford *et al.* [130] have used Method 2 to reduce Pt(II)-activated DNA with sodium borohydride. The result was formation of ~1 nm Pt nanoclusters on the DNA. Electroless plating with gold resulted in 4-6 nm gold deposits on the template. Gaps were present between the resultant nanoparticles, but the chains could still serve as precursors for nanowires or display interesting properties for quantum electronics behavior, which differs from that of most bulk metals, was ascribed to Seidel *et al.* [131] have

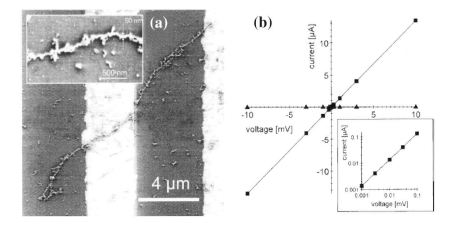

Figure 13. Palladium nanowires. (a) A single nanowire fabricated between two electrodes. Inset shows a nanowire with a diameter of ~ 50 nm. (b) I-V curve of a palladium nanowire. The resistance is 743 Ω, corresponding to a conductivity of 2×10^4 S cm^{-1}. The sample behaved as an insulator after the wire was severed. Reprinted from ref. 127 with permission. Copyright 2001 American Institute of Physics.

investigated effects of DNA activation on the heterogeneity of the nanowires. It was found that DNA was not metallized by Pt(0) without prior activation by Pt(II). By contrast, DNA activated with Pt(II) at 37 °C for 16 h was uniformly metallized by 3-5 nm Pt(0).

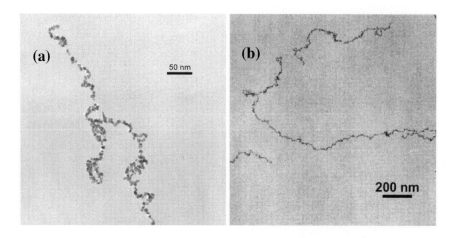

Figure 14. TEM micrographs of platinum nanoclusters on DNA. Pt(II)-DNA complexes in solution were treated with (a) dimethylaminoborane or (b) citrate. Reprinted from ref. 133 with permission. Copyright 2004 American Chemical Society.

The microscopic mechanism of platinum cluster nucleation on DNA templates has been studied by first-principle molecular dynamics simulations [132]. The authors found that Pt(II) complexes bound to DNA can form strong platinum-platinum bonds with free platinum complexes after a single reduction step, and may thus act as preferential nucleation sites. The hypothesis was corroborated by a series of experiments in which purely heterogeneous platinum growth on DNA was achieved. A long activation time was required to deposit Pt(0) nanoparticles able to initiate growth of nanowires with high coverage and heterogeneity. The activation process controlled the structure and heterogeneity of Pt(0) clusters on DNA after chemical reduction. Metal cluster necklaces of exceptional thinness and regularity were thus fabricated.

Seidel and co-workers have investigated conditions for growing chains of nanosized clusters of platinum on DNA templates [133]. The metallization procedure consists of just a few efficient steps (Method 2). A relatively long incubation period for complexation of dsDNA molecules and Pt(II) complexes was necessary to obtain template-directed formation of chains having thin and uniform clusters of complexes. Figure 14a shows uniform ~5 nm clusters of platinum on DNA. The clusters were acquired by treating Pt(II)-DNA with dimethylaminoborane. The platinum ion to base ratio was 65:1. Addition of citrate stabilized the metal clusters and yielded metallized chains of ~5 nm having increased structural uniformity (Figure 14b). Chemical reduction of the DNA/salt solution by citrate was done prior to the incubation step. In the absence of this chemical "activation," DNA acts a non-specific capping agent for the complexes and random formation of aggregates occurs. Base stacking along the DNA molecule was significantly distorted by complexation with Pt(II) but the molecule remained double-stranded. Systematic variation of the concentration of platinum salt, DNA, and reducing agent led to the conclusion that there is a moderately broad optimum concentration value for fabrication of comparable quality of clusters on DNA.

Copper has a low dielectric constant and is currently used for interconnection of devices in integrated circuits. Monson and Woolley have deposited copper onto surface-attached DNA to form nanowire-like structures that are ~3 nm tall [134]. DNA was first aligned on a silicon surface and then treated with aqueous $Cu(NO_3)_2$. Cu(II) associates with the DNA template by coulombic attraction (Method 2). Copper was then reduced by ascorbic acid to form a metallic sheath around the template. A more complete coating was obtained by repeating the Cu(II) and reduction steps. Control experiments involving treatments with aqueous NO_3^- or Mg^{2+} solutions showed no change in DNA height on exposure to reductant.

Becerril *et al.* [135] have described an "ionic surface masking" method to reduce off-template deposition in the fabrication of silver and copper nanorods. Non-specific background deposition of metal is low in this procedure. The template for the silver nanorods was single-stranded DNA, whereas that of the copper nanorods was double-stranded DNA. Alkali metal cations with high affinity for SiO_2 were used to passivate the silicon surface, creating a physical and electrostatic barrier against non-specific adsorption of Ag^+ or Cu^{2+} and subsequent deposition of metal. In silver nanorod synthesis there was a 51% reduction in the number of non-specifically deposited nanoparticles and an even greater decrease in their dimensions. With copper, there was a 74% decrease in the number of non-specifically deposited nanoparticles. The ability to fabricate metallic nanorods with ssDNA templates could be combined with direct surface hybridization of oligonucleotide-coupled, electronically active nanostructures at predetermined positions on single-stranded DNA, yielding nano-scale electronic circuits.

Table 2. Properties of Some DNA-templated Electrical Nanowires

Metal	Template	Method	Width (nm)	Resistivity (Ω-m)	Ref.
Ag	λ-DNA	2	~100	3.4×10^{-3}	99
	DNA nanoribbon	3	~ 43	2.4×10^{-6}	102
	DNA TAO nanotube	3	~40	$1.4\text{-}3.2 \times 10^{-5}$	103
	1D 3HB DNA	3	~30	$2.3\text{-}2.6 \times 10^{-6}$	104
	λ-DNA	3	~35	$1\text{-}2 \times 10^{-5}$	105
Au	λ-DNA	3	50-100	1.5×10^{-7}	100
	Calf thymus DNA	1	30-40	3×10^{-5}	110
	Calf thymus DNA	1	20-40	2×10^{-4}	111
Pd	λ-DNA	2	~50	5×10^{-7}	127
Pt	λ-DNA	2	1-5	NA	133
Cu	λ-DNA	2	~3	NA	134

As we have seen, conductive metallic nanoparticles or nanoclusters can be assembled on molecular DNA templates by direct assembly (Method 1), electroless plating (Method 2), and two-step metallization (Method 3) approaches. Table 2 summarizes the various electrical nanowires that have been made by these methods. In general, nanowire uniformity depends on DNA activation, the nanoscale nuclei, and the template molecules themselves. Branching structures, uneven interparticle distances, and background deposition can occur. Method 3 is especially advantageous for formation of uniform nanoclusters on DNA because the reducing agent is bound to templates prior to metallization. This method, however, has not been used to fabricate nanowires other than silver. The more flexible Method 2 has been used to fabricate nanowires of palladium, platinum, and copper, as well as cobalt and nickel (see next section). Improvement in confinement of deposit onto DNA is needed for further development of nanowires prepared by the template-based approach.

To conclude this section, synthetic DNA structures such as nanoribbons, TAO nanotubes, and 1D 3HB filaments seem more advantageous than λ DNA for fabricating highly uniform and conductive silver nanowires. The wider synthetic templates can probably bind more aldehyde than intrinsic B-DNA. As a result, more silver nanonuclei deposit on the DNA, resulting in increased uniformity of nanowire growth. The lowest resistivity achieved with these nanowires is ~2-3 × 10^{-6} Ω-m, about two orders of magnitude greater than the resistivity of bulk polycrystalline silver. Highly conductive gold and palladium nanowires seem promising for nanoelectronics development. Although there is no report on the electric measurement of DNA-templated nanowires made of platinum, it should be possible to prepare highly uniform and conductive nanowires with this metal. Ultra-thin and uniform-spaced platinum nanoparticles have been formed on DNA by tuning reaction conditions.

3.6. Magnetic Nanowires

Development of magnetic nanowires will be important for high-density memory storage devices and magnetic field sensors [26-28]. One-dimensional nanostructures of iron, cobalt, and nickel exhibit directional anisotropy, as do bimetallic nanowires of these elements. Remanence ratio and coercivity are greatly enhanced on the longer axis. Electrochemical deposition has been the main approach in magnetic nanowire fabrication. The quality of electrodeposited nanowires has depended strongly on pore uniformity and size distribution of the membrane templates. Here, we summarize attempts to fabricate magnetic nanowires by combining DNA-templating and Fe_3O_4 [138], Fe_2O_3 [139-140], $CoFe_2O_3$ [140], cobalt [141, 143], or nickel [142]. The methods used are direct assembly (Method 1) and electroless plating (Method 2) in aqueous solution. As in electrical nanowire fabrication, DNA bases and phosphate groups are the target binding sites for template functionalization with magnetic materials.

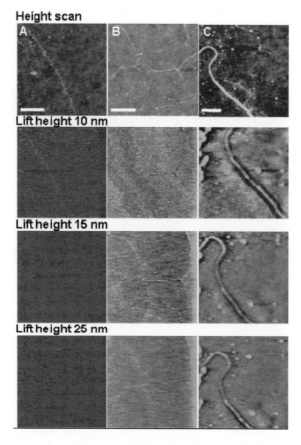

Figure 15. Magnetic force microscopy images of DNA-templated magnetic nanowires. Column A, lift height effect on 5 nm gold nanoparticles on DNA. Column B, MFM images of Fe_2O_3 nanoparticles on DNA at different heights. Column C, MFM images of $CoFe_2O_3$ nanoparticles on DNA at different heights. Scale bar is 500 nm. Reprinted from ref. 140 with permission. Copyright 2007, American Chemical Society.

Nyamjav *et al.* [138] have employed direct assembly (Method 1) to fabricate magnetic nanowires from Fe_3O_4. Positively charged Fe_3O_4 nanoparticles of 1.5-8 nm diameter were uniformly coated onto DNA in solution by electrostatic attraction. Magnetic properties of the resulting nanowires were measured by magnetic force microscopy (MFM). The results showed that the nanoparticles were densely packed onto the DNA templates. The force gradient between the MFM tip and the sample was detected by monitoring the frequency shift in locked phase. At a height of 25 nm, individual Fe_3O_4 nanoparticles in the topographic image appeared continuous, indicating magnetic interaction.

Kinsella *et al.* have studied how DNA coated with magnetic nanoparticles remains biologically active and accessible to the Bam HI restriction enzyme [139]. Long DNA molecules were coated with 4.1 nm diameter F_2O_3 nanoparticles by electrostatic attraction. Coated, stretched, and surface-bound DNA was then incubated with Bam HI, a restriction enzyme that specifically recognizes the based sequence GGATCC and cuts the DNA into two pieces. The authors showed that despite the presence of nanoparticles on the DNA, the enzyme was still able to recognize the cleavage site and effectively digest the assembly. The result would suggest that the binding affinity of the nanoparticles for DNA is not high.

Others have fabricated magnetic Fe_2O_3 and $CoFe_2O_3$ nanowires on DNA templates by direct assembly [140]. Pyrrolidinone-coated Fe_2O_3 having a diameter of ~4.1 nm and 3.4 nm-diameter $CoFe_2O_3$ nanoparticles were bound to λ DNA by electrostatic interaction. A superconducting quantum interference device indicated that both Fe_2O_3 and $CoFe_2O_3$ nanoparticles are superparamagnetic at room temperature. For a control, polylysine-coated gold nanoparticles were also assembled on DNA in the same way. The structure and magnetic properties of the nanowires thus formed was studied by MFM. Figure 15 shows a series of MFM images for the three types of nanowires at different heights. Columns A, B, and C show the phase shifts of gold, Fe_2O_3, and $CoFe_2O_3$ nanoparticles on DNA, respectively. The top row shows topography. The Fe_2O_3 column has a much stronger phase shift than the gold column, indicating a gradient in the stray field. As height increases, the phase shift becomes weaker. Phase shifts in the $CoFe_2O_3$ column are stronger than in the Fe_2O_3 column, because $CoFe_2O_3$ has a significantly higher saturation magnetization than Fe_2O_3. A signal is clearly detectable at a height of 25 nm. Multiple domains in the $CoFe_2O_3$ nanowires are apparent in column C, indicated by a difference in brightness. These measurements confirm that the deposits on DNA templates were in fact magnetic.

Heterogeneous nucleation and growth has also been used to deposit one type of metal nanoparticles on another type of metal. Gu *et al.* have reported the fabrication of Co [141] and Ni [142] nanowires on a DNA template by the catalysis of Pd nanoclusters having a diameter of several nanometers in a multi-step metallization process. The nanowires thus formed ranged from 10 nm to 30 nm in diameter. The effect immersing Pd(II):DNA complexes in a Co reducing bath was monitored by spectrophotometry. DNA activation by Pd(II) was essential for growing Co and Ni structures on the DNA templates. Uniformly distributed Pd(0) nanonuclei of 2–3 nm diameter on DNA templates catalyzed deposition of Co and Ni in a subsequent reaction. Immobilization of the DNA templates was important for high-quality electroless plating of the nanostructures. Quasi-parallel Co and Ni nano-arrays were prepared by combing DNA templates before metallization. The method could be developed to fabricate magnetic nanowires of iron, cobalt, nickel and other materials. The metallization process could probably be improved at each step, for example, by controlling the ratio of reactants, reaction time, and temperature. Study of the interaction between DNA molecules and nanoscale metal particles could be useful in the creation of novel hybrid structures.

3.7. Semiconductor Nanowires

Above, we briefly discussed semiconductor nanowire fabrication by VLS, LCG, and TBM. Here, we describe bottom-up assembly of 1D semiconductor chains by DNA templating. Although DNA-templated semiconductor nanowires are typically less regular than semiconductor nanowires fabricated by the other methods, the templating approach is more straightforward and less expensive than the others, and the resulting structures can be as thin as a few nanometers. Quantum confinement is important in this range; nanorods of CdSe, for example, show high photoluminance in polarized light. The focus in this section is on the fabrication process (Methods 1 and 2) and structure-associated optical properties of the nanowires.

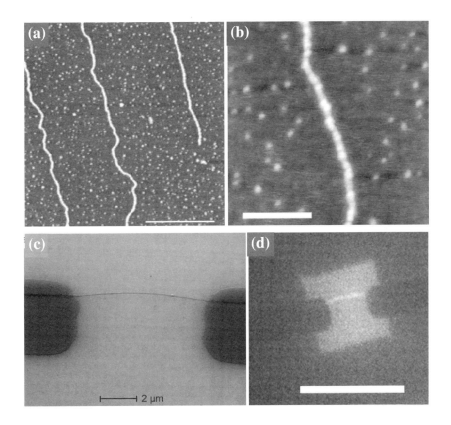

Figure 16. DNA-templated CdS nanowires. (a-b) AFM image of nanowires. Scale bare 500 nm in (a), 100 nm in (b). (c) SEM image of a single CdS nanowire bridging electrodes. (d) Luminance image of a nanowire between electrodes. Reprinted from ref. 147 with permission. Copyright 2007 Wiley-VCH Verlag.

Two years before the birth of DNA-templated electrical nanowires [99], Coffer *et al.* assembled semiconductor CdS nanoparticles on a 3455-bp plasmid DNA template (Method 2) [144]. Cd ions bound to DNA to form Cd(II)-DNA complexes, which were immobilized on a surface and then treated with hydrogen sulfide. CdS nanoparticles having a diameter of ~5 nm formed ring structures with a circumference of 1.2 µm, defined by the plasmid DNA template. The basic idea for constructing semiconductor nanostrucutures from DNA templates was thus demonstrated. Later, in work by Torimoto *et al.* [145], ~3 nm CdS nanoparticles modified with thiocholine to make them positively charged bound to the phosphate groups of DNA by electrostatic interaction (Method 1). Chains composed of densely packed CdS nanoparticles were found by transmission electron microscopy. The inter-particle distance was ~ 3.5 nm, corresponding to a length of 10 base pairs in B-DNA.

Dittmer and Simmel have reported a method of depositing nanoparticles of the *p*-type semiconductor CuS on DNA templates, both in solution and on a surface [146]. Cu(II) ions bound to DNA phosphate groups and bases. The Cu(II)-DNA complex solution was treated with the reducing agent Na_2S. 1-10 nm CuS nanoparticles with individual nanocrystals were obtained. The authors found that the use of DNA bundles as templates was helpful for increasing the density of CuS nanoparticles on DNA.

Better control over the synthesis DNA-templated CdS nanowires has been reported by Dong *et al.* [147]. The same chemical scheme as in ref. 143 was utilized: Cd(II) in a complex with DNA was reacted with Na_2S to produce CdS. The reaction was carried out in solution and on two different surfaces: mica and alkyl-coated silicon. It was found that nanowires on the silicon surface were irregular and aggregated due to the movement of DNA strands in the reaction. Figure 16a-b shows CdS nanowires with a diameter of 12-14 nm. The nanowires were prepared by incubating DNA solution and the reactants for 24 h at 4 °C and transferring them to a substrate. CdS nanoparticles prepared in solution were more uniform and more densely coated on DNA than on the surface. In the absence of DNA, CdS quantum dots had a peak emission at 650 nm. The DNA-CdS system, by contrast, had a peak emission of 520-540 nm in photoluminance spectra, suggesting that CdS had deposited mainly on the DNA templates. Figure 16d-c shows a single CdS nanowire stretched between electrodes. I-V curves were typically non-linear.

Bright emission in the luminescence image confirmed that the nanowire was made of CdS nanoparticles.

Elongated semiconductor CdSe nanorods would be useful for the development of microemitters of polarized light and micron-scale photosensors. Artemyev *et al.* have reported that DNA molecules can be used to build highly illuminant CdSe nanorods [148]. Positively charged CdSe/ZnS core-shell nanorods (22 nm in length and 4.5 nm in diameter) were bound to the phosphate groups of DNA immobilized on a Langmuir-Blodgett film. After 5 min incubation, the resultant DNA-CdSe nanorods had formed highly organized linear filaments. These structures extended to a length of 1 μm and a width of ~100 nm, suggesting that electrostatic interactions were important in the structure formation process. The detailed mechanism is unclear. Figure 17 shows the polarized micro-photoluminescence images of synthesized DNA-CdSe complexes following illumination with 488 nm laser light. The spectra show a narrow band centered ~580 nm. Intensity was much stronger for vertical (red) than horizontal polarization, confirming that sample filaments made of CdSe nanorods were uni-directionally luminescent. In a similar report, Stsiapura *et al.* [149] describe the assembly of highly-ordered fluorescent CdSe/ZnS quasi-nanowires, which was driven by electrostatic interaction between positively charged CdSe quantum dots and DNA phosphate groups. The fluorescent patterns are tunable by changing particle charge density, morphology, and stoichiometry relative to DNA.

An electrodeposition technique combined with the molecular recognition property of DNA has been developed by Sarangi *et al.* [150] for the synthesis of CdSe nanobeads and nanowires. Two single-stranded oligonucleotides, one with 30 guanine bases (poly-G30) and one with 30 cytosine bases (poly-C30), were used in this work. Cations of Cd(II) and $HSeO_2$(I) interacted with poly-G30 DNA electrostatically, and the resulting complex exhibited a net positive charge due to charge reversal. The complex migrated toward the cathode under an applied electric field. CdSe nanobeads having an average diameter of ~ 3 nm were found to deposit on the complexes by electroactive aggregation. The poly-C30:poly-G30-cation complex formed long filament-like structures under an applied field by electrodeposition and surface diffusion. Electron diffraction analysis revealed that such nano-filaments are made of single cubic crystal of CdSe with a width of ~4.0 nm. Quantum confinement in these CdSe nanobeads was demonstrated by a blue shift of 0.8 eV in optical absorption. The photoluminescence of the

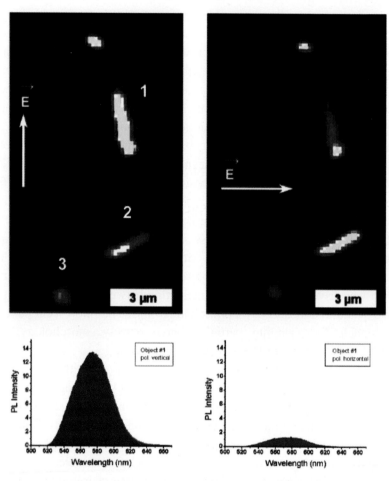

Figure 17. Room-temperature polarized micro-photoluminescence images of DNA-CdSe nanorod complexes. Top, false-color images of photoluminescence, intensity increasing from black to blue to yellow. Bottom, corresponding room-temperature photoluminescence spectra confirm the strong polarization of emission along the filament. Red spectrum, vertical polarization; blue spectrum, horizontal polarization. Reprinted from ref. 148 with permission. Copyright 2005 American Chemical Society.

DNA-CdSe nanostructures was blue-shifted and quenched in comparison with bare CdSe, suggesting a possible application in the detection of molecular recognition. This electrodeposition technique has also been used to fabricate HgTe nanowires in the catalysis of DNA molecules [151]. The experiments further confirmed that electrostatic interaction between DNA and cations plays an important role in nanowire growth.

3.8. Conducting Polymer Nanowires

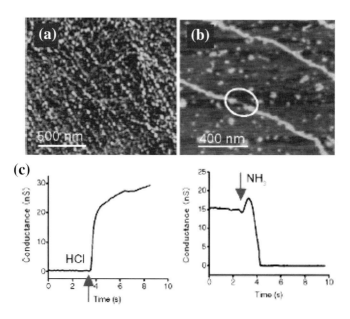

Figure 18. DNA-templated polyaniline nanowires. Fabrication at (a) pH 5 and (b) pH 3.2. (c) Response to the proton doping-undoping process. Left, conductance increases in addition of HCl vapor. Right, nanowires become non conductive on addition of NH₃. Reprinted from ref. 152 with permission. Copyright 2004 the American Chemical Society.

Electrical properties of nanowires made of π-conjugated polymers are unique in that the conductivity can be reversibly controlled by chemical methods [205-207]. The method used to fabricate conducting polymer nanowires of this type was electrochemical deposition, the target materials "filling" a micron- or nano-sized cylinder-like membrane under an applied field [208, 209]. Ma *et al.* have reported that polyaniline nanowires can be fabricated on a DNA molecular template [152]. Other nanowires made of π-conjugated polymers [153-156] have also been fabricated on DNA. Electrostatic interaction and π-π interaction between the target polymers and DNA are important in polymer nanoparticle deposition (Method 1). Monomer oxidation and protonic doping have been shown useful for increasing polymer conductivity. Here, we briefly review the nanoscale assembly of polymers on DNA and properties of the resulting nanowires.

In work on polyaniline nanowire fabrication DNA was stretched on silicon oxide by "molecular combing" [152]. Protonated aniline monomers (pK 9.4) were then attached to DNA and aligned by electrostatic interaction. The DNA-bound monomers were enzymatically polymerized in the presence of horseradish peroxidase and H_2O_2, becoming 0.8-1 nm-thick polyaniline nanowires. Morphology was found to have a strong dependence on pH. Polymerization of anilines on DNA was best at pH 4.0. At pH 3.2 or pH 5, by contrast, polymerization was incomplete or resulted in discontinuous nanoparticles, respectively (Figure 18a-b). Figure 18c shows the unique electrical behavior of polyaniline nanowires: in an acidic (doped) solution the nanowires are electrically conductive, whereas in a basic (undoped) solution they are insulating. These properties can be reversed by simple addition of acid or base.

The key step in polyaniline nanowire synthesis is polymerization of monomers bound to DNA. Nickels *et al.* [153] have reported that ammonium persulfate oxidation and photo-oxidation can be effective in aniline polymerization, in addition to enzymatic oxidation. Photo-oxidation involves the ruthenium trisbipyridinium complex, which absorbs at 450 nm and oxidizes anilines. It was found that persulfate worked best for nanowire fabrication, leading to ~5 nm polyaniline particles uniformly deposited on pre-immobilized DNA. Enzymatic oxidation, by contrast, led to extensive protein adsorption on the surface. Polyaniline nanowires made by photo-oxidation were less uniform than by other methods.

DNA-templated nanowires made of the conducting polymer polypyrrole (PPy) have been synthesized by electrostatic interaction between the target polymer and DNA [154]. The fabrication strategy resembled the assembly of DNA polyaniline nanowires. Pyrrole was oxidized and polymerized in the presence of iron chlorides. Fourier transform infrared spectra indicated a strong association between cationic PPy and anionic DNA in aqueous solution. Uniform and continuous 5 nm-wide PPy nanowires were found on the surface after stretching DNA/PPy complexes prepared in solution, whereas polymerization on DNA immobilized on a mica surface led to non-continuous nanowires. The nanoparticle coverage on DNA was low. A two-terminal I-V measurement showed a resistance of 843 MΩ for a 5.2 nm wide and 7 μm long PPy nanowire, corresponding to the bulk conductivity of the polymer. This value is 1000-fold higher than the conductivity of PPy nanowires synthesized by alumina template-based chemodeposition.

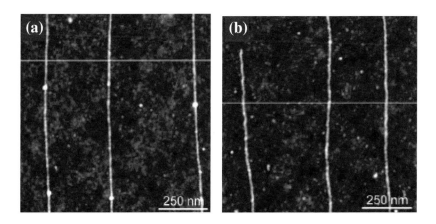

Figure 19. Redox reactions of PPhenaz-TMA/DNA nanowires. (a) PPhenaz-TMA nanowires oxidized by ammonium peroxodisulfate ($(NH_4)_2$-S_2O_8); (b) Au^{3+} reduced to Au(0) metal by PPhenaz-TMA. The height scale is 5 nm in both images. Reprinted from ref. 155 with permission. Copyright 2005 the American Chemical Society.

Above we saw that DNA stretching techniques are useful for fabricating highly-oriented nanowire arrays. DNA is usually covalently bonded on surface prior to being oriented, stabilizing the molecules for subsequent chemical reactions. Nakao *et al.* have synthesized highly-oriented arrays of nanowires made of the polymer polyphenazasiline with alkylammonium salts on the N atom (PPhenaz-TMA) [155]. See Figure 19. The π-π interaction allows strong attachment of π-conjugated units of the polymer to DNA base pairs. Two methods of fabrication of parallel PPhenaz-TMA nanowires have been demonstrated. In one, the polymer is attached in order to stretch and orient DNA templates. In the other, stretched DNA/PPhenaz-TMA complexes are prepared in solution. Either method can be used to prepare uniform arrays of nanowires with an average diameter of ~1.3 nm. Electrochemical *p*-doping of PPhenaz-TMA nanowires was done to increase electric conductivity. Oxidation of the nanowires by ammonium peroxodisulfate resulted in a height increase of 0.6-0.8 nm (Figure 19a), related to the size of a reduced SO_4 anion or a morphology change in the nanwories. After treatment of the nanowires with Au^{3+} solution, gold nanowires were produced with a height of 2.4 nm (Figure 19b). The unique modifiable electrical properties of synthetic polymer nanowires could be useful in nanoelectronics and nanosensor development.

DNA is also serving as a prototype template for building nanoelectronic devices by self-assembly. Electronic functionality is made possible by coordinating electronic polymer chains to DNA. Two methods of fabricating aligned and ordered DNA nanowires by complexation with conjugated polyelectrolytes (CPEs) have been described [156]. Complexes were formed either in solution prior to DNA stretching or after stretching on a surface. Molecular combing was used to stretch the complexes on patterned surfaces and naked DNA on PMMA. Various methods have been used to show the coordination of short CPE chains to the aligned DNA. Photoluminescent nanowire arrays were made of complexes of conjugated ploythiophene polyelectrolyte and DNA (CPE/DNA). A slight red-shift of fluorescence spectra of CPE/DNA relative to CPE was observed in solution, indicating association and absence of aggregation. Stretching the CPE/DNA nanowires with a PDMS stamp on a mica substrate resulted in a linear array of quasi-wire nanostructures, with up to 1.8 nm nanoparticles decorated on DNA chains. Assembly was also achieved by complexation of CPE to stretched and oriented DNA templates.

We have provided a brief review of DNA-templated conducting polymer nanowires. Fabrication of these structures is typically done by direct assembly of the positively charged polymer on the negatively charged backbone of DNA. Complexation can be done in solution or on a solid surface. The resulting polymer nanowires have a much lower conductivity than the metallic nanowires discussed earlier in this work, but the potential tunability of the electronic conductivity of polymer nanowires makes them attractive for the development of nanosensing devices. Moreover, polymer nanowires templated by DNA are very thin (often just a few nanometers) and very flexible, potentially important for fabrication of circuits on plastic substrates. Björk *et al.* [156] have pointed out that the superior mechanical properties of polymer nanowires could be an advantage over inorganic nanowires in the assembly of complex 3D nanostructures.

3.9. Applications

The creation and characterization of functional nanoscale devices templated on single DNA molecules is advancing nanotechnology. There are many interesting opportunities for further scientific study and technology development. Some potential applications of DNA nanowire technology are in the areas of molecular detection devices [157], nanoelectronics [158, 159], quantum interferences devices [160, 161], and nanoswitches [162, 163]. A DNA array detection method, for example, has been developed in which the binding of oligonucleotides "functionalized" with gold nanoparticles brings about a change conductivity following binding to a target probe [157]. Binding localizes the nanoparticles in an electrode gap. The deposition of gold facilitated by the nanoparticles bridges the gap, enabling a measurable change in conductivity. The method has been used to detect target DNA molecules at concentrations as low as 500 fM with high nucleotide sequence specificity ($\sim 10^5$:1). Highly conductive electric nanowires are promising in future nanocircuitry. Magnetic nanowires could be useful in the development of field sensors and storage media. Semiconductor nanowires have unique optical properties and could be used to develop sensors and optical devices. What follows is a brief discussion of recent progress in converting the ability to fabricate DNA-templated nanowires into applications.

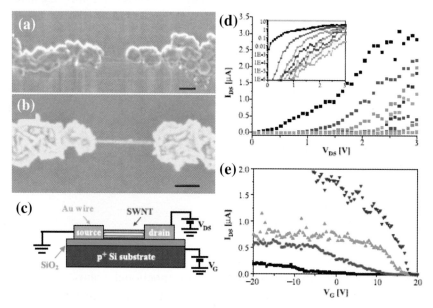

Figure 20. FET made of a DNA-templated carbon nanotube and a gold nanowire. (a-b) A carbon nanotube in contact with DNA-templated gold nanowires; scale bar is 100 nm. (c) Device scheme. (d) Drain-source current versus drain-source bias (V_{DS}) for different values of gate bias (V_G): -20 V (black), -15 V (red), -10 V (green), -5 V (blue), 0 V (cyan), $+5$ V (magenta), $+10$ V (yellow), $+15$ V (olive), $+20$ V (slate blue). Inset, same data on log scale. (e) Drain-source current versus V_G for different values of V_{DS}: -0.5 V (black), 1 V (red), 1.5 V (green), 2 V (blue). Reprinted from ref. 158 with permission. Copyright 2003 American Association for the Advancement of Science.

FETs are basic devices in integrated circuit design. Keren *et al.* [158] have developed a scheme for fabricating room-temperature FETs based on molecular recognition. A semiconducting single-wall carbon nanotube (SWNT) was self-assembled between two DNA-templated gold nanowires. The elements of the structure functioned respectively as active channel, source, and drain (Figure 20a-c). DNA scaffold molecules were used for the dual purpose of providing the "address" for precise localization of the SWNT and serving as the template for interconnecting the gold wires. Similar to the scheme in Figure 9a, ~250 nm long RecA nucleoprotein filaments were bound to specific sequences on an aldehyde-derivatized DNA template by homologous recombination. The RecA filament was then used to localize a streptavidin-functionalized SWNT: a primary antibody bound to RecA, and a biotin-conjugated secondary antibody bound to the primary antibody. Two-step metallization of DNA with silver and gold yielded functional FETs (Figure 20a-b). The gating polarity showed *p*-type conduction of the SWNT with saturation of the drain-source current for negative gate voltages (Figure 20d-e). Although drawbacks were found in the fabrication methodology and function of these FETs, the work nevertheless demonstrated a basic assembly strategy for precise localization of nanotubes. Such localization will be necessary for incorporating structures such as carbon nanotubes in molecular electronic devices.

Discrete three-branched metal nanostructures have been assembled by DNA templating [159]. Such structures are potentially useful for positioning semiconductor nanocrystals in three-terminal electronic devices and interconnecting such devices. Three 120-mer oligonucleotides with complementary end sequences were hybridized to form a ssDNA ring ~10 nm in diameter with three double-stranded DNA arms ~21 nm in length. Other ssDNA molecules were biotinylated and then hybridized to the single-stranded ring. A single streptavidin nanoparticle was thus localized to the center of the ring-arm DNA structures. The resulting DNA nanostructures were then selectively metallized with silver or copper. pH 2-6 and mild reducing conditions for metallization favored the formation of more uniform crystalline structures.

Metallization of DNA by sputtering has been used to fabricate superconducting nanowires [160]. DNA molecules were stretched between electrodes prepared by EBL with a spacing of ~150 nm (Figure 21a). The superconducting alloy $Mo_{21}Ge_{79}$ was then coated on the DNA templates. The resulting nanowires were only 5-15 nm thick (Figure 21b). Measurement of quantum interference has been made on two-nanowire devices, the wire separations ranging from 265 nm to 4050 nm. Figure 21c shows the change in temperature-dependent resistance at a separation distance of 595 nm in the presence and absence of a magnetic field. A transition in resistance was found to occur over a broad temperature range in the absence of the field. When the field was on, periodic oscillations in the resistance which differed from Little-Parks oscillations were readily detected at any temperature (Figure 21d). A quantitative explanation for the observed quantum interference phenomenon is that strong phase gradients are created in the leads by the applied magnetic field. This superconducting device could possibly be used as a local magnetometer or a superconducting phase gradiometer.

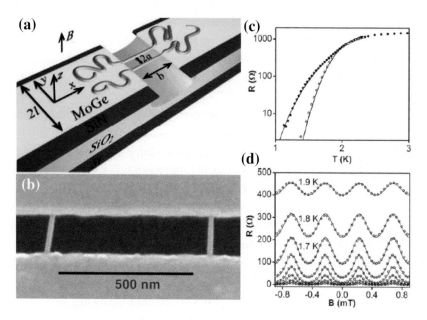

Figure 21. A DNA-templated two-nanowire device. (a) Schematic; (b) superconducting $Mo_{21}Ge_{79}$ coated on stretched DNA molecules; (c) resistance versus temperature in 0 T magnetic field (open symbols, lower curve) and in a 228 μT field (filled symbols, upper curve); (d) resistance versus magnetic field strength in the range 1.2-1.9 K. Reprinted from ref. 160 with permission. Copyright 2005 American Association for the Advancement of Science.

In similar work [161], homogeneous metallic nanowires with a diameter below 10 nm were fabricated by suspending DNA molecules and metallizing them by sputtering. A high dose, high current density focused electron beam from a transmission electron microscope was used to induce local crystallization, etching the nanowires. The resulting structures, which had metallic grains and constrictions at predefined locations, could possibly be used in a room-temperature single-electron tunneling device.

Molecular recognition and metallization of DNA have been used to build a 10-15 nm-wide protein-functionalized nanogaps which are potentially useful as nanoswitches [162-163]. A biotin molecule was attached at the center of a synthetic DNA template. The DNA-biotin complex was mixed with positively charged gold particles, and the resulting complex was mixed with streptavidin. The strong affinity of streptavidin for biotin led to the displacement of Au nanoparticles in the immediate vicinity of biotin but not from other locations on the DNA molecule. A gold enhancement process yielded a nanosize gap at the center of conductive gold nanowires. Exposure of these structures to biotin-modified gold nanoparticles resulted in a single gold nanoparticles localized at the gaps. The nanoscale switches thus fabricated could display novel and potentially useful electrical properties.

We have presented a few examples of potential applications of DNA nanowires. Future functional nanowires must be highly productive and reproducible. Achieving this goal will require robust fabrication strategies, supreme material properties, and cost-effective processes. Many fundamental questions need to be answered. DNA-based nanoassembly is very young; time is needed to develop market-driven applications. The exquisite structural and chemical properties of DNA make this molecular template interesting for the further exploration of its use in nanoelectronics, biological and chemical sensors development, micro-scale optical devices, and new molecule-assembled materials and systems.

4. Conclusion

Electrical nanowires and other kinds of functional nanowire and nanostructure have been created by the specific coupling of DNA template molecules and functional nanoparticles. Watson-Crick base pairing, negatively charged phosphate groups, and chelating nitrogenous bases, as well as gold-thiol coupling and biotin-streptavidin interactions, play important roles in DNA-templated nanowire fabrication. This review has discussed various methods to fabricate electrical, magnetic, semiconductor, and conducting polymer nanowires on a DNA template. The broader context, fabrication methods, nanowire properties, and applications of the technology have been addressed.

DNA nanowires are typically less ordered than nanowires fabricated by VLS or membrane-templated electrodeposition, due to the use of wet chemical processes at room temperature. Nevertheless, DNA-templated nanowires can be just a few nanometers thick. Moreover, precise control and localization is possible, and retention of molecular recognition has been demonstrated in some approaches. Templated approaches based on DNA are more consistent for nanostructure fabrication than ones involving membranes and electrodeposition. The typical DNA nanowire fabrication process does not require expensive equipment. DNA nanowires are suitable for bionanosystems development because DNA can be modified to bind many different chemical molecules, proteins and nanoparticles – key building blocks of future nanosystems.

We have summarized different approaches to stretching and positioning DNA templates on 2D substrates. Molecular combing is the most effective approach because it does not require modification of DNA, can be used with a wide range of surfaces, and is controlled by meniscus movement. Metallization is the key to realizing the potential of conductive DNA-templated nanowires in future nanoelectronics devices. We have described various methods and processes used to fabricate electrical nanowires of silver, gold, palladium, platinum, and copper. We have also discussed magnetic, semiconductor, and polymer nanowires. Direct assembly (Method 1) is based on electrostatic interaction for the assembly of target nanoparticles or ions on DNA templates. In electroless plating (Method 2), metal ions bind to the template, and the resulting complexes are treated with a reducing agent or plating bath. Two-step metallization (Method 3) involves the binding of a reducing agent (aldehyde) to the DNA template and treatment with silver ions. All three methods have been used to analyze the mechanisms of nucleation and growth which govern the uniformity and conductivity of the resulting nanowires.

Various electrical nanowires with diameters ranging from a few nanometers to a hundred nanometers have been fabricated on DNA templates. Most studies have involved silver or gold for metallization. Uniform silver nanowires with a diameter of 30-40 nm can be fabricated routinely by two-step metallization (Method 3). The lowest resistivity of silver nanowires achieved to date is 2-3 $\times$ 10^{-6} Ω-m, about two orders of magnitude higher than the resistivity of bulk polycrystalline silver. The most conductive DNA-templated nanowires were metallized with palladium; a resistivity of 1.5 $\times$ 10^{-7} Ω-m was achieved. Platinum nanowires can be as thin as 1-2 nm. Besides electrical nanowires, superconducting, magnetic, semiconductor, and conducting polymer nanowires are promising for the development of future functional nanosystems due to their unique material properties. It seems likely that DNA will increasingly be recognized as an important nanostructure, not only in biotechnology, but also in futuristic manufacturing processes in nanotechnology.

References

1. Q. Gu, C. Cheng, R. Gonela, S. Suryanarayanan, K. Dai and D. T. Haynie, Nanotechnology 17, R14 (2006).
2. W. A. Goddard, III, D. W. Brenner, S. E. Lyshevski and G. J. Iafrate, Handbook of Nanoscience, Engineering, and Technology, CRC Press, Boca Raton, Florida (2003).
3. International Technology Roadmap for Semiconductors (2005).
4. Quantum Information Science and Technology Roadmapping Project, University of California (2002).
5. W. Aspray, John von Neumann and the Origins of Modern Computing, Massachusetts Institute of Technology Press, Cambridge, Massachusetts (1990).
6. G. Meng, Y. J. Jung, A. Cao, R. Vajtai and P. M. Ajayan, Proc. Natl Acad. Sci. (USA) 102, 7074 (2005).
7. D. S. Goodsell, Bionanotechnology: Lessons from Nature, Wiley-Liss, New York (2004).
8. D. T. Haynie, L. Zhang, W. Zhao and J. S. Rudra, Nanomedicine:NBM 2, 150 (2006).
9. R. P. Feynman, Eng. Sci. (Caltech) 23, 5 (1960).
10. D. T. Haynie, Int. J. Nanomed. 2, 125 (2007).
11. J. M. Schnur, Science 262, 1669 (1993).

12. V. P. Roychowdhury, D. B. Janes and X. Wang, IEEE Trans. Elect. Dev. 43, 1688 (1996).
13. C. A. Mirkin, R. C. Mucie, and J. J. Storhoff, Nature 382, 607 (1996).
14. R. Elghanian, J. J. Storhoff and C. A. Mirkin, Science 277, 1078 (1997).
15. M. Mertig, R. Kirsch and W. Pompe, Appl. Phys. A 66, S723 (1998).
16. R. C. Mucic, J. J. Storhoff, C. A. Mirkin, and R. L. Letsinger, J. Am. Chem. Soc. 120, 12674 (1998).
17. R.R. Naik, S.E. Jones, C.J. Murray, J.C. McAuliffe, R.A. Vaia, M.O. Stone, Adv. Funct. Mater. 14, 25 (2004).
18. M. Knez, A.M. Bittner, F. Boes, C. Wege, H. Jeske, E. Maiβ, K. Kern, Nano Lett. 3, 1079 (2003).
19. W. Shenton, D. Pum, U. B. Sleytr and S. Mann, Nature 389, 585 (1997).
20. G. M. Chow, M. Pazirandeh, S. Baral and J. R. Campbell, Nanostruct. Mater. 2, 495 (1993).
21. Y. Huang, X. Duan, Y. Cui, L. Lauhon, K. Kim and C. M. Lieber, Science 294, 1313 (2001).
22. Y. Cui and C. M. Lieber, Science 291, 851 (2001).
23. X. Duan, Y. Huang, Y. Cui, J. Wang and C. M. Lieber, Nature 409, 66 (2001).
24. C. Niu, V. Sahi, J. Chen, J. W. Parce, S. Empedocles and J. L. Goldman, Nature 425, 274 (2003).
25. Y. Cui, Z. Zhong, D. Wang, W. U. Wang and C. M. Lieber, Nano Lett. 3, 149 (2003).
26. L. Piraux, J. George, J. Despres, C. Leroy, E. Ferain, R. Legres, K. Ounadjela and A. Fert, Appl. Phys. Lett. 65, 2484 (1994).
27. T. Thurn-Albrecht, J. Schotter, C. A. Kastle, N. Emley, T. Shibauchi, L. Krusin-Elbaum, K. Guarini, C. T. Black, M. T. Tuominen and T.P. Russell, Science 290, 2126 (2000).
28. K. Nielsch, R. B. Wehrspohn, J. Barthel, J. Kirschner and U. Gösele, Appl. Phys. Lett. 79, 1360 (2001).
29. R. S. Wagner and W. C. Ellis, Appl. Phys. Lett. 4, 89 (1964).
30. E. I. Givargizov, J. Cryst. Growth 31, 20 (1975).
31. Y. Wu and P. Yang, J. Am. Chem. Soc. 123, 3165 (2001).
32. A. M. Morales and C. M. Lieber, Science 279, 208 (1998).
33. X. Duan and C. M. Lieber, J. Am. Chem. Soc. 122, 188 (2000).
34. X. Duan and C. M. Lieber, Adv. Mat. 12, 298 (2000).
35. G. E. Possin, Rev. Sci. Instrum. 41, 772 (1970).
36. H. He and N. J. Tao, in Encyclopedia of Nanoscience and Nanotechnology edited H. S. Nalwa, American Scientific Publishers, Stevenson, California (2003) Vol. X, p. 1.
37. C. R. Martin, Science 266, 1961 (1994).
38. C. J. Brumlik and C. R. Martin, J. Am. Chem. Soc. 113, 3174 (1991).
39. C. J. Brumlik, V. P. Menon and C. R. Martin, J. Mater. Res. 9, 1174 (1994).
40. S. A. Sapp, B. B. Lakshmi and C. R. Martin, Adv. Mater. 11, 402 (1999).

41. M. Wirtz and C. R. Martin, Adv. Mat. 15, 455 (2003).
42. T. M. Whitney, J. S. Jiang, P. C. Searson and C. L. Chien, Science 261, 1316 (1993).
43. A. Blomdel, J. P. Meier, B. Doudin and J.-Ph. Ansermet, Appl. Phys. Lett. 65, 3019 (1994).
44. H. Zeng, R. Skomki, L. Menon, Y. Liu, S. Bandyopadhyay, and D. J. Sellmyer, Phys. Rev. B 65, 13426 (2002).
45. K. Liu, K. Nagodawithana, P. C. Searson and C. L. Chien, Phys. Rev. B 51, 7381 (1995).
46. A. Fert and L. Pirauxet, J. Magn. Magn. Mater. 200, 338 (1999).
47. M. Tanase, L. A. Bauer, A. Hultgren, D. M. Silevitch, L. Sun, D. H. Reich, P. C. Searson and G. J. Meyer, Nano Lett. 1, 155 (2001).
48. B. Lakshmi, P. K. Dorhout and C. R. Martin, Chem. Mater. 9, 857 (1997).
49. B. Lakshmi, C. J. Patrissi and C.R. Martin, Chem. Mater. 9, 2544 (1997).
50. C. R. Martin, in Handbook of Conducting Polymers 2nd edn edited J. R. Reynolds, T. Skotheim and R. Elsenbaumer, Marcel Dekker, New York (1997) p. 409.
51. C. R. Martin, Acc. Chem. Res. 28, 61 (1995).
52. C. R. Martin, Z. Cai, L. S. Van Dyke and W. Liang, Polym. Mater. Sci. Eng. 64, 204 (1991).
53. C. R. Martin, Z. Cai, L. S. Van Dyke and W. Liang, Polym. Prepr. 32, 89 (1991).
54. M. H. Huang, S. Mao, H. Feick, H. Yan, Y. Wu, H. Kind, E. Weber, R. Russo and P. Yang, Science 292, 1897 (2001).
55. J. C. Johnson, H. J. Choi, K. P. Knutsen, R. D. Schaller, P. Yang and R. J. Saykally, Nat. Mater. 1, 101 (2002).
56. X. Duan, Y. Huang, R. Agarwal and C. M. Lieber, Nature 421, 241 (2003).
57. H. Yan, R. He, J. Johnson, M. Law, R. J. Saykally and P. Yang, J. Am. Chem. Soc. 125, 4728 (2003).
58. J. C. Johnson, H. Yan, R. D. Schaller, L. H. Haber, P. Yang and R. J. Saykally, J. Phys. Chem. B 105, 11387 (2000).
59. M. Law, D. Sirbuly, J. Johnson, J. Goldberger, R. Saykally and P. Yang, Science 305, 1269 (2004).
60. M. C. McAlpine, R. S. Friedmann, S. Jin, K. H. Lin, W. U. Wang and C. M. Lieber, Nano Lett. 3, 1531 (2003).
61. H. Kind, H. Yan, M. Law, B. Messer and P. Yang, Adv. Mater. 14, 158 (2002).
62. F. Favier, E. Walter, M. Zach, T. Benter and R. M. Penner, Science 293, 2227 (2001).
63. J. Wang, M. S. Gudiksen, X. Duan, Y. Cui and C. M. Lieber, Science 293, 1455 (2001).
64. Y. Cui, Q. Wei, H. Park and C. M. Lieber, Science 293, 1289 (2001).
65. A. Maiti, J. A. Rodriguez, M. Law, P. Kung, J. R. McKinney and P. Yang, Nano Lett. 3, 1025 (2003).
66. J. Hahm and C. M. Lieber, Nano Lett. 4, 51 (2004).

67. Y. Huang, X. Duan, Q. Wei and C. M. Lieber, Science 291, 630 (2001).
68. M. S. Dresselhaus, in Chemistry, Physics and Materials Science of Thermoelectric Materials: Beyond Bismuth Telluride edited M. G. Kanatzidis, S. D. Mahanti, T. P. Hogan and M. F. Thorpe, Kluwer Academic/Plenum Publishers, New York (2003), p. 1.
69. K. T. Nam, D. Kim, P. J. Yoo, C. Y. Chiang, N. Meethong, P. T. Hammond, Y.-M. Chiang and A. M. Belcher, Science 312, 885 (2006).
70. C. Cheng, R. K. Gonela, Q. Gu and D. T. Haynie, Nano Lett. 5, 175 (2005).
71. C. Cheng and D. T. Haynie, Appl. Phys. Lett. 87, 263112 (2005).
72. C. Mao, D. J. Solis, B. D. Reiss, S. T. Kottmann, R. Y. Sweeney, A. Hayhurst, G. Georgiou, B. Iverson and A. M. Belcher, Science 303, 213 (2004).
73. S. Ju, A. Facchetti, Y. Xuan, J. Liu, F. Ishikawa, P. Ye, C. Zhou, T. J. Marks and D. B. Janes, Nat. Nanotech. 2, 378 (2007).
74. J. Xiang, A. Vidan, M. Tinkham, R. M. Westervelt and C. M. Lieber, Nat. Nanotech. 1, 208 (2006).
75. F. Patolsky, G. Zheng and C. M. Lieber, Nat. Protocols 1, 1711 (2006).
76. J. Xiang, W. Lu, Y. Hu, Y. Wu, H. Yan and C. M. Lieber, Nature 441, 489 (2006).
77. P. J. Pauzauskie, A. Radenovic, E. Trepagnier, H. Shroff, P. Yang and J. Liphardt, Nat. Mater. 5, 97 (2006).
78. G. Zheng, F. Patolsky, Y. Cui, W. U. Wang and C. M Lieber, Nat. Biotechnol. 23, 1294 (2005).
79. R. Beckman, E. J. Halperin, Y. Luo, J. E. Green and J. R. Heath, Science 310, 465 (2005).
80. Z. L. Wang and J. Song, Science 312, 242 (2006).
81. F. Patolsky, B. P. Timko, G. Yu, Y. Fang, A. B. Greytak, G. Zheng and C. M. Lieber, Science 313, 1100 (2006).
82. C. Yang, Z. Zhong and C. M. Lieber, Science 310, 1304 (2005).
83. Y. J. Doh, J. A. van Dam, A. L. Roest, E. P. A. M. Bakkers, L. P. Kouwenhoven and S. De Franceschi, Science 309, 5732 (2005).
84. D. J. Sirbuly, M. Law, P. Pauzauskie, H. Yan, A. V. Maslov, K. Knutsen, C. Ning, R. J. Saykally and P. Yang, *Proc. Natl Acad. Sci. USA* 102, 7800 (2005).
85. W. Lu, J. Xiang, B. P. Timko, Y. Wu and C. M. Lieber, Proc. Natl Acad. Sci. (USA) 102, 10046 (2005).
86. W. U. Wang, C. Chen, K. Lin, Y. Fang and C. M. Lieber, Proc. Natl Acad. Sci. (USA) 102, 3208 (2005).
87. R. S. Friedman, M. C. McAlpine, D. S. Ricketts, D. Ham and C. M. Lieber, Nature 434, 1085 (2005).
88. V. Schmidt and U. Gösele, Science 316, 698 (2007).
89. Y. Nakayama, P. J. Pauzauskie, A. Radenovic, R. M. Onorato, R. J. Saykally, J. Liphardt, and P. Yang, Nature 447, 1098 (2007).
90. S. Pramanik, C.-G. Stefanita, S. Patibandla, S. Bandyopadhyay, K. Garre, N. Harth and M. Cahay, Nat. Nanotech. 2, 216 (2007).

91. M. Law, L. E. Greene, J. C. Johnson, R. Saykally and P. Yang, Nat. Mater. 4, 455 (2005).

92. S. Kodambaka, J. Tersoff, M. C. Reuter and F. M. Ross, Science 316, 729 (2007).

93. Y. A. Gorby, S. Yanina, J. S. McLean, K. M. Rosso, D. Moyles, A. Dohnalkova, T. J. Beveridge, I. S. Chang, B. H. Kim, K. S. Kim, D. E. Culley, S. B. Ree, M. F. Romine, D. A. Saffarini, E. A. Hill, L. Shi, D. A. Elias, D. W. Kennedy, G. Pinchuk, K. Watanaba, S.-i. I. B. Logan, K. H. Nealson and J. K. Fredrickson, Proc. Natl Acad. Sci. (USA) 103, 11358 (2006).

94. N. C. Seeman, Annu. Rev. Biophys. Biomol. Struct. 27, 225 (1998).

95. L. M. Adleman, Science 266, 1021 (1994).

96. P. A. Alivisatos, K. P. Johnsson, X. Peng, T. E. Wilson, C. J. Loweth, M. P. Bruchez and P. G. Schultz, Nature 382, 609 (1996).

97. C. A. Mirkin, R. L. Letsinger, R. C. Mucic and J. J. Storhoff, Nature 382, 607 (1996).

98. C. J. Loweth, W. B. Caldwell, X. Peng, A. P. Alisisatos and P. G. Schultz, Angew. Chem. Int. Edn 38, 1808 (1999).

99. E. Braun, Y. Eichen, U. Sivan and G. Ben-Yoseph, Nature 391, 775 (1998).

100. K. Keren, M. Krueger, R. Gilad, G. Ben-Yoseph, U. Sivan and E. Braun, Science 297, 72 (2002).

101. K. Keren, R. S. Berman and E. Braun, Nano Lett. 4, 323 (2004).

102. H. Yan, S. H. Park, G. Finkelstein, J. H. Reif and T. H. LaBean, Science 301, 1882 (2003).

103. D. Liu, S.H. Park, J. H. Reif and T. H. LaBean, Proc. Natl Acad. Sci. (USA) 101, 717 (2004).

104. S. H. Park, R. Barish, H. Li, J. H. Reif, G. Finkelstein, H. Yan and T.H. LaBean, Nano Lett. 5, 693 (2005).

105. S. H. Park, M. W. Prior, T. H. LaBean and G. Finkelstein, Appl. Phys. Lett. 89, 033901 (2006).

106. M. Fischler, U. Simon, H. Nir, Y. Eichen, G. A. Burley, J. Gierlich, P. M. E. Gramlich and T. Carell, Small 3, 1049 (2007).

107. K. Ijiro, Y. Matsuo and Y. Hashimoto, Mol. Cryst. Liq. Cryst. 445, 207 (2006).

108. A. Kumar, M. Pattarkine, M. Bhadbhade, A. B. Mandale, K. N. Ganesh, S. S. Datar, C. V. Dharmadhikari and M. Sastry, Adv. Mater. 13, 341 (2001).

109. M. Sastry, A. Kumar, S. Datar, C. V. Dharmadhikari and K. N. Ganesh, Appl. Phys. Lett. 78, 2943 (2001).

110. O. Harnack, W. E. Ford, A. Yasuda and J. M. Wessels, Nano Lett. 2, 919 (2002).

111. A. Ongaro, F. Griffin, P. Beecher, L. Nagle, D. Iacopino, A. Quinn, G. Redmond and D. Fitzmaurice, Chem. Mater. 17, 1959 (2005).

112. H. Nakao, H. Shiigi, Y. Yamamoto, S. Tokonami, T. Nagaoka, S. Sugiyama and T. Ohtani, Nano Lett. 3, 1391 (2003).

113. T. Nishinaka, A. Takano, Y. Doi, M. Hashimoto, A. Nakamura, Y. Matsushita, J. Kumaki and E. Yashima, J. Am. Chem. Soc. 127, 8120 (2005).

114. Y. Weizmann, F. Patolsky, I. Popov and I. Willner, Nano Lett. 4, 787 (2004).

115. S. Beyer, P. Nickels and F. C. Simmel, Nano Lett. 5, 719 (2005).

116. Z. Deng, Y. Tian, S. H. Lee, A. E. Ribbe and C. Mao, Angew. Chem. Int. Ed. 44, 3582 (2005).

117. H. Li, S. H. Park, J. H. Reif, T. H. LaBean and H. Yan, J. Am. Chem. Soc. 9, 418 (2004).

118. D. H. Lee, S. J. Kim, S. Y. Heo and D. Jang, Appl. Phys. Lett. 87, 233103 (2005).

119. S. R. Quake and A. Scherer, Science 290, 1536 (2000).

120. F. Patolsky, Y. Weizmann, O. Lioubashevski and I. Willner, Angew. Chem. Int. Edn. 41, 2323 (2002).

121. L. Berti, A. Alessandrini and P. Facci, J. Am. Chem. Soc. 127, 11216 (2005).

122. H. J. Kim, Y. Roh and B. Hong, J. Vac. Sci. Tech. A 24, 1327 (2006).

123. J. S. Hwang, S. H. Hong, H. K. Kim, Y. W. Kwon, J. Jin, S. W. Hwang and D. Ahn, Jp. J. Appl. Phys. 44, 2623 (2005).

124. D. R. S. Cumming, A. D. Bates, B. P. Callen, J. M. Cooper, R. Cosstick, C. Geary, A. Glidle, L. Jaeger, J. L. Pearson, M. Proupín-Pérez and C. Xu, J. Vac. Sci. Tech. B **24**, 3196 (2006).

125. A. D. Bates, B. P. Callen, J. M. Cooper, R. Cosstick, C. Geary, A. Glidle, L. Jaeger, J. L. Pearson, M. Proupín-Pérez, C. Xu and D. R. S. Cumming, Nano Lett. 6, 445 (2006).

126. J. Richter, R. Seidel, R. Kirsch, M. Mertig, W. Pompe, J. Plaschke and H. K. Schackert, Adv. Mater. 12, 507 (2000).

127. J. Richter, M. Mertig, W. Pompe, I. Mönch and H. K. Schackert, Appl. Phys. Lett. 78, 536 (2001).

128. Z. Deng and C. Mao, Nano Lett. 3, 1545 (2004).

129. J. Richter, M. Mertig, W. Pompe and H. Vinzelberg, Appl. Phys. A 74, 725 (2002).

130. W. E. Ford, O. Harnack, A. Yasuda and J. M. Wessels, Adv. Mater. 13, 1793 (2001).

131. R. Seidel, M. Mertig and W. Pompe, Surf. Interf. Anal. 33, 151 (2002).

132. M. Mertig, L. C. Ciacchi, R. Seidel and W. Pompe, Nano Lett. 2, 841 (2002).

133. R. Seidel, L. C. Ciacchi, M. Weigel, W. Pompe and M. Mertig, J. Phys. Chem. B 108, 10801 (2004).

134. C. F. Monson and A. T. Woolley, Nano Lett. 3, 359 (2003).

135. H. A. Becerril, R. M. Stoltenberg, C. F. Monson and A. T. Woolley, J. Mater. Chem. 14, 611 (2004).

136. R. M. Stoltenberg and A. T. Woolley, Biomed Microdev 6, 105 (2004).

137. H. Kudo and M. Fujihira, IEEE Trans Nanotech 5, 90 (2006).

138. D. Nyamjav, J. M. Kinsella and A. Ivanisevic, Appl. Phys. Lett. 86, 093107 (2005).

139. J. M. Kinsella and A. Ivanisevic, J. Am. Chem. Soc. 127, 3276 (2005).

140. J. M. Kinsella and A. Ivanisevic, Langmuir 23, 3886 (2007).

141. Q. Gu, C. Cheng and D. T. Haynie, Nanotechnology 16, 1358 (2005).

142. Q. Gu, C. Cheng, S. Suryanarayanan, K. Dai and D. T. Haynie, Physica E 33, 92 (2006).
143. C. Lin, S. Tong, S. Lin, H. Chen, Y. Liu, H. Lin, W. Liu, S. Cheng and Y. Wang, MRS Proceedings 921E, T02-10 (2006).
144. J. L. Coffer, S. R. Bigham, X. Li, R. F. Pinizzotto, Y. G. Rho, R. M. Pirtle and I. L. Pirtle, Appl. Phys. Lett. 69, 3851 (1996).
145. T. Torimoto, M. Yamashita, S. Kuwabata, T. Sakata, H. Mori and H. Yoneyama, J. Phys. Chem. B 103, 8799 (1999).
146. W. U. Dittmer and F. C. Simmel, Appl. Phys. Lett. 85, 633 (2004).
147. L. Dong , T. Hollis, B. A. Connolly, N. G. Wright, B. R. Horrocks and A. Houlton, Adv. Mater. 19, 1748 (2007).
148. M. Artemyev, D. Kisiel, S. Abmiotko, M. N. Antipina, G. B. Khomutov, V. V. Kislov and A. A. Rakhnyanskaya, J. Am. Chem. Soc. 126, 10594 (2004).
149. V. Stsiapura, A. Sukhanova, A. Baranov, M. Artemyev, O. Kulakovich, V. Oleinikov, M. Pluot, J. H. M. Cohen and I. Nabiev, Nanotechnology 17, 581 (2006).
150. S. N. Sarangi, K. Goswami and S.N. Sahu, Biosens. Bioelectron. 22, 3086 (2007).
151. S. Rath, S. N. Sarangi and S. N. Sahu, J. Appl. Phys. 101, 074306 (2007).
152. Y. Ma, J. Zhang, G. Zhang and H. He, J. Am. Chem. Soc. 126, 7097 (2004).
153. P. Nickels, W. U. Dittmer, S. Beyer, J. P. Kotthaus and F. C. Simmel, Nanotechnology 15, 1524 (2004).
154. L. Dong, T. Hollis, S. Fishwick, B. A. Connolly, N. G. Wright, B. R. Horrocks and A. Houlton, Chem. Eur. J. 13, 822 (2007).
155. H. Nakao, H. Hayashi, F. Iwata, H. Karasawa, K, Hirano, S. Sugiyama and T. Ohtani, Langmuir 21, 7945 (2005).
156. P. Björk, A. Herland, I.G. Scheblykin and O. Inganäs, Nano Lett. 5, 1948 (2005).
157. S. J. Park, T. A. Taton and C. A. Mirkin, Science 295, 1503 (2002).
158. K. Keren, R. S. Berman, E. Buchstab, U. Sivan and E. Braun, Science 302, 1380 (2003).
159. H. A. Becerril, R. M. Stoltenberg, D. R. Wheeler, R. C. Davis, J. N. Harb and A. T. Woolley, J. Am. Chem. Soc. 127, 2828 (2005).
160. D. S. Hopkins, D. Pekker, P. M. Goldbart and A. Bezryadin, Science 308, 1762 (2005).
161. M. Remeika and A. Bezryadin, Nanotechnology 16, 1172 (2005).
162. A. Ongaro, F. Griffin, L. Nagle, D. Iacopino, R. Eritja and D. Fitzmaurice, Adv. Mater. 16, 1799 (2004).
163. B. Manning, A. D. Salvo, S. Stanca, A. Ongaro and D. Fitzmaurice, Proc. SPIE 5824, 93 (2005).
164. A. Bensimon, A. Simon, A. Chiffaudel, V. Croquette, F. Heslot and D. Bensimon, Science 265, 2096 (1994).
165. D. Bensimon, A.J. Simon, V. Croquette and A. Bensimon, Phys. Rev. Lett. 74, 4754 (1995).

166. J.-F. Allemand, D. Bensimon, L. Jullien, A. Bensimon and V. Croquette, Biophys. J. 73, 2064 (1997).

167. Z. Gueroui, C. Place, E. Freyssingeas and B. Berge, Proc. Natl Acad. Sci. (USA) 99, 6005 (2002).

168. H. Yokota, F. Johnson, H. Lu, R. M. Robinson, A. M. Belu, M. D. Garrison, B. D. Ratner, B. J. Trask and D. L. Miller, Nucleic Acids Res. 25, 1064 (1997).

169. K. Otobe and T. Ohtani, Nucleic Acids Res. 29, e109 (2001).

170. X. Michalet, R. Ekong, F. Fougerousse, S. Rousseaux, C. Schurra, N. Hornigold, M. Slegtenhorst, J. Wolfe, S. Povey, J.S. Beckmann and A. Bensimon, Science 277, 1518 (1997).

171. J. Li, C. Bai, C. Wang, C. Zhu, Z. Lin, Q. Li and E. Cao, Nucleic Acids Res. 26, 4785 (1998).

172. J. H. Kim, W. X. Shi and R. G. Larson, Langmuir 23, 755 (2007).

173. H. Nakao, M. Gad, S. Sugiyama, K. Otobe and T. Ohtani, J. Am. Chem. Soc. 125, 7162 (2003).

174. J. Guan and L. J. Lee, Proc. Natl Acad. Sci. (USA) 102, 18321 (2005).

175. J. M. Schurr and S. B. Smith, Biopolymers 29, 1161 (1999).

176. S. B. Smith and A. J. Bendich, Biopolymers 29, 1167 (1999).

177. R. M. Zimmermann and E. C. Cox, Nucleic Acid Res. 22, 492 (1994).

178. M. Washizu and O. Kurosawa, IEEE T. Ind. Appl. 26, 1165 (1990).

179. M. Washizu, O. Kurusawa, I. Arai, S. Suzuki and N. Shimamoto, IEEE T. Ind. Appl. 31, 447 (1995).

180. T. Yamamoto, O. Kurosawa, H. Kabata, N. Shimamoto and M. Washizu, IEEE T. Ind. Appl. 36, 1010 (2000).

181. F. Dewarrat, M. Calame and C. Schönenberger, Single Mol. 3, 189 (2002).

182. S. Suzuki, T. Yamanashi, S.-i. Tazawa, O. Kurosawa and M. Washizu, IEEE T. Ind. Appl. 34, 75 (1998).

183. V. Namasivayam, R. G. Larson, D. T. Burke and M. A. Burns, Anal. Chem. 74, 3378 (2002).

184. W. A. Germishuizen, C. Wälti, R. Wirtz, M. B. Johnston, M. Pepper, A. G. Davies and A. P. J. Middelberg, Nanotechnology 14, 896 (2003).

185. N. Kaji, M. Ueda and Y. Baba, Biophys. J. 82, 335 (2002).

186. T. T. Perkins, S. R. Quake, D. E. Smith and S. Chu, Science 264, 822 (1994).

187. T. T. Perkins, D. E. Smith, R. G. Larson and S. Chu, Science 268, 83 (1995).

188. S. B. Smith, Y. Cui and C. Bustamante, Science 271, 795 (1996).

189. H. Yokota, J. Sunwoo, M. Sarikaya, G. van den Engh and R. Aebersold, Anal. Chem. 71, 4418 (1999).

190. J. Y. Ye, K. Umemura, M. Ishikawa and R. Kuroda, Anal. Biochem. 281, 21 (2000).

191. A. Bezryadin, A. Bollinger, D. Hopkins, M. Murphey, M. Remeika and A. Rogachev, in Dekker Encyclopedia of Nanoscience and Nanotechnology edited J. A. Schwarz, C. I. Contescu and K. Putyera (Eds.), Marcel Dekker, New York (2004), p. 3761.

192. J. K. N. Mbindyo, B. R. Reiss, B. R. Martin, C. D. Keating, M. J. Natan and T. E. Mallouk, Adv. Mater. 13, 249 (2001).

193. G. Maubach, A. Csáki, D. Born and W. Fritzsche, Nanotechnology 14, 546 (2003).

194. G. Maubach and W. Fritzsche, Nano Lett. 4, 607 (2004).

195. J. Zhang, Y. Ma, S. Stachura and H. He, Langmuir 21, 4180 (2005).

196. D. Porath, A. Bezryadin, S. d. Vries and C. Dekker, Nature 403, 635 (2000).

197. H. W. Fink and C. Schönenberger, Nature 398, 407 (1999).

198. S. O. Kelley and J. K. Barton, Science 283, 375 (1999).

199. E. Meggers, M. E. Michel-Beyerle and B. Giese, J. Am. Chem. Soc. 120, 12950 (1998).

200. Y. Okahata, T. Kobayashi, K. Tanaka and M. Shimomura, J. Am. Chem. Soc. 120, 6165 (1998).

201. P. M. Takahara, A. C. Rosenzweig and S. J. Lippard, Nature 377, 649 (1995).

202. H. Huang, L. Zhu and P. B. Hopkins, Science 270, 1842 (1995).

203. G. B. Onoa, G. Cervantes, V. Moreno and M. J. Prieto, Nucleic Acid Res. 26, 1473 (1998).

204. J. Duguid, V. A. Bloomfield, J. Benevides and G. J. Thomas, Biophys. J. 65, 1916 (1993).

205. S. A. Chen and G. W. Hwang, J. Am. Chem. Soc. 117, 10055 (1995).

206. A. F. Diaz and J. A. Logan, J. Electroanal. Chem. 111, 111 (1980).

207. H.S.O. Chan, S.C. Ng, and P.K.H. Ho, J. Am. Chem. Soc. 117, 8517 (1995).

208. C. G. Wu and T. Bein, Science 264, 1757 (1994).

209. Z. Cai, J. Lei, W. Liang, V. Menon and C. R. Martin, Chem. Mater. 3, 960 (1991).

CHAPTER 6

SOLUTION-BASED SYNTHESIS OF ORIENTED ONE-DIMENSIONAL NANOMATERIALS

Jun Liu[1] and Guozhong Cao[2]

[1]Pacific Northwest National Laboratory, Richland, WA 99352, E-Mail: jun.liu@pnl.gov, [2]Department of Materials Science and Engineering, University of Washington, Seattle, WA 98195, E-mail: gzcao@u.washington.edu

Oriented nanostructures with controlled architecture from nano- to macro- length scales have potentials for wide range of applications. Although one dimensional nanostructured materials (1DNMs) have been extensively reported, there have been few papers devoted to systematic discussions of the principles and applications of oriented 1 DNMs from low temperature, solution phases. The importance of this approach has been recently highlighted in some significant breakthroughs in energy conversion devices and in controlling the physical and chemical properties of surfaces. In this paper, we review two major approaches for solution synthesis of oriented 1DNMs on different substrates, with or without templates. The low temperature solution based synthesis is complementary to the widely used gas phase reaction routes, but may allow us to reduce the cost for large scale fabrication, improve processing reliability, simplify the procedure for complicated shapes and geometries, and provide the opportunities to better control the experimental parameters and systematically fine tune the resultant microstructures. We will discuss the fundamental principles of the driving forces for template filling the templates, and the requirement of understanding of the nucleation and growth for growing 1DNMs. For the template-based synthesis, three general methods are used: electrochemical deposition, electrophoretic deposition, the template filling, in junction with multiple templates including anodized alumina membrane, radiation track-etched polymer membranes, nanochannel array glass, radiation track-etched mica, mesoporous materials, porous silicon by electrochemical etching of silicon wafer, zeolites, carbon nanotubes, etc. The template approaches have been applied to metals, ceramics, polymers, and more complex compositions. For the templateless approach, both seedless and seeded growth methods have been successfully used for oxide microrods, microtubes, nanowires, nanotubes, and polymeric materials. The flexibility of this approach to systematically control the sizes of the 1DNMs, and the ability to form complex oriented nanostructures has been demonstrated.

1. Introduction: New Frontiers in One-Dimensional Nanomaterials

Recently, one-dimensional nanomaterials (1DNMs) such as oriented nanowires have attracted wide attention due to their unique nanoscale anisotropic properties and their potentials in micro- and nanodevices. Varies aspects of nanomaterials synthesis and properties were extensively reviewed in a special issue by Advanced Materials published on one-dimensional nanostructures [1]. The objective of this paper is to review the strategies for growing oriented 1DNMs, in particular solution based approaches. This paper is motivated by some very important development in using 1DNMs for energy conversion and for controlling surface properties.

1.1. Nanowire Array Based Nano-Piezoelectric Devices

Integrating nanomaterials and devices to perform specific functions is one of the most significant challenges in nanotechnology. Up to date, most nanodevices are based on manipulating single nanoparticles and nanowires. Although oriented nanoarrays present golden opportunities for large scale fabrication of nanodevices, there are few successful examples.

Wang and Song first reported an important breakthrough in piezoelectric nanogenerators based on ZnO nanowire arrays [2]. In this landmark work, [0001] oriented ZnO nanowire arrays were grown on α–Al_2O_3 substrates. The ZnO nanowires were deflected by a conductive atomic force microscope (AFM) tip (Figure 1, left panel). This deformation produced a couple piezoelectric and semiconducting responses in the ZnO. The bending and strain field resulted a charge separation, which was detected as electrical currents in the AFM tip. The nanogenerators based on the piezoelectric properties of ZnO arrays are expected to have potential for powering remote sensors, biomedical devices and many other optoelectronic devices because many different kind of mechanical energy can be used to generate electricity.

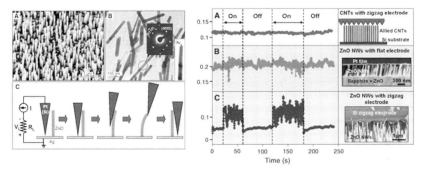

Figure 1. Converting mechanical energy into electricity using ZnO nanoarrays. Left panel: Electricity generated by AFM tips. Right panel: Electricity generated by saw shaped Si electrodes. Notice the electricity was only generated with ZnO using the zig-zag Si electrode. [Z. L. Wang, J. Song, Science, 2006, 312, 242. X. Wang, J. Song, Z. L. Wang, Science, 2007, 316, 102.]

Although in the AFM based device, the power generation efficiency can be as high as 17 to 30%, the manufacture and operation of the device are challenging. To solve this problem, Wang et al. replaced the AFM tip with a saw-shaped zig-zag Si electrode (Figure 1, right panel) [3]. The new device became much more practical, and also significantly increased the total current that can be generated.

1.2. Dye Sensitized Solar Cells

Solar energy is considered the ultimate solution to the energy challenge due to the fast depletion of fossil fuels. Currently the commercial solar energy technology is based on Si semiconductors, and the low efficiency and high cost have slowed wide spread applications. In early 1990s, a new class of dye sensitized solar cells (DSSCs) is reported and showed surprisingly high efficiency of over 4% considering that very inexpensive TiO_2 is used as the bulk of the photovoltaic cells [4]. The DSSCs are similar to a traditional electrochemical cell and are made of a few major components: a nanoporous semiconducting electrodes made of sintered TiO_2 nanoparticles and a dye molecule (bipyridine metal complex). Under photoexcitation, the dye molecules generate electrons and holes and inject the electrons into the TiO_2 semiconductors. The excited dye cations are reduced to the neutral ground state by a liquid electrolyte (iodide/triiodide redox-active couple dissolved in an organic solvent). The triiodide to iodide cycle is completed by drawing the electrons from the counter electrode. Because of the simplicity of the device and the low cost of the TiO_2 (which is essentially the same materials used in paints), the new DSSCs have great potential for large scale applications.

However, DSSCs still face significant challenges. First, even though the cost of the electrodes is low, the cost of the dye molecules is too higher. New, very inexpensive dye molecules that can efficiently absorb the sun light in the visible range are desired. Second, the long term stability, reliability and the cell operation needs to be significantly improved. Finally, the longstanding efficiency of 10% needs to be increased. Many approaches have been investigated to increase the efficiency by developing dyes with more efficient and broader spectral response, by increasing the open circuit voltage through manipulating the band gaps of the semiconductors and the redox agents, and significantly, by increasing the diffusion length of the electrons in the semiconductors.

The DSSCs depends on the high surface area nanocrystalline oxides as the anode for current collection. The large surface area and small crystalline sizes are required because of the need to anchor a large amount of dye molecules on the semiconductor surfaces. However, the diffusion of the electrons in the nanocrystalline materials is limited by the slow diffusion through different grains and by the trap states on the grain boundaries. Law et al. introduced a new concept by replacing the nanocrystalline films with oriented, lone, and high density ZnO nanowires prepared from solution seeded synthesis [5]. The high surface area is favorable for trapping the dye molecules, and the electron transport in oriented nanowires should be orders of magnitude faster than percolation in polycrystalline films. This approach produced a full sun efficiency of 1.5%, which is believed to be limited by the total available surface areas of the arrays (Figure 2). More recently, the same group reported that applying a thin crystalline TiO_2 coating, the efficiency can be increased to 2.25% [6]. The increase is attributed to the passivation of the surface trap sites and the energy barrier to repel the electrons from the surfaces.

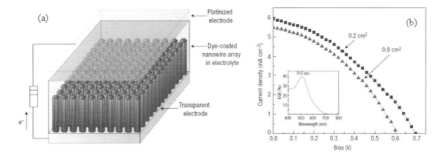

Figure 2. ZnO nanowires based DSSCs. (a) Schematic diagram of the cell. (b) Current density as a function of bias. The inset is the wavelength dependent quantum efficiency. [M. Law, L. L. Greene, J. C. Johnson, R. Saykally, P. Yang, Nature Materials, 2005, 4, 455.]

1.3. Superhydrophobic Surfaces (SHSs)

Surface wetting is one of the most fundamental phenomena in nature. It has been long realized that the wetting behavior is related to the surface roughness [7,8]. In plants such as lotus leaves, the self-cleaning and self-repairing mechanism depends on the fine nanostructures on the leaf surfaces that produce wetting angles as high as 160° (thus giving rise to the terminology superhydrophobic surfaces, or SHSs) [9]. There has been great interest in mimicking the SH phenomenon for practical applications. Not surprisingly, coatings of arrays of oriented nanowires become a very attractive candidate for such applications [10]. If the oriented nanowire arrays are considered as a porous film which can not be penetrated by the liquid, the measured apparent wetting angle θ_{eq} can be related to the true wetting angle θ on a smooth surface by the following equation [11]:

$$\cos\theta_{eq} = \Phi_s(\cos\theta + 1) - 1 \tag{1}$$

where Φ_s is the fraction of the area of the liquid-solid in touch, as estimated using the following expression:

$$\Phi_s = \begin{cases} \sigma \times \dfrac{\pi}{4} d^2 & \text{for needles} \\ \sigma \times \dfrac{3\sqrt{3}}{8} d^2 & \text{for rods} \end{cases} \tag{2}$$

here σ is the needle/rod number aerial density, and d represents the needle tip diameter or the maximum rod diameter.

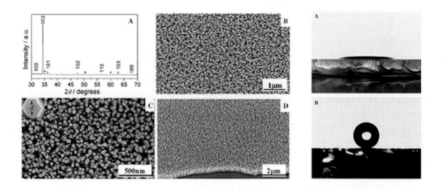

Figure 3. ZnO nanoarrays prepared from solution approaches and the wetting behavior of untreated and treated surfaces. Left panel: XRD pattern and SEM images of the ZnO arrays. Right panel: wetting of the untreated surface (top) and nonwetting of the octadecanethiol treated surface (bottom). [M. Guo, P. Deao, S. Cai, Thin Solid Films, 2007, 515, 7162.]

There are numerous examples of SHSs from nanowire arrays. Figure 3 shows an example of ZnO nanoarrays and the wetting behavior towards water [12]. The ZnO is hydrophilic and normally the wetting angle is close to 0°. However, after the ZnO surface is coated with hydrophobic molecules, the wetting angles increases to more than 160°. Significantly, the wetting behavior also depends on the microstructure of the nanowire arrays, which can only be systematically controlled using the solution based approach. It is worthwhile to point out that in semiconducting or conducting nanowire systems, the wetting angles can be tuned over a wide range by the application of an electrical voltage [13]. This provides means to externally control and manipulate the surface.

1.4. Synthesis of 1DNMs

In the literature, a large number of 1DNMs have been reported. A comprehensive coverage of the references in this area is beyond the scope of this paper, but in general most 1DNMs fall in the categories of (1) elemental, (2) inorganic compound, (3) polymeric, and (4) biomolecular. Among the elemental 1DNMs, carbon nanotubes (CNTs) [14], both single-walled (SWCNT) and multi-walled (MWCNT), are extensively studied. Other than the CNTs, an incomplete list of 1DNMs reported include Au [15], Ag [16], Pd [1], Pt [1], Si [17], Ge [18], Se [19], Te [20], W [21], Pb [22], Fe [23], Cu [24], and Bi [25]. About twice as many inorganic compound types of 1DNMs have been prepared, including nanowires, nanorods, and nanotubes of ZnO [26], In_2O_3 [27], Ga_2O_3 [28], GeO_2 [29], MgO [30], ZnS [31], CdS [32], CdSe [33], Ag_2Se [34], CdTe [35], GaN [36], GaAs [37], GaP [38], InP [39], InAs [40], SiC [41], WO_3 [42], SnO_2 [43], CuO [44], TiO_2 [45], MnO_2 [46], PbS [47], etc. In comparison, the synthesis of polymeric 1DNMs is more challenging. Only few examples of polymeric 1DNMs have been reported [48]. There has been an increasing interest in biomolecular 1DNMs, including DNA molecular nanowircs [49], peptide nanotubes [50], and other protein based nanofibers and nanotubes [51]. The biomolecular 1DNMs were either directly used for novel device applications, or were investigated as template to form new functional materials.

The papers in the Advanced Materials special issue discussed many methods for growing 1DNMs [1], including anisotropic crystal growth, gas phase reaction, templated growth, selective capping, self-assembly, etc. 1DNMS of materials with highly anisotropic crystalline structures can be prepared directly from either the solution phase or gas phase. Direct solution synthesis has been applied to Se and other chalcogens with controllable dimensions [19,20]. 1DNMs with less anisotropic crystalline structures were usually prepared with the help of a template structures, such as nanoporous membranes [52], mesoporous channels [53], or polymer chains [54]. Gas phase reaction is another commonly used method for 1DNMs [55]. Besides nanowires, nanobelts [56] and nanorings [57] have also been reported.

Most of the 1DNMs have been made in a randomly oriented fashion. 1DNMs with controlled orientation and architecture are highly desirable for many applications ranging from chemical and biological sensing and diagnosis, microelectronic devices and interconnects, energy conversion and storage (photovoltaic cells, batteries and capacitors, and hydrogen storage devices), catalysis, optical emission, display and data storage. For example, oriented nanowires are required for optical emission and display. Oriented nanostructures are also critical for improving the speed and sensitivity of sensing devices, and for rapid charging and discharging of energy storage devices. Catalytic applications not only require high surface area and high porosity, but favorable orientations for controlled kinetics and surface reactions. Therefore, growing oriented 1DNMs is a critical step for "bottom-up" approaches towards functional nanodevices.

One of the most widely used methods for preparing oriented 1DNMs is through gas phase reaction via a vapor-liquid-solid (VSL) mechanism [58]. In this mechanism, catalyst nanoparticles are first deposited on the substrate. The catalyst nanoparticle is melted and form an alloy with one the reacting elements in the vapor phase. The 1D nanowires are nucleated from the nanoparticles. The sizes of the nanowires are confined by the size of the catalyst particles. The VLS methods have been successfully used to prepare oriented carbon nanotubes [59,60], oriented ZnO nanowires [61,62], and many other materials.

Gas phase reaction is attractive for materials that are stable at relatively high temperatures. Compared to gas phase synthesis, solution based synthesis of 1DNMs received less attention. Trentler et al. developed a solution-liquid-solid (SLS) method similar to the VLS method [63]. In SLS method, the reaction is performed at a relatively low temperature in an organic solvent. The precursor is generated by decomposing organometallic compound. In general these techniques have not been applied to oriented 1DNMs. In this paper, we are interested in low temperature (below 100°C) aqueous solution based synthesis of oriented 1DNMs. We believe that such approaches will allow us to reduce the cost for large scale fabrication, improve processing reliability, and simplify the procedure for complicated shapes and geometries. The low temperature solution based synthesis conditions will also provide the opportunities to better control the experimental parameters and systematically fine tune the resultant microstructures.

2. Solution Approaches for Making Oriented 1DNM's

Two solution based approaches have been investigated in literature, one with, and the other without the use of a template. Template-based synthesis is a very general and versatile method and has been used in fabrication of nanorods, nanowires, and nanotubules of polymers, metals, semiconductors, and oxides. A variety of templates with nanosized channels have been explored.

In addition to the desired pore or channel size, morphology, size distribution and density of pores, template materials must meet other requirements. First, the template materials must be compatible with the processing conditions. For example, an electrical insulator is required for a template to be used in electrochemical deposition. Except for the template directed synthesis, template materials should be chemically and thermally stable during the synthesis and following processing steps. Secondly, depositing materials or solutions must wet the internal pore walls. Thirdly, for the synthesis of nanorods or nanowires, the deposition should start from the bottom or one end of the template channels and proceed from one side to another. However, for the growth of nanotubules, the deposition should start from the pore wall and proceed inwardly. Inward growth may result in the pore blockage, so that should be avoided in the growth of "solid" nanorods or nanowires. Kinetically, enough surface relaxation permits maximal packing density, so a diffusion-limited process is preferred. Other considerations include the easiness of releasing the nanowires or nanorods from the templates and easiness of handling during the experiments.

The template approach is easy to apply to a wide range of materials. A more challenging approach is to prepare oriented 1DNMS without using a template. The templateless approach is attractive because it eliminates a processing step. In the templated approach, the template needs to be carefully removed, and in many cases the nanowires show tendency to aggregate after template removal. Besides, the templated approach is difficult to apply to complex geometries, such as on textured surfaces, or within a confined space like a microfluidic channel.

However, developing a templateless approach requires a good understanding of the materials growth (nucleation and crystallization in terms of crystalline materials) and careful control of the experimental conditions. A general, new mechanism has yet to be developed for aligning the nanostructures without the guidance of a template. In order for this approach to be successful, the following events must be studied and understood: (1) nucleation, (2) growth, (3) formation of 1DNMs and mechanism of alignment, and (4) interaction at interfaces.

3. Template-Based Approach

The template approach has been extensively investigated to prepare free-standing, non-oriented 1DNMs and oriented 1DNMs. The most commonly used and commercially available templates are anodized alumina membrane [64], radiation track-etched polymer membranes [65]. Other membranes have also used as templates such as nanochannel array glass [66], radiation track-etched mica [67], and mesoporous materials [68], porous silicon by electrochemical etching of silicon wafer [69], zeolites [70] and carbon nanotubes [71,72]. Among the commonly used template, alumina membranes with uniform and parallel porous structure are made by anodic oxidation of aluminum sheet in solutions of sulfuric, oxalic, or phosphoric acids [64,73]. The pores can be arranged in a regular hexagonal array, and densities as high as 10^{11} pores/cm^2 can be achieved [74]. Pore size ranging from 10 nm to 100 µm can be made [64,75]. Polycarbonate membranes are made by bombarding a nonporous polycarbonate sheet, typically 6 to 20 µm in thickness, with nuclear fission fragments to create damage tracks, and then chemically etching these tracks into pores [67]. In these radiation track etched membranes, pores have a uniform size as small as 10 nm, though randomly distributed. Pore densities can be as high as 10^9 pores/cm^2.

3.1. Electrochemical Deposition

Electrochemical deposition, also known as electrodeposition, can be understood as a special electrolysis resulting in the deposition of solid material on an electrode. This process involves (1) oriented diffusion of charged reactive species (typically positively charged cations) through a solution when an external electric field is applied, and (2) reduction of the charged growth species at the growth or deposition surface which also serves as an electrode. In general, electrochemical deposition is only applicable to electrical conductive materials such as metals, alloys, semiconductors, and electrical conductive polymers, since after the initial deposition, electrode is separated from the depositing solution by the deposit and the electrical current must go through the deposit to allow the deposition process to continue. In industry, electrochemical deposition is widely used in making metallic coatings, a process also known as electroplating [76]. When deposition is confined inside the pores of template membranes, nanocomposites are produced. If the template membrane is removed, nanorods or nanowires are prepared.

When a solid immerses in a polar solvent or an electrolyte solution, surface charge will be developed. A surface oxidation or reduction reaction occurs at the interface between an electrode and an electrolyte solution, accompanied with charge transfer across the interface, until equilibrium is reached. For a given system, the electrode potential or surface charge density is described by the Nernst equation:

$$E = E_0 + \frac{RT}{n_i F} \ln(a_i) \tag{3}$$

where E_0 is the standard electrode potential, or the potential difference between the electrode and the solution, when the activity, a_i of the ions is unity, F, the Faraday's constant, R, the gas constant, and T, temperature. When the electrode potential is more negative (higher) than the energy level of vacant molecular orbital in the electrolyte solution, electrons will transfer from the electrode to the solution, accompanied with dissolution or reduction of electrode as shown in Figure 4a [77]. If the electrode potential is more positive (lower) than the energy level of the occupied molecular orbital, the electrons will transfer from the electrolyte solution to the electrode, and the deposition or oxidation of electrolyte ions on the electrode will proceed simultaneously as illustrated in Figure 4b [77]. The reactions stop when equilibrium is achieved.

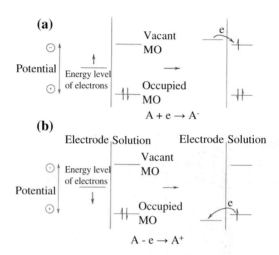

Figure 4. Representation of (a) reduction and (b) oxidation process of a species A in solution. The molecular orbitals (MO) of species A shown are the highest occupied MO and the lowest vacant MO. As shown, these correspond in an approximate way to the Eo's of the A/A- and A+/A couples, respectively. [A.J. Bard and L.R. Faulkner, Electrochemical Methods, John Wiley & Sons, New York, 1980.]

When an external electric field is applied to two dissimilar electrodes, electrode potentials can be changed so that electrochemical reactions at both electrodes and the electrons flow from a more positive electrode to a more negative electrode. This process is called electrolysis, which converts electrical energy to chemical potential. This process is widely used for applications of energy storage and materials processing. The system used for the electrolysis process is called electrolytic cell; in such a system the electrode connected to the positive side of the power supply is an anode, at which an oxidation reaction takes place, whereas the electrode connected to the negative side of the power supply is a cathode, at which a reduction reaction proceeds, accompanied with deposition. Sometimes, electrolytic deposition is therefore also called cathode deposition. In an electrolytic cell, it is not necessary that anode dissolves into the electrolytic solution and deposit is the same material as cathode.

Which electrochemical reaction takes place at an electrode (either anode or cathode) is determined by the relative electrode potentials of the materials present in the system. Noble metals are often used as an inert electrode in electrolytic cells.

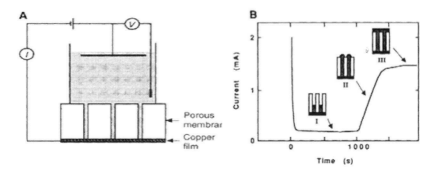

Figure 5. Common experimental set-up for the template-based growth of nanowires using electrochemical deposition. (a) Schematic illustration of electrode arrangement for deposition of nanowires. (b) Current-time curve for electrodeposition of Ni into a polycarbonate membrane with 60 nm diameter pores at –1.0 V. Insets depict the different stages of the electrodeposition. [T.M. Whitney, J.S. Jiang, P.C. Searson, and C.L. Chien, Science 261, 1316 (1993).]

Electrochemical deposition has been explored in the fabrication of nanowires of metals, semiconductors and conductive polymers. The growth of nanowires of conductive materials is a self-propagating process [78]. Once little fluctuation yields the formation of small rods, the growth of rods or wires will continue, since the electric field and the density of current lines between the tips of nanowires and the opposing electrode are greater than that between two electrodes due to a shorter distance. Therefore the growth species will be more likely deposit onto the tip of nanowires, resulting in continued growth. However, this method is not widely used in practice for nanowires since it is difficult to control the growth. Therefore, templates with desired channels are normally used for the growth of nanowires in electrochemical deposition. Figure 5 illustrates the common set-up for the template-based growth of nanowires using electrochemical deposition [79]. Template is attached onto the cathode, which is subsequently brought into contact with the deposition solution. The anode is placed in the deposition solution parallel to the cathode. When an electric field is applied, cations diffuse toward and reduce at the cathode, resulting in the growth of nanowires inside the pores of template. This figure also schematically shows the current density at different deposition stages when a constant electric field is applied. Possin [69] prepared various metallic nanowires by electrochemical deposition inside pores of radiation track-etched mica. Williams and Giordano [80] grew silver nanowires with diameters below 10 nm. The potentiostatic electrochemical template synthesis yielded different metal nanowires, including Ni, Co, Cu and Au with nominal pore diameters between 10 and 200 nm and the nanowires were found true replicas of the pores [81]. Whitney et al. [79] fabricated the arrays of nickel and cobalt nanowires by electrochemical deposition of the metals into track-etched-templates. Single crystal antimony nanowires have been grown by Zhang et al. in anodic alumina membranes using pulsed electrodeposition [82]. Single crystal and polycrystalline superconducting lead nanowires were also prepared by pulse electrodeposition [83]. The growth of single crystal lead nanowires required a greater departure from equilibrium conditions (greater overpotential) than the growth of polycrystalline ones. Semiconductor nanorods by electrodeposition include CdSe and CdTe synthesized by Klein et al in anodic alumina templates [84], and Schönenberger et al. [85] have made conducting polyporrole electrochemically in porous polycarbonate. Figure 6 shows SEM and TEM images and XRD spectrum of metal nanowires grown by electrochemical deposition in templates [82].

Hollow metal tubules can also be prepared using electrochemical deposition [86,87]. For growth of metal tubules, the pore walls of the template need to be chemically derivatized first so that the metal will preferentially deposit onto the pore walls instead of the bottom electrode. Such surface chemistry of the pore walls is achieved by anchoring silane molecules. For example, the pore surface of an anodic alumina template were covered with cyanosilanes, subsequent electrochemical deposition resulted in the growth of gold tubules [88].

An electroless electrolysis process has also been applied in the fabrication of nanowires or nanorods [86,89,90,91]. Electroless

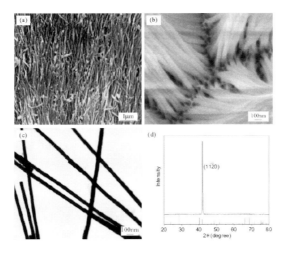

Figures 6. (a) Field-emission SEM image of the general morphology of the antimony nanowire array. (b) Field emission SEM showing the filling degree of the template and height variation of the nanowires. (c) TEM image of antimony nanowires showing the morphology of individual nanowires. (d) XRD pattern of the antimony nanowire array; the sole diffraction peak indicates the same orientation of all the nanowires. [Y. Zhang, G. Li, Y. Wu, B. Zhang, W. Song, and L. Zhang, Adv. Mater. 14, 1227 (2002).]

deposition is actually a chemical deposition and involves the use of a chemical agent to plate a material from the surrounding phase onto a template surface [92]. The significant difference between electrochemical deposition and electroless deposition is that in the former, the deposition begins at the bottom electrode and the deposited materials must be electrically conductive, whereas the latter method does not require the deposited materials to be electrically conductive and the deposition starts from the pore wall and proceeds inwardly. Therefore, in general, electrochemical deposition results in the formation of "solid" nanorods or nanowires of conductive materials, whereas the electroless deposition often grows hollow fibrils or nanotubes. For electrochemical deposition, the length of nanowires or nanorods can be controlled by the deposition time, whereas in electroless deposition the length of the nanotubes is solely dependent on the length of the deposition channels or pores, which often equal to the thickness of membranes. Variation of deposition time would result in a different wall thickness of nanotubules. An increase in deposition time leads to a thick wall and a prolonged deposition may form a solid nanorods. However, a prolonged deposition time does not guarantee the formation of solid nanorods. For example, the polyaniline tubules never closed up, even with prolonged polymerization time [93].

Nanotubes are commonly observed for polymer materials, even using electrochemical deposition, in contrast to "solid" metal nanorods or nanowires. It seems deposition or solidification of polymers insides template pores starts at the surface and proceeds inwardly. Martin [94] proposed to explain this phenomenon by the electrostatic attraction between the growing polycationic polymer and anionic sites along the pore walls of the polycarbonate membrane. In addition, although the monomers are highly soluble, the polymerized form is completely insoluble. Hence, there is a solvophobic component, leading to the deposition at the surface of the pores [95,96]. In the final stage, the diffusion of monomers through the inner pores becomes retarded and monomers inside the pores are quickly depleted. The deposition of polymer inside the inner pores stops and the entrance becomes corked. Figure 7a shows SEM images of such polymer nanotubes [97].

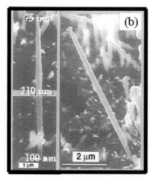

Figure 7. (a) SEM images of polymer nanotubes. [L. Piraux, S. Dubois, and S. Demoustier-Champagne, Nucl. Instrum. Methods Phys. Res. B131, 357 (1997).] (b) SEM images of non-uniformly sized metal nanowires grown in polycarbonate membranes by electrochemical deposition. [C. Schönenberger, B.M.I. van der Zande, L.G.J. Fokkink, M. Henny, C. Schmid, M. Krüger, A. Bachtold, R. Huber, H. Birk, and U. Staufer, J. Phys. Chem. B 101, 5497 (1997).]

Although many research groups have reported of growth of uniformly sized nanorods and nanowires grown on polycarbonate template membranes, Schönenberger et al. [85] reported that the channels of carbonate membranes were not always uniform in diameter. They grew metal, including Ni, Co, Cu, Au and polyporrole nanowires using polycarbonate membranes with nominal pore diameters between 10 and 200 nm by an electrolysis method. From both potentiostatic study of growth process and SEM analysis of nanowire morphology, they concluded that the pores are in general not cylindrical with a constant cross section, but are rather cigar-like. For the pores with a nominal diameter of 80 nm, the middle section of the pores is wider by up to a factor of 3. Figure 7b shows some such non-uniformly sized metal nanowires grown in polycarbonate membranes by electrochemical deposition [85].

3.2. Electrophoretic Deposition

The electrophoretic deposition technique has been widely explored, particularly for deposition of ceramic and organoceramic materials on cathode from colloidal dispersions [98,99,100]. Electrophoretic deposition differs from electrochemical deposition in several aspects. First, the deposit by electrophoretic deposition method needs not to be electrically conductive. Secondly, nanosized particles in colloidal dispersions are typically stabilized by electrostatic or electrosteric mechanisms. As discussed in the previous section, when dispersed in a polar solvent or an electrolyte solution, the surface of nanoparticles develops an electrical charge via one or more of the following mechanisms: (1) preferential dissolution or (2) deposition of charges or charged species, (3) preferential reduction or (4) oxidation, and (5) adsorption of charged species such as polymers. Charged surfaces will electrostatically attract oppositely charged species (typically called counter-ions) in the solvent or solution. A combination of electrostatic forces, Brownian motion and osmotic forces would result in the formation of a so-called double layer structure, and schematically illustrated in Figure 8.

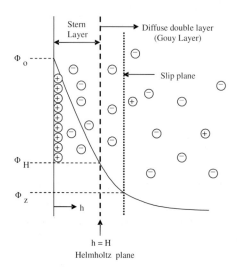

Figure 8. Schematic illustrating electrical double layer structure and the electric potential near the solid surface with both Stern and Gouy layers indicated. Surface charge is assumed to be positive.

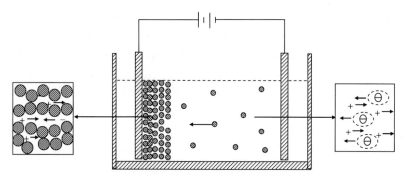

Figure 9. Schematic showing the electrophoresis. Upon application of an external electric field to a colloidal system or a sol, the constituent charged nanoparticles or nanoclusters are set in motion in response to the electric field, whereas the counter-ions diffuse in the opposite direction.

The figure depicts a positively charged particle surface, the concentration profiles of negative ions (counter-ions) and positive ions (surface-charge-determining-ions), and the electric potential profile. The concentration of counter-ions gradually decreases with distance from the particle surface, whereas that of charge-determining ions increases. As a result, the electric potential decreases with distance. Near to the particle surface, the electric potential decreases linearly, in the region known as the Stern layer. Outside of the Stern layer, the decrease follows an exponential relationship, and the region between Stern layer and the point where the electric potential equals zero is called the diffusion layer. Together, the Stern layer and diffusion layer are called the double layer structure in the classic theory of electrostatic stabilization.

Upon application of an external electric field to a colloidal system or a sol, the constituent charged particles are set in motion in response to the electric field, as schematically illustrated in Figure 9 [104]. This type of motion is referred to as electrophoresis. When a charged particle is in motion, some of the solvent or solution surrounding the particle will move with it, since part of the solvent or solution is tightly bound to the particle. The plane that separates the tightly bound liquid layer from the rest of the liquid is called the slip plane. The electric potential at the slip plane is known as the zeta-potential. Zeta-potential is an important parameter in determining the stability of a colloidal dispersion or a sol; a zeta potential larger than about 25 mV is typically required to stabilize a system [101]. Zeta potential is determined by a number of factors, such as the particle surface charge density, the concentration of counter-ions in the solution, solvent polarity and temperature. The zeta potential, ζ, around a spherical particle can be described as [102]:

$$\zeta = \frac{Q}{4\pi\varepsilon_r a(1+\kappa a)} \tag{4}$$

$$\text{with} \quad \kappa = \left(\frac{e^2 \sum n_i z_i^2}{\varepsilon_r \varepsilon_0 kT} \right)^{\frac{1}{2}}$$

where Q is the charge on the particle, a is the radius of the particle out to the shear plane, ε_r is the relative dielectric constant of the medium, and n_i and z_i are the bulk concentration and valence of the ith ion in the system, respectively. It is worthwhile to note that a positively charged surface results in a positive zeta potential in a dilute system. A high concentration of counter ions, however, can result in a zeta-potential of the opposite sign.

The mobility of a nanoparticle in a colloidal dispersion or a sol, μ, is dependent on the dielectric constant of the liquid medium, ε_r, the zeta potential of the nanoparticle, ζ, and the viscosity, η, of the fluid. Several forms for this relationship have been proposed, such as the Hückel equation [102]:

$$\mu = \frac{2\varepsilon_r \varepsilon_0 \zeta}{3\pi\eta} \tag{5}$$

Double layer stabilization and electrophoresis are extensively studied subjects. Readers may find additional detailed information in books on sol-gel processing [103,104,105] and colloidal dispersions [102,106].

Electrophoretic deposition simply uses oriented motion of charged particles in a electrical field to grow films or monoliths by enriching the solid particles from a colloidal dispersion or a sol onto the surface of an electrode. If particles are positively charged (more precisely speaking, having a positive zeta potential), then the deposition of solid particles will occur at the cathode. Otherwise, deposition will be at the anode. At the electrodes, surface electrochemical reactions proceed to generate or receive electrons. The electrostatic double layers collapse upon deposition on the growth surface, and the particles coagulate. There is not much information on the deposition behavior of particles at the growth surface. Some surface diffusion and relaxation is expected. Relatively strong attractive forces, including the formation of chemical bonds between two particles, develop once the particles coagulate. The films or monoliths grown by electrophoretic deposition from colloidal dispersions or sols are essentially a compaction of nanosized particles. Such films or monoliths are porous, i.e., there are voids inside. Typical packing densities, defined as the fraction of solid (also called green density) are less than 74%, which is the highest packing density for uniformly sized spherical particles [107]. The green density of films or monoliths by electrophoretic deposition is strongly dependent on the concentration of particles in sols or colloidal dispersions, zeta-potential, externally applied electric field and reaction kinetics between particle surfaces. Slow reaction and slow arrival of nanoparticles onto the surface would allow sufficient particle relaxation on the deposition surface, so that a high packing density is expected.

Many theories have been proposed to explain the processes at the deposition surface during electrophoretic deposition. Electrochemical process at the deposition surface or electrodes is complex and varies from system to system. However, in general, a current exists during electrophoretic deposition, indicating reduction and oxidation reactions occur at electrodes and/or deposition surface. In many cases, films or monoliths grown by electrophoretic deposition are electric insulators. However, the films or monoliths are porous and the surface of the pores would be electrically charged just like the nanoparticle surfaces, since surface charge is dependent on the solid material and the solution. Furthermore, the pores are filled with solvent or solution that contains counter-ions and charge-determining ions. The electrical conduction between the growth surface and the bottom electrode could proceed via either surface conduction or solution conduction.

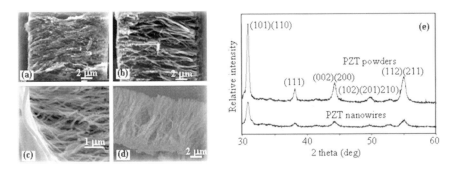

Figure 10. a) SEM micrograph of nanorods and b) X-ray diffraction spectra of Pb(Zr,Ti)O3 nanorods grown by template-based sol-gel electrophoretic deposition. [S. J. Limmer, S. Seraji, M. J. Forbess, Y. Wu, T. P. Chou, C. Nguyen, and G.Z. Cao, Adv. Mater. 13, 1269 (2001).]

Limmer et al. [108,109,110] combined sol-gel preparation and electrophoretic deposition in the growth of nanorods of various complex oxides. In their approach, conventional sol-gel processing was applied for the synthesis of various sols. By appropriate control of the sol preparation, nanometer particles with desired stoichiometric composition were formed, electrostatically stabilized by adjusting to an appropriate pH and uniformly dispersed in the solvent. When an external electric field is applied, these electrostatically stabilized nanoparticles will respond and move towards and deposit on either cathode or anode, depending on the surface charge (more precisely speaking, the zeta potential) of the nanoparticles. Using radiation tracked etched polycarbonate membranes with an electric field of ~1.5 V/cm, they have grown nanowires with diameters ranging from 40 to 175 nm and a length of 10 μm corresponding to the thickness of the membrane. The materials include anatase TiO_2, amorphous SiO_2, perovskite structured $BaTiO_3$ and $Pb(Ti,Zr)O_3$, and layered structured perovskite $Sr_2Nb_2O_7$. Nanorods grown by sol electrophoretic deposition are polycrystalline or amorphous. One of the advantages of this technique is the ability to synthesize complex oxides and organic-inorganic hybrids with desired stoichiometric composition; Figure 10 shows the SEM micrographs and XRD spectra of $Pb(Zr,Ti)O_3$ nanorods [108]. Another advantage is the applicability for variety of materials. Other materials, such as nanorods of SiO_2, TiO_2, $Sr_2Nb_2O_7$ and $BaTiO_3$ [109], have been prepared.

Wang et al. [111] used electrophoretic deposition to form nanorods of ZnO from colloidal sols. ZnO colloidal sol was prepared by hydrolyzing an alcoholic solution of zinc acetate with NaOH, with a small amount of zinc nitrate added to act as a binder. This solution was then deposited into the pores of anodic alumina membranes at voltages in the range of 10-400 V. It was found that lower voltages led to dense, solid nanorods, while higher voltages caused the formation of hollow tubules. The suggested mechanism is that the higher voltages cause dielectric breakdown of the anodic alumina, causing it to become positively charged as the cathode. Electrostatic attraction between the ZnO nanoparticles and the pore walls then leads to tubule formation.

Miao et al. [112] prepared single crystalline TiO_2 nanowires by template-based electrochemically induced sol-gel deposition. Titania electrolyte solution was prepared using a method developed by Natarajan and Nogami [113], in which Ti powder was dissolved into a H_2O_2 and NH_4OH aqueous solution and formed TiO^{2+} ionic clusters. When an external electric field was applied, TiO^{2+} ionic clusters diffused to cathode and underwent hydrolysis and condensation reactions, resulting in deposition of nanorods of amorphous TiO_2 gel. After heat treatment at

240°C for 24 hr in air, nanowires of single crystal TiO$_2$ with anatase structure and with diameters of 10, 20, and 40 nm and lengths ranging from 2 to 10 µm were synthesized. However, no axis crystal orientation was identified. The formation of single crystal TiO$_2$ nanorods here is different from that reported by Martin's group [114]. Here the formation of single crystal TiO$_2$ is via crystallization of amorphous phase at an elevated temperature, whereas nanoscale crystalline TiO$_2$ particles are believed to assemble epitaxially to form a single crystal nanorods. Epitaxial agglomeration of two nanoscale crystalline particles has been reported [115], though no large single crystals have been produced by assemble nanocrystalline particles. Figures 11a and 11b shows the micrograph and XRD spectra of single crystal nanorods of TiO$_2$ grown by template-based electrochemically induced sol-gel deposition [112].

Figure 11. (a) and (b) of single crystal nanorods of TiO2 grown by template-based electrochemically induced sol-gel deposition. [Z. Miao, D. Xu, J. Ouyang, G. Guo, Z. Zhao, and Y. Tang, Nano Lett. 2, 717 (2002).]. (c) and (d) Oxide nanorods made by filling the templates with sol-gels: (a) ZnO and (b) TiO2. [B.B. Lakshmi, P.K. Dorhout, and C.R. Martin, Chem. Mater. 9, 857 (1997).]

3.3. Template Filling

Direct template filling is the most straightforward and versatile method in preparation of nanorods and nanotubules. Most commonly, a liquid precursor or precursor mixture is used to fill the pores. There are several concerns in the template filling. First of all, the wetability of the pore wall should be good enough to permit the penetration and complete filling of the liquid precursor or precursor mixture. For filling at low temperatures, the surface of pore walls can be easily modified to be either hydrophilic or hydrophobic by introducing a monolayer of organic molecules. Second, the template materials should be chemically inert. Thirdly, control of shrinkage during solidification is required. If adhesion between the pore walls and the filling material is weak, or solidification starts at the center, or from one end of the pore, or uniformly throughout the rods, solid nanorods are most likely to form. However, if the adhesion is very strong, or the solidification starts at the interfaces and proceeds inwardly, hollow nanotubules are most likely to form.

3.3.1. Colloidal dispersion filling

Martin and his co-workers [114,116] have studied the formation of various oxide nanorods and nanotubules by simply filling the templates with colloidal dispersions. Colloidal dispersions were prepared using appropriate sol-gel processing. The filling of the template was to place a template in a stable sol for a various period of time. The capillary force is believed to drive the sol into the pores, when the surface chemistry of the template pores were appropriate modified to have a good wetability for the sol. After the pores were filled with sol, the template was withdrawn from the sol and dried prior to firing at elevated temperatures. The firing at elevated temperatures served two purposes: removal of template so that free standing nanorods can be obtained and densification of the sol-gel-derived green nanorods. This is a very versatile method and can be applied for any material, which can be made by sol-gel processing. However, the drawback is the difficult to ensure the complete filling of the template pores. Figures 11c and 11d show SEM micrographs of TiO$_2$ and ZnO nanorods made by filling the templates with sol-gel [114].

It is known that the typical sol consists of a large volume fraction of solvent up to 90% or higher [103]. Although the capillary force may ensure the complete filling of colloidal dispersion inside pores of the template, the amount of the solid filled inside the pores can be very small. Upon drying and subsequent firing processes, a significant amount of shrinkage would be expected. However, the results showed that the amount of shrinkage is small when compared with the size of the template pores. The results indicated that there are some unknown mechanisms, which enrich the concentration of solid inside pores. One possible mechanism could be the diffusion of solvent through the membrane, leading to the enrichment of solid along the internal surface of template pores, a process used in ceramic slip casting [117]. The observation of formation of nanotubules (as shown in Figure 12 [114]) by such a sol filling process may imply such a process is indeed present. However, considering the fact that the templates typically were emerged into sol for just a few minutes, the diffusion through membrane and enrichment of solid inside the pores must be a rather rapid process. It is also noticed that the nanorods made by template filling are commonly polycrystalline or amorphous. The exception was found, when the diameter of nanorods is smaller than 20 nm, single crystal TiO$_2$ nanorods were made [114].

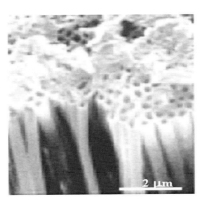

Figure 12. Hollow nanotubes formed by incomplete filling of the template. [B.B. Lakshmi, P.K. Dorhout, and C.R. Martin, Chem. Mater. 9, 857 (1997).]

3.3.2. Melt and solution filling

Metallic nanowires can also be synthesized by filling a template with molten metals [118, 26]. One example is the preparation of bismuth nanowires by pressure injection of molten bismuth into the nanochannels of an anodic alumina template [119]. The anodic alumina template was degassed and immersed in the liquid bismuth at 325°C (T_m = 271.5°C for Bi), and then high pressure Ar gas of ~300 bar was applied to inject liquid Bi into the nanochannels of the template for 5 hours. Bi nanowires with diameters of 13 –110 nm and large aspect ratios of several hundred have been obtained. Individual nanowires are believed to be single crystal. When exposed to air, bismuth nanowires are readily to be oxidized. An amorphous oxide layer of ~ 4 nm in thickness was observed after 48 hours. After 4 weeks, bismuth nanowires of 65 nm in diameter were found to be totally oxidized. Nanowires of other metals, In, Sn, and Al, and semiconductors, Se, Te, GaSb, and Bi_2Te_3 were prepared by injection of melt liquid into anodic alumina templates [37].

Polymeric fibrils have been made by filling a monomer solution, which contain the desired monomer and a polymerization reagent, into the template pores and then polymerizing the monomer solution [120,121,122,123]. The polymer preferentially nucleates and grows on the pore walls, resulting in tubules at short deposition times, as discussed previously in the growth of conductive polymer nanowires or nanotubules by electrochemical deposition and fibers at long times. Cai et al. [124] synthesized polymeric fibrils using this technique.

Similarly, metal and semiconductor nanowires have been synthesized through solution techniques. For example, Han et al. [125] have synthesized Au, Ag and Pt nanowires in mesoporous silica templates. The mesoporous templates were filled with aqueous solutions of the appropriate metal salts (such as $HAuCl_4$), and after drying and treatment with CH_2Cl_2 the samples were reduced under H_2 flow to convert the salts to pure metal. Nanowires and nanorods of both metal and oxides were carefully studied by Liu et al. [126] using high resolution electron microscopy and electron energy loss spectroscopy techniques. A sharp interface only exists between noble metal nanowires and the matrix. For magnetic nickel oxide, a core shell nanorod structure (Figure 13) was observed containing a nickel oxide core and a thin nickel silicate shell. The magnetic properties of the aligned nickel oxide were found to be significantly different from nickel oxide nanopowders due to the alignment of the nanorods. Chen et al filled the pores of a mesoporous silica template with an aqueous solution of Cd and Mn salts, dried the sample, and reacted it with H_2S gas to convert to (Cd,Mn)S [127]. Ni $(OH)_2$ nanorods have been grown in carbon-coated anodic alumina membranes by Matsui et al [128], by filling the template with ethanol $Ni(NO_3)_2$ solutions, drying, and hydrothermally treating the sample in NaOH solution at 150°C.

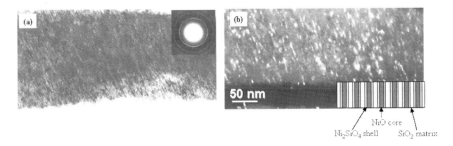

Figure 13. Core NiO nanowires using mesoporous silica as the template. (a) Bright field TEM image. The insert is a selected area electron diffraction pattern. (b) Dark field TEM image. The insert is the core-shell model based on TEM and extended fine structure electron energy loss spectroscopy. [J. Liu, G. E. Fryxell, M. Qian, L.-Q. Wang, Y. Wang, Pure and Applied Chemistry, 2000, 72, 269-279, 2000.]

3.3.3. Centrifugation

Template filling of nanoclusters assisted with centrifugation force is another inexpensive method for mass production of nanorod arrays. Figure 14 shows SEM images of lead zirconate titanate (PZT) nanorod arrays with uniformly sizes and unidirectional alignment [129]. Such nanorod arrays were grown in polycarbonate membrane from PZT sol by centrifugation at 1500 rpm for 60 minutes. The samples were attached to silica glass and fired at 650°C in air for 60 minutes. Nanorod arrays of other oxides including silica and titania have also been grown in this method. The advantages of centrifugation include its applicability to any colloidal dispersion systems including those consisting of electrolyte-sensitive nanoclusters or molecules. However, in order to grow nanowire arrays, the centrifugation force must be larger than the repulsion force between two nanoparticles or nanoclusters.

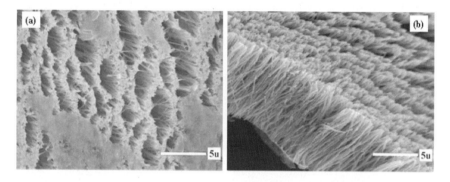

Figure 14. SEM images of the top view (a) and side view (b) of lead zirconate titanate (PZT) nanorod arrays grown in polycarbonate membrane from PZT sol by centrifugation at 1500 rpm for 60 min. Samples were attached to silica glass and fired at 650 C in air for 60 min. [T.L. Wen, J. Zhang, T.P. Chou, and G.Z. Cao, Adv. Mater. (2003).]

3.4. Converting from Consumable Templates

Nanorods or nanowires can also be synthesized using consumable templates [130], though the resultant nanowires and nanorods are in general not ordered to form an aligned array. Nanowires of compounds can be synthesized or prepared using a template-directed reaction. First nanowires or nanorods of constituent element is prepared, and then reacted with chemicals containing desired element to form final products. Gates et al. [131] converted single crystalline trigonal selenium nanowires into single crystalline nanowires of Ag_2Se by reacting with aqueous $AgNO_3$ solutions at room temperature. Nanorods can also be synthesized by reacting volatile metal halide or oxide species with formerly obtained carbon nanotubes to form solid carbide nanorods with diameters between 2 and 30 nm and lengths up 20 μm [132,133]. ZnO nanowires were prepared by oxidizing metallic zinc nanowires [134]. Hollow nanotubules of MoS_2 of ~ 30 μm long and 50 nm in external diameter with wall thickness of 10 nm were prepared by filling a solution mixture of molecular precursors, $(NH_4)_2MoS_4$ and $(NH_4)_2Mo_3S_{13}$ into the pores of alumina membrane templates. Then template filled with the molecular precursors was heated to an elevated temperature and the molecular precursors thermally decomposed into MoS_2 [135]. Certain polymers and proteins were also reported to have used to direct the growth of nanowires of metals or semiconductors. For example, Braun et al. [136] reported a two-step procedure to use DNA as a template for the vectorial growth of a silver nanorods of 12 μm in length and 100 nm in diameter. CdS nanowires were prepared by polymer-controlled growth [137]. For the synthesis of CdS nanowires, cadmium ions were well distributed in a polyacrylamide matrix. The Cd^{2+} containing polymer was treated with thiourea (NH_2CSNH_2) solvothermally in ethylenediamine at 170°C, resulting in degradation of polyacrylamide. Single crystal CdS nanowires of 40 nm in diameter and up to 100 μm in length with a preferential orientation of [001] were then simply filtered from the solvent.

4. Templateless 1DNMs Synthesis

4.1. General Theory of Nucleation and Growth

In contrast to the template synthesis, the conditions for growing oriented 1DNMs without a template are much less understood. In order to develop a general templateless approach, we need to understand the nucleation and crystal growth in different solubility regions. According to the classic theory of nucleation and growth [138], the free energy of forming stable nuclei on a substrate is determined by four factors: the degree of supersaturation S, the interfacial energy between the particle (c) and the liquid (l) σ_{cl}, the interfacial energy between the particle and the substrate (s) σ_{cs}, and the interfacial energy between the substrate and the liquid σ_{sl}:

$$\Delta G = -RT \ln S + \sigma_{cl} + (\sigma_{cs} - \sigma_{sl})\ A_{cs} \qquad (6)$$

where A_{cs} is the surface area of the particle.

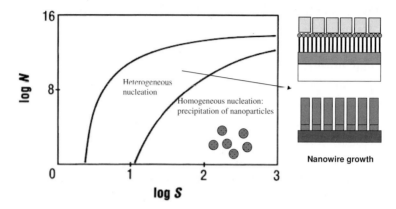

Figure 15. Solubility diagram reflected as number of nuclei generated in the solution as a function of the degree of supersaturation. The region (heterogeneous nucleation) for growing oriented nanostructured films and 1DNMs is indicated, either using self-assembled monolayers (functional groups) or using seeded growth techniques. Nanoparticles are precipitated in the homogeneous nucleation region.

Figure 15 is a schematic plot of the number of nuclei (N) as a function of degree of supersaturation (S), which is related to the concentration of the precursor and the solubility in the solution. Figure 16 suggests several regions for crystal growth. At a high concentration or a high temperature, homogeneous nucleation is dominant and precipitation is the bulk solution is the main mechanism. The region must be avoided if controlled crystal growth is desired. The region slightly above the solubility line is the heterogeneous nucleation region. In this region heterogeneous nucleation on a substrate dominates and therefore it is possible to grow uniform nanostructured films and oriented1DNMs. Unfortunately, most solution synthesis is carried out at too high a concentration so that undesirable precipitation dominates. As a result, oriented nanostructures were difficult to form.

From Equation (4) and Figure 16, we can use the following rules to design a generalized solution synthesis method for oriented nanostructures:

(1) Controlling the solubility of the precursors and the degree of supersaturation so that massive precipitation is not the dominating

reaction. Similar to VLS growth, the degree of supersaturation has a large effect on the growth behavior. A low degree of supersaturation promotes heterogeneous growth of 1DNMs, while a high degree of supersaturation favors bulk growth in the solution. Experimentally this is accomplished by reducing the reaction temperature, and by reducing the precursor concentrations as much as possible, while at the same time ensuring nucleation and growth can still take place. The formation of the new materials is characterized by the increase in cloudiness of the solution, which can be monitored by light scattering or turbidity measurement. A rapid increase in cloudiness is an indication of rapid precipitation and should be avoided.

(2) Reducing the interfacial energy between the substrate and the particle. In most cases, the activation energy for crystal growth on a substrate is lower that that required to create new nuclei from the bulk solution. Still, chemical methods can be used to further lower the interfacial energy. For example, self-assembled monolayers containing surface active groups were used to promote heterogeneous nucleation. The self-assembled monolayers had at least two functions: reducing the surface tension, and stimulate nucleation of specific crystalline phases. A "biomimetic" approach was developed to prepare oriented nanostructured ceramic films [138]. Highly oriented nanorods of FeOOH and other materials were reported [139].

(3) Controlling of crystal growth. Most crystalline materials, such as ZnO, have anisotropic crystalline structures and specific growth habits. If the growth along certain directions is much faster than other directions, nanowires or nanorodes can be produced.

During crystal growth, different facets in a given crystal have different atomic density, and atoms on different facets have a different number of unsatisfied bonds (also referred to as broken or dangling bonds), leading to different surface energy. Such a difference in surface energy or the number of broken chemical bonds leads to different growth mechanisms and varied growth rates. According to Periodic Bond Chain (PBC) theory developed by Hartman and Perdok [140], all crystal facets can be categorized into three groups based on the number broken periodic bond chains on a given facets: flat surface, stepped surface, and kinked surface. The number of broken periodic bond chains can be understood as the number of broken bonds per atom on a given facets in a simplified manner.

The PBC theory suggests different growth rate and behavior on different facets [141,142]. For example, in a simple cubic crystal, according to the PBC theory, {100} faces are flat surfaces (denoted as F-face) with one PBC running through one such surface, {110} are stepped surfaces (S-face) that have two PBCs, and {111} are kinked surfaces (K-face) that have three PBCs. For {110} surfaces, each surface site is a step or ledge site, and thus any impinging atom would be incorporated wherever it adsorbs. For {111} facets, each surface site is a kink site and would irreversibly incorporate any incoming atom adsorbed onto the surface. For both {110} and {111} surfaces, the above growth is referred to as a random addition mechanism and no adsorbed atoms would escape back to the vapor phase. Both {110} and {111} faces have faster growth rate than that of {100} surface in a simple cubic crystal. In a general term, S-faces and K-faces have a higher growth rate than F-faces. For both S and K faces, the growth process is always adsorption limited, since the accommodation coefficients on these two type surfaces are unity, that all impinging atoms are captured and incorporated into the growth surface. For F faces, the accommodation coefficient varies between zero (no growth at all) and unity (adsorption limited), depending on the availability of kink and ledge sites.

These theories help explain why some facets in a given crystal grow much faster than others. Facets with fast growth rate tend to disappear, i.e., surfaces with high surface energy will disappear. In a thermodynamically equilibrium crystal, only those surfaces with the lowest total surface energy will survive as determined by the Wulff plot [143,144]. Therefore, the formation of high aspect ratio nanorods or nanowires entirely based on different growth rates of various facets is limited to materials with special crystal structures. In general, other mechanisms are required for the continued growth along the axis of nanorods or nanowires, such as defect-induced growth, impurity-inhibited growth, or physical confinement. Xia and many other groups investigated using capping agent to force the growth of long nanowires [145]. Sun et al [141] using Ag nanoparticles as the seeds, and poly(vinyl

Liang *et al* [173] reported a direct electrochemical synthesis of oriented nanowires of polyaniline (PANI), a conducting polymer with its backbone conjugated by phenyl and amine groups, from solutions using no templates. The entire synthesis involves not only the electropolymerization of aniline ($C_6H_5NH_2$) but also the in situ electrodeposition, in concert with a growth of the PANI nanowires in an oriented array. The experimental design is on such basis that, in theory, the rate of electropolymerization (or nanowires) is related to the current density. Therefore, it is possible to control the nucleation and the polymerization rate by adjusting the current density.

Large arrays of oriented polymer nanowires were produced on a wide range of substrates, including metal, glass, oxide, carbon, microbead, etc. The electrolyte solutions contained 0.5 M aniline and 1.0 M perchloric acid. Several steps are creatively used in this synthesis, each with a different current density (Figure 20a). The first step, with a higher current density, provides a dense and uniformed array of PANI nanoparticles as nanoseeds on substrate. In the subsequent steps, the current density was reduced step-by-step to suppress the creation of new polymer particles, and to promote the growth of the polymers from the existing polymer seeds on the substrate. Upon doing so, only those nanowires, with their orientations perpendicular to the substrate surface, would be allowed to grow into the long-and-thin nanowires, forming a dense oriented array (Figures 20b to 20e). These oriented conductive nanowires have thin tips, facing upward on substrate, ~50-70 nm in diameter for each. These tips are well separated with an ample space around each nanowire, good for other chemical species to easily reach the nanowire surfaces in sensing applications. On the other hand, a one-step electrochemical deposition resulted in a mixture of misaligned nanowires and nanodots with different diameters.

The step-wise electrochemical synthetic strategy, through the separation of the nucleation and the growth processes, is in line with that for growing oriented ZnO nanowires. Therefore, the polymer nanowires can also be prepared by textured surface and be patterned.

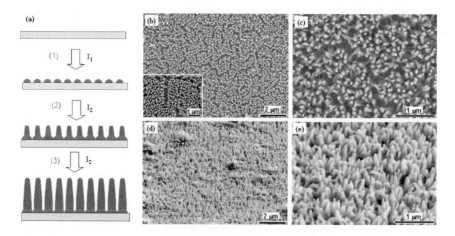

Figure 20. Large arrays of oriented and patterned ZnO nanowires using seeded growth. (a) A Low magnification SEM image of large arrays of ZnO. (b) A SEM image with a tilted sample showing good alignment. (c) ZnO microrods without seeds. (d) Preferred ZnO nanorod growth with stamped circular pattern of ZnO nanoseeds. Reproduced from Reference 144 with permission from Nature Materials. [L. Liang, J. Liu, C. F. Jr. Windisch, G. J. Exarhos, Y. Lin, Angew. Chemie. Int. Ed. 2002, 41, 3665.]

Since the conducting PANI is electrochemically active, the cyclic voltammetry (CV) is used for characterization on a pair of redox CV peaks characteristic for the PANI. The CV results show that the oriented sample has the highest peak intensities for the pair, nearly ten times that for a sample from the one-step synthesis, and three times that for a sample with interconnected and misoriented 1DNMs. This difference implies that the oriented nanowires would have the highest surface areas that electrolytes can access, suggesting a high efficiency and sensitivity desired in making sensing devices.

Polarized Fourier transform infrared (FTIR) spectroscopy was used for characterizing these oriented PANI nanowires. The difference in the spectra between the p-polarized mode (perpendicular to the substrate surface), and an s-polarized mode (parallel to the substrate surface) suggested that the polymer molecules are also aligned within the nanowires. Near phenyl planes are vertically aligned on their edges and all PANI backbones are parallel to the substrate surface–like a molecular fence. This alignment implies that these PANI nanowires could be semicrystalline in structure. Based on this result, novelties in melting point, yielding behavior, toughness, elasticity, and electrical and thermal conductivities can be expected.

4.4. Sequential Nucleation and Growth

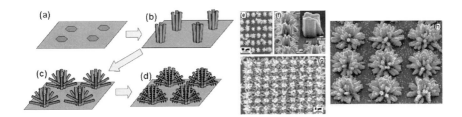

Figure 21. Strategies for growing complex hierarchical structures by multi-step, sequential nucleation and growth. (a) Planting seeds of micropatterns. (b) First generation nanorods. (c) Second generation structures. (d) Third generation structures. (e) to (g) Large arrays of micropatterned second generation structure. (h) Arrays of micropatterned third generation structures. [T. L. Sounart, J. Liu, J. A. Voigt,J. W. P Hsu, E. D. Spoerke, Z. Tian,Y. Jian, Advanced Functional Materials, 16, 335-344, 2006.]

Mutil-step, seeded growth of simple one dimensional oriented nanocrystalline films and multi-stage growth of more complicated nanostructures are vanguards for a process to systematically assemble complex, hierarchical crystal architectures. Several groups have recently used more than one synthesis step to nucleate new oriented nanocrystals on crystals formed in a previous reaction step. In gas-phase synthesis, Dick et al. [174] first synthesized GaP and InP semiconductor nanowires, and then sequentially reseeded the nanowire surfaces to produce tree-like nanostructures. In solution-phase synthesis, similar ZnO [175] and ZnO/TiO$_2$-based [176] composite nanostructures were produced by multi-step precipitation of powders in bulk aqueous solutions. We have investigated multi-step, aqueous nucleation and growth methods that have produced higher-order, hierarchical films of several important minerals [177]. In the hierarchical growth method, the first step is to create nucleation sites, via processes such as seed deposition or micropatterning of SAMs (Figure 21). Oriented nanocrystals grow from these nucleation sites, and in subsequent reaction steps, new crystals nucleate and grow on the crystals produced in previous stages. This provides the capability to build a diverse range of complex nanostructures from various primary subunits that can be tuned by the growth chemistry in each step.

Higher-order ZnO crystal structures assembled from rod- or needle-shaped primary subunits have been synthesized with a sequential nucleation and growth process. Primary ZnO rods are first prepared on micropatterned substrate (Figure 21b). During the second step, new crystals grow on the surfaces of the primary rods when bifunctional diaminoalkane molecules are added to the solution (Figure 21c). In the third step, secondary crystals are "healed" to the hexagonal prismatic shape, and in the fourth step, additional branch crystals nucleate from the secondary structures to form unusual wagon wheel-like crystals (Figure 21d). Therefore, with multiple reaction steps, new nucleation sites and a variety of large supercrystal structures can be created. Morphological variations between the structures, such as the size, population density, and shape of the rods in each stage, can be precisely controlled with the solution chemistry in each step.

The combination with top-down approach is most attractive for generating hierarchical structures. Two-dimensional patterns of oriented nanocrystals can be created by modifying the spatial distribution of the interfacial energy on a substrate. For example, Aizenberg et al. [178,179,180] investigated the combination of SAMs and soft lithography (microstamping or microcontact printing) to prepare spatially controlled micropatterns of calcite crystals on a surface with precisely controlled location, nucleation density, size, orientation, and morphology. Mineral nucleation was favored on acid-terminated regions, but suppressed in methyl-terminated regions where the influx of nutrients was maintained below saturation. We applied similar microcontact printing techniques to grow oriented ZnO nanorods on patterned substrates [160]. Extended microarrays of carboxyl-terminated alkylthiols were printed on electron beam evaporated silver films. When the patterned silver substrates were placed in aqueous zinc nitrate solutions, oriented ZnO nanorods formed on the bare silver surface, but not on the surface covered by the carboxylic acid groups. Using this approach, we were able to make patterned lines, dots, and a variety of structures, and control the density and the spacing to micron scales (Figure 18e).

We produced micropatterns of hierarchical ZnO nanorod clusters by diamine-induced sequential nucleation and growth on micropatterned primary crystals. Secondary growth produced flower-like crystals from new crystal growth on the top face of the primary rods and side branches formed on the edge. Figures 21e to 21g show arrays of ordered flower-like ZnO structures that formed during secondary growth on a micropattern of oriented primary rods with the top and side view, and Figures 21g show a densely packed arrays of similar structures in which the secondary crystals are almost connected. Additional growth steps with diamine produce a pattern of tertiary structures with fine-branched crystals, such as those shown in Figure 21h. Some of these small tertiary subunits also nucleate sparsely on the substrate off the pattern, but it is remarkable how well the substrate is protected by only a monolayer through four reaction stages conducted over the course of several days. The length, morphology, and population density of the tertiary subunits are tunable with the reaction conditions as discussed previously. Thus, by combining top-down micropatterning techniques with bottom-up chemical synthesis control, complex tertiary "cactus-like" crystals can be tuned in structure and organized spatially on a substrate.

The key to the sequential nucleation and growth process is the ability to induce secondary nucleation of new, oriented branched crystals on primary ZnO. Without the diamine molecules, new nucleation is not observed after the primary stage, and repeated crystal growth simply increases the primary crystal size. This is a very unusual, important phenomenon that does not occur in typical crystal growth, but is not limited to ZnO [181]. In typical growth, different crystalline planes in a given crystal have different atomic density, and atoms on different facets have a different number of unsatisfied bonds, leading to different surface energies for different facets. During crystal growth, crystals will epitaxially grow larger [141,142]. Facets with a fast growth rate tend to disappear, i.e., surfaces with high surface energy will disappear. On a thermodynamically equilibrated crystal, those surfaces with the lowest total surface energy will survive, as determined by the Wulff Plot

[143,144]. However, such nucleation and growth theory cannot explain the secondary nucleation phenomena.

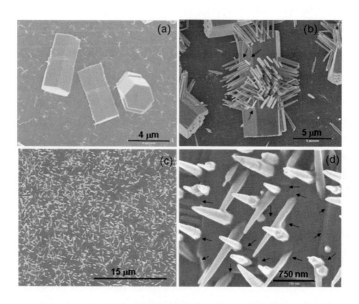

Figure 22. SEM images of secondary nucleation on larger primary ZnO crystals and on (100) ZnO substrates. (a) Primary ZnO crystals grown with 50 mg/L citrate in the first growth stage. (b) Secondary ZnO crystal from renucleation on the primary crystals in second-stage growth. (c, d) Branch nucleation on ($1\bar{1}00$) ZnO substrate. [T. L Sounart, J. Liu, James A. Voigt, M. Huo, E. D. Spoerke, B. Mckenzie, J. Am. Chem. Soc., ASAP article, 2007]

The concentration of diamine is critical. Secondary nucleation only occurs within a critical diamine concentration range [182]. Careful study of the different diamines suggests that the secondary nucleation may be related to the dissolution of ZnO surfaces. The critical diamine concentration might be the concentration required to cause light surface dissolution (etching) of the primary ZnO crystal during the approach to the 60°C incubation temperature, thus creating nucleation sites during supersaturation conditions at 60°C. To understand where the renucleation takes place on the ZnO surface, we studied the nucleation of ZnO nanorods on larger primary ZnO crystals, as well as on single crystalline ZnO ($1\bar{1}00$) prismatic surfaces. Figure 22a shows a typical hexagonal bi-crystal with a joining middle plane due to the polar nature of ZnO. The ($1\bar{1}00$) surface is smooth except for the middle plane. Figure 22b shows the morphology with the growth of the branched crystals. Notice that many of the branched crystals are concentrated in the central region. Careful examination reveals a high density of line defects under the branched crystals in this central area. We believe that the line defects were created by the early dissolution event, and served as pin-points for the secondary nucleation to occur. Branch-growth on single crystalline ZnO (Figures 22c and 22d) further supports that the new crystals are mostly formed on the defect sites of the {$1\bar{1}00$} surfaces, such as edges and holes, as indicated by the arrows in Figure 22d. Therefore, our results support the hypothesis that the CNC may be the concentration at which light surface etching forms nucleation sites, and the upper critical concentration is bound by the concentration to cause complete dissolution of the ZnO crystals [182].

The mechanism of branch nucleation driven by solubility is also consistent with the report of secondary nucleation of ZnO using an entirely different chemistry, viz. NaOH solutions [183]. However, controlled tunable branch nucleation has only been demonstrated to date with diamines. The bifunctional diamine molecules might play additional roles in promoting branch nucleation. Kanaras et al. [184] observed that bifunctional phosphonic acids promoted more branching of CdSe and CdTe nanocrystals than monofunctional phosphonic acids, and suggested that the bifunctional molecules may be important because they yield higher local acid concentrations at the molecular level than the monofunctional equivalent.

5. Concluding Remarks

This review has mainly covered two major approaches for solution synthesis of oriented 1DNMs on different substrates, one with templates and the other using nanoseeds. Much of this review is to focus on the novelties in synthetic approach and detailed developments in methodology.

For all the synthetic successes highlighted by the template-based solution approach, multiple templates are used. They are anodized alumina membrane, radiation track-etched polymer membranes, nanochannel array glass, radiation track-etched mica, mesoporous materials, porous silicon by electrochemical etching of silicon wafer, zeolites, carbon nanotubes, etc. In contrast, only three general methods for growing the 1DNMs are used in all these syntheses: electrochemical deposition, electrophoretic deposition, and template filling, depending on the nature of each 1DNM. The template based approach is straight forward and applicable to a wide range of materials, including metal, polymers, oxides, and more complex compositions.

In the work highlighted by the templateless approach, the main challenge is to control the nucleation, crystal growth, and growth kinetics. The templateless approach is complementary to the template approach, may simplify the procedure, and is applicable to both simple and complicated geometries. Both seedless and seeded growth methods have been successfully used for oxide microrods, microtubes, nanowires, nanotubes, and polymeric materials. The flexibility of this technique to systematically control the sizes of the 1DNMs, and the ability to form complex oriented nanostructures has been demonstrated. However, there technique is still at an infancy stage and has not been applied to a wide range of materials.

We hope that this review highlighted the importance progresses made in the synthesis of oriented 1DNMs. These materials will find applications for energy, environment, electronics, and biomedicine related applications. The future challenge is the understanding of the performance requirements of the specific applications, and developing materials designing and synthesis rules for the nanomaterials to meet these requirements.

Acknowledgements

The authors thank the support from PNNL's Laboratory-Directed Research and Development Program and by the Office of Basic Energy Sciences (BES), U. S. Department of Energy (DOE). PNNL is a multiprogram laboratory operated by Battelle Memorial Institute for the

Department of Energy under Contract DE-AC05-76RL01830. Guozhong Cao wants to acknowledge the support from NSF, AFOSR, DOE, and WTC. The authors also thank the Dr. Zhengrong (Ryan) Tian (University of Arkansas) and Dr. Qifeng (Jeff) Zhang (University of Washington) for their assistance in manuscript preparation.

References

1. Y. Xia, P. Yang, Y. Sun, Y. Wu, B. Mayers, B. Gates, Y. Yin, F. Kim, H.Yan, *Adv. Mater.* **2003**, *15*, 353.
2. Z. L. Wang, J. Song, *Science*, **2006**, 312, 242.
3. X. Wang, J. Song, Z. L. Wang, *Science*, *2007*, 316, 102.
4. a) B. O'Regan, M. Grätze, *Nature*, **1991**, 252, 737. b) M. Grätze, *Nature*, **2000**, 403, 363.
5. M. Law, L. L. Greene, J. C. Johnson, R. Saykally, P. Yang, *Nature Materials*, **2005**, 4, 455.
6. M. Law, L. E. Greene, A. Radenovic, T. Kuykendall, J. Liphardt, P. Yang, *J. Phys. Chem. B*, **2006**, 110, 22652.
7. R. N. Wenzel, J. Phys. Colloid Chem., **1949**, 53, 1466.
8. A. B. D. Cassie, S. Baxter, *Trans. Farad. Soc.*, **1944**, 40, 546.
9. W. Barthlott, C. Neinhuis, *Planta*, **1997**, 202, 1.
10. a) X. Wu, L. Zheng, D. Wu, *Langmuir*, **2005**, 21, 2665. b) C. Journet, S. Moulinet, C. Ybert, S. T. Purcell, L. Bocquet, L, *Europhys. Lett.*, **2005**, 71, 104-109. c) L. Feng, S. Li, Y. Li, H. Li, L. Zhang, J. Shai, Y. Song, B. Liu, D. Zhu, D., *Adv. Mater.*, **2002**, 14,1857. d) H. Li, X. Wang, Y. Song, Y. Liu, Q. Li, L. Jiang, D. Zhu, *Angew. Chem. Int. Ed.*, **2001**, 40, 1743. e) E. Hosono, S. Fujihara, I. Honma, H. Zhou, H., *J. Am. Chem. Soc.*, **2005**, 127, 13458.
11. C. R. Kessel, S Granick, *Langmuir*, **1999**, 7, 532.
12. M. Guo, P. Deao, S. Cai, *Thin Solid Films*, **2007**, 515, 7162.
13. N. Verplanck, E. Galopin, J.-C. Camart, V. Thomy, *Nano Letters*, **2007**, 7, 813.
14. S. Iijima, *Nature*, *1991*, 354, 56.
15. a) J. H. Song, Y. Wu, B. Messer, H. Kind. P. Yang, *J. Am. Chem. Soc.* **2001**, *123*, 10397. b) S. Link, Z. L. Wang, M. A. El-Sayed, *J. Phys. Chem. B* **2000**, *104*, 7867. c) Z. J. Wang, M. B. Mohemed, S. Link, M. A. El-Sayed, *Surf. Sci.* **1999**, *440*, L809. d) G. Bogel, H. Meekes, P. Bennema, D. Bollen, *J. Phys. Chem. B* **1999**, *103*, 7577. e) N. R. Jana, L. Gearheart, C. J. Murphy, *J. Phys. Chem. B* **2001**, *105*, 4065. f) N. R. Jana, L. Gearheart, C. J. Murphy, *Chem. Comm.* **2001**, 617.
16. a) Y. Sun, B. T. Mayers, T. Xia, *Nano Lett.* **2002**, 2, 481. b) Y. Sun, Y. Yin, B. Mayers, T. Herricks, Y. Xia, *Chem. Mater.* **2002**, *14*, 4736. c) Y. Sun, Y. Xia, *Adv. Mater.* **2002**, *14*, 833. d) Y. Sun, B. Gates, B. Mayers, Y. Xia, *Nano Lett.* **2002**, 2,

165. f) X. Wen. S. Yang, *Nano Lett.* **2002**, *2*, 451. g) N. R. Jana, L. Gearheart, C. J. Murphy, *Adv. Mater.* **2001**, *13*, 1389.

17.   a) T. I. Kamins, R. S. Williams, D. P. Basile, H. T. J. S. Harris, *J. Appl. Phys.* **2001**, *89*, 1008. b) T. Hanrath, B. A. Korgel, *Avd. Mater.* **2003**, *15*, 437. c) X. Lu, T. Hanrath, K. P. Johnson, B. A. Korgel, *Nano Lett.* **2003**, *3*, 93. T. Hanrath, B. A. Korgel, *J. Am. Chem. Soc.* **2001**, *124*, 1424. d) J. D. Holmes, K. P. Johnson, R. C. Doty, B. A. Korgel, *Science*, **2000**, *287*, 1471.

18.   A. M. Morales, C. M. Lieber, *Nano Lett.* **2002**, *2*, 487.

19.   a) B. Gates, B. Mayers, B. Cattle, Y. Xia, *Adv. Funct. Mater.* **2002**, *12*, 219. b) B. Gates, Y. Xia, *J. Am. Chem. Soc.* **2000**, *122*, 12582.

20.   a) B. Mayers, Y. Xia, *J. Mater. Chem.* **2002**, *12*, 1875. b) B. Mayers, Y. Xia, *Adv. Mater.* **2002**, *14*, 279. c) M. Mo, J. Zeng, X. Liu, W. Yu, S. Zhang, Y. Qian, *Adv. Mater.* **2002**, *14*, 1658. d) F. Fievet, J. P. Lagier, M. Figlaz, *MRS Bull.* **1989**, Dec., 29.

21.   Y. Li, X. Li, Z.-X. Deng, B. Zhou, S. Fan, J. Wang, X. Sun, *Angew. Chem. Int. Ed.* **2002**, *41*, 333.

22.   G. Yi, W. Schwarzecher, *Appl. Phys. Lett.* **1999**, *74*, 1747.

23.   S-J. Park, S. Kim, S. Lee, Z. G. Khim, K. Char, T. Hyeon, *J. Am. Chem. Soc.* **2000**, *122*, 8581.

24.   W. S. Yun, J. J. Urban, Q. Gu, H. Park, *Nano Lett.* **2002**, *2*, 447.

25.   A) Z. Zhang, D.hlman, M. S. Dresselhaus, J. Y. Ying, *Chem. Mater.* **1999**, *11*, 1659. b) Z. Zhang, J. Y. Ying, M. S. Dresselhaus, *J. Mater. Res.* **1998**, 13, 1745.

26.   a) M. Huang, S. Mao, H. Feick, H. Yan, Y. Wu, H. Kind, E. Weber. R. Russo, P. Yang, *Science*, **2001**, *292*, 1897. b) M. Huang, Y. Wu, H. Feick, N. Tran, E. Weber. P. Yang, *Adv. Mater.* **2001**, *13*, 113. c) Z. W. Pan, Z. R. Dai, Z. L. Wang, *Science*, **2001**, *291*, 1947.

27.   C. H. Liang, G. W. Meng, Y. Lei, F. Phillipp, L. D. Zhang, *Adv. Mater.* **2001**, *13*, 1330.

28.   a) X. C. Wu, W. H. Song, W. D. Huang, M. H. Pu, B. Zhao, Y. P. Sun, J. J. Du, *Chem. Phys. Lett.* **2000**, *328*, 5. b) H. Z. Zhang, Y. C. Kong, Y. Z. Wang, X. Du, Z. G. Bai, J. J. Wang, D. P. Yu, Y. Ding, Q. L. Hang, S. Q. Feng, *Solid State Commun.* **1999**, *109*, 677. c) C. H. Liang, G. W. Wang, Y. W. Wang, L. D. Zhang, S. Y. Zhang, *Appl. Phys. Lett.* **2001**, *78*, 3202.

29.   Z. G. Bai, D. P. Yu, H. Z. Zhang, Y. Ding, Y. P. Wang, X. Z. Gai, Q. L. hang, G. C. Xiong, S. Q. Feng, *Chem. Phys. Lett.* **1999**, *303*, 311.

30.   a) P. D. Yang, C. M. Lieber, *Science*, **1996**, 273, 1836. b) Y. Yin, G. Zhang, Y. Xia, *Adv. Funct. Mater.* **2002**, *12*, 293.

31.   a) D. Zhang, L. Qi, H. cheng, J. Ma, *J. Coll. Surf. Sci.* **2002**, *246*, 413. b) J. Xu, Y. Li, *J. Coll. Surf. Sci.* **2003**, *259*, 275. c) C. Lan, K. Hong, W. Wang, G. Wang, *Solid State Commun.* **2003**, *125*, 455. c) X. Chen, H. Xu, N. Xu, F. Zhao, W. Lin,

G. Lin, Y. Fu, Z. Huang, H. Wang, M. Wu, *Inorg. Chem.* **2003**, *42*, 3100. d) Z. Qiao, G. Xie, H. Tao, Z. Nie, Y. Lin, X. Chen, *J. Solid State Chem.* **2002**, *166*, 49.

32. a) Y. Jun, S.-M. Lee, N.-J. Kang, J. Cheon, *J. Am. Chem. Soc.* **2001**, *123*, 5150. b) J. H. Zhan, X. G. Yang, D. W. Wang, S. D. Li, Y. Xie, Y. Xia, Y. T. Qian, *Adv. Mater.* **2000**, *12*, 1348.

33. a) X. Peng, L. Manna, W. Yang, J. Wickham, E. Scher, A. Kadavanish, A. P. Alivisatos, *Nature*, **2000**, *404*, 59. b) L. Manna, E. Scher, A. P. Alivisatos, *J. Am. Chem. Soc.* **2000**, *122*, 12700. c) Z. A. Peng, X. Peng, *J. Am. Chem. Soc.* **2001**, *123*, 1389. d) Q. Yang, K. B. Tang, C. R. Wang, Y. T. Qian, S. Y. Zhang, *J. Phys. Chem. B* **2002**, *106*, 9227.

34. a) B. Gates, Y. Wu, Y. Yin, P. Yang, Y, Xia, *J. Am. Chem. Soc.* **2001**, *123*, 11500. b) B. Gates, B. Mayers, Y. Wu, Y. Sun, B. cattle, P. Yang, Y, Xia, *Adv. Funct. Mater.* **2002**, *12*, 679.

35. Z. Tang, N. A. Kotov, M. Giersig, *Science*, **2002**, *297*, 237.

36. a) Y. Huang, X. Duan, Y. Cui, J. Lauhon, K.-H. Kim, C. M. Lieber, *Science*, **2001**, *294*, 1313. b) J. Goldberger, R. He, Y. Zhang, S. Lee, H. Yan, H.-J. Choi, P. Yang, *Nature,* **2003**, *422*, 599.

37. X. Duan, J. Wang, C. M. Lieber, *Appl. Phys. Lett.* **2000**, *76*, 1116.

38. M. S. Gudiksen, C. M. Lieber, *J. Am. Chem. Soc.* **2000**, *122*, 8801.

39. M. S. Gudiksen, J. Wang, C. M. Lieber, *J. Phys. Chem. B* **2001**, *106*, 4062.

40. Y. Xie, P. Yan, J. Lu, W. Z. Wang, Y. T. Qian, *Chem. Mater.* **1999**, *11*, 2619.

41. A) H. Dai, W. Wong, Y. Z. Lu, S. Fan, C. M. Lieber, *Nature*, **1995**, *375*, 769. b) Q. Y. Lu, J. Q. Hu, K. B. Tang, Y. T. Qian, G. Zhou, X. M. Liu, J. S. Zhu, *Appl. Phys. Lett.* **1999**, *75*, 507.

42. G. Gu, B. Zheng, W. Q. Han, S. Roth, J. Liu, *Nano Lett.* **2002**, *2*, 849.

43. a) R. R. He, M. Law, R. Fan, F. Kim, P. Yang, *Nano Lett.* **2002**, *2*, 1109. b) Z. R. Dai, J. L. Gole, J. D. Stout, Z. L. Wang, *J. Phys. Chem. B* **2002**, *106*, 1274. c) Z. R. Dai, Z. W. Pan, Z. L. Wang, *J. Phys. Chem. B* **2002**, *106*, 902. d) H. Zhang, A. C. Dohnalkova, C. Wang, J. S. Young, E. C. Buck, Z. L. Wang, *Nano Lett.* **2002**, *2*, 105

44. X. Jiang, T. Herricks, Y. Xia, *Nano Lett.* **2002**, *2*, 1333.

45. a) T. Kasuga, M.Hiramatsu, A. Hoson, T.Sekino, K. Niihara, *Langmuir*, **1998**, *14*, 3160. b) T. Kasuga, M. Hiramatsu, A. Hoson, T. Sekino, K. Niihara, *Adv. Mater.* **1999**, *11*, 1307. c) S. L. Zhang, J. F. Zhou, Z. J. Zhang, Z. L. Du, A. V. Vorontsov, Z. S. Jin, *Chinese Sci. Bull.* **2000**, *45*, 1533. d) D. S. Seo, J. K. Lee, H. Kim, *J. Crystal Growth* **2001**, *229*, 428. e) Q. H. Zhang, L. A. Gao, J. Sun, , S. Zheng, *Chem. Lett.* **2002**, *2*, 226. f) Q. H. Zhang, L. Gao, S. Zheng, J. Sun, *Acta Chim. Sinica* **2002**, *60*, 1439. g) C. H. Lin, S. H. Chien, J. H. Chao, C. Y. Sheu, Y. C. Cheng, Y. J. Huang, C. H. Tsai, *Catalysis Lett.* **2002**, *80*, 153. h) G. H. Du, Q. Chen, R. C. Che, Z. L. Yuan, L.-M. Peng, *Applied Phys. Lett.* **2001**, *79*, 3702. i) Q. Chen,

H. Du, S. Zhang, L.-M. Peng, *Acta Cryst.* **2002**, B58, 587. j) B. D. Yao, Y. F. Chan, X. Y. Zhang, W. F. Zhang, Z. Y. Yang, N. Wang, *Applied Phys. Let.* **2003**, *82*, 281.

46. J. Liu, J. Cai, Y. Son, Q. Gao, S. L. Suib, M. Aindow, *J. Phys. Chem. B* **2002**, *106*, 9761.

47. D. B. Yu, D. B. Wang, Z. Y. Meng, J. Lu, Y. T. Qian, *J. Mater. Chem.* **2002**, *12*, 403.

48. a) J. Doshi, D. H. Reneker, *J. Electrost.* **1999**, *35*, 151. b) D. H. Renek, A. L. Yarin, H. Fong, S. Koombhongse, *J. Appl. Phys.* **2000**, *87*, 4531. c) C. Jerome, R. Jerome, *Angew. Chem.* **1998**, *110*, 2639; *Angew. Chem. Int. Ed.* **1998**, *37*, 2488. e)

49. A. Turberfield, review in *Physics World*, **2003**, 16, 43.

50. (a) C. H. Gorbiz, *Current Opinions in Solid State & Materials Science*, **2002**, 6, 109. (b) M. Reches, E. Gazit, *Science*, **2003**, 300, 625.

51. J. Howard, *Mechanics of Motor Proteins and Cytoskeleton*, Sinauer Associates, Sunderland, Massachusetts **2001**.

52. C. R. Martin, *Science* **1994**, *34*, 266.

53. M. Sasaki, M. Osada, N. Higshimoto, T. Yamamato, A. Fukuoda, M. Ichikawa, *J. Molecular Catalysis A-Chemical*, **1999**, *141*, 223.

54. a) Y. Sun, B. Gates, B. Mayers, Y. Xia, *Nnao Lett.* **2002**, *2*, 165. b) Y. Sun, Y. Xia, *Adv. Mater.* **2002**, *14*, 833. c) Y. Sun, B. T. Mayers, T. Herricks, Y. Xia, *Chem., Mater.* **2002**, *14*, 4736.

55. Gas phase reaction

56. Z. W. Pan, Z. R. Dai, Z. L. Wang, *Science* **2001**, *291*, 1947.

57. X. Y. Kong, Y. Ding, R. Yang, Z. L Wang, *Science*, **2004**, 303, 1348.

58. R. S. Wagner, W. C. Ellis, *Appl. Phys. Lett.* **1964**, *4*, 89.

59. W. Z. Li, S. S. Xie, L. X. Qian, B. H. Chang, B. S. Zou, W. Y. Zhou, R. A. Zhao, G. Wang, *Science* **1996**, *36*, 274.

60. Z. F. Ren, Z. P. Huang, J. W. Xu, J. H. Wang, P. Bush, M. P. Siegel, P. N. Provencio, *Science* **1998**, *282*, 1150.

61. M. H. Huanh, S. Mao, H. Feick, H. Yan, Y. Wu, H. Kind, E. Webber, R. Russo, P. Yang, *Science* **2001**, *292*, 1897.

62. Y. Wu, H. Yan, M. Huang, B. Messer, J. H. Song, P. Yang, *Chem. Eur. J.* **2002**, *8*, 1261.

63. T. J. Tentler, K. M. Hickman, S. C. Geol, A. M. Viano, P. C. Gibbons, W. E. Buhro, *Science* **1995**, *270*, 1791.

64. R.C. Furneaux, W.R. Rigby, A.P. Davidson, *Nature* **1989**, *337*, 147.

65. R.L. Fleisher, P.B. Price, R.M. Walker, *Nuclear Tracks in Solids*, University of California Press, Berkeley **1975**.

66. R.J. Tonucci, B.L. Justus, A.J. Campillo, C.E. Ford, *Science* **1992**, *258*, 783.

67. G.E. Possin, *Rev. Sci. Instrum.* **1970**, *41*, 772.

68. C. Wu T. Bein, *Science* **1994**, *264*, 1757.

69. S. Fan, M.G. Chapline, N.R. Franklin, T.W. Tombler, A.M. Cassell, H. Dai, *Science* **1999**, *283*, 512.
70. P. Enzel, J.J. Zoller, T. Bein, *Chem. Commun.* **1992**, 633.
71. C. Guerret-Piecourt, Y. Le Bouar, A. Loiseau, H. Pascard, *Nature* **1994**, *372*, 761.
72. P.M. Ajayan, O. Stephan, P. Redlich, C. Colliex, *Nature* **1995**, *375*, 564.
73. A. Despic, V.P. Parkhuitik, *Modern Aspects of Electrochemistry* Vol. 20, Plenum, New York **1989**.
74. D. AlMawiawi, N. Coombs, M. Moskovits, *J. Appl. Phys.* **1991**, *70*, 4421.
75. C.A. Foss, M.J. Tierney, C.R. Martin, *J. Phys. Chem.* **1992**, *96*, 9001.
76. J.B. Mohler, H.J. Sedusky, *Electroplating for the Metallurgist, Engineer and Chemist*, Chemical Publishing Co., Inc. New York **1951**.
77. A.J. Bard, L.R. Faulkner, *Electrochemical Methods*, John Wiley & Sons, New York **1980**.
78. F.R.N. Nabarro, P.J. Jackson, in *Growth and Perfection of Crystals* (eds. R.H. Doremus, B.W. Roberts, D. Turnbull), John Wiley, New York **1958**.
79. T.M. Whitney, J.S. Jiang, P.C. Searson, C.L. Chien, Science **1993**, 261, 1316.
80. W.D. Williams, N. Giordano, *Rev. Sci. Instrum.* **1984**, *55*, 410.
81. B.Z. Tang, H. Xu, *Macromolecules* **1999**, *32*, 2569.
82. Y. Zhang, G. Li, Y. Wu, B. Zhang, W. Song, L. Zhang, Adv. Mater. **2002**, 14, 1227.
83. G. Yi, W. Schwarzacher, *Appl. Phys. Lett.* **1999**, *74*, 1746.
84. J.D. Klein, R.D. Herrick, II, D. Palmer, M.J. Sailor, C.J. Brumlik, C.R. Martin, *Chem. Mater.* **1993**, 5, 902.
85. C. Schönenberger, B.M.I. van der Zande, L.G.J. Fokkink, M. Henny, C. Schmid, M. Krüger, A. Bachtold, R. Huber, H. Birk, U. Staufer, *J. Phys. Chem.* **1997**, *B 101*, 5497.
86. C.J. Brumlik, V.P. Menon, C.R. Martin, *J. Mater. Res.* **1994**, *268*, 1174.
87. C.J. Brumlik, C.R. Martin, *J. Am. Chem. Soc.* **1991**, *113*, 3174.
88. C.J. Miller, C.A. Widrig, D.H. Charych, M. Majda, *J. Phys. Chem.* **1988**, *92*, 1928.
89. C.G. Wu, T. Bein, *Science* **1994**, *264*, 1757.
90. P.M. Ajayan, O. Stephan, P. Redlich, *Nature* **1995**, *375*, 564.
91. W. Han, S. Fan, Q. Li, Y. Hu, *Science* **1997**, *277*, 1287.
92. *Electroless Plating: Fundamentals and Applications*, (eds. J.B. Hajdu, G.O. Mallory), American Electroplaters and Surface Finishers Society, Orlando **1990**.
93. C.R. Martin, *Chem. Mater.* **1996**, *8*, 1739.
94. C.R. Martin, *Science* **1994**, *266*, 1961.
95. C.R. Martin, *Adv. Mater.* **1991**, *3*, 457.
96. J.C. Hulteen, C.R. Martin, *J. Mater. Chem.* **1997**, *7*, 1075.
97. L. Piraux, S. Dubois, S. Demoustier-Champagne, *Nucl. Instrum. Methods Phys. Res.* **1997**, *B131*, 357.
98. I. Zhitomirsky, *Adv. Colloid Interf. Sci.* **2002**, *97*, 297.
99. O.O. Van der Biest, L.J. Vandeperre, *Annu. Rev. Mater. Sci.* **1999**, *29*, 327.

100. P. Sarkar, P.S. Nicholson, *J. Am. Ceram. Soc.* **1996**, *79*, 1987.

101. J. S. Reed, *Introduction to the Principles of Ceramic Processing*, John Wiley & Sons, New York **1988**.

102. R. J. Hunter, *Zeta Potential in Colloid Science: Principles and Applications*, Academic Press, London **1981**.

103. C. J. Brinker, G. W. Scherer, *Sol-Gel Science: the Physics and Chemistry of Sol-Gel Processing*, Academic Press, San Diego **1990**.

104. A.C. Pierre, *Introduction to Sol-Gel Processing*, Kluwer, Norwell **1998**.

105. J. D. Wright, N. A.J.M. Sommerdijk, *Sol-Gel Materials: Chemistry and Applications*, Gordon and Breach, Amsterdam **2001**.

106. D. H. Everett, *Basic Principles of Colloid Science, the Royal Society of Chemistry*, London **1988**.

107. W. D. Callister, *Materials Science and Engineering: An Introduction*, John Wiley & Sons, New York **1997**.

108. S. J. Limmer, S. Seraji, M. J. Forbess, Y. Wu, T. P. Chou, C. Nguyen, G.Z. Cao, *Adv. Mater.* **2001**, *13*, 1269.

109. S. J. Limmer, S. Seraji, M. J. Forbess, Y. Wu, T. P. Chou, C. Nguyen, G.Z. Cao, *Adv. Func. Mater.* **2002**, *12*, 59.

110. S.J. Limmer, G.Z. Cao, *Adv. Mater.* **2003**, *15*, 427.

111. Y.C. Wang, I.C. Leu, M.N. Hon, *J. Mater. Chem.* **2002**, *12*, 2439.

112. Z. Miao, D. Xu, J. Ouyang, G. Guo, Z. Zhao, Y. Tang, *Nano Lett.* **2002**, *2*, 717.

113. C. Natarajan, G. Nogami, *J. Electrochem. Soc.* **1996**, *143*, 1547.

114. B.B. Lakshmi, P.K. Dorhout, C.R. Martin, *Chem. Mater.* **1997**, *9*, 857.

115. R.L. Penn, J.F. Banfield, *Geochim. Cosmochim. Ac.* **1999**, *63*, 1549.

116. B.B. Lakshmi, C.J. Patrissi, C.R. Martin, *Chem. Mater.* **1997**, *9*, 2544.

117. J.S. Reed, Introduction to Principles of Ceramic Processing, Wiley, New York **1988**.

118. C.A. Huber, T.E. Huber, M. Sadoqi, J.A. Lubin, S. Manalis, C.B. Prater, *Science* **1994**, *263*, 800.

119. Z. Zhang, D. Gekhtman, M.S. Dresselhaus, J.Y. Ying, *Chem. Mater.* **1999**, *11*, 1659.

120. W. Liang, C.R. Martin, *J. Am. Chem. Soc.* **1990**, *112*, 9666.

121. S.M. Marinakos, L.C. Brousseau, III, A. Jones, D.L. Feldheim, *Chem. Mater.* **1998**, *10*, 1214.

122. P. Enzel, J.J. Zoller, T. Bein, *Chem. Commun.* **1992**, 633.

123. H.D. Sun, Z.K. Tang, J. Chen, G. Li, *Solid State Commun.* **1999**, *109*, 365.

124. Z. Cai, J. Lei, W. Liang, V. Menon, C.R. Martin, *Chem. Mater.* **1991**, *3*, 960.

125. Y.J. Han, J.M. Kim, G.D. Stucky, *Chem. Mater.* **2000**, *12*, 2068.

126. J. Liu, G. E. Fryxell, M. Qian, L.-Q. Wang, Y. Wang, *Pure and Applied Chemistry*, **2000**, *72*, 269-279, 2000.

127. L. Chen, P.J. Klar, W. Heimbrodt, F. Brieler, M. Fröba, *Appl. Phys. Lett.* **2000**, *76*, 3531.
128. K. Matsui, T. Kyotani, A. Tomita, *Adv. Mater.* **2002**, *14*, 1216.
129. T. Wen, J. Zhang, T.P. Chou, S. J. Limmer, G.Z. Cao, submitted to *Chem. Mater.*
130. C.G. Wu, T. Bein, *Science* **1994**, *264*, 1757.
131. B. Gates, Y. Wu, Y. Yin, P. Yang, Y. Xia, *J. Am. Chem. Soc.* **2001**, *123*, 11500.
132. H. Dai, E.W. Wong, Y.Z. Lu, S. Fan, C.M. Lieber, *Nature* **1995**, *375*, 769.
133. E.W. Wong, B.W. Maynor, L.D. Burns, C.M. Lieber, *Chem. Mater.* **1996**, *8*, 2041.
134. Y.Li, G.S. Cheng, L.D. Zhang, *J. Mater. Res.* 2000, *15*, 2305.
135. C.M. Zelenski, P.K. Dorhout, *J. Am. Chem. Soc.* **1998**, *120*, 734.
136. E. Braun, Y. Eichen, U. Sivan, G. Ben-Yoseph, *Nature* **1998**, *391*, 775.
137. J. Zhan, X. Yang, D. Wang, S. Li, Y. Xie, Y. Xia, Y. Qian, *Adv. Mater.* **2000**, *12*, 1348.
138. B. C. Bunker, P. C. Rieke, B. J. Tarasevich, A. A. Campbell, G. E. Fryxell, G. L. Graff, L. Song, J. Liu, J. W. Virden, *Science* **1994**, *264*, 48.
139. J. Liu, A. Y. Kim, L. Q. Wang, B. J. Palmer, Y. L. Chen, P. Bruinsma, B. C. Bunker, G. J. Exarhos, G. L. Graff, P. C. Rieke, G. E. Fryxell, J. W. Virden, B. J. Tarasevich, L. A. Chick, *Advances in Colloidal and Interface Science* **1996**, *69*, 131.
140. P. Hartman and W.G. Perdok, *Acta Cryst.* **1955**, *8*, 49.
141. P. Hartman, *Z. Kristallogr.* **1965**, *121*, 78.
142. P. Hartman, *Crystal Growth: An Introduction*, North Holland, Amsterdam **1973**.
143. C. Herring, *Structure and Properties of Solid Surfaces*, University of Chicago, Chicago **1952**.
144. W.W. Mullins, *Metal Surfaces: Structure Energetics and Kinetics*, The American Society of Metals, Metals Park **1962**.
145. A) Y. Sun, B. Gates, B. Mayers, Y. Xia, *Nnao Lett.* **2002**, *2*, 165. b) Y. Sun, Y. Xia, *Adv. Mater.* **2002**, *14*, 833. c) Y. Sun, B. T. Mayers, T. Herricks, Y. Xia, *Chem., Mater.* **2002**, *14*, 4736.
146. L. Vayssieres, N. Beermann, S. E. Lindguist, A. Hagfeldt, *Chem Mater.* **2001**, *13*, 233.
147. L. Vayssieres, L. Rabenberg, A. Manthiram, Nano Letters, **2002**, *2*, 1393.
148. L. Vayssieres, K. Keis, S-E. Linquist, A. Hagfelt, *J. Phys. Chem. B* **2001**, *105*, 3350.
149. Y. Wu, H. Yan, M. Huang, B. Messer, J. H. Song, P. Yang, *Chem. Eur. J.* **2002**, *8*, 1261.
150. L. Vayssieres, K. Keis, A. Hagfelt, S-E. Linquist, *Chem. Mater.* **2001**, 13, 4395.
151. L. Vayssieres, *Adv. Mater.* **2003**, *15*, 464.
152. Z. Tian, J. A. Voigt, J. Liu, B. Mckenzie, M. J. Mcdermott, J. Am. Chem. Soc. 2002, 124, 12954.

153. J. Liu, Y. Lin, L. Liang, J. A. Voigt, D. L. Huber, Z. R. Tian., E. Coker, B. Mckenzie, M. J. Mcdermott, Chem.: A European J. 2003, 9, 604.

154. L. E. Greene, M. Law, J. Goldberger, F. Kim, J. C. Johnson, Y. Zhang, R. J. Saykally, P. Yang, *Angew, Chem. Int. Ed.* **2003**, *42*, 3031.

155. Z. R. Tian, J. A. Voigt, J. Liu, B. Mckenzie, M. J. Mcdermott, R. T. Cygan, L. J. Criscenti, *Nature Materials*, 2003.

156. Z. R. Tian, J. A. Voigt, J. Liu, B. McKenzie, H. F. Xu, J. Am. Chem. Soc. 2003, 125, 12384.

157. (a) T. Kasuga, M. Hiramatsu, A. Hoson, T. Sekino, K. Niihara, *Langmuir*, **1998**, *14*, 3160. (b) T. Kasuga, M. Hiramatsu, A. Hoson, T. Sekino, K. Niihara, *Advanced Mater.* **1999**, *11*, 1307.

158. (a) S. L. Zhang, J. F. Zhou, Z. J. Zhang, Z. L. Du, A. V. Vorontsov, Z. S. Jin, *Chinese Sci. Bull.* **2000**, *45*, 1533. (b) D. S. Seo, J. K. Lee, H. J. Kim, *Crystal Growth* **2001**, *229*, 428. (c) Q. H. Zhang, L. A. Gao, J. Sun, S. Zheng, *Chem. Lett.* **2002**, *2*, 226. (d) Q. H. Zhang, L. Gao, S. Zheng, J. Sun, *Acta Chim. Sinica* **2002**, *60*, 1439. (e) C. H. Lin, S. H. Chien, J. H. Chao, C. Y. Sheu, Y. C. Cheng, Y. J. Huang, C. H. Tsai, *Catalysis Lett.* **2002**, *80*, 153. (f) G. H. Du, Q. Chen, R. C. Che, Z. L. Yuan, L.M. Peng, *Applied Phys. Lett.* **2001**, *79*, 3702.

159. (a) B. D. Yao, Y. F. Chan, X. Y. Zhang, W. F. Zhang, Z. Y. Yang, N. Wang, *Applied Phys. Let.* **2003**, *82*, 281. (b) Q. Chen, W. Zhou, G. Du, L. M. Peng, *Adv. Mat.*, **2002**, *14*, 2008. (c) Q. Chen, G. H. Du, S. Zhang, L. M. Peng, *Acta Cryst. B* **2002**, *58*, 587.

160. J. W. P Hsu, Z. R. Tian, N. C. Simmons, C. M. Matzke, J. A. Voigt, J. Liu, *Nano Letters*, **2005**, *83*, 2005.

161. *Advanced in Crystal Growth Research* (eds. K. Sato, Y. Furukawa, K. Nakajima,) Weizmann Inst. Sci., Israel **2001**.

162. J. P. Jolivet, M. Gzara, Mazieres, J. Lefebvre, *J. Colloid Interface Sci.* **1985**, *107*, 429.

163. A. López-Macipe, J. Gómez-Morales, R. Rodríguez-Clemente, *J. Colloid Interface Sci.* **1998**, *200*, 114.

164. P. C. Hidber, T. J. Graule, L. J. Gauckler, *J. Am. Ceram. Soc.* **1996**, *79*, 1857.

165. S.Biggs, P. J. Scales, Y. K. Leong, T. W. Healy, *J. Chem. Soc.-Faraday Trans.* 1995, *91*, 2921.

166. C. Liu, P. M. Huang, *Soil. Sci. Soc. Am. J.* **1999**, *63*, 65.

167. Z. R. Tian, Z. R., J. A. Voigt, J. Liu, J., B., Mckenzie, M. J. Mcdermott,. *J. Am. Chem. Soc.* **2002**, *124*, 12954-1295.

168. A. G. MacDiarmid, *Rev. Modern Physics*, **2001**, *73*, 701.

169. K. Doblhofer, K. Rajeshwar, *Handbook of Conducting Polymers*, Chapter 20, Marcel Dekker, Inc., New York **1998**.

170. J. Doshi, D. H. Reneker, *J. Electros.* **1999**, *35*, 151.

171. D. H. Renek, A. L. Yarin, H. Fong, S. Koombhongse, *J. Appl. Phys.* **2000**, *87*, 4531.

172. C. Jérôme, R. Jérôme, *Angew. Chem. Int. Ed.* **1998**, *37*, 2488.

173. L. Liang, J. Liu, C. F. Jr. Windisch, G. J. Exarhos, Y. Lin, *Angew. Chemie. Int. Ed.* **2002**, *41*, 3665.

174. K. A. Dick, K. Deppert, M. Larsson, T. Martensson, W. Seifert, L. Wallenberg, L. Samuelson, *Nat. Mater.* **2004**, *3*, 380.

175. X. Gao, Z. Zheng, H. Zhu, G. Pan, J. Bao, F. Wu, D. Song, *Chem. Commun.* **2004**, *12*, 1428.

176. H. Yang, H. Zeng, *J. Am. Chem. Soc.* **2005**, *127*, 270.

177. T. L. Sounart, J. Liu, J. A. Voigt,J. W. P Hsu, E. D. Spoerke, Z. Tian,Y. Jian, *Advanced Functional Materials*, **2006**, 16, 335.

178. J. Aizenberg, A. Black, G. Whitesides, *Nature* **1999**, *398*, 495.

179. J. Aizenberg, *J. Cryst. Growth* **2000**, *211*, 143.

180. J. Aizenberg, *Adv. Mater.* **2004**, *16*, 1295.

181. P. Hartman, W. G. Perdok,. *Acta Cryst.* **1955**, *8*, 49.

182. T. L. Sounart, J. Liu, J. A. Voigt, M. Huo, E. D. Spoerke, B. Mckenzie, J Am Chem Soc, ASAP article, 2007.

183. X. Gao, Z. Zheng, H. Zhu, G. Pan, J. Bao, F. Wu, D. Song, D. *Chem. Commun.* 2004, *12*, 1428.

184. A. G. Kanaras, S. Carsten, H. Liu, P. Alivisatos, *Nano Lett.* **2005**, *5*, 2164.

CHAPTER 7

ONE- AND TWO-DIMENSIONAL ASSEMBLIES OF NANOPARTICLES: MECHANISMS OF FORMATION AND FUNCTIONALITY

Nicholas A. Kotov [1] and Zhiyong Tang [2]

[1]*Department of Chemical Engineering, University of Michigan, Ann Arbor, Michigan 48109-2136, [2]National Center for Nanoscience and Technology, Beijing 100080, China, Email: kotov@umich.edu; zytang@nanoctr.cn*

Nanoparticle (NP) assemblies have attracted many scientific and technological attentions due to their abilities to bridge between nano-scale objects and macro-scale world. Among those, one-dimensional (1D) and two-dimensional (2D) NP assemblies arouse the scientists' interests due to their high efficiencies of managing the flow direction of electronic, photonic, and magnetic information in the NP devices. Additionally, 1D and 2D assemblies of NP, i.e. chains and sheets, can significantly help the scientists in understanding of a number of biological processes and fundamental quantum effects of nanoscale systems. However, the difficulties to prepare anisotropic 1D and 2D NP assemblies are obvious since both the apparent shape and structure of single composed NPs are isotropic. Such a big challenge drives the scientist to explore the anisotropic interaction between NPs and then fabricate their 1D and 2D assemblies with the desirable functions. This review summarizes the recent progress on the research of 1D and 2D NP assemblies. The formation mechanisms of 1D and 2D NP assemblies are introduced, and the functionality and application of 1D and 2D NP assemblies are discussed. Current problems underlying the fundamental research and practical applications of NPs are also addressed.

1. Introduction

The unique electrical, optical, magnetic, and catalytic properties of nanoparticles (NPs) encourage the scientists to seek their possible applications in devices.[1-3] As discussed in our previous reviews,[4] current researches on NPs can be classified into two main trends. The first one involves manipulation and exploration of *single* NP in devices reaching the limit of miniaturization possible for electronic and photonic circuits.[5] The second trend is focused on application of NP/polymer composites as macro-scale thin films producing a new generation of currently used devices, such as light-emitting diodes and solar cells.[6] In our opinion, successful practical implementation of these devices will most likely come sooner for NP thin films rather than for those made from single nanocolloids because of simpler processing, and easier interfacing with current technologies.[4]

Overall, the NP thin films and similar assemblies can be classified in three categories: one dimensional (1D), two dimensional (2D), and three dimensional (3D) systems. The preparation of 3D NP assembly was traced back to the early of 1990s,[7-11] and several conclusive and influential reviews were published in the past 10 years.[3, 12-14] So the review on works of 3D NP assemblies is intentionally excluded in this chapter, and the authors can read the Refs. 3, 12-14 to obtain the details on 3D NP assemblies. As a comparison, anisotropic 1D and 2D assemblies of NPs face more challenges, and the increasing numbers of research groups are involving such projects. In part 2 and part 3 of this review the self-assembly mechanism and functionality of 1D NP assemblies are discussed, respectively. Because our previous review summarizes the early development of 1D NP assemblies, only recent works are introduced. The self-assembly mechanism and functionality of 2D NP assemblies are elucidated in part 4 and part 5, respectively. Finally, the problems and promises of 1D and 2D assemblies are briefly discussed in part 6.

2. Formation mechanism of 1D NP assembly

The formation of 1D NP assemblies is based on either external templates or intrinsic anisotropy in NPs. The external templates are linear nanoscale materials including polyelectrolyte,[15-17] inorganic nanotubes (NTs) and nanowires (NWs),[18-23] or biomolecules.[24-28] 1D NP assemblies are fabricated when NPs spontaneously adsorbed onto above linear templates upon the physical/chemical/biological interactions. Contrastively, the formation mechanism of 1D NP assemblies induced by the intrinsic anisotropy is much more complex, and the corresponding studies are becoming more attractive. The below parts summarize the recent works on the anisotropy-induced 1D NP assemblies.

2.1 Origins of the NP anisotropy

The anisotropy of NPs determines whether NPs spontaneously align into 1D structure. The anisotropy of NPs arises from either magnetic/electric dipole moments inside NPs[29-33] or inhomogeneous distribution of stabilizers on NP surface.[34, 35] The anisotropy inside magnetic NPs has been known for a long time, for example, 1D structure of maghemite (γ-Fe$_2$O$_3$) NPs induced by magnetic dipole attraction between NPs have been observed in terrestrial magnetotactic bacteria, such as the MV-1 strain and *Magnetospirillum magnetotacticum* (MS-1) strain.[36, 37] On the contrary, the origins of electric dipole moments inside varying types of NPs are still ambiguous and remain in controversy, especially considering the non-existence of asymmetry crystalline structures inside most metal, metal oxide, or semiconductor NPs.

Our group studied the origin of a permanent electric dipole moment in CdS NPs with a symmetry cubic crystal lattice by simulation.[38] Figure 1A shows a prototypical tetrahedral CdS NP with 84 cadmium atoms and 123 sulfur atoms as the base for calculations. We suggested that, while a variety of potential defects on the NP surfaces were possible, a probable deviation from the ideal geometric shape of a tetrahedron was the truncation of apexes, which led to lowering of the surface energy associated with the NPs. Therefore, from an atomic perspective, NPs with truncated apexes could possess some thermodynamic advantage and, in fact, could be quite abundant in polar solvents. Accordingly, a systematic progression of tetrahedral NPs with gradually varying degree and placement of truncation was evaluated in our work.

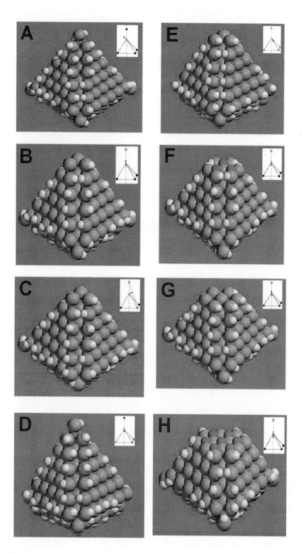

Figure 1. (A) Regular tetrahedral CdS NP with S-H terminal groups. The molecule is depicted using a space-filling representation: S (teal); Cd (green); H (gray). Starting from the (a) base NP, (B) one, (C) two, (D) three, and (E) four corners were truncated to obtain NPs denoted as C0T0, C1T1, C2T1, C3T1, and C4T1, respectively. Starting from C1T1, the truncated clusters (F) C1T2, (G) C1T3, and (H) C1T4 were obtained. The inset in each figure depicts the direction of the dipole moment schematically, pointing from the positive end toward the negative end. Filled/unfilled circles represent untruncated/truncated corners of the tetrahedron. (Taken from Ref. 38).

The simplest case in this series was a NP in which one of the four corners of the regular tetrahedron was modified by deleting a Cd and S atom. The resulting NP, C1T1, is depicted in Figure 1B. It should be noted that three sulfur atoms from the layer below the truncated apex were exposed. Parts C, D, and E of Figure 1 showed NPs C2T1, C3T1, and C4T1 from the same progression that were similarly truncated at two, three, and four corners, respectively. Starting from C1T1 (Figure 1B), an increasing number of atomic layers were removed from a single corner to study the effect of the degree of truncation. The three exposed sulfur atoms at the truncated corner in C1T1 were deleted to obtain C1T2, which exposed three Cd atoms (Figure 1F). Deleting these three Cd atoms resulted in C1T3, which, in turn, exposed 7 S atoms (Figure 1G). The last model system in this family, C1T4, was obtained by removing these 7 S atoms (Figure 1H). Here, the labels assigned to the various asymmetric molecules had a general format CmTn, where m was the number of truncated corners and n was the number of layers removed from a corner.

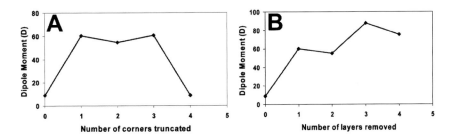

Figure 2. Variation of the DM of a CdS nanocrystal as (A) the number of truncated corners increases (C0T0, C1T1, C2T1, C3T1, C4T1 progression) and (B) the number of layers removed from a single corner of a regular tetrahedron (C0T0, C1T1, C1T2, C1T3, C1T4 progression). (Taken from Ref. 38).

The software package Spartan (Wavefunction Inc., Irvine, CA) was used to calculate the net dipole moment inside the CdS NPs, and the results were plotted in Figure 2. Truncation of one of the corners increased the dipole moment dramatically (Figure 2A), and NPs with two or three truncated corners did not reveal further change and the dipole moment remained around 60 D. The dipole moment was produced due to the asymmetry in electron distribution caused by the atoms in existing (nontruncated) corners. When all corners were present, the polarity vectors were compensated. As soon as one corner was missing, the vectorial sum of three other contributions gave rise to a strong dipole moment. Similar explanation was applied to the truncation of different layers from a single corner of a regular tetrahedron. The removal of the first layer produced a remarkable dipole moment, whereas the change of dipole moment was slight for subsequent removal of the additional layers (Figure 2B). In one word, the crystalline defects of NPs with the symmetry crystal structure lead to production of large, permanent electric dipole moment of NPs.

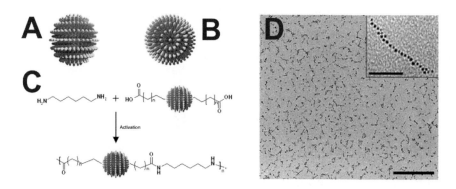

Figure 3. From rippled particles to NP chains. Idealized drawing of (A) a side view and (B) a top view of a rippled particle showing the two polar defects that must exist to allow the alternation of concentric rings. (C) Schematic depiction of the chain formation reaction. (D) TEM images of chains that compose the precipitate obtained when MUA pole-functionalized rippled NPs are reacted with DAH in a two-phase reaction. Scale bars 200 nm, inset 50 nm. (Taken from Ref. 40).

Besides defects, the inhomogeneous distribution of stabilizers also leads to the anisotropy of NPs. Stellacci and coworkers recently shown that mixtures of thiol molecules, which on flat gold surfaces separated into randomly distributed domains, formed ordered alternating phases (ripples) when assembled on surfaces with a positive Gaussian curvature, such as the core of NPs.[39] Moreover, these types of domains profoundly demarcated the two diametrically opposed singularities at the particle poles, where the rings collapsed into points (Figure 3A and 3B).[40] The authors proposed that, in the case of a self-assembled ligand shell, the polar singularities manifested themselves as defect points, that was, sites at which the ligands must assume a nonequilibrium tilt angle. Ligands at the poles, being not optimally stabilized by intermolecular interactions with their neighbors, should be the first molecules to be replaced in place-exchange reactions (Figure 3C). Therefore, the anisotropy inside NPs could be easily produced by accurately controlling the ratios of the mixed stabilizers.

2.2 Preparation of 1D NP assemblies upon the anisotropy

The next question is how anisotropy inside NPs induces their 1D self-assembly. We studied 1D self-assembly process of NP by Monte Carlo computer simulation.[41] Figures 4A-4C represent the simulated distribution of the potential of the electric field around the NPs. Here, three trimers of different sized NPs with the same net charge and dipole moment were presented overlaid on the field potential. It could be seen that if another similarly charged and polar particle would approach the chain, the gradient of the field would be the smallest along the long axis of the trimer from the side of the opposite charged end of the dipole. This could be best seen for the smallest particles where only the rightmost one was accessible without bumping into the high potential area (coded as white color in Figure 4). So, the formation of 1D structures proceeded in the solution by a step-by-step process where either chains grew by sweeping individual particles or dimers as they diffused in the solution, approaching them with the "right" side, or small chain fragments of 2 to maybe 4 constituents were formed from the individual NPs and then bound together to form large aggregates.

Similar inducement effect was experimentally confirmed by Talapin, Murray, and coworkers.[42] They reported the solution-based synthesis of crystalline PbSe NWs by oriented attachment of collections of single NPs that attached and fused along identical crystal faces forming oriented chains (Figures 5A-5F). In the crude fractions taken during the one-step PbSe NW synthesis, the authors observed long (1-30 μm)

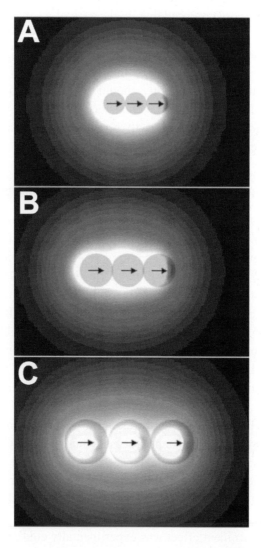

Figure 4. Distribution of electric field potential around NPs of different sizes. The net charge and dipole are constant for the three sizes. The actual sizes are arbitrary and are provided only for the visualization of the difference in the field distribution between equal dipoles at different separations. (Taken from Ref. 41).

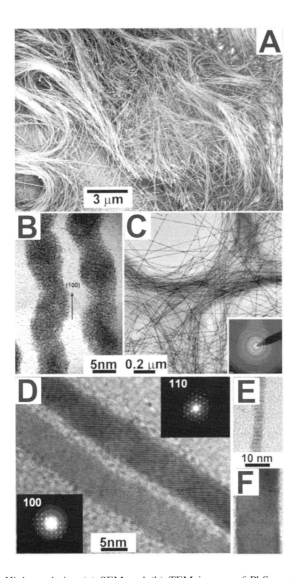

Figure 5. High-resolution (a) SEM and (b) TEM images of PbSe nanowires grown in solution in the presence of oleic acid. (c) Overview and (d-f) high-resolution TEM images of PbSe nanowires formed in the presence of oleic acid and n-tetradecylphosphonic acid. Selected area electron diffraction from a film of PbSe nanowires (inset to c) and single nanowires imaged along the (100) and (110) zone axes (insets to d). The diameter of PbSe nanowires can be tuned from (e) ~4 nm to (f) ~18 nm. (Taken from Ref. 42).

nanowires and isolated spherical (or cubic) NPs, while dimers and oligomers were very rarely observed (Figure 5A). The authors concluded that the assembly and fusion of the first two PbSe NPs was the rate-limiting step in NW formation. Most likely, NW growth was an avalanche-like process driven by a combination of the increasing anisotropy and the scaling of the dipole moment of the growing PbSe NW as the number of attached NPs increased.

1D NP assemblies can also be obtained by taking advantage of anisotropy from the inhomogeneous distribution of stabilizers.[40] As schemed in Figure 3C, the divalent 1,6-diaminohexane was used to combine stabilizers of 11-mercaptoundecanoic acid at the two poles of Au NPs, and then the 1D assemblies of Au NPs were prepared (Figure 3D). Such an assembly process is very similar to a polymerization reaction to synthesize nylon.

3. Functionality of 1D NP assembly

As summarized in our previous review, functionality of 1D NP assembly has led to their broad application in optoelectronics, magnetic devices, and sensors.[4] Herein, two recent examples are given to show a new application of 1D NP assemblies in energy storage and conversion. Belcher and coworkers used the linear and genetically engineered M13 bacteriophage viruses as the templates to synthesize 1D NP assemblies.[43] As schemed in Figure 6A, there were two types of p8 proteins randomly to package onto the virus progeny: intact p8 proteins of E4 viruses and engineered p8 proteins containing the gold-binding peptide motif. Incubation of the engineered M13 bacteriophage viruses with an Au NP solution resulted in 1D arrays of Au NPs bound to the gold-binding peptides distributed among p8 proteins (Figure 6B). Subsequently, Co_3O_4 was nucleated and grown via the tetraglutamate functionality, resulting in 1D hybrid nano-structures of Au NPs spatially interspersed within the Co_3O_4 NWs (Figure 6C). The electrochemical properties of the hybrid Au-Co_3O_4 nanowires were evaluated by using galvanostatic cycling and cyclic voltammetry, and the specific capacity of the hybrid was estimated to be at least 30% greater than that of pure Co_3O_4 NWs (Figure 6D and 6E). The incorporation of Au NPs could improve electronic conductivity to the Co_3O_4 NPs in 1D nanostructures, and thus improved battery capacity. Furthermore, the authors claimed that such functionality strategy of 1D NP assemblies allowed for the growth and assembly of other functional nanomaterials for applications such as photovoltaic devices, high–surface area catalysts, and supercapacitors.

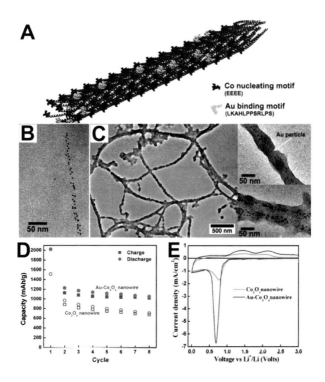

Figure 6. (A) Visualization of the genetically engineered M13 bacteriophage viruses. P8 proteins containing a gold-binding motif (yellow) were doped by the phagemid method in E4 clones, which can grow Co_3O_4. (B) TEM images of the assembled gold NPs on the virus. Control experiments showed that gold NPs were bound by the gold-specific peptides. (C) TEM image of hybrid NWs of Au nanoparticles/Co_3O_4. (D) Specific capacity of hybrid Au-Co_3O_4 NWs. Half cell with Li electrode was cycled at a rate of C/26.5. Virus mass was subtracted and the mass of active materials such as Co_3O_4 and Au was counted. The capacity of virus-directing Co3O4 NWs without Au NPs was also compared. (E) Cyclic voltammograms of hybrid Au-Co_3O_4 and Co_3O_4 nanowires at a scanning rate of 0.3 mV/s. (Taken from Ref. 43).

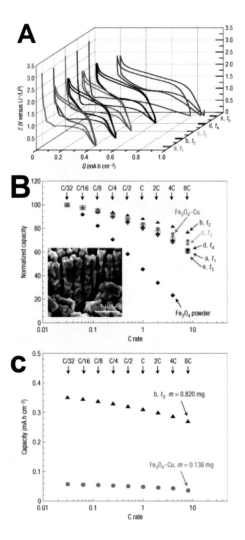

Figure 7. (A) potential-capacity profiles for the as-prepared copper-supported Fe₃O₄ deposits galvanostatically cycled at a rate of 1 Li+/2 h versus Li. (B) Rate capability plots for the five Fe₃O₄ deposits on Cu nanostructured electrodes, compared with a Fe₃O₄ deposit denoted ' Fe₃O₄–Cu' grown on a planar Cu foil electrode. (C) For comparison, the normalized capacity (mA h cm-2 of geometrical surface area) is plotted versus rate for our optimized Fe₃O₄ based Cu-nanostructured electrode and a Fe₃O₄-based Cu planar electrode. Inset: scanning electron micrograph of the 1D Cu-supported Fe₃O₄ nanostructures. (Taken from Ref. 44).

Simon and coworkers prepared the vertically-oriented Cu nanorods (NRs) arrays by using porous alumina oxide as the templates, followed by coating 1D Fe_3O_4 NP assemblies onto Cu NRs by electrodeposition.[44] The resulting Cu NR electrodes, differentiated by their degree of Fe_3O_4 NP coverage, were characterized for their performance as electrodes in Li half cells which were cycled in a galvanostatic mode at a rate of 1 Li^+ per 2 h (Figure 7A). The self-supported Fe_3O_4/Cu 1D nanostructured electrodes were evaluated for their rate capability by using signature curves (Figure 7B). Excellent rate capability was observed for all of the nanostructured electrodes compared with the Fe_3O_4 powder cell because they recovered 80% of their total capacity at an 8C rate. Furthermore, Figure 7C revealed the benefit of having a nanostructured current collector as opposed to a planar one in terms of power density, because the current scales with the amount of Fe_3O_4 deposited allowing total discharge at comparable capacity rates. It was very impressive that the 1D hybrid NP/NR assembly electrodes had the capability of increasing the power density by a factor of six.

4. Formation mechanism of 2D NP assembly

One of the most universal and simplest ways to achieve 2D NP assemblies is to organize NPs at two-phase interfaces, such as gas-solid, liquid-solid, gas-liquid or liquid-liquid where the interfaces as a template. Other templates to be used for producing 2D NP assemblies include proteins and biomolecules. There is a big challenge to devolop the templateless methods to prepare 2D NP assemblies, and until very recently, the first example was reported by our goup.

4.1 2D assembly of NPs produced at the immiscible interface

Despite the simplicity, some of the best organized 2D NP superstructures were obtained by the forces during liquid evaporation and other interfacial solid-liquid interactions at nanoscale.[8, 45-47] Ordered NP mono- and multilayers in a closely packed arrangement were produced from a NP suspension upon the solvent evaporation or adsorption of NPs on charged substrates provided that the NPs are highly uniform in diameter. The ordered domains can cover relatively large areas in the micron scale. On the other hand, their deliberate modulation of the fluid dynamics can make NPs assemblies in surprisedly, highly ordered layers and networks even for fairly polydispersed colloids.[8] A complex interplay of wetting dynamics, capillary forces, interface instabilities, and the spontaneous formation of complex patterns is implicated in their formation.[48, 49] The predictive power of the nanoscale liquid dynamics was demonstrated by deriving the formation of characteristic bands and rings on the basis of linear stability analysis and numerical simulations, which revealed the dependence of particle distribution on equilibrium film separation distance, initial packing concentration, rate of evaporation, and NP surface activity.[50]

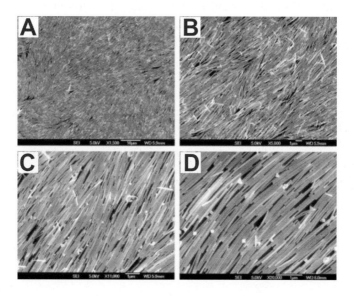

Figure 8. Scanning electron microscopy images (at different magnifications) of the silver NW monolayer deposited on a silicon wafer. (Taken from Ref. 59).

Highly ordered 2D arrays of magnetic particles were obtained by the evaporation of aqueous solutions on octadecyltrichlorosilane stamped surfaces.[51] The droplets of water containing metal salts prefer the hydrophilic surface of untreated silica and were confined by surface forces. The metal salt residue could be subsequently transformed into ferrites or other compounds. The resolution of 70-460 nm in X and Y of these arrays enabled addressing each magnetic island as an individual memory cell. Similar result can be achieved by, so called, nanosphere lithography, when a mono- or multilayer of uniform latex spheres serves as a mask through which a metal can be plated on the surface.[52]

A 2D organization method related to fluid dynamics is the assembly of NPs on surfaces of copolymers, which form intricate surface patterns due to nanoscale phase separation of the polymer blocks with hydrophilic/hydrophobic balance.[53] Being cast from organic solution, NPs adsorb preferentially on the low surface tension areas,[54] and often form aggregates with pronounced fractal dimensionality.[55]

Except for at the solid-liquid interface, the 2D NP assemblies are also produced either at the gas-liquid or the liquid-liquid interface.

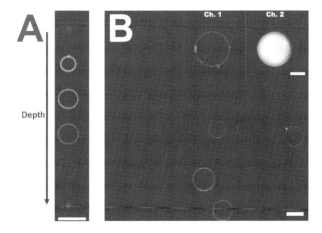

Figure 9. (A) Fluorescence confocal microscope images of a water droplet in toluene in which CdSe nanoparticles are suspended. (B) Confocal microscope images of water droplets of different sizes. Scale bar, 20 μm. (Taken from Ref. 64).

One of the most powerful techniques of 2D organization of NPs at the gas-liquid interface is Langmuir-Blodgett (LB) deposition which combines the advantages of the self-organization of NPs and operator controlled pattern organization. It has very well established reputation for nanoscale organic films and was extended to NP systems in 1994 by

Kotov *et al.*[56] and Daboussi *et al.*[57] Kotov *et al.* dissolved dodecylbenzenesulfonic acid stabilized CdS NPs in chloroform solvent, and then the chloroform solution of CdS NPs were spread at the interface of water and air to form a stable monolayer. Subsequently, LB monolayer films were prepared by compressing the NP monolayer to a certain surface pressure. Such LB NP monolayer films were easily transferred to varying types of the solid substrates by immersion and subsequent extraction of the substrates from the subphase.[56] Besides the spherical NPs, LB technique is also used to prepare 2D assemblies of anisotropic NWs.[58] Yang and coworkers synthesized the 1-hexadecanethiol stabilized Ag NWs and dispersed onto a water surface of the LB trough.[59] Under compression, Ag NWs organized into 2D monolayer at the air-water interface. Most interestingly, the compressed Ag NW monolayer exhibits remarkable alignment parallel to the trough barrier. Figure 8 shows scanning electron microscopy (SEM) images of NW monolayer transferred onto a silicon wafer. The Ag NWs were aligned side-by-side over large areas, resembling a nematic two-dimensional ordering of a liquid crystal. The authors illustrated that these aligned NW monolayers were readily used as surface-enhanced Raman spectroscopy substrates for molecular sensing with high sensitivity and specificity. These metallic layers were observed to exhibit giant local electromagnetic field enhancement, particularly due to NWs with sharp tips and noncircular cross-sections. Lieber and coworkers prepared hierarchical parallel nanowire arrays by LB techniques and used them as masks to define nanometer pitch lines in $10 * 10$ μm^2 arrays repeated with a 25 μm array pitch over centimeter square areas.[60] The authors suggested that this nanolithography method represented a highly scalable and flexible pathway for defining nanometer scale lines on multiple length scales and thus had substantial potential for enabling the fabrication of integrated nanosystems. As summarized, LB technique remains one of the most popular approaches to the preparation of remarkable 2D superstructures with potential practical applications in nanoscale photonics and electronics. Here we should note that insufficient mechanical strength and susceptibility to environmental effects resulted in significant decrease of interest to LB 2D assemblies of NPs after initial enthusiasm. Although in many cases this concern is justified, there are a number of ways to make fairly robust LB films, which can be further improved with inclusion of nanocolloids.[12, 61-63]

Russell and coworkers realized the 2D self-assembly of NPs at the liquid-liquid interface.[64-66] In order to create an immiscible liquid-liquid interface, a small amount of water was added into toluene solution of tri-*n*-octylphosphine oxide (TOPO)-stabilized CdSe NPs.[64] As shown in Figure 9A, ~20-µm-diameter water droplet appeared in toluene, and the fluorescence of the droplet should arise from the CdSe NPs. Each image in Figure 9A was an optical cross-section taken at 2.7-µm intervals in depth through the droplet. Together, these data showed that the droplet was spherical and that the nanoparticles segregated to the toluene-water interface to form a shell and stabilize the droplet. As a comparison, the fluorescence of water droplets containing sulforhodamine-B dye was also imaged (inset in Figure 9B). Channel 1 (green) showed the fluorescence from the CdSe nanoparticles (detection, 525 nm) and channel 2 (red) showed the fluorescence from the sulforhodamine-B dye dissolved in the water (detection, 585 nm). Evidently, the NPs were aggregated into monolayer at the interface of water and oil, whereas the dye was homogenously dispersed in the whole water droplets. The more details of the distribution of CdSe NPs on the droplet surfaces were achieved by detailed reflection interference contrast microscopy studies on dispersed droplets, coupled with atomic force and electron microscopy studies on dried droplets. All results revealed that CdSe spontaneously formed a 2D monolayer at the liquid-liquid interface.

4.2 2D assembly of NPs produced by the biological templates

Self-assembly of NPs in organized 2D systems can also be facilitated by proteins and other biomolecules. Considering the popularity of hybrids of biotechnology and nanotechnology, the recent appearance of multiple studies in this area becomes quite reasonable. Willner and coworkers proposed the NP-biomolecule systems as one of the most promising techniques for programmed assemblies of nanostructures.[67] This notion is substantiated by the record performance of NP-based detection and imaging procedures, which can significantly advance the experimental methods of life science.[68-70]

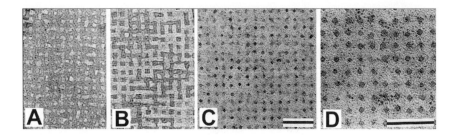

Figure 10. Transmission electron micrographs of the chemically modified and gold(III)chloride treated S-layer lattice of Bacillus sphaericus CCM2177 under increasing electron doses. (A) A coherent film of fine grainy gold precipitates is found under low electron dose conditions. (B and C) Upon increase of the electron dose regularly arranged monodisperse gold clusters are formed in the pore region of the S-layer. Bar for A to C, 50 nm. (D) Frequently the square shaped gold particles were rotated by 45° with respect to the base vectors of the S-layer lattice. Scale bar, 50 nm. (Taken from Ref. 72).

Several groups utilized the ability of proteins to self-organize in 2D lattices. One of the best examples of such superstructures are proteins from bacterial surface layers, *i.e.* S-layers, recrystalizing in vitro into sheets and tube-shaped protein crystals with typical dimensions in the micrometer range.[71] Pum and coworkers investigated the formation of Au NPs on monolayers of the thiol-modified S-layer protein from *Bacillus sphaericus* CCM2177.[72] After treated with $HAuCl_4$ solution, the S-layer lattice was exposed under low electron dose conditions. Figures 10A-10D demonstrate that upon increase of the electron dose the contiguous gold coating disappeared and clearly visible monodisperse gold clusters were formed in the pore region of the S-layer. Finally, a 2D square superlattice of uniform 4 to 5 nm sized gold particles with 12.8 nm repeat distance was fabricated (Figure 10D). The NP arrays on S-layers displayed a particle density above $6*10^{11}$ cm^{-2}. Similar film geometries were obtained with genetically engineered hollow double-ring protein chaperonin, which has either 3 nm or 9 nm apical pores surrounded by chemically reactive thiols.[73] The periodic solvent-exposed thiols within chaperonin templates were used to size-selectively bind and organize either gold (1.4, 5 or 10nm) or CdSe-ZnS semiconductor (4.5 nm) quantum dots into arrays. The lattices with pronounced alignment motif in NP can be made from viral proteins.[74]

The 2D crystalline layers of ferritin and apoferritin were mostly used for the preparation of 2D arrays of iron-containing magnetic particles.[75] These proteins are attractive candidate for the biological approach to NP organization because 8 nm internal cavity of apoferritin can be reconstituted with a variety of non-native inorganic cores, such as magnetite, iron sulfide, manganese oxides, and cadmium sulfide. Encapsulation of the inorganic phase within the protein cage restricts the length scale of particle-particle interactions and prevents direct physical contact, particle fusion, and growth within the organized phase. Ferritin-based arrays could have important applications in magnetic storage and nanoelectronic devices.[76] Thus, Co/Pt NP arrays were prepared within apoferritin S-layer. By varying the annealing conditions the coercivities at 500-8000 Oe were achieved. Electrical testing of NP films shows they are capable of sustaining recording densities greater than 12.6 Gbits/in^2.[77]

4.3 Preparation of 2D assembly of NPs upon the aniosotropy

In the previously considered examples of self-assembly of NP in 2D superstructures there was always a template, either a substrate or interface, to help NP to remain in 2D. Template methods for the preparation of 2 NP assemblies have several intrinsic disadvantages, for instance, the substrates or interfaces may have a considerable effect on either optoelectric or magnetic properties of the resulting 2D NP assemblies. Although post-synthetic physical or chemical treatment can help removing the templates, possible morphological and structural alternations of 2D NP assemblies from post-treatment are detrimental for their application.

Very recently, we demonstrated for the first time that NPs spontaneously formed 2D monolayer in single-phase aqueous solutions.[78] Upon partial removal of stabilizers, 2-(dimethylamino)ethanethiol (DMAET) stabilized CdTe NPs with positive charges were shown to form monomolecular sheets after kept under ambient condition for 1 month (Figure 11A). The 2D network of NP in the sheets provided them substantial mechanic strength compared to other monolayer films, which was likely due to partial merging of the

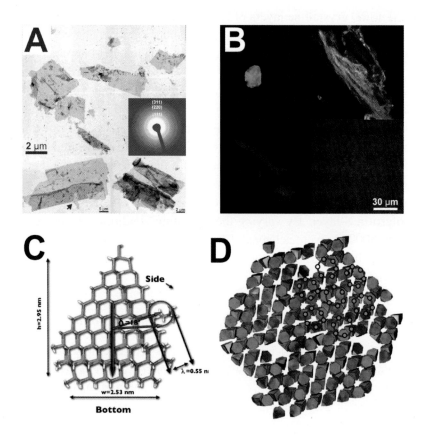

Figure 11. (A) TEM images of free-floating films of CdTe NPs. Insert: electron diffraction pattern obtained from the films. (B) Fluorescence images of self-assembled sheets of green-emitting NPs with a diameter of ~2.4 nm (left upper), yellow-emitting NPs with a diameter of ~3.6 nm (right upper), red-emitting NPs with a diameter of ~5.0 nm (left lower), and red-emitting NPs before self-assembly demonstrating only disordered aggregation on glass substrates (right lower). (C) Side view of films obtained by mesoscale simulation with N = 480 and = 0.13. The system size shown here has been doubled for clarity. (D) Face view of a separate sheet in C. The basic structure within the sheet is rings composed of six NPs. (Taken from Ref. 78).

crystal lattices of NP as well the large Van del Waals forces among the NP aggregates in 2D monolayer. There was a substantial optical activity of CdTe remaining in such layers, which could be detected by fluorescence microscopy (Figure 11B). Furthermore, the photoemission of the films was the same over the entire sheet regardless of size, which reflected the equal degree of quantization of the NPs forming the free-floating film. It should be pointed out that both the shape and structure of 2D NP assemblies were similar to those of 2D S-layer protein film.

The formation mechanism of 2D NP assemblies was explored by studying the multiple interactions, i.e. hydrophobic attraction, dipole attraction, and electrostatic repulsion, among the NPs.[79] The electrostatic potential between two NPs U_{ij} was modeled by

$$U_{ij}\left(r_{ij}\right) = \frac{q_i q_j}{4\pi\varepsilon_0 \varepsilon r_{ij}} e^{-kr_{ij}} C_0^2 + \frac{q_i \mu_j \cos\theta_j + q_j \mu_i \cos\theta_i}{4\pi\varepsilon_0 \varepsilon r_{ij}^2} e^{-kr_{ij}} C_0 C_1$$

$$+ \frac{\mu_i \mu_j}{4\pi\varepsilon_0 \varepsilon r_{ij}^3} \left\{ \cos\theta_i \cos\theta_j \left[2 + kr_{ij} + \left(kr_{ij}\right)^2 \right] \right.$$

$$+ \sin\theta_i \sin\theta_j \cos\left(\phi_i - \phi_j\right) \left[1 + kr_{ij} \right] \left. \right\} e^{-kr_{ij}} C_1^2 \qquad (1)$$

which was proposed by Phillies for polyelectrolyte colloids and proteins in dilute solution.

In the above equation,

$$C_0 = \frac{e^{ka}}{1 + ka}$$

$$C_1 = \frac{3e^{ka}}{2 + 2ka + \left(ka\right)^2 + \varepsilon_0\left(1 + ka\right)/\varepsilon} . \qquad (2)$$

Here q_i and q_j were the net charge carried by NP i and NP j, μ_i and μ_j were dipole moments, r_{ij} was the distance between two NPs, θ_i and θ_j were angles of the dipole vector with respect to the vector connecting the centers of the NPs, where $0 < \theta < \pi$, φ_i and φ_j were dihedral angles describing the relative rotation of dipoles, where $0 < \varphi < 2\pi$, $1/k$ was the Debye screening length which was set to be 2.5nm, ε_0 was the permittivity of vacuum, ε was the effective permittivity of the solvent (water), and a was the radius of a NP. The dipole moment was set to be 100 Debye as suggested by our quantum calculations. The net charge q was set to be $+3e$ as determined by experimental measurements. MC simulations in the canonical (*NVT*) ensemble were performed in a cubic box with periodic boundary conditions implemented in all three

Cartesian directions. A MC step was defined as N (number of NPs) attempts at moving the particles by either translation or rotation. The simulations began from a disordered state which was obtained by running 3 - 5 million MC steps with attractive energy $\xi = 0.0$ and $T = 25°C$. ξ was then gradually increased from a low value (disordered state) to a high value at which ordered films were formed. We investigated three different systems: $N = 20$ at volume fraction $\phi = 0.13$, $N = 30$ at $\phi = 0.12$, and $N = 60$ at $\phi = 0.13$. 20 independent runs were performed for the first two systems and 4 independent runs were performed for the third system. All of them demonstrated that CdTe NPs self-assembled into 2D monolayer films in solution (Figure 11C and 11D). The simulation results confirmed that all three types interactions between NPs, including hydrophobic attraction, dipole attraction, and electrostatic repulsion, were necessary for the formation of 2D NP assemblies.

Potentially, there can be other NP dispersions with specifically designed surface functionalities which afford the spontaneous organization of NP in sheets based on their preferential attraction to each other. In fact, we also found that DMAET-stabilized CdSe NP spontaneously organized into 2D monolayer in solution.[78] Moreover, somewhat similar self-organization of 90 nm triangular prisms from Ag in 2D sheets was also observed by Lin and coworkers.[80] The crystal lattices in the Ag platelets in the superstructure were aligned producing an intricately interconnected network.

5. Functionality of 2D NP assembly

A number of interesting prototype devices of functional 2D NP assemblies have been demonstrated on their basis such as memory structures,[81, 82] command surfaces,[83] single electrical transistors,[84] non-linear optical elements,[85] magnetic storage units,[86] and sensors.[67] Such a broad spectrum of applications is identical to other 3D films from NPs because the charge transfer in 2D films of NPs and other properties follow the same regularities as for other 3D NP assemblies. So only a couple of unique applications based on 2D NP assemblies are illustrated in this part.

5.1 Optical property of 2D NP assemblies

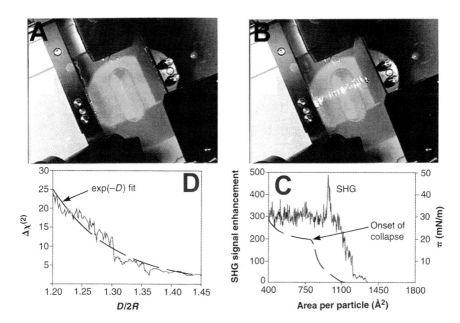

Figure 12. Photographs taken during compression of a Langmuir monolayer of butanethiol-capped Ag NPs, before (A) and just after (B) the metal-insulator transition. (C) Surface pressure of the Langmuir film (π) and SHG signal enhancement as a function of area per particle for hexanethiol-capped particles. The SHG has been normalized to the signal observed from the uncompressed monolayer, indicating that the enhancement originates from interparticle coupling. (D) Change in $|\chi(2)|$ versus D/2R. (A and B are taken from Ref. 87, and C and D taken from Ref. 88).

The optical property of 2D NP assemblies was easily tuned by controlling the inter-distances between NPs.[87] Heath and coworkers prepared 2D LB monolayer of alkylthiol-capped Ag NPs at the air-water interface. The inter-distances between NPs in the 2D monolayer were tuned by applied different surface pressures.[88] As shown in Figure 12A, before compression the 2D film showed a red color due to a strong optical resonance among Ag NPs. Upon compression, the middle part of 2D phase took on the appearance of a continuous, shiny film, which was analogous to the bulk Ag metal (Figure 12B). The optical change of 2D

NP assemblies was further investigated by studying the second-harmonic generation (SHG) response. An enhancement factor of ~ 500 was observed for the p-polarized SHG via the increase of surface pressure (Figure 12C), and the tremendous enhancement in SHG signal strength should originate from inter-NP coupling upon compression. A sharp discontinuity in the SHG data for the Ag NP films (Figure 12C) was that the energy gain from electron delocalization became sufficient to overcome the site (single-particle) charging energy. The strong inter-NP coupling was also confirmed by exponential change in $|\chi^{(2)}|$ versus $D/2R$. (Figure 12D), here $|\chi^{(2)}|$, D, and R were the square root of the SHG sign, the center-to-center particle separation distance, and a particle radius, respectively. All optical changes pointed to an insulator-to-metal transition of 2D NP assemblies with the decrease of inter-distances.

5.2 Optoelectrical response of 2D NP assemblies

The unique optoelectrial property of 2D NP assemblies has a potential application for electronic devices. Bawendi, Bulović and coworkers fabricated a 2D trioctylphosphine oxide-coated CdSe/ZnS core-shell NP monolayer atop a hole transporting N,N'-diphenyl-N,N'-bis(3-methylphenyl)-(1,1'-biphenyl)-4,4'-diamine (TPD) layer by taking advantage of the spontaneous phase separation of the mixture via a single spin-casting process onto the ITO substrates (Figure 13A).[89] Then, an electroluminescent device was prepared by thermal evaporation of a 40-nm-thick film of tris-(8-hydroxyquinoline)aluminium onto CdSe/ZnS core-shell NP monolayer, followed by a 1-mm-diameter, 75-nm-thick Mg:Ag (10:1 by mass) cathode with a 50-nm Ag cap (Figure 13A). The external quantum efficiency of such a device exceeded $\eta = 0.4\%$ for a broad range of device luminances (from 5 to 2,000 cd m^{-2}), peaking at $\eta = 0.52\%$ at 10 mA cm^{-2} (Figure 13B). The brightness of 100 cd m^{-2} was achieved at current density $J = 5.3$ mA cm^{-2}, voltage $V = 6.1$ V, corresponding to a luminescence efficiency of 1.9 cd A^{-1}. At 125 mA cm^{-2}, the brightness of the device was 2,000 cd m^{-2}, which corresponded to 1.6 cd A^{-1}, and a 25-fold improvement over the best previously reported NP-based light-emitting diode (LED) result (Figure 13B). Very recently, the authors prepared the white LEDs with a broad spectral emission generated by electroluminescence from a mixed-monolayer of red, green, and blue emitting CdSe/ZnS NP assemblies.[90] The white LED exhibited external quantum efficiencies of 0.36% coordinates of (0.35, 0.41) at video brightness, and color rendering index of 86 as compared to a 5500 K blackbody reference. Accordingly, the dramatic improvement of the brightness as well the broad spectrum makes 2D NP assemblies to become a good candidate in the future luminescent device.

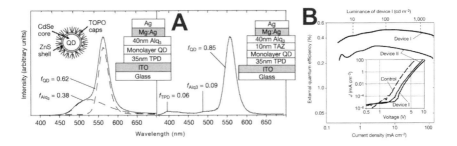

Figure 13. (A) Electroluminescence spectra and structures for two QD-LEDs, devices I and II. (B) External quantum efficiency versus current density for the two devices shown in A. Taken from Ref. 89.

6. Problems and promises of 1D and 2D NP assemblies

Although 1D and 2D NP assemblies are ubiquitous in natural biology, the research work on their preparation and functionality are still in the burgeon stage. There are plenty of unknown phenomena awaiting exploration and many problems requiring resolution. Besides, current reports on new application of functional NP assemblies are seldom, and the burst development of this area can be expected in the near future.

6.1 Formation mechanism of 1D and 2D NP assemblies

Recent progresses on studying the anisotropy of NPs, arising from either the crystalline structure or the distribution of stabilizers, have been helping the scientists to understand the self-assembly mechanism of 1D and 2D NP assemblies.[4] However, all above studies are based on the qualitative analysis. For example, theoretical calculation showed the dipole moments inside NPs are 50-200 D,[91, 92] which were strong enough to overcome the thermal energy at the room temperature and drove the 1D self assembly of NPs. Unfortunately, until now the absence of the suitable measurement methods prevents the scientists to obtain the direct and quantitative data on the NP dipole moments, and as well the accumulated effect of dipole moments during formation of the 1D NP assemblies.

The same problems also trouble the scientists who are studying the anisotropy of NPs from inhomogeneous distribution of stabilizers on their surface.[39] The absence of the characterization techniques makes them impossible for direct observation of stabilizer distribution on NP surfaces, as well the evolution process of stabilizer during formation of the 1D NP assemblies.

The 2D self assembly mechanism of NPs is more complex and elusive. Actually, there is only one report on 2D templateless self assembly of NPs by taking advantage of the triple folds of interactions between NPs, such as hydrophobic attraction, dipole attraction, and electrostatic repulsion.[78] The future work should target at understanding the general formation mechanism of 2D NP assemblies, for instance, except for above three types, are there any other types of interactions may lead to formation of 2D NP assemblies? Other urgent studies include quantitative investigation on the interaction strengths between different types of forces, which result in the production of 2D NP assemblies.

6.2 Functionality of 1D and 2D NP assemblies

The mechanism study of 1D and 2D NP assemblies should serve for design of functional assemblies and application of them in our life. So, in some extent the studies on functionality of 1D and 2D are paramount for their future development.

One of tendencies on 1D NP assemblies is to fabricate the functional and binary structures. As discussed in Part 3, the 1D hybrid assemblies of metal oxide and metal NPs have shown an intriguing application promise in the field of energy storage and conversion.[43, 44] Certainly, the hybrid and application of 1D assemblies are not limited to above two types of NPs. As examples, the hybrid assemblies of semiconductor and metal NPs can be expected to have prominent functionality in the field of solar cells or photodecomposition of organic pollutants due to the integration of effective light harvesting of semiconductor NPs and good conductivity of metal NPs. The 1D hybrid assemblies of semiconductor and magnetic NPs may also lead to their application as spintronics.

As shown in part 5, the single type of NPs have been assembled into 2D films and used as the optoelectronics.[89, 90] There are no any reports on preparation and application of 2D hybrid assemblies with binary NPs. Undoubtedly, future work on preparation of 2D assemblies containing different types of NPs will extend their application to a broad area, such as functional membranes, solar cells, sensors, and etc.

Acknowledgements

The authors thank 100-talent program of Chinese Academy of Sciences (ZYT), NSF (NAK), AFOSR (NAK), and DARPA (NAK) for the financial support of this research.

References

1. A. P. Alivisatos. *Science* **271**, 933-937 (1996).
2. A. P. Alivisatos. *Journal of Physical Chemistry* **100**, 13226-13239 (1996).
3. C. B. Murray, C. R. Kagan, M. G. Bawendi. *Annual Review of Materials Science* **30**, 545-610 (2000).
4. Z. Y. Tang, N. A. Kotov. *Advanced Materials* **17**, 951-962 (2005).
5. D. L. Feldheim, C. D. Keating. *Chemical Society Reviews* **27**, 1-12 (1998).
6. J. H. Fendler. *Chemistry of Materials* **8**, 1616-1624 (1996).
7. C. B. Murray, C. R. Kagan, M. G. Bawendi. *Science* **270**, 1335-1338 (1995).
8. L. Motte, F. Billoudet, E. Lacaze, J. Douin, M. P. Pileni. *Journal of Physical Chemistry B* **101**, 138-144 (1997).
9. C. J. Kiely, J. Fink, M. Brust, D. Bethell, D. J. Schiffrin. *Nature* **396**, 444-446 (1998).
10. C. T. Black, C. B. Murray, R. L. Sandstrom, S. H. Sun. *Science* **290**, 1131-1134 (2000).
11. E. V. Shevchenko, D. V. Talapin, N. A. Kotov, S. O'Brien, C. B. Murray. *Nature* **439**, 55-59 (2006).

12. C. P. Collier, T. Vossmeyer, J. R. Heath. *Annual Review of Physical Chemistry* **49**, 371-404 (1998).

13. A. L. Rogach, D. V. Talapin, E. V. Shevchenko, A. Kornowski, M. Haase, H. Weller. *Advanced Functional Materials* **12**, 653-664 (2002).

14. M. P. Pileni. *Journal of Physics-Condensed Matter* **18**, S67-S84 (2006).

15. A. Kiriy, S. Minko, G. Gorodyska, M. Stamm, W. Jaeger. *Nano Letters* **2**, 881-885 (2002).

16. R. Sheparovych, Y. Sahoo, M. Motornov, S. M. Wang, H. Luo, P. N. Prasad, I. Sokolov, S. Minko. *Chemistry of Materials* **18**, 591-593 (2006).

17. S. H. Yu, H. Colfen. *Journal of Materials Chemistry* **14**, 2124-2147 (2004).

18. P. M. Ajayan, S. Iijima. *Nature* **361**, 333-334 (1993).

19. D. Ugarte, T. Stockli, J. M. Bonard, A. Chatelain, W. A. de Heer. *Applied Physics a-Materials Science & Processing* **67**, 101-105 (1998).

20. S. Banerjee, S. S. Wong. *Nano Letters* **2**, 195-200 (2002).

21. C. J. Murphy, T. K. San, A. M. Gole, C. J. Orendorff, J. X. Gao, L. Gou, S. E. Hunyadi, T. Li. *Journal of Physical Chemistry B* **109**, 13857-13870 (2005).

22. M. Grzelczak, M. A. Correa-Duarte, V. Salgueirino-Maceira, M. Giersig, R. Diaz, L. M. Liz-Marzan. *Advanced Materials* **18**, 415-+ (2006).

23. J. Lee, P. Hernandez, J. Lee, A. O. Govorov, N. A. Kotov. *Nature Materials* **6**, 291-295 (2007).

24. M. G. Warner, J. E. Hutchison. *Nature Materials* **2**, 272-277 (2003).

25. E. Braun, K. Keren. *Advances in Physics* **53**, 441-496 (2004).

26. C. B. Mao, D. J. Solis, B. D. Reiss, S. T. Kottmann, R. Y. Sweeney, A. Hayhurst, G. Georgiou, B. Iverson, A. M. Belcher. *Science* **303**, 213-217 (2004).

27. Y. Huang, C. Y. Chiang, S. K. Lee, Y. Gao, E. L. Hu, J. De Yoreo, A. M. Belcher. *Nano Letters* **5**, 1429-1434 (2005).

28. T. Scheibel, R. Parthasarathy, G. Sawicki, X. M. Lin, H. Jaeger, S. L. Lindquist. *Proceedings of the National Academy of Sciences of the United States of America* **100**, 4527-4532 (2003).

29. Z. Y. Tang, N. A. Kotov, M. Giersig. *Science* **297**, 237-240 (2002).

30. Z. Y. Tang, B. Ozturk, Y. Wang, N. A. Kotov. *Journal of Physical Chemistry B* **108**, 6927-6931 (2004).

31. Y. Volkov, S. Mitchell, N. Gaponik, Y. P. Rakovich, J. F. Donegan, D. Kelleher, A. L. Rogach. *Chemphyschem* **5**, 1600-1602 (2004).

32. Y. Lalatonne, J. Richardi, M. P. Pileni. *Nature Materials* **3**, 121-125 (2004).

33. A. M. Cao, J. S. Hu, H. P. Liang, W. G. Song, L. J. Wan, X. L. He, X. G. Gao, S. H. Xia. *Journal of Physical Chemistry B* **110**, 15858-15863 (2006).

34. J. Polleux, N. Pinna, M. Antonietti, M. Niederberger. *Advanced Materials* **16**, 436-+ (2004).

35. M. Niederberger, H. Colfen. *Physical Chemistry Chemical Physics* **8**, 3271-3287 (2006).

36. R. E. Dunin-Borkowski, M. R. McCartney, M. Posfai, R. B. Frankel, D. A. Bazylinski, P. R. Buseck. *European Journal of Mineralogy* **13**, 671-684 (2001).

37. M. M. Walker, T. E. Dennis, J. L. Kirschvink. *Current Opinion in Neurobiology* **12**, 735-744 (2002).

38. S. Shanbhag, N. A. Kotov. *Journal of Physical Chemistry B* **110**, 12211-12217 (2006).

39. A. M. Jackson, J. W. Myerson, F. Stellacci. *Nature Materials* **3**, 330-336 (2004).

40. G. A. DeVries, M. Brunnbauer, Y. Hu, A. M. Jackson, B. Long, B. T. Neltner, O. Uzun, B. H. Wunsch, F. Stellacci. *Science* **315**, 358-361 (2007).

41. A. Y. Sinyagin, A. Belov, Z. N. Tang, N. A. Kotov. *Journal of Physical Chemistry B* **110**, 7500-7507 (2006).

42. K. S. Cho, D. V. Talapin, W. Gaschler, C. B. Murray. *Journal of the American Chemical Society* **127**, 7140-7147 (2005).

43. K. T. Nam, D. W. Kim, P. J. Yoo, C. Y. Chiang, N. Meethong, P. T. Hammond, Y. M. Chiang, A. M. Belcher. *Science* **312**, 885-888 (2006).

44. L. Taberna, S. Mitra, P. Poizot, P. Simon, J. M. Tarascon. *Nature Materials* **5**, 567-573 (2006).

45. Z. L. Wang. *Advanced Materials* **10**, 13-+ (1998).

46. J. Tang, G. L. Ge, L. E. Brus. *Journal of Physical Chemistry B* **106**, 5653-5658 (2002).

47. E. Rabani, D. R. Reichman, P. L. Geissler, L. E. Brus. *Nature* **426**, 271-274 (2003).

48. H. Haidara, K. Mougin, J. Schultz. *Langmuir* **17**, 1432-1436 (2001).

49. H. Haidara, K. Mougin, J. Schultz. *Langmuir* **17**, 659-663 (2001).

50. M. R. E. Warner, R. V. Craster, O. K. Matar. *Journal of Colloid and Interface Science* **267**, 92-110 (2003).

51. Z. Y. Zhong, B. Gates, Y. N. Xia, D. Qin. *Langmuir* **16**, 10369-10375 (2000).

52. S. P. Li, W. S. Lew, Y. B. Xu, A. Hirohata, A. Samad, F. Baker, J. A. C. Bland. *Applied Physics Letters* **76**, 748-750 (2000).

53. I. A. Ansari, I. W. Hamley. *Journal of Materials Chemistry* **13**, 2412-2413 (2003).

54. G. Kim, M. Libera. *Macromolecules* **31**, 2569-2577 (1998).

55. L. S. Li, J. Jin, S. Yu, Y. Y. Zhao, C. X. Zhang, T. J. Li. *Journal of Physical Chemistry B* **102**, 5648-5652 (1998).

56. N. A. Kotov, F. C. Meldrum, C. Wu, J. H. Fendler. *Journal of Physical Chemistry* **98**, 2735-2738 (1994).

57. B. O. Dabbousi, C. B. Murray, M. F. Rubner, M. G. Bawendi. *Chemistry of Materials* **6**, 216-219 (1994).

58. P. D. Yang. *Nature* **425**, 243-244 (2003).

59. A. Tao, F. Kim, C. Hess, J. Goldberger, R. R. He, Y. G. Sun, Y. N. Xia, P. D. Yang. *Nano Letters* **3**, 1229-1233 (2003).

60. D. Whang, S. Jin, Y. Wu, C. M. Lieber. *Nano Letters* **3**, 1255-1259 (2003).

61. I. A. Greene, F. X. Wu, J. Z. Zhang, S. W. Chen. *Journal of Physical Chemistry B* **107**, 5733-5739 (2003).

62. G. Schmid. *Advanced Engineering Materials* **3**, 737-743 (2001).
63. G. Brezesinski, H. Mohwald. *Advances in Colloid and Interface Science* **100**, 563-584 (2003).
64. Y. Lin, H. Skaff, T. Emrick, A. D. Dinsmore, T. P. Russell. *Science* **299**, 226-229 (2003).
65. E. Glogowski, J. B. He, T. P. Russell, T. Emrick. *Chemical Communications*, 4050-4052 (2005).
66. Y. Lin, A. Boker, H. Skaff, D. Cookson, A. D. Dinsmore, T. Emrick, T. P. Russell. *Langmuir* **21**, 191-194 (2005).
67. I. Willner, A. N. Shipway, B. Willner. In *Molecules as Components of Electronic Devices*, pp. 88-105 (2003).
68. D. R. Larson, W. R. Zipfel, R. M. Williams, S. W. Clark, M. P. Bruchez, F. W. Wise, W. W. Webb. *Science* **300**, 1434-1436 (2003).
69. R. F. Service. *Science* **310**, 1132-1134 (2005).
70. M. M. C. Cheng, G. Cuda, Y. L. Bunimovich, M. Gaspari, J. R. Heath, H. D. Hill, C. A. Mirkin, A. J. Nijdam, R. Terracciano, T. Thundat, M. Ferrari. *Current Opinion in Chemical Biology* **10**, 11-19 (2006).
71. W. Shenton, D. Pum, U. B. Sleytr, S. Mann. *Nature* **389**, 585-587 (1997).
72. S. Dieluweit, D. Pum, U. B. Sleytr. *Supramolecular Science* **5**, 15-19 (1998).
73. R. A. McMillan, C. D. Paavola, J. Howard, S. L. Chan, N. J. Zaluzec, J. D. Trent. *Nature Materials* **1**, 247-252 (2002).
74. S. W. Lee, S. K. Lee, A. M. Belcher. *Advanced Materials* **15**, 689-692 (2003).
75. I. Yamashita. *Thin Solid Films* **393**, 12-18 (2001).
76. M. Li, K. K. W. Wong, S. Mann. *Chemistry of Materials* **11**, 23-+ (1999).
77. J. Hoinville, A. Bewick, D. Gleeson, R. Jones, O. Kasyutich, E. Mayes, A. Nartowski, B. Warne, J. Wiggins, K. Wong. *Journal of Applied Physics* **93**, 7187-7189 (2003).
78. Z. Y. Tang, Z. L. Zhang, Y. Wang, S. C. Glotzer, N. A. Kotov. *Science* **314**, 274-278 (2006).
79. Z. L. Zhang, Z. Y. Tang, N. A. Kotov, S. C. Glotzer. *Nano Lett.* **7**, 1670-1675 (2007).
80. J. Y. Chang, J. J. Chang, B. Lo, S. H. Tzing, Y. C. Ling. *Chemical Physics Letters* **379**, 261-267 (2003).
81. P. Poddar, T. Telem-Shafir, T. Fried, G. Markovich. *Physical Review B* **66** (2002).
82. S. Paul, C. Pearson, A. Molloy, M. A. Cousins, M. Green, S. Kolliopoulou, P. Dimitrakis, P. Normand, D. Tsoukalas, M. C. Petty. *Nano Letters* **3**, 533-536 (2003).
83. J. Ortegren, K. D. Wantke, H. Motschmann, H. Mohwald. *Journal of Colloid and Interface Science* **279**, 266-276 (2004).
84. S. W. Chen, R. S. Ingram, M. J. Hostetler, J. J. Pietron, R. W. Murray, T. G. Schaaff, J. T. Khoury, M. M. Alvarez, R. L. Whetten. *Science* **280**, 2098-2101 (1998).

85. S. Bernard, N. Felidj, S. Truong, P. Peretti, G. Levi, J. Aubard. *Biopolymers* **67**, 314-318 (2002).
86. S. H. Sun, C. B. Murray, D. Weller, L. Folks, A. Moser. *Science* **287**, 1989-1992 (2000).
87. G. Markovich, C. P. Collier, S. E. Henrichs, F. Remacle, R. D. Levine, J. R. Heath. *Accounts of Chemical Research* **32**, 415-423 (1999).
88. C. P. Collier, R. J. Saykally, J. J. Shiang, S. E. Henrichs, J. R. Heath. *Science* **277**, 1978-1981 (1997).
89. S. Coe, W. K. Woo, M. Bawendi, V. Bulovic. *Nature* **420**, 800-803 (2002).
90. P. O. Anikeeva, J. E. Halpert, M. G. Bawendi, V. Bulovic. *Nano Lett.* (2007).
91. M. Shim, P. Guyot-Sionnest. *Journal of Chemical Physics* **111**, 6955-6964 (1999).
92. E. Rabani. *Journal of Chemical Physics* **115**, 1493-1497 (2001).

CHAPTER 8

SYNTHESIS OF POROUS POLYMERS USING SUPERCRITICAL CARBON DIOXIDE

Colin D. Wood* and Andrew I. Cooper

Donnan and Robert Robinson Laboratories, Department of Chemistry, University of Liverpool, Liverpool, L69 3BX, UK
E-mail: C.D.Wood@liverpool.ac.uk

Supercritical carbon dioxide is the most extensively studied supercritical fluid (SCF) medium for polymerization reactions and organic transformations. This can be attributed to a list of advantages ranging from solvent properties to practical environmental as well as economic considerations. Aside from these gains, CO_2 finds particularly advantageous application in the synthesis and processing of porous materials and as such, this will be the subject of this review.

1. Introduction

Supercritical carbon dioxide ($scCO_2$) has been promoted recently as a sustainable solvent because it is nontoxic, nonflammable, and naturally abundant.[1] $scCO_2$ has been shown to be a versatile solvent for polymer synthesis and processing,[2-5] and it has been exploited quite widely for the preparation of porous materials:[6] for example, $scCO_2$ has been used for the production of microcellular polymer foams,[7,8] biodegradable composite materials,[9] macroporous polyacrylates,[10-13] and fluorinated microcellular materials.[14]

Although CO_2 is attractive in a range of different applications a significant technical barrier remains in that it is a relatively weak solvent: important classes of materials which tend to exhibit low solubility in $scCO_2$ include polar biomolecules, pharmaceutical actives, and high-molecular weight polymers.[1,2,6,15-18] Until recently, the only polymers found to have significant solubility in CO_2 under moderate conditions ($<100°C$, <400 bar) were amorphous fluoropolymers[1] and, to a lesser extent, polysiloxanes.[17] Therefore, the discovery of inexpensive CO_2-soluble materials or "CO_2-philes" has been an important challenge.[19-21] Herein, we will focus on our work looking at applying carbon dioxide in the synthesis of porous polymers and recently developed CO_2-soluble hydrocarbon polymers and surfactants.

2. Porous materials and supercritical fluids

Porous materials are used in a wide range of applications, including catalysis, chemical separations, and tissue engineering.[6] However, the synthesis of these materials is often solvent intensive. Supercritical carbon dioxide as an alternative solvent for the synthesis of functional porous materials can circumvent this issue as well as affording a number of specific physical, chemical, and toxicological advantages. For example, energy intensive drying steps are required in order to dry porous materials whereas the transient "dry" nature of CO_2 overcomes these issues. Pore collapse can occur in certain materials when removing conventional liquid solvents; this can be avoided using supercritical fluids (SCFs) because they do not possess a liquid-vapour interface. Porous structures are important in biomedical applications (*e.g.*, tissue engineering) where the low toxicity of CO_2 offers specific advantages in terms of minimizing the use of toxic organic solvents. In addition, the wetting properties and low viscosity of CO_2 offers specific benefits in terms of surface modification.

A number of new approaches have been developed in the past few years for the preparation of porous materials using supercritical fluids (SCF).[6,22,23,24,25] Current routes include foaming,[9,26-29] CO_2-induced phase separation,[30,31] reactive-[10-13] and nonreactive[14,32] gelation of CO_2 solutions, nanoscale casting using supercritical CO_2,[33,34] and CO_2-in-water (C/W) emulsion templating.[22-25,35,36] Each of these methods uses a different mechanism to generate the porosity in the material. In the following sections we will discuss some of the methods that we have developed.

3. CO_2 as a pressure-adjustable template/porogen

$scCO_2$ has been used for the formation of permanently porous crosslinked poly(acrylate) and poly(methacrylates) monoliths using $scCO_2$ as the porogenic solvent.[10-12] Materials of this type[37,38] are useful in applications such as high-performance liquid chromatography, high-performance membrane chromatography, capillary electrochromatography, microfluidics,[39] and high-throughput bioreactors.[40] In this process, no organic solvents are used in either the synthesis or purification. It is possible to synthesize the monoliths in a variety of containment vessels, including chromatography columns and narrow-bore capillary tubing. Moreover, the variable density associated with SCF solvents was exploited in order to "fine-tune" the polymer morphology. The apparent Brunauer-Emmett-Teller (BET) surface area and the average pore size of the materials varied substantially for a series of crosslinked monoliths synthesized using $scCO_2$ as the porogen over a range of reaction pressures.[12] This approach was also extended to synthesize well-defined porous, cross-linked beads by suspension polymerization, again without using any organic solvents.[13] The surface area of the of the beads could be tuned over a broad range (5-500 m^2/g) simply by varying the CO_2 density. The ability to fine tune the polymer morphology in materials of this type can be rationalized by considering the variation in solvent quality as a function of CO_2 density and the resulting influence on the mechanism of nucleation, phase separation, aggregation, monomer partitioning, and pore formation.[41]

Recently, an entirely new approach to preparing porous materials was developed by templating the structure of *solid* CO_2 by directional freezing.[42] In this process, a liquid CO_2 solution was frozen in liquid nitrogen unidirectionally. The solid CO_2 was subsequently removed by direct sublimation to yield a porous, solvent-free structure with no additional purification steps. Other CO_2-soluble actives could be incorporated into the porous structure uniformly (Figure 1). This was demonstrated by dispersing oil red uniformly in the aligned porous sugar acetates.[42] This method differs fundamentally from the other CO_2-based techniques[9,10,12-14,22-36] and offers the unique advantage of generating materials with aligned pore structures. Materials with aligned microstructures and nanostructures are of interest in a wide range of applications such as organic electronics,[43] microfluidics,[44] molecular filtration,[45] nanowires,[46] and tissue engineering.[47]

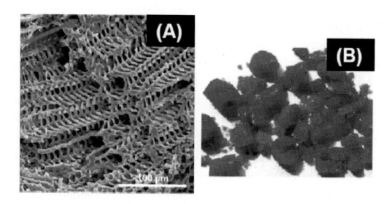

Figure 1. A) Porous sugar acetate with an aligned structure prepared by directional freezing of a liquid CO_2 solution. B) Oil Red O, a CO_2-soluble dye, could be uniformly dispersed in this aligned porous material.[42]

In addition to producing materials with aligned porosity, there are a number of additional advantages associated with this new technique. The method avoids the use of any organic solvents, thus eliminating toxic residues in the resulting material. The CO_2 can be removed by simple sublimation, unlike aqueous-based processes where the water must be removed by freeze-drying.[48-50] Moreover, the method can be applied to relatively nonpolar, water-insoluble materials. These aligned porous structures may find numerous applications, for example, as biomaterials. Aligned porous materials with micrometer-sized pores are of importance in tissue engineering where modification with biological cells is required. We are particularly interested in the use of such porous materials as scaffolds for aligned nerve cell growth. This latter application will be greatly facilitated by the recent development of biodegradable CO_2-soluble hydrocarbon polymers as potential scaffold materials.[51]

4. Polymer solubility in CO_2

One of the fundamental issues that one must consider when implementing CO_2 for polymer synthesis or processing is polymer solubility. As mentioned, CO_2 is a weak solvent and there has been considerable research effort focused on discovering inexpensive biodegradable CO_2-soluble polymers from which inexpensive CO_2-soluble surfactants, ligands, and phase transfer agents could be developed. However, it is very difficult to predict which polymer structures would be CO_2-soluble, despite attempts to rationalize specific solvent-solute interactions by using *ab initio* calculations.[52] Only a few examples of CO_2-soluble polymers currently exist and, as such, there are a limited number of "design motifs" to draw upon. Moreover, it is clear that polymer solubility in CO_2 is influenced by a large number of interrelated factors[17] such as specific solvent-solute interactions,[20,52-54] backbone flexibility,[20,53,55] topology,[55] and the nature of the end-groups.[55] Given the current limits of predictive understanding, the discovery of new CO_2-soluble polymers might be accelerated using parallel or 'high-throughput' methodology. The synthetic approaches for such a strategy are already well in place; for example, a growing number of methods exist whereby one may synthesize and characterize polymer libraries.[56] By contrast, there are no examples of techniques for the rapid, parallel determination of solubility for libraries of materials in $scCO_2$ or other SCFs. The conventional method for evaluating polymer solubility in SCFs is cloud point measurement,[17,20,53,55] which involves the use of a variable-volume view cell. This technique is not suitable for rapid solubility measurement and would be impractical for large libraries of materials.

A number of research groups have synthesised 'CO_2-philic' fluoropolymers or silicone-based materials for use as steric stabilisers in dispersion polymerization[5,57-60] as phase transfer agents for liquid–liquid extraction,[61] as supports for homogeneous catalysis,[62,63] and as surfactants for the formation of water / CO_2 emulsions and microemulsions.[64,65] Unfortunately, the high cost of fluorinated polymers may prohibit their use on an industrial scale for some applications. Fluoropolymers also tend to have poor environmental degradability, and this could negate the environmental advantages associated with the use of $scCO_2$. The lack of inexpensive CO_2-soluble polymers and surfactants is a significant barrier to the future implementation of this solvent technology.[66]

Inexpensive poly(ether carbonate) (PEC) copolymers have been reported to be soluble in CO_2 under moderate conditions.[20] Similarly, sugar acetates are highly soluble and have been proposed as renewable CO_2-philes.[52,67] Such materials could, in principle, function as CO_2-philic building blocks for inexpensive ligands and surfactants, but this potential has not yet been realized and numerous practical difficulties remain. For example, CO_2 solubility does not in itself guarantee performance in the various applications of interest. Effective surfactants, in particular, tend to require specific asymmetric topologies such as diblock copolymers.[64,68] This in turn necessitates a flexible and robust synthetic methodology to produce well-defined architectures for specific applications.

5. High throughput solubility measurements in CO_2

We reported a new method which allows for the rapid parallel solubility measurements for libraries of materials in supercritical fluids.[69] The technique was used to evaluate the solubility of a mixed library of 100 synthetic polymers including polyesters, polycarbonates, and vinyl polymers. It was found that poly(vinyl acetate) (PVAc) showed the highest solubility in CO_2, the anamolously high solubility of PVAc has been shown previously.[66] This method is at least 50 times faster than other techniques in terms of the rate of useful information that is obtained and has broad utility in the discovery of novel SCF-soluble ligands, catalysts, biomolecules, dyes, or pharmaceuticals for a wide range of materials applications.

6. Inexpensive and Biodegradable CO_2-Philes

Poly(vinyl acetate) (PVAc) is an inexpensive, high-tonnage bulk commodity polymer which, unlike most vinyl polymers, is moderately biodegradable and has been used in pharmaceutical excipient formulations. PVAc has also been shown to exhibit anomalously high solubility in CO_2 with respect to other vinyl hydrocarbon polymers,[66] although the polymer is soluble only at relatively low molecular weights under conditions of practical relevance (P < 300 bar, T < 100 °C).

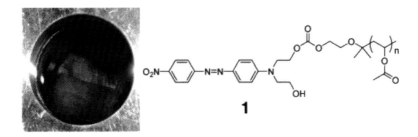

Figure 2. Photograph showing the dissolution of an OVAc-functionalized dye, 1, in CO_2 (200 bar, 20 °C, 0.77 wt %).[51]

We presented a simple and generic method for producing inexpensive and biodegradable polymer surfactants for $scCO_2$ for solubilization, emulsification, and related applications.[51] In this method, the terminal hydroxyl group of a poly(vinyl acetate) (PVAc) oligomer is transformed into an imidazole ester by reaction with carbonyl diimidazole (CDI). This route has a number of advantages. First, the OVAc imidazolide intermediate can be isolated, purified, and then coupled with a wide range of alcohols (or amines) to produce a variety of structures. Second, the route introduces a carbonate linkage that may further enhance CO_2 solubility[53,70] and could also improve the biodegradability of the resulting materials. To illustrate the use of OVAc as a solubilizing group, an organic dye, Disperse Red 19 (DR19), was functionalized with OVAc (Mn) 1070 g/mol, (Mw) 1430 g/mol to produce **1** (Figure 2). The stoichiometry of the reaction was controlled such that one OVAc chain was attached to each DR19 molecule, as confirmed by GPC and 1H NMR. DR19 itself had negligible solubility in CO_2 up to pressures of 300 bar/25 °C (no color was observed in the CO_2 phase). By contrast, the functionalized dye, **1**, was found to be soluble in CO_2 (100-200 bar) at least up to concentrations of around 1 wt % (Figure 2). This suggested that OVAc has potential as a less expensive and more biodegradable replacement for the highly fluorinated materials used previously to solubilize species such as dyes, catalysts, proteins, and nanoparticles in CO_2.[1,4,5,60-62,64,65,71,72] Another important area in $scCO_2$ technology is the formation of water-in-CO_2 (W/C) and CO_2-in-water (C/W) emulsions and microemulsions.[65,73-76] The same CDI route was used to couple OVAc with poly(ethylene glycol) monomethyl ethers (HO-PEG-OMe) and poly(ethylene glycol) diols (PEG) to produce diblock (Figure 4, **2a**) and triblock (**2b**) copolymers, respectively (Figure 3).[51]

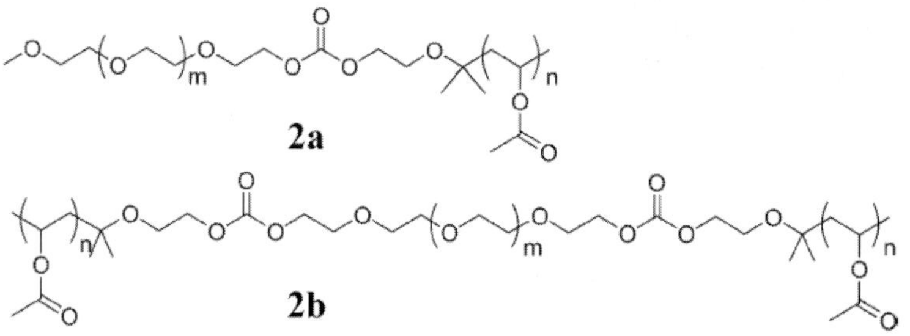

Figure 3. Structures of CO_2-philic surfactants for C/W emulsion formation: OVAc-b-PEG diblock polymer (2a) and OVAc-b-PEG-b-OVAc triblock polymer (2b).[51]

These polymers were found to be useful surfactants. For example, an OVAc-b-PEG-OVAC triblock surfactant was found to emulsify up to 97 v/v % C/W emulsion which was stable for at lease 48 h. An OVAc-b-PEG diblock copolymer was used to form a 90 v/v % C/W emulsion. Materials of this type were used to template SCF emulsions and will be discussed in detail in the following section.

7. Templating of supercritical fluid emulsions

Emulsion templating is useful for the synthesis of highly porous inorganic,[77-80] and organic materials.[81-83] In principle, it is possible to access a wide range of porous hydrophilic materials by reaction-induced phase separation of concentrated oil-in-water (O/W) emulsions. A significant drawback to this process is that large quantities of a water-immiscible oil or organic solvent are required as the internal

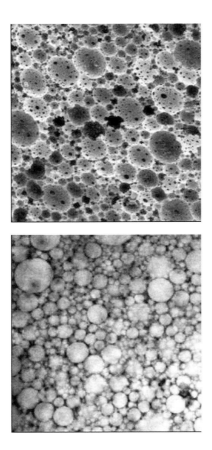

Figure 4. Emulsion-templated crosslinked polyacrylamide materials synthesized by polymerization of a high-internal phase CO_2-in-water emulsion (C/W HIPE). (a) SEM image of sectioned material. (b) Confocal image of same material, obtained by filling the pore structure with a solution of fluorescent dye. As such (a) shows the "walls" of the material while (b) show the "holes" formed by templating the $scCO_2$ emulsion droplets. Both images = 230 μm x 230 μm. Ratio of CO_2/aqueous phase = 80:20 v/v.d Pore volume = 3.9 cm^3/g. Average pore diameter = 3.9 μm. Adapted from Butler et al. [35]

phase (usually >80 vol.%). In addition, it is often difficult to remove this oil phase after the reaction. Based on studies concerning SCF emulsion stability and formation,[74] we have developed methods for templating high internal phase CO_2-in-water (C/W) emulsions (HIPE) to generate highly porous materials in the absence of any organic solvents – only water and CO_2 are used.[35] If the emulsions are stable one can generate low-density materials (~ 0.1 g/cm^3) with relatively large pore volumes (up to 6 cm^3/g) from water-soluble vinyl monomers such as acrylamide and hydroxyethyl acrylate. Figure 4 shows a crosslinked polyacrylamide material synthesized from a high internal phase C/W emulsion, as characterized by SEM and confocal microscopy (scale = 230 μm x 230 μm). Comparison of the two images illustrates quite clearly how the porous structure shown in the SEM image is templated from the C/W emulsion (as represented by the confocal microscopy image of the pores). In general, the confocal image gives a better measure of the CO_2 emulsion droplet size and size distribution immediately before gelation of the aqueous phase. Initially we used low molecular weight (Mw ~ 550 g/mol) perfluoropolyether ammonium carboxylate surfactants to stabilize the C/W emulsions[35] but as discussed there are some practical disadvantages of using these surfactants in this particular process such as cost and the surfactant is non-degradable. It was subsequently shown that it is possible to use inexpensive hydrocarbon surfactants to stabilize C/W emulsions and that these emulsions can also be templated to yield low-density porous materials.[36] In this study it was shown that all of the problems associated with the initial approach could be overcome and it was possible to synthesize C/W emulsion-templated polymers at relatively modest pressures (60 – 70 bar) and low temperatures (20 °C) using inexpensive and readily available hydrocarbon surfactants. Moreover, we demonstrated that this technique can in principle be extended to the synthesis of emulsion-templated HEA and HEMA hydrogels that may be useful, for example, in biomedical applications.[84-86]

As mentioned, we demonstrated a simple and generic method for producing inexpensive, functional hydrocarbon CO_2-philes for solubilization, emulsification, and related applications.[51] This approach was extended and water-soluble diblock and triblock surfactant architectures were accessed and it was found that both types of structure could stabilize highly concentrated C/W emulsions. A detailed investigation into the factors affecting the C/W emulsion stability was carried out in order to utilize these optimized emulsions to generate materials with significantly increased levels of porosity.[22] This new method is a simple and generic method for producing inexpensive and biodegradable polymer surfactants for use in supercritical CO_2. Low molecular weight (M-w < 7000 g/mol) hydroxyl-terminated poly(vinyl acetate) (PVAc-OH) was synthesized using optimized reaction conditions and isopropyl ethanol (IPE) as the chain transfer agent. Oligomeric PVAc-OH (OVAc-OH, M-w < 3000 g/mol) was then obtained by supercritical fluid fractionation. The OVAc-OH species was converted to the imidazole ester by reaction with carbonyl diimidazole (CDI) and CO_2-soluble surfactants were produced by coupling these reactive blocks with poly(ethylene glycol) methyl ethers or poly(ethylene glycol) diols. The surfactants were found to be extremely effective in the production of stable CO_2-in-water (C/W) emulsions (Figure 5), which were then used as templates to produce emulsion-templated materials with unprecedentedly high levels of porosity for materials produced by this route. It was shown that these hydrocarbon surfactants can outperform perfluorinated species in applications of this type. The synthetic methodology also allows fine-tuning of the hydrophilic-CO_2-philic balance to suit different applications. Surfactants of this type may find a range of additional uses in emulsion technology, particularly where biodegradability of the hydrophobic segment is required.

Figure 5. Optical image of C/W using OVAC(1070)-b-PEG(2000) surfactants at varying CO_2: H_2O volume ratios.[22]

We recently presented a new methodology to produce highly porous cross-linked hydrogel materials by templating concentrated CO_2-in-water (C/W) emulsions.[23] Poly(vinyl alcohol) (PVA), blended PVA/PEG, and naturally derived chitosan materials were produced via this route.

Using the PVAc-based surfactants discussed above the C/W emulsions were sufficiently stable for templating to occur and for open-cell porous materials to be produced, as shown in Figure 6. It was observed that the internal structure was uniformly porous and consisted of a skeletal replica of the original C/W HIPE. The pore structure was highly interconnected and there were open pores on the surface that were connected to the interior (Figure 6a). The diameter of the macropores was found to be in the range 3–15 μm. The technique can be carried out at moderate temperatures and pressures (25 degrees C, < 120 bar) using inexpensive hydrocarbon surfactants such as PVAc-based block copolymers which are composed of biodegradable blocks. This methodology opens up a new solvent-free route for the preparation of porous biopolymers, hydrogels, and composites, including materials which cannot readily be produced by foaming.

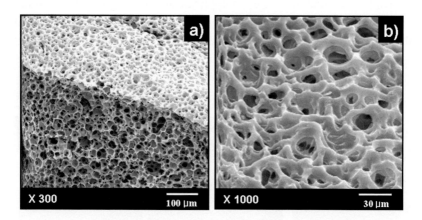

Figure 6. Electron micrographs of open-cell porous PVA hydrogel produced from C/W emulsions in the presence of PVAc-based surfactant. a) internal and surface pore structures; b) showing surface morphology with higher magnification.[23]

A number of limitations were apparent in our initial C/W emulsion templating approach; namely that the PFPE surfactant was expensive and nondegradable, reaction pressures were high (250-290 bar), and reaction temperatures were elevated (50-60 °C).[36] We have since extended our methodology to produce highly porous cross-linked PVA materials, blended PVA/PEG, and naturally derived chitosan by the gelation of C/W HIPEs. Moreover, we have shown that this technique can be carried out at much lower temperatures and pressures (25 °C, < 120 bar) using inexpensive, biodegradable hydrocarbon surfactants such as PVAc-based block copolymers. Our methodology opens up a new solvent-free route for the preparation of porous biopolymers, hydrogels, and composites, including materials which cannot readily be produced by foaming. We plan to use this knowledge in future studies to develop highly porous materials, and to achieve fine control over porous structure by tuning the CO_2 density for a number of applications, particularly those in which organic solvent residues pose a problem.

8. Conclusions

In general, CO_2 is an attractive solvent alternative for the synthesis of polymers because it is 'environmentally friendly', non-toxic, non-flammable, inexpensive, and readily available in high-purity from a number of sources. Product isolation is straightforward because CO_2 is a gas under ambient conditions, removing the need for energy intensive drying steps. It offers the potential of reducing organic solvent usage in the production of a range of materials. This is particularly advantageous in processes where large volumes of organic solvents are used such as the production of porous materials. Moreover, the discovery of inexpensive, functional hydrocarbon CO_2-philes has opened up opportunities for solubilization, emulsification, and related applications. It has been recently shown that these structures can outperform perfluorinated analogues in specific applications.

Acknowledgements

We thank EPSRC for financial support (EP/C511794/1) and the Royal Society for a Royal Society University Research Fellowship (to A.I.C).

References

1. J.M. DeSimone, Science. 297, 799 (2002).
2. A. I. Cooper, J. Mater. Chem. 10, 207-234 (2000).
3. A. I. Cooper, Adv. Mater. 13, 1111 (2001).
4. J. M.; DeSimone, Z. Guan, C.S. Elsbernd, Science 257, 945 (1992).
5. J.M. DeSimone, E.E. Maury, Y.Z. Menceloglu, J.B. McClain, T.J. Romack, J.R. Combes, Science 265, 356 (1994).
6. A.I Cooper, Adv. Mater. 15, 1049 (2003).
7. K.L. Parks, E.J. Beckman, Polym. Eng. Sci. 36, 2404 (1996).
8. K.L Parks, E.J. Beckman, J. Polym. Eng. Sci. 36, 2417 (1996).
9. S.M. Howdle, M.S. Watson, M.J. Whitaker, V.K. Popov, M.C. Davies, F.S. Mandel, J.D. Wang, K.M. Shakesheff, Chem. Commun. 109 (2001).
10. A.I. Cooper, A.B. Holmes, Adv. Mater. 11, 15, 1270 (1999).
11. A.I. Cooper, C.D. Wood, A.B. Holmes, Ind. Eng. Chem. Res. 39, 12, 4741 (2000).
12. A.K. Hebb, K. Senoo, R. Bhat, A.I. Cooper, Chem. Mat. 15, 10, 2061 (2003).
13. C.D. Wood, A.I. Cooper, Macromolecules 34, 1, 5 (2001).
14. C. Shi, Z. Huang, S. Kilic, J. Xu, R.M. Enick, E.J. Beckman, A.J. Carr, R.E. Melendez, A.D, Hamilton, Science 286, 1540 (1999).
15. J. L. Kendall, D. A. Canelas, J. L. Young, and J. M. DeSimone, Chem. Rev. 99, 543 (1999).
16. H. M. Woods, M. Silva, C. Nouvel, K. M. Shakesheff, S. M. Howdle, J. Mater. Chem. 14, 1663 (2004).
17. C. F. Kirby, M. A. McHugh, Chem. Rev. 99, 565 (1999).
18. P. G. Jessop, W. Leitner, Chemical Synthesis Using Supercritical Fluids; Wiley-VCH: Weinheim, (1999).
19. E. J. Beckman, Chem. Comm. 1885 (2004).
20. T. Sarbu, T. J. Styranec, E. J. Beckman Ind. Eng. Chem. Res. 39, 4678 (2004).
21. J. Eastoe, A. Paul, S. Nave, D. C. Steytler, B. H. Robinson, E. Rumsey, M. Thorpe, R. K. Heenan, J. Am. Chem. Soc. 123, 988 (2001).
22. Bien Tan, Jun-Young Lee, and Andrew I. Cooper Macromolecules 40, 1945 (2007).
23. Jun-Young Lee, Bien Tan, and Andrew I. Cooper Macromolecules 40, 1955 (2007).
24. C. Palocci, A. Barbetta, A.L Grotta, M. Dentini Langmuir 23, 8243 (2007).
25. S. Partap, I. Rehman, J.R. Jones, J.A. Darr, Adv. Mater. 18, 501 (2006).
26. S. K. Goel, E. J. Beckman, Polymer 34, 1410 (1993).
27. S. Siripurapu, Y. J. Gay, J. R. Royer, J. M. DeSimone, R. J. Spontak, S. A. Khan, Polymer, 43, 5511 (2002).
28. B. Krause, G. H. Koops, N. F. A. van der Vegt, M. Wessling, M. Wubbenhorst, J. van Turnhout, Adv. Mater. 14, 1041 (2002).

29. S. Siripurapu, J. M. DeSimone, S. A. Khan, R. J. Spontak, Adv. Mater. 16, 989 (2004).
30. H. Matsuyama, H. Yano, T. Maki, M. Teramoto, K. Mishima, K. J. Matsuyama,. Membr. Sci. 194, 157 (2001).
31. H. Matsuyama, A. Yamamoto, H. Yano, T. Maki, M. Teramoto, K. Mishima, K. J. Matsuyama, Membr. Sci. 204, 81 (2002).
32. F. Placin, J. P. Desvergne, F. J. Cansell, Mater. Chem. 10, 2147. (2000).
33. H. Wakayama, H. Itahara, N. Tatsuda, S. Inagaki, Y. Fukushima, Chem. Mater. 13, 2392 (2001).
34. Y. Fukushima, H. Wakayama, J. Phys. Chem. B 103, 3062. (1999).
35. R. Butler, C. M. Davies, A. I. Cooper, Adv. Mater. 13, 1459 (2001).
36. R. Butler, I. Hopkinson, A. I. Cooper, J. Am. Chem. Soc. 125, 14473 (2003).
37. F. Svec, J. M. J. Fréchet, Science. 273, 205 (1996).
38. F. Svec, J. M. J. Fréchet, Ind. Eng. Chem. Res. 38, 34 (1998).
39. C. Yu, M. C. Xu, F. Svec, and J. M. J. Fréchet, J. Polym. Sci. A, Polym. Chem.40, 755 (2002).
40. M. Petro, F. Svec and J. M. J. Fréchet, Biotechnol. Bioeng. 49, 355 (1996).
41. D. C. Sherrington Chem. Comm. 21, 2275 (1998).
42. H. Zhang, J. Long, A. I. Cooper, J. Am. Chem. Soc. 127, 13482 (2005).
43. H. Gu, R. Zheng, X. Zhang, B. Xu, Adv. Mater. 16, 1356 (2004).
44. S. R. Quake, A. Scherer, Science 290, 1536 (2000).
45. A. Yamaguchi, F. Uejo, T. Yoda, T. Uchida, Y. Tanamura, T. Yamashita, N. Teramae, Nat. Mater. 3, 337 (2004).
46. R. Adelung, O. C. Aktas, J. Franc, A. Biswas, R. Kunz, M. Elbahri, J. Kanzow, U. Schuramann, F. Faupel, Nat. Mater. 3, 375 (2004).
47. C. Y. Xu, R. Inai, M. Kotaki, S. Ramakrishna, Biomaterials 25, 877 (2004).
48. W. Mahler, M. F. Bechtold, Nature 285, 27 (1980).
49. S. R. Mukai, H. Nishihara, H. Tamon, Chem. Comm. 874 (2004).
50. H. Nishihara, S. R. Mukai, D. Yamashita, H. Tamon, Chem.Mater. 17, 683 (2005).
51. B. Tan, A. I. Cooper, J. Am. Chem. Soc. 127, 8938 (2005).
52. P. Raveendran, S. L. Wallen, J. Am. Chem. Soc 124, 12590 (2002).
53. T. Sarbu, T. Styranec, E. J. Beckman, Nature 405, 165 (2000).
54. S.G. Kazarian, M. F. Vincent, F. V. Bright, C. L. Liotta, C. A. Eckert, J. Am. Chem. Soc 118, 1729 (1996).
55. C. Drohmann, E. J. Beckman, Journal of Supercritical Fluids 22, 103 (2002).
56. R. Hoogenboom, M. A. R. Meier, U. S. Schubert, Macromolecular Rapid Comm. 24, 16. (2003).
57. K. A. Shaffer, T. A. Jones, D. A. Canelas, J. M. DeSimone, S. P. Wilkinson, Macromolecules 29, 2704 (1996).
58. C. Lepilleur, E. J. Beckman, Macromolecules 30, 745 (1997).
59. A. I. Cooper, W. P. Hems, A. B. Holmes, Macromolecules 32, 2156 (1999).
60. P. Christian, S. M. Howdle, D. J. Irvine, Macromolecules 33, 237 (2000).

61. A. I. Cooper, J. D. Londono, G. Wignall, J. B. McClain, E. T. Samulski, J. S. Lin, A. Dobrynin, M. Rubinstein, A. L. C. Burke, J. M. J. Fréchet, J. M. DeSimone, Nature 389, 368 (1997).

62. W. P. Chen, L. J. Xu, Y. L. Hu, A. M. B. Osuna, J. L Xiao, Tetrahedron 58, 3889 (2002).

63. I. Kani, M. A. Omary, M. A. Rawashdeh-Omary, Z. K. Lopez-Castillo, R. Flores, A. Akgerman, J. P. Fackler, Tetrahedron 58, 3923 (2002).

64. K. P. Johnston, Current Opinion in Colloid & Interface Science 5, 351 (2000).

65. K. P. Johnston, K. L. Harrison, M. J. Clarke, S. M. Howdle, M. P. Heitz, F. V. Bright, C. Carlier, T. W. Randolph, T. W. Science 271, 624. (1996).

66. Z. Shen, M. A. McHugh, J. Xu, J. Belardi, S. Kilic, A. Mesiano, S. Bane, C. Karnikas, E. J. Beckman, R. Enick, Polymer 44, 1491 (2003).

67. P. Raveendran, S.L. Wallen, J. Am. Chem. Soc. 124, 7274 (2002).

68. D.A. Canelas, D.E. Betts, J.M. DeSimone, Macromolecules 29, 2818 (1996).

69. C. L. Bray, B. Tan, C. D. Wood, A. I. Cooper, J. Mater. Chem. 15, 456, (2005).

70. B. Tan, H. M. Woods, P. Licence, S. M. Howdle, A. I. Cooper, Macromolecules 38, 1691 (2005).

71. M. A. Carroll, and A. B. Holmes, Chem. Comm. 1395 (1998).

72. P. G. Shah J. D. Holmes, R. C. Doty, K. P. Johnston, B. A. Korgel, J. Am. Chem. Soc. 122, 4245 (2000).

73. S. R. P. da Rocha, P. A. Psathas, E. Klein, K. P. Johnston, J. Coll.Int. Sci. 239, 241. (2001).

74. C. T. Lee, P. A. Psathas, K. P. Johnston, J. deGrazia, T. W. Randolph, Langmuir, 15, 6781 (1999).

75. J. L. Dickson B. P. Binks, K. P. Johnston, Langmuir 20, 7976 (2004).

76. J. L. Dickson, P. G. Smith, V. V. Dhanuka, V. Srinivasan, M. T. Stone, P. J. Rossky, J. A. Behles, J. S. Keiper, B. Xu, C. Johnson, J. M. DeSimone, K. P. Johnston, Ind. Eng. Chem. Res. 44, 1370 (2005).

77. A. Imhof, D. J. Pine, Adv. Mater. 10, 697 (1998).

78. P. Schmidt-Winkel, W. W. Lukens, P. D. Yang, D. I, Margolese, J. S. Letlow, J. Y. Ying, G. D. Stucky, Chem. Mater. 12, 686 (2000).

79. V. N. Manoharan, A. Imhof, J. D. Thorne, D. J. Pine. Adv. Mater. 13, 447 (2001).

80. H. Zhang, G. C. Hardy, M. J. Rosseinsky, A. I. Cooper, Adv. Mater. 15, 78 (2003).

81. N. R. Cameron, and D. C. Sherrington. Adv. Polym. Sci. 126, 163 (1996).

82. W. Busby, N. R. Cameron, C. A. B. Jahoda, Biomacromolecules. 2, 154 (2001).

83. H. Zhang, and A. I. Cooper, Chem. Mater. 14, 4017 (2002).

84. J. Song E. Saiz, C. R. Bertozzi, J. Am. Chem. Soc. 125, 1236 (2003).

85. Y. Luo, P. D. Dalton, M. S. Shoichet, Chem. Mater. 13, 4087 (2001).

86. F. Chiellini, R. Bizzari, C. K. Ober, D. Schmaljohann, T. Y. Yu, R. Solaro, and E. Chiellini, E. Macromol. Rapid. Comm. 22, 1284 (2001).

CHAPTER 9

HIERARCHICAL MACRO-MESOPOROUS OXIDES AND CARBONS: TOWARDS NEW AND MORE EFFICIENT HIERARCHICAL CATALYSIS

Alexandre Léonard[1], Aurélien Vantomme[2] and Bao-Lian Su*

*Laboratory of Inorganic Materials Chemistry (CMI), Groupe de Chimie des Nanomatériaux (GCNM), The University of Namur (FUNDP), 61 rue de Bruxelles, B-5000 Namur, Belgium, * Corresponding author. Tel : 32 81 72 45 31, Fax : 32 81 72 54 14, email : bao-lian.su@fundp.ac.be, 1. Chargé de Recherches, Fonds National de la Recherche Scientifique, Belgium. 2. Present address : Total Petrochemicals Research Feluy, Polyolefins Catalysis Department, Zoning 1C, B-7181 Feluy, Belgium.*

Hierarchy is a nature-inspired concept that has made its appearance in the field of materials synthesis. Regarding the high potential of meso- (and micro-) porous oxides in heterogeneous catalysis, the accessibility to their active sites has to be improved. This is rendered possible by introducing a second larger porosity level, i.e. by creating hierarchically porous structures. Such materials are expected to enhance the catalytic performance comparing to single-sized oxides and to broaden the spectrum of applications. After introduction the concept of "Hierarchical Catalysis", this review article gives a survey of the work accomplished during the past years in the conception of oxides with multiple porosities. The developed methodologies are described and discussed for silica, aluminosilicates and (bi-) metallic oxides, giving a glance at the techniques available for the synthesis of these given oxides and opening up perspectives for the conception of new unprecedented compositions with hierarchical pore structure. A special attention has been paid to our new discovery on the self-formation phenomenon of porous hierarchy, an ultime powerful and quite simple synthesis method on the basis of the chemistry of metal oxides that conducts to the generation of macro-meso-(or micro-)porous single or binary oxides materials with versatile chemical compositions. The catalytic applications of these structures are also addressed by emphasizing the benefit of the multi-scale porosity.

1. Introduction

Hierarchy is essential to nature and is encountered anywhere in our environment.[1] Trees for instance exhibit a hierarchical structure going from the large-sized stem to the small branches that support the leaves. Our respiratory system as well as our blood circulation network also consist of large vessels connected to ever smaller capillaries [2-3]. The underlying reason for the existence of hierarchy is mainly diffusion. Indeed, the nutrition of leaves in a tree is performed by very small branches, allowing for a high exchange surface; the same is true for oxygenation of blood or oxygen transport to individual cells in our body. Nevertheless, to ensure an efficient transport of nutrients, oxygen, etc, large starting stems or vessels are essential to allow a fast and efficient diffusion. To make short, nature tells us that efficient processes can only exist by the combination of high surface areas and good diffusion properties, i.e. by creating *hierarchical systems.*

How can high surface areas be achieved ? Either by reducing the size of items or by introducing porosity into solid bodies. In fact, porosity is also a concept from nature if we consider the beautiful porous structures of diatoms or the efficient transport of water and ions through cell membranes. Scientists have been recognizing for a long time the high potential of porous materials in processes such as catalysis, adsorption and separation [4-5]. That is why numerous research projects have been devoted to the synthesis, comprehension of formation mechanisms and applications of mostly unimodal porous materials.

The first major breakthrough of porous materials occurred when scientists recognized the high potential of zeolites, highly crystalline aluminosilicate *micro*porous materials with homogeneous openings in the nanometre range, inducing a deep revolution in petroleum refinery industry. Lots of efforts have since been devoted to the characterization and to the development of new structures with tuneable compositions aiming industrial applications [6-7]. The porosity of zeolites is induced by single molecules or solvated cations as structure-directing agents. Thus, in order to achieve materials with larger pore sizes, Mobil Oil researchers turned towards supramolecular assemblies as templates in the beginning of the 90's of the last century. Aluminosilicate *meso*porous phases with surface areas reaching 1000 m^2/g and pore sizes of 2-3 nm were prepared by polymerization of inorganic sources around micelles formed by the self-assembly of surfactant molecules [8-9]. This pathway was further extended to the use of amphiphilic block copolymers for micelle formation, leading to larger-pore sized *meso*porous materials [10-11] and even swelling agents have been employed, reaching pore sizes up to 30 nm [12]. Tremendous work has been performed in the domain of *meso*porous materials since the first preparations and complete characterizations [13-15]. In particular, many efforts are devoted to diversifying the chemical composition of the frameworks, to fine-tuning the pore sizes and the three-dimensional stacking, to introducing defined functionalities, to improving their thermal and boiling water stability and to understanding their formation mechanism. A review article by the groups of Sanchez and Patarin very well illustrates the advances in *meso*structured materials synthesis performed during their first 10 years of existence [16]. The resulting application potential of the different *meso*porous compositions have also been reviewed in detail [17-20].

In order to still increase the pore sizes, studies have also been performed to create *macro*porous materials, with pore sizes larger than 50 nm (IUPAC classification). Such materials are indeed of interest in a range of potential applications such as catalysis, catalyst supports, cell-immobilization, optics and chemical filtration and separation [21]. The most direct pathway makes use of templates such as polymeric or silica spheres around which inorganic precursors can polymerize, giving rise to a rich variety of *macro*porous inorganic compositions [22-30]. Other strategies encompass the use of polymeric hydrogels, emulsions or even bacteria...[31-36] Monolithic silica gels with a bicontinuous three-dimensional morphology have successfully been prepared by a method based on the phase separation between an organic polymer and silica oligomers [37].

Despite the huge amount of work described above and as mentioned in the first lines of this manuscript, nature tells us that *hierarchy* is essential. A major breakthrough thus consists in crossing the borders of unimodal porous structures and creating materials with *hierarchical porosity*. Indeed, the introduction of secondary larger pores in *meso*porous catalysts has already shown considerable enhancement of the diffusion of reactants and products [38-39]. Other than bimodality in *meso*porous systems, more interesting catalysts would consist of a *macro*porous array where the pores are separated by *meso*porous walls. The association of a high dispersion of active sites, achieved by confinement in *meso*pores, with efficient mass transfer and reduced diffusion limitations of guest species, thanks to large *macro*porous channels, would lead to more active catalysts, especially when using bulky molecules or viscous systems. In particular, the aim is to create materials, with a rich diversity in compositions, that possess adjustable and well-defined macropores separated by walls bearing interconnected and tuneable mesopores. The great challenge is thus open, for each chemical composition, to control individually different length scaled pore sizes of such materials without affecting the structural regularity.

In this paper, review the state of the art in the synthesis of oxides and carbons with hierarchical porosity. After introducing the concept of "Hierarchical Catalysis" in view of the huge potential of oxides with multi-scale porosity, in the first part, we will describe and discuss the synthesis methodologies that have been developed in the preparation of hierarchical silica, aluminosilicates, (bi-)metallic oxides and carbons. We focus on the existing procedures to give insight into the available synthesis routes for given materials and to stimulate research towards new unprecedented compositions with hierarchical pore structure. A special accent was put on our recent discovery concerning the self-formation phenomenon to program porous hierarchy on the basis of the power of the chemistry of metal oxides. This very simple synthesis route can be applicable to the synthesis of a large variety of materials with different chemical compositions and is industrializable in large scale. The last part is devoted to the application potential with a description of some developed reactions that clearly demonstrate the superiority of hierarchical materials on unimodal structures.

2. Introduction of the concept of "hierarchical catalysis" [40]

In the chemical industries, catalytic processes usually occur as a sequence of different steps, i.e. the sequential coupling for instance of pretreatment, chemical conversion into valuable products and purification of resulting product mixtures. In fact, one desired product from a precise reagent is often prepared by a multiple steps reaction with the production of a series of intermediates, where each step requires a defined catalyst with one precise porosity and one specific functionality. In the future, facing the actual problems of global warming and ever-growing raw materials and energy consumption, there is an evident need for developing new processes that could ideally be realized with minimal intermediary steps. This suggests that ideally, the multiple steps reaction should be realized in one only reactor without any intermediary separation processes. Such a goal could only be attained by the use of hierarchically porous materials. Indeed, the integration of multiple levels of porosities combined with desired functionalities inside one single body could potentially allow for the successive realization of the complete reaction from starting reagents to final desired products. This means that the product of one reaction can be the reagent for the next reaction without separation and purification processes that would inherently occur by the sieving capacity of the adjusted pore sizes. This concept, called "Hierarchical Catalysis", would thus allow for the integration of multifunctional processes on the basis of a hierarchical single nanocatalyst concept. The advantages of such a concept are evident: reduction of the number of steps involved in a chemical conversion, thus a reduction of energy consumption, less waste products, enhanced performances and increased operational safety.

This ideal concept thus clearly brings to the fore the huge potential of porous compositions with well-defined multiple porosities and also underlines that efforts have still to be made in order to develop ever new synthesis procedures to make this dream coming true.

Table 1. Methods employed for the synthesis of different hierarchical macro-mesoporous oxides and carbons

Macrotemplating Method	Composition	Ref.
Polymeric spheres	SiO_2	[26], [27], [29], [41]-[53]
	Organosilica	[51]-[52]
	Al-O-Si	[102]
	ZrO_2	[26]
	TiO_2	[26], [41], [118]-[119]
	TiO_2-Ta_2O_5	[153]
	TiO_2-ZrO_2	[153]
Polymer foam	SiO_2	[59]-[64]
	ZrO_2	[117]
	TiO_2	[123]-[125]
	TiO_2-SiO_2	[154]
Microorganisms	SiO_2	[66]-[72]
Natural structures	ZrO_2	[121]-[122]
	$CeZrO_2$	[120]
	TiO_2	[122]
"Natural" Molecules	SiO_2	[73]-[76]
Emulsions / bubbles	SiO_2	[35], [77]-[84]
	Al-O-Si	[103]
	YSZ	[150]-[151]
Phase separation	SiO_2	[37], [85]-[93]
	TiO_2	[126]-[127]
Inorganic Salt	SiO_2	[94]
	Nb_2O_5	[135]
Ice-templating	SiO_2	[95]-[96]
	Al-O-Si	[95], [104]
Fusion of spheres	SiO_2	[97]
Spontaneous route	Al-O-Si	[106]-[107], [110]
	ZrO_2	[40], [105], [128]-[129]
	TiO_2	[40], [130]-[131], [131]
	Nb_2O_5	[40]
	Y_2O_3	[40]
	Ta_2O_5	[40]
	ZrO_2-SiO_2	[155]
	ZrO_2-TiO_2	[155]
	YSZ	[155]
	AlPO	[155]
	TiPO	[156]
Nanocasting of preformed macro-mesoporous silica	Co_3O_4	[137]
	SnO_2	[137]
	MnO_2	[137]
	Mn_2O_3	[137]
	C	[60], [84], [172]-[181]

3. Conception of hierarchically porous materials

In this part, we will review the work accomplished until now in the preparation of hierarchical macro-meso (or micro-)porous oxides and carbons. The different synthesis methodologies will be addressed and discussed first for silica, then for aluminosilicates, (bi-) metallic oxides and carbons. A summary of the preparation strategies with the corresponding achieved chemical compositions is given in *table 1*.

3.1. Hierarchical macro-mesoporous silica

3.1.1. Micromolding by spheres and other "hard" macrotemplates

The most widespread method was reported first in 1998 and implies the joint use of both the synthesis methods of macro- and mesoporous materials. The combination between colloidal crystal templating and surfactant-assisted mesostructuration is a very efficient way for the construction of ordered macro-mesoporous architectures [27]. The method consists in aggregating colloidal uniformly-sized latex spheres in a regular array and to infiltrate the voids between them by a silica precursor mixed with a surfactant or block-copolymer micellar solution. After ageing, condensation and polymerization, the surfactant and latex spheres are removed by solvent extraction or calcinations under flowing air in order to create an open porous framework [26, 41-44] (Figure 1). By this method, the specific surface area of a macroporous silica could be increased from 200 to more than 1300 m^2/g thanks to the addition of the mesopores in the wall structures [29].

This dual templating method has further widely been developed in order to improve the structural quality of the materials. In particular, the size and composition (latex, polystyrene, …) of the colloidal spheres can be tuned as well as the precursor gel that leads to the mesostructure. That is why macroporous materials with hexagonal or cubic ordered mesoporous walls could successfully be prepared by this method. Strategies were also developed by adding polymer spheres to the mixed surfactant-silica solutions and have led to MCM-48-type cubic materials containing uniform macropores [45]. The addition of the spheres to a

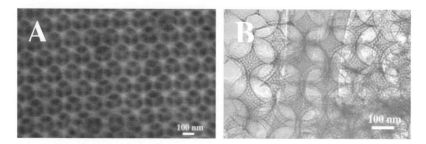

Figure 1. SEM picture showing the regular macropore array templated by colloidal spheres (A) and TEM image showing that the framework is made of ordered mesopores (B). *From: [38]: P. Yang, T. Deng, D. Zhao, P. Feng, D. Pine, B.F. Chmelka, G.M. Whitesides and G.D. Stucky, Science 282, 2244 (1998)*

surfactant micellar solution *before* the inorganic source has also led to highly structured macro-mesoporous MCM-41 and 48 [46-47].

The same procedures were also applied to the preparation of macro-microporous materials, i.e. macrostructures with zeolitic walls [48-51]. Going further than bimodality, hierarchical silica with a trimodal porous system and a high structural regularity were synthesized by using jointly polystyrene spheres to create macropores and an "ionic liquid" (1-hexadecyl-3-methylimidazolium-chloride) as well as a block copolymer to form very regular mesopores of 3 and 11 nm in diameter respectively [52]. Moreover, it is well-known that mesoporous materials prepared via a block copolymer as surfactant contain a substantial amount of micropores in addition to their mesopores, due to the structure of the copolymer [10]. By employing the dual templating pathway, porous silica could be prepared with 4 levels of porosity, namely ordered macropores induced by the polymer spheres, uniformly-sized windows connecting these macropores and micro- and mesopores originating from the structuring of silica precursors around block- copolymer micelles [53].

This colloidal templating method is also very advantageous for the synthesis of hybrid organic-inorganic hierarchical silica. Indeed, instead of adding one only inorganic source, co-codensation can be carried out by using jointly an alkoxysilane and an organosiloxane $R'Si(OR)_3$ or a bis(organosiloxane) $(RO)_3Si-R'-Si(OR)_3$ to introduce organic functionalities either attached to the surface of the pores or located as linkers in the silicate framework [54-55]. Such hybrid structures with macropores and a broad mesopore size distribution were used as support for polyoxometalate clusters and exhibited a good catalytic activity in the epoxydation of cyclooctene by hydrogen peroxide at room temperature [56].

All of the exposed strategies strongly suggest that the colloidal templating technique is a very versatile pathway for the creation of a rich variety of bi-(or multi-)modal porous silica. Indeed, it is very easy to tune the size of the macropores by choosing the appropriate diameter of polymer spheres, which are, for a large range, commercially available. The mesopores are also adjustable, by choosing surfactants with variable chain lengths, formation of micelles by block copolymers or even by the use of swelling agents like alkanes or TMB [57-58]. The structural diversity can also be enriched by combining the microsphere templating with the large series of existing zeolites nanocrystals.

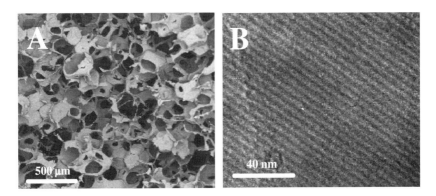

Figure 2. SEM image of the polyurethane foam-templated silica (A) and TEM image of the porous structure (B). *From: [56]: S. Alvarez and A.B. Fuertes, Mater. Lett. 61, 2378 (2007)*

A somewhat different but fundamentally similar approach consists of impregnating a macrocellular polymeric foam by the inorganic species. A commercial polyurethane foam was impregnated with a mixture of surfactant and silica precursor, leading to final silica monoliths that are exact replicas of the polymer foam (i.e. with macropores) but with extra mesopores [59] (Figure 2). Coating of prefabricated foams was also carried out by using monolithic polystyrene foams and the resulting materials had cellular macropores of 0.3 to 2 µm in diameter, interconnected by windows and separated by walls containing highly ordered mesopores of 5.1 nm in size [60]. Preformed mesoporous silica nanoparticles were also used as building blocks for coating a polyurethane foam giving raise, after mineralization, to a monolithic silica replica of the starting foam [61]. The same strategy was employed by using polymer membranes like cellulose acetate and polyamide that were infiltrated by a surfactant-silica solution following by gelling and mineralization [62-64]. Deposition of mesoporous silica onto macroporous silicon oxycarbide (SiOC) foams results in a ceramic body with hierarchical macro-mesoporosity, since a uniform mesoporous coating with a highly ordered cubic structure was deposited on the walls of the macropores, giving raise to a significant increase in the surface area of the starting ceramic foam [65].

The rich variety in available macrocellular foams ensures a wide choice in macropore sizes that can be achieved via this method, just alike the polymer spheres. Whether colloidal spheres, membranes or foams are used, all the materials result from a simple construction principle based on the (inverted) replication of a mould with appropriate macrosizes.

3.1.2. Microorganisms and biological molecules as macrotemplates

In view of all the preparation procedures described until yet, it is evident that the larger openings in the materials all result from the condensation of an inorganic source around a macrotemplate that has a size of several hundreds of nanometers. Nature is also made of submicron-sized items, thus potential macrotemplates, an idea that has been exploited by several research groups and in particular by the group of Mann *et al.*

The first example we would like to describe here is in fact a combination between micromolding and utilization of biologicals. The first step consists in fabricating an inverted carbon replica of diatoms, a microorganism that displays a very regular micron-sized porosity, by nanocasting [66]. The voids of the carbon were then filled up by a surfactant-silica mixture, leading, after mineralization and carbon elimination, to a "replica of the replica", or "nanoduplicate" [67-68]. *In fine*, the silica wall of the diatom was replaced by an ordered network of mesopores while the macroporous array of the starting microorganism was maintained thanks to the nanoreplication process. Alike the unimodal macroporous materials prepared with spheres, this process results in a significant increase in specific surface area.

The use of living organisms as macrotemplates is further illustrated by coaligned multicellular filaments of bacteria that were involved in the growth of porous silica. A template-directed precursor gel of mesoporous MCM-41 was infiltrated into the voids between the filaments, giving rise, after calcination, to a bimodal network composed of mineralized biologically-induced macrochannels surrounding surfactant-templated mesopores [69]. Hierarchically ordered silica replicas of wood cellular structures were also successfully synthesized by infiltrating the voids between the cells by a surfactant-templated mesoporous silica precursor. This procedure results in a bimodal network made of macro-openings coated by highly ordered mesopores, as for the bacterial filaments-directed synthesis [70]. Insect wings as well as intact pollen grains have been employed following the same strategy [71-72].

Finally, "natural" chemicals can also successfully be employed as macrostructure-directing agents. In that way, starch gels and polysaccharides like dextran or chitosan can act as macrotemplates for the creation of macropores after silica mineralization [73-75]. Polypeptide-based triblock copolymers have been sysnthesized to act as templates for the formation of hierarchically structured silica [76]. The silica prepared by using the polypeptide poly(l-phenylalanine)-*b*-poly(ethylene glycol)-*b*-poly(l-phenylalanine) (Phe$_7$–PEG$_{135}$–Phe$_7$) as template shows an ordered mesostucture with supermicropores and interconnected layers of macropores, originating respectively from the templating action of the poly(l-phenylalanine) and the poly(ethylene glycol).

3.1.3. Emulsions and bubbles as macrotemplates

Another very convenient methodology for the synthesis of hierarchically porous silica consists in using microemulsions as templates for macopores. Like for the "hard" spheres, emulsions can be prepared as to produce droplets with tuneable sizes and have successfully been employed in the preparation of macroporous silica [35]. Replacing the silica source by supramolecular surfactant-silica assemblies then leads to a hierarchically porous structure.

The first synthesis by this pathway was reported by Sen who prepared meso-cellular foams along with macro-cellular foams by an emulsion made of oil droplets dispersed in a continuous water phase and stabilized by an ionic surfactant (CTMABr) [77-78]. Comparing to the "hard-sphere" templating, the macropore sizes are much less homogeneous, due to the softness and deformability of the emulsion droplets and their low resistance to stirring. High surface area silica monoliths with hierarchical porosity have been prepared by Carn and coworkers by using concentrated emulsions and micellar templates. The textural properties could be adjusted by varying the pH of the continuous aqueous phase, by controlling the emulsification or by tuning the volume fraction of the dispersed phase [79].

This synthesis strategy has been extended to air-liquid interfacial foams. Hierarchically porous silica with very large macropores have been prepared by silicification of metastable non-ionic polyethylene oxide air-liquid foams formed by strong stirring [80]. Nevertheless, the macropore sizes were quite difficult to control as the formation of the foam was largely dependent on the stirring conditions of the reaction medium. A better foaming could be induced by bubbling air in a surfactant solution, allowing a better control on the sizes and shapes of the formed bubbles [81]. Also, the spraying of a solution containing surfactant-silica assemblies leads to controllable macro-mesoporous foams [82]. Nevertheless, in all those cases, the structural regularity of the mesopores was quite difficult to control. That is why, based on the emulsion route, by using an oil in water emulsion formed by decaline ($C_{10}F_{18}$) droplets stabilized by a fluorinated surfactant, materials with macropores of 1-3 µm in size separated by a hexagonally ordered large-pore mesoporous framework has been prepared by Blin *et al.* in 2006 [83]. A further increase in structural quality was achieved by Sun *et al.* who prepared ordered hierarchical macroporous silica with ordered large mesoporous wall structures by carefully tuning the emulsification of the oil phase and the self-assembly of the surfactant template [84]. Large mesopores of 15 nm in diameter were obtained whereas the sizes of the macropores were comprised between 200 and 250 nm.

As for the hard-sphere or "biological" templating procedure, emulsions appear to be very suitable for the synthesis of highly porous hierarchical silica. Nevertheless, more physico-chemical variables seem to be involved for reaching the final structure in contrast to the static and hard character of the colloidal spheres or of the foams and membranes, especially regarding the homogeneity of the macropores. It is also important to see that the emulsion remains intact upon addition of the inorganics and that the surfactant-silica assemblies, precursors to mesoporous arrays, are not disrupted upon rapid stirring for instance. That is why only few materials bearing both macropores and *ordered* mesopores have been reported [83-84].

The main advantage in the emulsion-derived preparation pathway certainly relies on the ease of elimination of the templating agent, comparing with the required calcinations to remove the polymer spheres or the biological organics. This is a key point in the synthesis of (transition-)metallic oxides, which usually display quite a poor thermal stability and risk to collapse upon heat treatment due to the crystallization of the walls.

3.1.4. Macro-mesoporous silica monoliths by polymerization-induced phase separation

As mentioned in the introductive part of this manuscript, porous oxide materials with co-continuous macropores can be formed by a chemically induced liquid-liquid phase separation [37, 85-86]. This technique is based on the chemical instability created by the polymerization of network-forming inorganic precursors, leading to the formation of biphasic morphologies that are frozen by gelling. The addition of a water-soluble polymer such as poly(acrylic acid) further induces phase separation due to its incompatibility with the growing inorganic polymers, playing thus an assisting role in the phase separation to form microstructured arrays. The removal of fluid phase containing the polymer then *in fine* leads to a macroporous oxide framework

(Figure 3). If a surfactant or a block-copolymer is added to the synthesis mixture, an additional homogeneous mesoporous channel array will be formed, leading to hierarchical macro-mesoporous structures, resulting from the combination of phase-separation technique and micelle-directed mesopore creation. The success of this idea was first demonstrated by the group of Nakanishi who prepared co-continuous silica gels containing mesopores templated by a block-copolymer [87]. It was shown that the disordered mesoporous array was independent of the micrometric structure. Hierarchically porous silicas with disordered mesopores were also prepared by the same group by using a cationic surfactant [88]. The achievement of partially ordered mesopores in these silica was made by Shi *et al.* who used a triblock copolymer that served both as mesostructure templating agent and as phase separating species [89] Macroporous silica with highly ordered mesopores were finally prepared by Nakanishi *et al.*, first with 1,2-bis (trimethoxysilyl)ethane, then tetramethoxysilane as silica source and with a triblock copolymer as surfactant combined with TMB as micelle-swelling agent. The presence of the latter turned out to be necessary for reaching a 2-D hexagonal framework [90-91]. Phase separation was also employed by the group of Huesing *et al.* with the aim of preparing highly ordered mesostructures [92-93]. These authors took into consideration that the alcohol molecules released during silicon alkoxides hydrolysis and polymerization often destroy the lyotropic surfactant phases, impeding the long-range order of the final materials. That is why the phase-separation route was combined with the use of different glycol-modified silanes as silica source and block copolymers as surfactants, leading to a hierarchical silica network with macropores and periodically arranged mesopores. Moreover, this pathway does not require the presence of an additional phase separation polymer as described above.

The phase separation induced by the polymerization of silica in the presence of polymers is a very convenient method for the preparation of hierarchical macro-mesoporous silica, especially for the conception of monoliths. Nevertheless, care has to be taken if ordered mesopores are required. The sizes and shapes of the macrostructure can be tuned depending on the polymer used as well as on the relative composition of the starting reaction mixture [37].

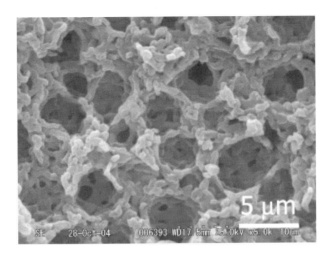

Figure 3. SEM image of a macroporous framework prepared by phase separation. *From: [88]: T. Amatani, K. Nakanishi, K. Hirao and T. Kodaira, Chem. Mater. 17, 2114 (2005)*

When comparing the macro-mesoporous materials created by the above-cited methods with the ones obtained by phase separation, the latter can be seen as a new and different class of structures. Indeed, the polymer sphere templating route affords materials with interconnected spherical macroholes, which can be highly ordered in the three dimensions of space. Biological patterning gives rise to macrovoids that depend on the shapes and the sizes of the used microorganisms, with generally less regularity in spatial ordering. Foams and bubbles are quite difficult to be controlled due to the involved physico-chemical conditions, implying that the homogeneity of the macropores and their interconnectivity is hard to adjust. Here, the architecture obtained via the phase separation pathway can be described as a co-continuous or bi-continuous morphology, made of a three-dimensional continuous silica network and a continuous porous system. These materials should be of increased interest in catalytic reactions, chromatography and separation processes due to the continuous three-dimensional macropore system that will sharply reduce diffusion limitations [2, 93].

3.1.5. Macro-mesoporous silica prepared by other original routes

Following the ideas of templating and phase separation and keeping in mind that macroporous diatoms form in saltwater, macro-mesoporous silica membranes were synthesized by the group of Stucky by a novel multiphase process of acid-catalyzed silica sol-gel chemistry in the presence of inorganic salts and self-assembling block copolymers [94]. In this mixture, the silica self-assembles into an ordered mesostructure upon interaction with the block copolymer as described in the synthesis of pure mesoporous SBA-15 materials [10]. In the presence of an electrolyte however, this self-assembly will take place at the interface of inorganic salt solution droplets, leading *in fine* to a structure with hierarchical porosity. The sizes of the macropores can be adjusted as a function of the chemical nature of the electrolyte as well as the size of the droplets that can be changed by regulating the evaporation rate of the solvent. Independently, the size of the mesopores can be tuned by choosing the appropriate block copolymer or by adding micelle swelling agents like TMB for instance.

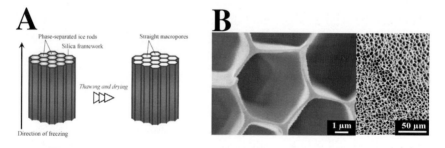

Figure 4. Growth mechanism (A) and SEM images of macrostructures grown by ice-crystal templating (B). *From : [92] : H. Nishihara, S.R.Mukai, D. Yamashita and H. Tamon, Chem. Mater. 17, 683 (2005)*

A very original method based on the use of ice crystals was developed by Nishihara *et al.* [95-96]. Ordered macroporous silica honeycomb-like arrays were prepared by freezing freshly gelled precursor hydrogels under conditions where pseudo-steady state growth of ice crystals can proceed. These polygonal ice rods act as macrotemplates for the growth of the silica structure and the size of the final macropores can be adjusted by controlling the freezing conditions (Figure 4). Moreover, these materials show a hierarchical array of pores since the walls separating the macropores contain disordered mesopores as obtained in classical sol-gel silica. Nevertheless, the sizes of the mesopores can be tuned upon adjustment of the pH of the starting sol or by a post-synthesis hydrothermal treatment, which can lead to materials with specific surface areas reaching 600 m^2/g and quite homogeneous mesopores of 4-5 nm in diameter [96]. This method is very convenient since the removal of the template (ice) spheres is achieved by simple thawing and drying of the material, ensuring the complete structural integrity of the porous network during post-treatments like calcination. A very interesting issue would be to control the spatial stacking of the mesopores like in the surfactant-assisted templating in order to increase further the specific surface area, to sharpen the pore size distributions and to orient the structure in the desired symmetry.

A more "physical" method towards macro-mesoporous hierarchical silica was developed by Vasiliev *et al.*, who prepared macro-mesoporous silica monoliths by partial fusion of mesoporous silica spheres by using the pulsed current processing (PCP) method that can achieve very rapid temperature increases [97]. Mesoporous silica spheres with different sizes ranging from 0.5 to 12 µm were first prepared by evaporation-driven surfactant-templating in microdroplets. The resulting microsized spheres were then subjected to a rapid temperature increase with compressive stress. This leads to interconnection of the spheres, with creation of voids between them. The hierarchical porous character of the prepared monoliths thus results from the surfactant-templated mesopores and the interspherical voids stemming from the packing and connection of the spheres during PCP. The size of the macrovoids could further be tuned by varying the size of the silica microspheres. Moreover, it has been shown that the structural integrity of the mesopores was maintained.

This method is quite interesting for the production of porous monoliths but it will not bring more porosity than a conventional stacking of mesostructured spheres. Besides, the diffusion properties will still be limited by the length of the mesochannels, thus the size of the spherical particles.

3.2. Hierarchical macro-mesoporous metal oxides

The first part of this publication was devoted to the description of the different preparation pathways that are available for the synthesis of macro-mesoporous silica. Our aim was to describe the different techniques available towards such hierarchical structures and to highlight the properties of the final materials. Nevertheless, pure silica are known to be of little use as intrinsic catalysts because of their quite inert character, but lot of research is carried out to use them as supports. Indeed advantage can be taken from the very high specific surface areas, tunable pore sizes and chemical stability in order to reach very high dispersions of active sites to heterogenize catalytic reactions.

In the aim of preparing intrinsic catalysts, numerous research projects have focused on the application of the described preparation pathways to single or even bi-metallic oxides. The transposition of the synthesis procedures is however not straightforward since it is well known that metallic alkoxides as inorganic precursors are usually highly reactive in aqueous solutions, but a careful control of the preparation conditions has led to hierarchically porous aluminosilicates, transition metallic and bimetallic oxides.

3.2.1. Hierarchical macro-mesoporous aluminosilicates

Porous aluminosilicate compositions can find many applications in catalytic processes. Catalytic cracking of heavy petroleum feedstocks and fine chemistry acid-synthesis are only the main examples for this [98]. The existing processes could significantly be improved if the low-surface area "amorphous" aluminosilicates, often employed as catalysts or supports, could be replaced by homogeneous hierarchical macro-mesoporous structures.

During the last years, a huge amount of work has been accomplished in the synthesis of hierarchical micro-mesoporous aluminosilicates, in particular mesoporous structures with zeolitic walls [99-101], in order to increase the hydrothermal stability of mesoporous structures. However, despite their evident potential in catalysis, only few studies report about the conception of macro-mesoporous aluminosilicates.

Micromolding with latex spheres as templates for macropores and a cationic surfactant for mesoporosity creation was applied in 2001 by the group of Gundiah to fabricate macro-mesoporous aluminosilicate composites [102]. The material is made of an ordered array of macropores of 150 nm in diameter separated by mesoporous walls. The mesopores are disordered but account for a very high specific surface area of the material up to 1000 m^2/g. The authors estimate that about 20% of the Al are located in framework positions and that most of the Al is present as Al_2O_3 in composite form with SiO_2.

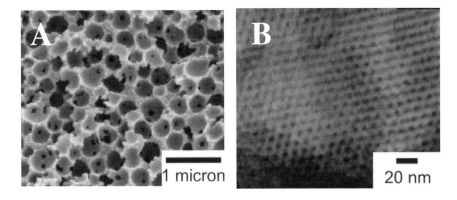

Figure 5. SEM image of the hierarchically porous aluminosilicate (A) and TEM image of the porous structure (B). *From: [100]: J.J. Chiu, D.J. Pine, S.T. Bishop and B.F. Chmelka, J. Catal. 221, 400 (2004)*

Stable monolithic hierarchical macro-mesoporous aluminosilicates Al-SBA-15 were reported by Chmelka *et al.* Their synthesis procedure consists in preparing an oil-in-water emulsion, where the oil is PDMS (poly(dimethylsiloxane)). The aqueous continuous phase consists of amphiphilic block copolymer in combination with an aluminosilica sol [103] and the emulsion oil droplets have a characteristic size of several hundred nanometers. The polycondensation of the hydrolyzed silica and alumina precursor species in interaction with the amphiphilic block copolymers yields a mesoscopically ordered aluminosilica/block copolymer mesophase (Figure 5). This procedure allows the direct and independent control of macro- and mesopore dimensions, so that the final pore structures can be tailored to different diffusion and reaction conditions.

Hierarchical aluminosilicate monoliths could also be prepared by the original ice-crystals templating procedure. The macropores form honeycomb-like structures with Al atoms homogeneously dispersed throughout the samples by forming Al-O-Si bonds [95, 104].

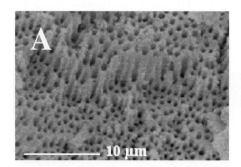 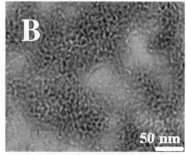

Figure 6. SEM image of a hierarchically porous aluminosilicate prepared by the spontaneous pathway (A) and TEM picture showing the mesoporous structure of the walls (B).

In our group, we developed a new strategy towards the formation of hierarchical aluminosilicates, based on our recent discovery concerning a self-formation phenomenon of porous hierarchy without need of any external templates such as surfactants or polymeric spheres [105-107]. We demonstrated that macro-mesoporous structures can form spontaneously without any external templating molecule, suggesting that the dual pore system may result from the polymerization chemistry of the inorganic sources in solution [107]. The materials are made of tubular macrochannels with openings ranging from 0.5 to 2 µm (Figure 6A), separated by wormhole-like disordered mesopores of about 4 nm in diameter and specific surface areas reaching 600 m^2/g (Figure 6B). The regularity in size of the macrochannels is demonstrated by the cross-section observed by TEM. When the same syntheses were carried out in the presence of a non-ionic surfactant, the same dual pore network was obtained with a higher proportion of Al atoms in tetrahedral framework positions and more regularity within the mesopores. However, it is clear that the surfactant does not play any role in the creation of the macrochannels. From this important statement, we proposed a synthesis scheme implying the rapid mineralization of the inorganics around water-alcohol channels that form due to the hydrodynamic flow of the solvent and where the mesopores represent voids between the rapidly formed aluminosilicate nanoparticles [40, 106-110]. In that sense, the formation of macrochannels only relies on the hydrolysis and polycondensation kinetics of the highly reactive inorganic source together with the behaviour of the solvents. The proposed mechanism can be described in detail as follows: The addition of the highly reactive Al source in the aqueous solution leads to a rapid formation of droplets with a solid aluminosilicate shell. The spontaneous hydrolysis and polymerization of the inorganic alkoxides then proceed inwards the droplets, generating more and more alcohol molecules that gather together leading to the formation of larger water/alcohol macrochannels inside the structure. The alcohol molecules suddenly released by the hydrolysis reaction can be considered as "self-formed porogene molecules" to generate large macrochannels. Meanwhile, the polymerization leads to the formation of mesoporous aluminosilicate nanoparticles and the self-aggregation of these nanoparticles in turn gives rise to interparticular mesoporosity. This reaction continues until all the inorganics have been "used up" and the substantial amount of water/alcohol entrapped inside the droplets creates high pressures that cause their bursting and splitting. The fragments of these initial droplets are the particles that can be observed by SEM (Figure 7).

If this mechanism is true, it is evident that the formation of macro-mesoporous hierarchical oxides is dependent on the inorganic source employed. This means that the precursor should be a metal alkoxide with a large electronegativity difference between the metal and the alkoxides in order to achieve high reactivity and thus fast polymerization rates. This mechanism, based on the control of the hydrolysis and polymerization kinetics of the inorganic sources in solution, is a powerful tool to fabricate hierarchical macro-mesoporous materials without needing complex templating procedures. A detailed and comprehensive explication of this self-formation phenomenon of porous hierarchy has been described in our recent papers [40, 111].

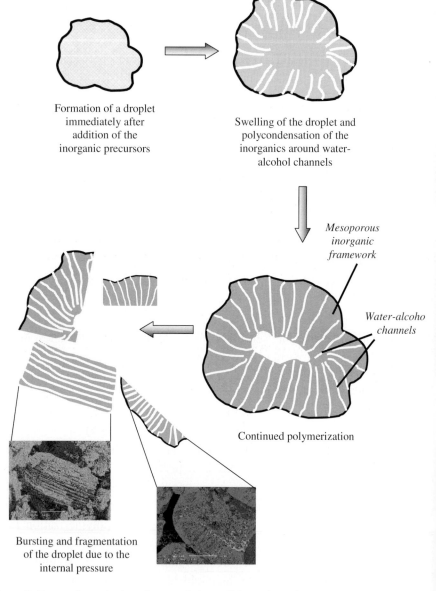

Formation of a droplet
immediately after
addition of the
inorganic precursors

Swelling of the droplet and
polycondensation of the
inorganics around water-
alcohol channels

*Mesoporous
inorganic
framework*

*Water-alcoho
channels*

Continued polymerization

Bursting and fragmentation
of the droplet due to the
internal pressure

Figure 7. Proposed synthesis scheme of the self-formation phenomenon of porous hierarchy.

3.2.2. Hierarchical macro-mesoporous ZrO_2 and TiO_2

Zirconium oxide porous structures are particularly interesting because of their high thermal stability, being potentially applicable as catalyst supports, adsorbents, heavy duty membranes and chemical sensors [112-115]. Titanium oxide is the object of a lot of attention, due to its high potential in photocatalysis and as catalyst support [116].

Hierarchically porous ZrO_2 were synthesized by using millimeter-sized polyacrylamide beads that contain macropores and serve as scaffolds for the preparation of porous inorganic oxides. The obtained materials contain mesopores (diameters 2-5 nm) resulting from the use of a block-copolymer (PEG-PPG-PEG) and large macropores of around 5-10 μm [117]. Colloid crystals (polystyrene spheres) together with surfactant templating agents (block-copolymers) have also led to the preparation of hierarchical macro-mesoporous TiO_2 [41]. The inverted opals obtained by latex-sphere templating also show a hierarchical character of pores [118]. An original preparation route of titania with multiple porosity based on polystyrene-beads templating was described by Dionigi *et al.* [119] Very homogeneous porous bimodal frameworks with preservation of macropore order and narrowly dispersed mesopores created by controlling the anatase nanocrystal size (6 nm) were obtained upon removing the template by evaporation in an inert atmosphere. It was shown that the polystyrene beads were particularly suitable as templates, being evaporated in the temperature range of anatase existence.

The natural templates such as wood, eggshells and other natural cellulose substances have been employed in the conception of hierarchically porous ZrO_2 and TiO_2, although the accent was put more on the replication of the hierarchical structure than on the porous character of the final materials [120-122]. The material prepared from the eggshell is constituted of interwoven microtubes with diameters of less than 1 μm, the walls of which contain mesopores of about 7 nm in size, resulting from the stacking of ZrO_2 nanocrystals [123].

Combined emulsions or air/liquid foams with surfactant templating could be transposed from silica to zirconia and titania, leading to high quality porous frameworks made of well-defined macroporous structures containing uniform mesopore distributions [123-125]. For instance, mesoporous titania with a hierarchical interior macroporous structure could be prepared in a one-step synthesis in the presence of a surfactant that combines conventional mesostructure templating with a reverse micelle approach to stabilize a water-in-oil microemulsion. The materials contain anatase, a sponge-like macroporous structure, mesopores sizes of 5.2 nm and a specific surface area of about 200 m^2/g [123].

The phase separation method was also successfully utilized in the aim of fabricating TiO$_2$ thin films with hierarchical pore structures made of homogeneous macropores (0.1-2 µm) and mesopores (3.-4 nm) [126-127].

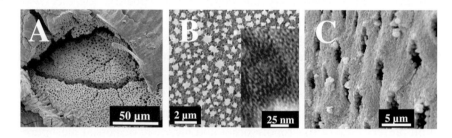

Figure 8. SEM (A) and TEM (B) micrographs illustrating the macrochannels, their cross-section and the mesoporous walls of the macro-mesoporous ZrO$_2$ materials and SEM image of the hierarchical TiO$_2$ (C). Both materials are prepared by the spontaneous route.

Finally, on the basis of our discovery of the self-formation phenomenon described in section 3.2.1. and which was successfully applied to the preparation of macro-mesoporous aluminosilicates, titania and zirconia featuring hierarchical macro-mesoporosities have been obtained [40, 105, 111, 128-129]. For example, ZrO$_2$ particles synthesised via this route are mainly tens of micrometers in size with a regular array of parallel or funnel-like shaped macrochannels of 300-800 nm in diameter (depending on the alkoxide used) (Figure 8A). The walls separating the macropores additionally exhibit accessible mesochannels (around 2.0 nm) with a wormhole-like array (Figure 8B). High surface area hierarchical macro-*micro*porous titanium oxide materials have also been prepared via this self-formation phenomenon. Their structure is made of a regular array of funnel-like sinusoidal macrochannels separated by an agglomeration of microporous TiO_2 nanoparticles (Figure 8C). Like for zirconia, this hierarchical structure gives rise to quite a high surfaces area (400-470 m^2/g). Similar procedures with surfactant [130] and without [131] have successfully been employed for the development of photocatalysts based on macro-mesoporous hierarchical titania.

3.2.3. Hierarchical macro-mesoporous Nb_2O_5, Ta_2O_5, Y_2O_3, Co_3O_4, SnO_2, MnO_2 and Mn_2O_3

Niobium oxides are known to be very active for reactions that require a strong Brønsted acid catalyst and are also excellent catalyst supports in HDS processes [132-134]. A few years ago, Antonelli showed that the combination of an amine template with niobium ethoxide followed by a treatment in aqueous NaCl leads to a spontaneous formation of a hierarchical framework [135]. The amine is supposed to play a double role in this strategy, first as template to direct the formation of mesopores and second in a cooperative macroscopic assembly of the inorganics in the presence of alcohols and salts. The hierarchically porous Nb_2O_5 have macropore sizes in the 200-300 nm range and mesopores around 2 nm.

Figure 9. SEM images of the hierarchically porous Nb_2O_5 (A), Ta_2O_5 (B) and Y_2O_3 (C) showing the homogeneous macroporous structure.

Following our previous experience in the synthesis of macro-mesoporous materials on the basis of the self-formation phenomenon of porous hierarchy, we recently tried to generalize this route to synthesize other compositions. For instance, by using niobium ethoxides, this simple synthesis protocol leads to the formation of Nb_2O_5 particles with a regular array of parallel macropores with diameters ranging from 0.3 to 10 (Figure 9A) [40]. Moreover, high magnification TEM images of the macropore walls reveal accessible micropores, which do not result from any surfactant templating in opposition to the mesopores generated by the amine-templating procedure in the pathway proposed by Antonelli [135]. The same strategy of self-formation has successfully led to similar results for tantalum oxides (Figure 9B).

Yttrium oxides possess unique properties, the main ones being their higher melting temperature than many other well-known oxides, a wide energy bandgap, high values of electrical resistivity and electric strength [136]. The self-formation strategy was applied to the preparation of hierarchically porous Y_2O_3 by a controlled polymerisation of yttrium butoxide in aqueous media [40]. The hierarchical Y_2O_3 particles possess macrochannels of 1-10 µm in diameter separated by walls formed of mesotructured nanoparticles, giving a supplementary interparticular mesoporosity centered at 30 nm (Figure 9C).

Hierarchically porous cobalt oxide (Co_3O_4), tin oxide (SnO_2) and manganese oxide (MnO_2 or Mn_2O_3) monoliths were prepared by an original route based on the nanocasting of preformed hierarchical silica. SiO_2 monoliths that contain macropores with adjustable size in the range of 0.5-30 µm as well as mesopores which can be altered between 3 and 30 nm are used as molds and are impregnated with a metal salt solution, which is subsequently decomposed to a metal oxide by heat treatments to form a SiO_2/MeO_x composite. Finally, the silica part can be removed by leaching in either NaOH or hydrofluoric acid. The final replicas exhibit both macropores and mesopores [137]. As no shrinkage could be observed for any of the replicas, the authors claim that by changing the morphology of the parent silica, it is possible to control the morphology of the replicas. During the first stage of formation, the macropores of the silica are filled with the salt solution and, when the solvent evaporates, the metal oxide precursors diffuse into the mesopores, which means that, at the micrometer scale, the final metal oxides are positive replicas of the starting silica. At the mesoporous level however, the textural mesopores are a negative replica of the parent silica [137].

3.2.4. Hierarchical ZnO

Zinc oxide has gained increased interest during the past years as a low-cost semiconductor with important catalytic, electrical, optoelectronic and photoelectrochemical properties [138-139]. *Porous* ZnO nanostructures could open up the way towards dye-sensitized photovoltaic cells, hydrogen storage and secondary batteries. Several methods towards porous zinc oxides have been developed. For instance, Jiu *et al.* reports the fabrication of mesoporous ZnO with various pore diameters by using a Zinc inorganic salt together with a copolymer gel to create the nanopores [140]. Macroporous ZnO was prepared by Liu *et al.* by the macrosphere templating method, i.e. by filling the interstices between close-packed polystyrene spheres with a ZnO sol made of Zinc acetate as precursor [141]. The final materials exhibit a highly ordered array of macropores upon removal of the polymeric template by calcination. In our group we developed a self-assembly pathway towards hierarchically porous zinc oxides by using surface-modified colloidal ZnO nanocrystallites as building blocks and P123 copolymers as templates [142]. To reach the final porous framework, the ZnO nanocrystallites are first functionalized on their surface by taurine (2-aminoethanesulfonic acid) in order to adjust the interactions between the nanoparticles and the copolymer template. After coalescence, ethanol extraction and calcinations, the final materials exhibit a hierarchical array of pores made of small and large mesopores. The combined characterizations by porosimetry and electron microscopies clearly prove the existence of interconnected small mesopores (~2.7 nm) within the individual ZnO nanocrystallites and large mesopores (~19 nm) originating from regular interparticular voids. These materials have been tested in the photocatalytic degradation of phenol under ambient conditions and show a superior activity compared to TiO_2 nanoparticles (PC-500). The formation mechanism is discussed in detail especially in terms of surface energy of the particles [142].

3.3. Hierarchical macro-mesoporous bimetallic oxides

The synthesis of bimetallic oxides with hierarchical porosity represents an important issue for the development of new catalytic applications. Indeed, titania and zirconia-based mixed oxides for instance already find applications in catalysis and solid oxide fuel cells [143-144]. It is also known that the catalytic efficiency of metal oxides can be improved by doping with a second metal or by combining it with a second metal oxide [145]. As an example, it has been shown that TiO_2 doped with ZrO_2 or In_2O_3 was more efficient in the photocatalytic decomposition of salicylic acid and 2-chlorophenol comparing with the sole titanium dioxide [146-147]. This increase in activity has been attributed to an increase in surface area and decrease in rutile phase formation due to the presence of the second oxide. Yttria-stabilized zirconia (YSZ) also turn to be very important especially in the development of solid oxide fuel cells [148-149]. Despite their evident potential, few works have been accomplished in the conception of mixed bimetallic oxides, especially regarding hierarchical porosity [150]. Unimodal mesoporous yttria-zirconia and ceria-zirconia thin films with an ordered network of channels exhibiting a high thermal stability have been prepared by evaporation-induced self-assembly of metal chlorides templated by block copolymers [151]. A recent review encompasses the potential of micro- and mesoporous amorphous mixed oxides for the design of new catalysts [152].

Hierarchically porous mixed oxides M/Ti (M = Zr or Ta) have been prepared by Wang *et al.* by cohydrolysis of 2 metallic precursors in the presence of polystyrene spheres [153]. The resulting materials exhibit ordered macropores and homogeneous wall compositions with Zr or Ta uniformly dispersed in the TiO_2 anatase framework. Moreover, the materials possess a mesoporous wall structure with BET surface areas larger than those of the corresponding pure oxides. The TiO_2/Ta_2O_5 and TiO_2/ZrO_2 exhibited higher photocatalytic than that of the solitary oxides in the photodegradation of 4-nitrophenol (4-NP) and rhodamine B (RB). Both the photocatalytic reactions confirmed that the presence of the second metal oxide in the titania framework resulted in enhanced photocatalytic activity compared with the pure titania framework, due to the homogeneous dispersion of the second metal inside the framework and the two levels of structural porosity.

The sequential coating of cellulose acetate, cellulose nitrate, polyamide, polyethersulfone and polypropylene membranes was carried out to prepare a TiO_2 coated bimodal macro-mesoporous silica [154].

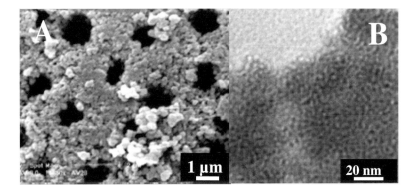

Figure 10. SEM image of the hierarchically porous bimetallic TiO_2-ZrO_2 prepared via the spontaneous route in the presence of a surfactant (A) and TEM micrograph of the wall revealing the mesopores (B).

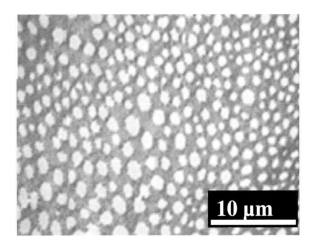

Figure 11. SEM image of a hierarchically porous aluminophosphate prepared via the spontaneous route in the presence of a surfactant.

Once again, our synthesis strategy based on the self-formation phenomenon of porous hierarchy demonstrated its simplicity and superiority in the preparation of a series of hierarchical macro-mesoporous binary mixed metal oxides. Titania–zirconia, titania–alumina, alumina–zirconia, zirconia–silica, alumina–silica, aluminophosphates, silicoaluminophosphates and titanium phosphates were prepared by using mixed alkoxide solutions [155-156] (Figures 10 and 11). In comparison with the meso–macrostructured single metal oxides, the introduction of a secondary oxide leads to a significant improvement of the structural and textural properties of the resultant materials, with a homogeneous distribution of the components and higher surface areas than the solitary metal oxides, as also stated by Wang [153]. Moreover, not only the mesopore sizes, but also the macropore sizes of binary metal oxides could be tailored and controlled by the variation of the relative molar ratios of the metal precursors. The thermal stability of the binary oxide compositions could also be enhanced significantly. These meso-macrostructured binary oxide compositions should be significant for the use as advanced functional materials, especially in the catalysis applications and in the concretization of the "hierarchical catalysis" concept.

3.4. Hierarchical carbon-based materials

It is well-known that porous carbons are promising candidates in a wide variety of applications including water and air purification, adsorption, catalysis, electrodes and energy storage [157-159]. This huge potential results from their high surface areas and pore volumes, their chemical inertness and their mechanical stability. However, periodically arranged fullerenes and single-walled nanotubes are held together via weak van der waals interactions and thus cannot be considered as systems with permanent ordered porosity [160]. That is why Ryoo developed a very original route towards mesoporous carbons in 1999 [161]. The employed pathway is based on nanocasting, i.e. the replication of existing mesoporous silica into negative carbon replicas upon infiltration of porosity by a carbon precursor and its subsequent polymerization. Many studies have since been devoted to the nanoreplication method and various starting sacrificial porous materials ("exotemplates") have been employed [162]. A general view on the replication procedure is given in a recent review article by Vinu *et al.* [163] Moreover, "true mesoporous carbons" were reported by Zhao *et al.* who used surfactant (block-copolymer) templating of a carbon precursor (thermopolymerizable polymer) [164-165]. In order to improve accessibility and diffusion efficiency, much interest has recently been devoted to carbons with a hierarchical pore system [84, 166]. Silica particles aggregates [167], polymeric foams [59], sponge-like and co-continuous macro-mesoporous silica (obtained via phase separation) [168] have already been used successfully as templates for creating new hierarchically porous carbons. In each case, the macropores are positive replica of the sacrificial exotemplate whereas the mesoporous part consists of a negative replica. Nevertheless, form all these studies, the achievement of hierarchically porous carbons featuring mesopores and *regular* macrochannels still seems difficult to achieve. Facing this problem, we investigated the possibility to use the macro-mesoporous metal oxides prepared via the self-formation phenomenon as exotemplates for the synthesis of carbons with multiples porosities. By fine-tuning the amount of carbon precursor, it has proven to be possible to make a negative replica of the mesoporous walls of these materials

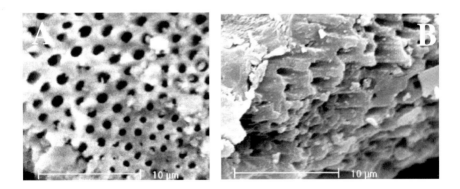

Figure 12. SEM images of a hierarchically porous aluminosilicate (A) and its corresponding carbon replica (B).

(macro-mesoporous zirconia, aluminosilicates, titania, niobia,...), leaving the macrochannels empty. After elimination of the exotemplate, hierarchical macro-mesoporous carbons made of tubular macrochannels separated by bimodal mesoporous walls could successfully be obtained (Figure 12) [111].

In addition to silica, metal oxides and bimetallic oxides, carbons with multiple porosities still widen the application potential in heterogeneous catalysis. Furthermore, if they are used as supports, macro-mesoporous carbons will further concretize the concept of hierarchical catalysis introduced in the first part of this paper.

4. Emerging catalytic applications of hierarchically porous materials

As mentioned in the introductive part of this manuscript, materials with hierarchical porosity exhibit a more accessible framework than their monomodal counterparts. Macropores separated by mesoporous walls combine two essential features for catalytic applications, namely high accessible surface areas and pore volumes with high diffusion rates and reduced transport limitations. That is why such materials could be expected to be more active when employed in the catalytic processes developed until yet for single mesoporous structures. In the following part, we will give a survey on the existing and potential catalytic applications of macro-mesoporous silica, aluminosilicates, (bi-)metallic oxides and carbons.

4.1. Macro-mesoporous silica in catalysis

Porous SiO_2 networks do not contain inherent catalytically active sites and silica does not possess variable oxidation states like the transition metals. In fact, they are quite inert and used mainly as a packing material in HPLC columns, showing often much higher performances for separation and sensing than conventional columns packed with particles [169-171]. Nevertheless, the surface of silica networks comprises generally a substantial amount of silanol Si-OH groups, which can serve as anchoring points for catalytically active species, rendering possible the heterogenization of homogeneous catalysis.

Acid catalysis could be envisioned for reactions already carried out on mesoporous modified silica in the presence of bulky molecules. Modifications include the grafting of heteropolyacids for acid cracking or liquid phase esterification reactions [172-173], grafting, impregnation or ion-exchange of Al species to carry out Friedel-Crafts alkylations, acetalizations, Diels-Alder reactions, Beckmann rearrangement, Aldol condensation,... Redox catalysis on mesoporous materials has also been studied extensively in literature. Again, it implies the modification of all-silica structures by transition metals like Ti, Zr, V, Cr, Mn, Fe, ... leading to potential oxidation catalysts of bulky molecules. The well-defined mesopores accompanied by their high specific surface area in hierarchical macro-mesoporous silica are also excellent candidates for the immobilization and dispersion of catalytically active nanoparticles. As examples, MoS_2 and Co-Mo for HDS and HDN processes [174], Ni, Pt, Rh, Ir, Ru as hydrogenation catalysts [175], Pd as hydrodehalogenation of polluting VOC [176], Au for the oxidation of CO and H_2 [177], $Mn_2(CO)_{10}$ for the total oxidation of propene [178], ... Well-defined catalytic structures can also be grafted in the mesopores, macropores or both in hierarchical silica like for instance enzymes [179] or organometallic complexes [180-181] or even more complex species like transition-metal substituted polyoxometallate clusters $[Co^{II}(H_2O)PW_{11}O_{39}]^{5-}$ for epoxydation reactions [182]. A very detailed review article written by Taguchi and Schüth describes in detail the main advances in the preparation methods and application of ordered mesoporous materials in catalysis [19]. The described reactions in this contribution could potentially be realized in a more efficient way in hierarchical macro-mesoporous silica, especially when large reactants and products are involved. Furthermore, the functions conferred to mesoporous silica that are described in a review article by Fryxell, could also be starting points for the design of new macro-mesoporous functional catalysts [20].

The one-pot synthesis of hierarchically porous materials with organic functionalities, either attached at the interior of the pores or as integral part of the walls, also opens up large perspectives. It has indeed been shown that a dye (2,4-dinitrophenylamine) could be directly incorporated in the wall structure by co-condensation of a silica source and the dye-coupled organosilane in the presence of both macro- and meso-porogens. This method can be extended to other organosilane moieties for the development of sensors, depollutants and catalysts [55-56].

This description is of course not exhaustive but considering the huge application potential of all-silica porous materials such as zeolites, mesoporous materials and 3-DOM, a combination of pores sized and shaped at different sizes within one single body could significantly broaden up the application domain due to the unique properties of the architectures and render real the concept of hierarchical catalysis that implies multiple reaction steps to proceed successively within the different pore levels of one single material.

4.2. Macro-mesoporous aluminosilicates in catalysis

The development of mesoporous MCM-41 type structures by Mobil in 1992 was triggered by the need of large-pore sized aluminosilicates to be used in FCC catalysis in addition to zeolites. Moreover, the FCC catalyst is a concrete example of hierarchical structure in use in industry and involves the formation of a composite made of a USY zeolite mixed with a macroporous matrix, usually "amorphous" low-surface area silica, alumina or silica/alumina with clay. The introduction of mesoporosity in the USY zeolite by vapor steaming treatment can facilitate the secondary catalytic cracking whereas the more oriented cracking and the fine rearrangement of cracked molecules take place in the supercages of USY zeolite. However, this micro–meso–macroporous hierarchy is obtained by artificial mixture of different components containing pre-defined porosities [17]. A more interesting pathway towards the synthesis of such a new FCC catalyst would be the direct synthesis of macrochannels that bear meso and/or micropores in their walls, via the synthesis methods described previously [98-107].

More generally, acid/base catalysis in cracking, isomerization and fine chemical synthesis involving bulky molecules could be considered to be feasible in hierarchically porous aluminosilicate structures, provided the acid sites are strong enough [17, 19]. Indeed, it is well-known that the amorphous character as well as the thickness of the walls in mesoporous materials strongly decreases the acid strength of the sites in comparison with zeolites. That is why a large deal of work has been carried out on the fabrication of mesoporous materials with zeolitic walls and it would be very exciting to continue this trend in realizing hierarchical macro-mesoporous aluminosilicates with the same zeolitic character [48-51, 99-101].

4.3. Macro-mesoporous single and bi-metallic oxides in catalysis

Porous ZrO_2 is a widely used material in catalysis as support. With supported palladium for instance, it has shown its high quality in the vapour phase phenol hydrogenation reaction and with vanadia and gold for the complete benzene oxidation [183-184].

Porous TiO_2 is intensively used as support and also as a very efficient photocatalyst. The high potential of TiO_2 has been reviewed in detail, highlighting advantages such as complete mineralization of pollutants, use of solar or near-UV light, operation at room temperature and low cost [185-186].

The beneficial effect of a hierarchical pore structure on photocatalytic activity was demonstrated by Zhang *et al.* Mesoporous Titania spheres were prepared via a sonochemical approach in the presence of a block copolymer surfactant. The stacking of the spheres leads to the presence of a substantial amount of textural interstitial meso- or macroporosity. Tested in the photodegradation of n-pentane in air, the materials show 50% higher photocatalytic activity compared to commercial P25 and 15% higher than a sample with less textural porosity [187]. Though these materials are not real hierarchical structures in the sense that the larger openings simply result from voids between nanoparticles, this example clearly demonstrates the beneficial effect of the presence of larger pores in addition to the mesopores to provide more efficient transport for the reactant molecules to reach the active sites on the walls of the small mesopores. The superior activity of a hierarchically porous titania prepared following the spontaneous route in the presence of surfactant was also demonstrated by Wang *et al.* [130]. It was shown that the photodegradation of ethylene into carbon dioxide and water was much more efficient than with commercial P25. This superior activity was attributed to several factors. First, the macrochannels act as a light-transfer path for introducing the incident photon flux onto the inner surface of mesoporous TiO_2 in the core of the structure. This allows light waves to penetrate deep inside the photocatalyst, making it a more efficient light harvester so that the effective light-activated surface area can be significantly enhanced. Secondly, the hierarchical architecture allows for an effective transport of reactants, helping in overcoming the intradiffusional resistance to mass transport present in a typical unimodal mesoporous titania. Sponge-like macro-mesoporous titania prepared by hydrothermal treatment of precipitates of tetrabutyl titanate have also been tested in the gaseous photocatalytic oxidative decomposition of acetone [131]. The hierarchically porous structures are shown to be beneficial for enhancing the adsorption efficiency of light and the flow rate of the gas molecules. The results indicate that, in the absence of macrochannels, there was about a 10% drop in photocatalytic activity. High surface area macro-mesoporous TiO_2 prepared via the spontaneous surfactant-assisted route by our group were also tested as catalytic supports of Pd for volatile organic compounds (VOCs) oxidation [188]. The Pd impregnated catalyst was found to be powerful for total oxidation of toluene and chlorobenzene.

In the same study, a hierarchically porous mixed oxide TiO_2-ZrO_2 was also tested as support for Pd and compared to ZrO_2 and TiO_2. The results show that the production of polychlorinated PhClx (with x = 2–6) compounds that occurs on the titania significantly decreases on the bimetallic oxide, showing that the nature of the support plays also a key role in the orientation of the reaction.

Hierarchical porous ZnO made of small and large mesopores were tested by our group in the photodegradation of phenol and compared with the most popular photocatalyst, commercial TiO_2 nanoparticles P-500 [142]. Under the same experimental conditions, the porous ZnO nanoparticles show a superior activity to TiO_2 nanoparticles, commercial ZnO powder and nanopowder ZnO. Compared to titanium dioxide, the higher activity of ZnO can be attributed to the larger fraction of absorbed UV light whereas the unique porous structure is responsible for the better performance in comparison to commercial ZnO (nano-)powder. This statement again puts in light the evident advantages of a hierarchically porous structure.

4.4. Macro-mesoporous carbons as high-potential supports

As for the other compositions with multiple levels of porosities, the applications are not well developed yet. Nevertheless, the applications developed by using only mesoporous carbons could be extended to macro-mesoporous carbons with potentially higher efficiency due to diffusion improvement. For instance, nanocasted mesoporous carbons CMK have been explored as potential supports for Pt and Pd in the liquid phase hydrogenation of nitrobenzene to p-aminophenol and showed better conversion and selectivity than conventional carbons [189]. The very high specific surface areas and homogeneous pore sizes of mesoporous carbons have also been explored for achieving a very high dispersion of Pt that showed its superior performance in a fuel cell setup [159]. Mesoporous carbons have been coated by polymers to create a high mechanical strength composite with high electric conductivity for electrode purposes [190]. All these applications could be extended towards using hierarchical macro-mesoporous carbons that will certainly improve the overall efficiency due to the adjunction of the second pore level. Further potential applications include adsorption to remove volatile organic compounds, odorous molecules or dioxins and furans from air [111].

5. Conclusions and outlook

Hierarchy is a very important nature-inspired feature that is essential to life. "Hierarchical Catalysis" is a concept that should definitely be concretized since it brings us back to natural phenomena, which, since ever, take advantage of hierarchy to fulfil natural processes very efficiently. In fact, the best efficiency in catalysis could be attained by realizing successive chemical conversion steps from starting reagents to final products within one single catalyst by taking advantage of the multiple scaled pore systems. At each level, a given reaction could take place leading to products that are the reactants of the next catalytic centre without needing intermediary separation and purification that would inherently occur by the sieving capacity of the adjusted pore sizes.

Though not concretized yet, such systems will be reality in a near future. Indeed, as shown all along in this review paper, many efforts have been carried out in the conception of catalysts or catalyst supports with multiple levels of porosity embedded within one single body. Though not exhaustive, this paper has tried to shed some light on the currently available synthetic procedures for the conception of new hierarchically porous catalysts with different compositions. Various preparation pathways have been developed for silica and work is currently being carried out to transpose them to other compositions like transition metal oxides, mixed oxides and carbons.

In particular, we have highlighted the innovative ideas that have emerged to tailor the different levels of pore sizes in order to create structures that act as efficiently as natural ones. Macropores in solid bodies can be created by using macrotemplates such as colloidal (polymeric) spheres, polymer or ceramic foams, bioorganisms and "natural" molecules, emulsions and droplets, inorganic salts and ice crystals and, in with metallic alkoxides, via a spontaneous formation route. In any chosen case, the aim is to find a template with sizes adapted for the replication into macropores, macrovoids or macrochannels and, by selecting the appropriate template, the size of the final macropores can be adjusted in a wide range from 0.1 to 10 µm. Well-defined ordered mesopores are known to be obtained via surfactant-templating, implying the self-assembly of inorganic precursors in the presence of micelles that delimit mesopores. By applying such a dual templating method, both the macro- and the mesoporosity can be independently adjusted. The macroporous structures can also be designed as monoliths, thin films or spheres. Such materials are expected to significantly broaden the field of application of mesoporous materials since the active sites, usually highly dispersed in mesopores, will be more easily reached through the macropores that enhance diffusion and reduce mass-transport limitations.

The survey of literature indicates that major work has been accomplished in the design of hierarchical silica but much less on other oxides. A very important issue would be to continue the investigations towards the development of other compositions, in particular transition metal oxides and mixed oxides. These possess indeed more a wide application range, especially in catalysis. Regarding this, research has to be carried out to synthesize and understand the underlying formation mechanisms of thermally stable oxide catalysts and catalyst supports. In that way, we believe that the self-formation route widely explored by our group is a very powerful tool in the conception of hierarchically porous oxides. Indeed, the careful choice of the inorganic precursors regarding their reactivity allows for the rapid and simple synthesis of homogeneous macrotubular oxides separated by meso-(or micro-)porous walls. These high-quality materials have also turned out to be very interesting exotemplates for the formation of hierarchically porous carbons.

Further work has also to be realized regarding the catalytic applications of materials with multiple porosities. It would indeed be interesting to start using the hierarchical structures in catalytic processes already developed with pure mesoporous materials in order to improve their efficiency. New catalytic reactions implying still larger molecules and that were difficult to realize with existing materials, should also be developed. Finally, the feasibility of successive reaction steps inside one single catalyst should be investigated in detail in order to fabricate "Hierarchical Catalysts" that combine high efficiency with minimal energy consumption.

Acknowledgements

Alexandre Léonard thanks the FNRS (Fonds National de la Recherche Scientifique, Belgium) for a "Chargé de Recherches" fellowship. Financial supports from a European Interreg III program (F.W-2.1.5) and the Wallonia region are greatly acknowledged. This work was realized in the framework of PAI-IUAP network P6/17.

References

1. P. Fratzl and R. Weinkamer, Progr. Mater. Sci. *In press.* (2007).
2. M.O. Coppens, J.H. Sun and T. Maschmeyer, Catal. Today 69, 331 (2001).

3. J.H. Sun, Z. Shan, T. Maschmeyer and M.O. Coppens, Langmuir 19, 8395 (2003).
4. M.E. Davis, Nature 417, 813 (2002).
5. A. Stein, Adv. Mater. 15, 763 (2003).
6. J.B. Nagy, P. Bodart, I. Hannus and I. Kiricsi, *Synthesis, Characterization and Use of Zeolitic Microporous Materials*; DecaGen Ltd: Hungary 159 (1998).
7. J. Weitkamp, Solid State Ionics 131, 175 (2000).
8. C.T. Kresge, M.E. Leonowicz, W.J. Roth, J.C. Vartuli and J.S. Beck, Nature 359, 710 (1992).
9. J.S. Beck, J.C. Vartuli, W.J. Roth, M.E. Leonowicz, C.T. Kresge, K.D. Schmitt, C.T.W. Chu, D.H. Olson, E.W. Sheppard, S.B. McCullen, J.B. Higgins and J.L. Schlenker, J. Am. Chem. Soc. 114, 10834 (1992).
10. D. Zhao, Q. Huo, J. Feng, B.F. Chmelka and G.D. Stucky, J. Am. Chem.. Soc. 120, 6024 (1998).
11. D. Zhao, J. Feng, Q. Huo, N. Melosh, G.H. Fredrickson, B.F. Chmelka and G.D. Stucky, Science 279, 548 (1998).
12. J. Fan, C. Yu, J. Lei, Q. Zhang, T. Li, B. Tu, W. Zhou and D. Zhao, J. Am. Chem. Soc. 127, 10794 (2005).
13. T. Linssen, K. Cassiers, P. Cool and E.F. Vansant, Adv. Coll. Interf. Sci. 103, 121 (2003).
14. B.L. Su, A. Léonard and Z.Y. Yuan, C.R. Chimie 8, 713 (2005).
15. A. Léonard, A. Vantomme, C. Bouvy, N. Moniotte, P. Mariaulle and B.L. Su, Nanopages 1, 1 (2006).
16. G.J.A.A. Soler-Illia, C. Sanchez, B. Lebeau and J. Patarin, Chem. Rev. 102, 4093 (2002)
17. A. Corma, Chem. Rev. 97, 2373 (1997).
18. A. Tuel, Microporous Mesoporous Mater. 27, 151 (1999).
19. A. Taguchi and F. Schüth, Microporous Mesoporous Mater. 77, 1 (2005).
20. G.E. Fryxell, Inorg. Chem. Commun. 9, 1141 (2006).
21. W. Yue, R.J. Park, A.N. Kulak and F.C. Meldrum, J. Cryst. Growth 294, 69 (2006).
22. B.T. Holland, C.F. Blanford and A. Stein, Science 281, 538 (1998).
23. O.D. Velev, T.A. Jede, R.F. Lobo and A.M. Lenhoff, Nature 389, 447 (1998).
24. O.D. Velev, T.A. Jede, R.F. Lobo and A.M. Lenhoff, Chem. Mater. 10, 3597 (1998).
25. J.E.G.J. Wijnhoven and W. Vos, Science 281, 802 (1998).
26. B.T. Holland, C.F. Blanford, T. Do and A. Stein, Chem. Mater 11, 795 (1999).
27. M. Antonietti, B. Berton, C. Göltner and H.P. Hentze, Adv. Mater. 10, 154 (1998).
28. Y. Xia, B. Gates, Y. Yin and Y. Lu, Adv. Mater. 12, 693 (2000).
29. A. Stein and R.C. Schroden, Curr. Opin. Solid State Mater. Sci. 5, 553 (2001).
30. S.H. Park and Y. Xia, Adv. Mater. 10, 1045 (1998).
31. R.A. Caruso, M. Giersig, F. Willig and M. Antonietti, Langmuir 14, 6333 (1998).
32. R.H. Jin and J.J. Yuan, J. Mater. Chem. 15, 4513 (2005).
33. A. Imhof and D.J. Pine, Nature 389, 948 (1997)

34. A. Imhof and D.J. Pine, Adv. Mater. 11, 311 (1999).

35. R. Ravikrishna, R. Green and K. Valsaraj, J. Sol Gel Sci. Technol. 34, 111 (2005).

36. S.A. Davis, S.L. Burkett, N.H. Mendelson and S. Mann, Nature 385, 420 (1997).

37. A. Yachi, R. Takahashi, S. Sato, T. Sodesawa, K. Oguma, K. Matsutani and N. Mikami J. Non Cryst. Solids 351, 331 (2005).

38. N. Tsubaki, Y. Zhang, S. Sun, H. Mori, Y. Yoneyama, X. Li and K. Fujimoto, Catal. Commun. 2, 311 (2001).

39. J. Yu, J.C. Yu, M.K.P. Leung, W. Ho, B. Cheng, X. Zhao and J. Zhao, J. Catal. 217, 69 (2003).

40. A. Vantomme, A. Léonard, Z.Y. Yuan and B.L. Su, Colloids Surf. A: Physicochem. Eng. Aspects 300, 70 (2007).

41. P. Yang, T. Deng, D. Zhao, P. Feng, D. Pine, B.F. Chmelka, G.M. Whitesides and G.D. Stucky, Science 282, 2244 (1998).

42. Q. Luo, L. Li, B. Yang and D. Zhao, Chem. Lett. 378 (2000).

43. Y.S. Yin and Z.L. Wang, Microsc. Microanal. 5, 818 (1999).

44. S. Fujita, H. Nakano, M. Ishii, H. Nakamura and S. Inagaki, Microporous Mesoporous Mater. 96, 205 (2006).

45. S. Vaudreuil, M. Bousmina, S. Kaliaguine and L. Bonneviot, Adv. Mater. 13, 1310 (2001).

46. C. Danumah, S. Vaudreuil, L. Bonneviot, M. Bousmina, S. Giasson and S. Kaliaguine, Microporous Mesoporous Mater. 44-45, 241 (2001).

47. C.G. Oh, Y.Y. Baek and S.K. Ihm, Adv. Mater. 17, 270 (2005).

48. Y.J. Wang, Y. Tang, Z. Ni, W.M. Hua, W.L. Yang, X.D. Wang, W.C. Tao and Z. Gao, Chem. Lett. 510 (2000).

49. L. Huang, Z. Wang, J. Sun, L. Miao, Q. Li, Y. Yan and D. Zhao, J. Am. Chem. Soc. 122, 3530 (2000).

50. S. Shimizu, Y. Kiyozumi and F. Mizukami, Chem. Lett. 403 (1996).

51. B.T. Holland, L. Abrams and A. Stein, J. Am. Chem. Soc. 121, 4308 (1999).

52. D. Huang, T. Brezesinski and B. Smarsly, J. Am. Chem. Soc. 126, 10534 (2004).

53. T. Sen, G.J.T. Tiddy, J.L. Casci and M.W. Anderson, Angew. Chem. Int. Ed. 42, 4649 (2003).

54. A. Stein, B.J. Melde and R.C. Schroden, Adv. Mater. 12, 1403 (2000).

55. B. Lebeau, C.E. Fowler, S. Mann, C. Farcet, B. Charleux and C. Sanchez, J. Mater. Chem. 10, 2105 (2000).

56. R.C Schroden, C.F. Blanford, B.J. Melde, B.J.S. Johnson and A. Stein, Chem. Mater. 13, 1074 (2001).

57. J.L. Blin, C. Otjacques, G. Herrier and B.L. Su, Langmuir 16, 4229 (2000).

58. J.L. Blin, G. Herrier, C. Otjacques and B.L. Su, Stud. Surf. Sci. Catal. 129, 57 (2000).

59. S. Alvarez and A.B. Fuertes, Mater. Lett. 61, 2378 (2007).

60. H. Maekawa, J. Esquena, S. Bishop, C. Solans and B.F. Chmelka, Adv. Mater. 15, 591 (2003).

61. L. Huerta, C. Guillern, J. Latorre, A. Beltran, D. Beltran and P. Amoros, Solid State Sci. 7, 405 (2005).

62. R.A. Caruso and J.H. Schattka, Adv. Mater. 12, 1921 (2000).

63. R.A. Caruso and M. Antonietti, Adv. Funct. Mater. 12, 307 (2002).

64. P.R. Giunta, R.P. Washington, T.D. Campbell, O. Steinbock and A.E. Stiegman, Angew. Chem. Int. Ed. 43, 1505 (2004).

65. S. Costacurta, L. Biasetto, E. Pippel, J. Woltersdorf and P. Colombo, J. Am. Ceram. Soc. 90, 2172 (2007).

66. X. Cai, G. Zhu, W. Zhang, H. Zhao, C. Wang, S. Qiu and Y. Wei, Eur. J. Inorg. Chem. 18, 3641 (2006).

67. H. Yang and D. Zhao, J. Mater. Chem. 15, 1217 (2005).

68. J. Parmentier, C. Vix-Guterl, S. Sadallah, M. Reda, M. Illescu, J. Werckmann and J. Patarin, Chem. Lett. 32, 262 (2003).

69. S.A. Davis, S.L. Burkett, N.H. Mendelson and S. Mann, Nature 385, 420 (1997).

70. L.G. Wang, Y. Shin, W.D. Samuels, G.J. Exarhos, I.L. Moudrakovski, V.V. Tersikh and J.A. Rippmeester, J. Phys. Chem. B 107, 13793 (2003).

71. G. Cook, P.L. Timms and C. Göltner-Spickermann, Angew. Chem. Int. Ed. 42, 557 (2003).

72. S.R. Hall, H. Bolger and S. Mann, Chem. Commun. 2784 (2003).

73. B. Zhang, S.A. Davis and S. Mann, Chem. Mater. 14, 1369 (2002).

74. D. Walsh, L. Arcelli, T. Ikoma, J. Tanaka and S. Mann, Nature Mater. 2, 386 (2003).

75. V. Pedroni, P.C. Schultz, M.E.G. de Ferreira and M.A. Morini, Colloid. Polym. Sci. 278, 964 (2000).

76. Y. Liu, Z. Shen, L. Li, P. Sun, X. Zhou, B. Li, Q. Jin, D. Ding and T. Chen, Microporous Mesoporous Mater. 92, 189 (2006).

77. T. Sen, G.J.T. Tiddy, J.L. Casci and M.W. Andersen, Chem. Commun. 2182 (2003).

78. T. Sen, G.J.T. Tiddy, J.L. Casci and M.W. Andersen, Microporous Mesoporous Mater. 78, 255 (2005).

79. F. Carn, A. Colin, M.F. Achard, H. Deleuze, E. Sellier, M. Birot and R. Backov, J. Mater. Chem. 14, 1370 (2004).

80. S.A. Bagshaw, Chem. Commun. 767 (1999).

81. F. Carn, A. Colin, M.F. Achard, H. Deleuze, Z. Zaadi and R. Backov, Adv. Mater. 16, 140 (2004).

82. K. Suzuki, K. Ikari and H. Imai, J. Mater. Chem.. 13, 1812 (2003).

83. J.L. Blin, R. Bleta, J. Ghanbaja and M.J. Stébé, Microporous Mesoporous Mater. 94, 74 (2006).

84. J. Sun, D. Ma, H. Zhang, X. Bao, G. Weinberg and D. Su, Microporous Mesoporous Mater. 100, 356 (2007).

85. K. Nakanishi, J. Porous Mater. 4, 67 (1997).

86. K. Nakanishi, Bull. Chem. Soc. Jpn, 79, 673 (2006).

87. Y. Sato, K. Nakanishi, K. Hirao, H. Jinnai, M. Shibayama, Y.B. Melnichenko and G.D. Wignall, Coll. Surf. A: Physicochem. Eng. Aspects 187, 117 (2001).

88. K. Nakanishi, Y. Sato, Y. Ruyat and K. Hirao, J. Sol-Gel Sci. Technol. 26, 567 (2003).

89. Z.G. Shi, Y.Q. Feng, L. Xu, S.L. Da and Y.Y. Ren, Microporous Mesoporous Mater. 68, 55 (2004).

90. K. Nakanishi, Y. Kobayashi, T. Amatani, K. Hirao and T. Kodaira, Chem. Mater. 16, 3652 (2004).

91. T. Amatani, K. Nakanishi, K. Hirao and T. Kodaira, Chem. Mater. 17, 2114 (2005).

92. N. Huesing, C. Raab, V. Torma, A. Roig and H. Peterlik, Chem. Mater 15, 2690 (2003).

93. D. Brandhuber, V. Torma, C. Raab, H. Peterlik, A. Kulak and N. Huesing, Chem. Mater. 17, 4262 (2005).

94. D. Zhao, P. Yang, B.F. Chmelka and G.D. Stucky, Chem. Mater. 11, 1174 (1999).

95. H. Nishihara, S.R. Mukai, D. Yamashita and H. Tamon, Chem. Mater. 17, 683 (2005).

96. S.R. Mukai, H. Nishihara and H. Tamon, Catal. Surv. Asia, 10, 161 (2006).

97. P.O. Vasiliev, Z. Shen, R.P. Hodgkins and L. Bergström, Chem. Mater. 18, 4933 (2006).

98. A. Corma, C. Martinez and L. Sauvenaud, Catal Today *In press.* (2007).

99. Y. Liu, W. Zhang and T.J. Pinnavaia, J. Am. Chem. Soc. 122, 8791 (2000).

100. D.T. On and S. Kaliaguine, Angew. Chem. Int. Ed. 40, 3248 (2001).

101. Z.T. Zhang, Y. Han, F.S. Xiao, S.L. Qiu, L. Zhu, R.W. Wang, Y. Yu, Z. Zhang, B.S. Zou, Y.Q. Wang, H.P. Sun, D.Y. Zhao and Y. Wei, J. Am. Chem. Soc. 123, 5014 (2001).

102. G. Gundiah, Bull. Mater. Sci. 24, 211 (2001).

103. J.J. Chiu, D.J. Pine, S.T. Bishop and B.F. Chmelka, J. Catal. 221, 400 (2004).

104. H. Nishihara, S.R. Mukai, Y. Fujii, T. Tago, T. Masuda and H. Tamon, J. Mater. Chem. 16, 3231 (2006).

105. J.L. Blin, A. Léonard, Z.Y. Yuan, L. Gigot, A. Vantomme, A.K. Cheetham and B.L. Su, Angew. Chem. Int. Ed. 42, 2872 (2003).

106. A. Léonard, J.L. Blin and B.L. Su, Chem. Commun. 2568 (2003).

107. A. Léonard and B.L. Su, Chem. Commun. 1674 (2004).

108. W. Deng, M.W. Toepke and B.H. Shanks, Adv. Funct. Mater. 13, 61 (2003).

109. A. Collins, D. Carriazo, S.A. Davis and S. Mann, Chem. Commun. 568 (2004).

110. A. Léonard and B.L. Su, Colloids Surf. A: Physicochem. Eng. Aspects 300, 129 (2007).

111. B.L. Su, A. Vantomme, L. Surahy, R. Pirard and J.P. Pirard, Chem. Mater. 19, 3325 (2007)

112. Y. Amenomiya, Appl. Catal. 30, 57 (1987).

113. K.D. Dobson and A.J. McQuillan, Langmuir 13, 3392 (1997).

114. L. Shi, K.C. Tin and N.B. Wong, J. Mater. Sci. 34, 3367 (1999).

115. E.M. Logothetis, Chem. Sensor Technol. 3, 89 (1991).

116. A. Fujishima and K. Honda, Nature 238, 37 (1972).

117. H. Zhang, G.C. Hardy, Y.Z. Khimyak, M.J. Rosseinsky and A.I. Cooper, Chem. Mater. 16, 4245 (2004).

118. M.C. Carbajo, E. Enciso and M.J. Torralvo, Coll. Surf. A: Physicochem. Eng. Aspects 293, 72 (2007).

119. C. Dionigi, G. Calestani, T. Ferraroni, G. Ruani, L.F. Liotta, A. Migliori, P. Nozar and D. Palles, J. Coll. Interf. Sci. 290, 201 (2005).

120. A.S. Deshpande, I. Burgert and O. Paris, Small 2, 994 (2006).

121. D. Yang, L. Qi and J. Ma, J. Mater. Chem. 13, 1119 (2003).

122. J. Huang and T. Kunitake, J. Am. Chem. Soc. 125, 11834 (2003).

123. Z.Y. Yuan, T.Z. Ren and B.L. Su, Adv. Mater. 15, 1462 (2003).

124. F. Carn, A. Colin, M.F. Achard, H. Deleuze, C. Sanchez and R. Backov, Adv. Mater. 17, 62 (2005).

125. F. Carn, M.F. Achard, O. Babot, H. Deleuze, S. Reculusa and R. Backov, J. Mater. Chem. 15, 3887 (2005).

126. S. Ito, S. Yoshita and T. Watanabe, Bull. Chem. Soc. Jap. 73, 1933 (2000).

127. M.C. Fuertes and G.J.A.A. Soler-Illia, Chem. Mater. 18, 2109 (2006).

128. Z.Y. Yuan, A. Vantomme, A. Léonard and B.L. Su, Chem. Commun. 1558 (2003).

129. A. Vantomme, Z.Y. Yuan and B.L. Su, New J. Chem. 28, 1083 (2004).

130. X. Wang, J.C. Yu, C. Ho, Y. Hou and X. Fu, Langmuir 21, 2552 (2005).

131. J. Yu, L. Zhang, B. Cheng and Y. Su, J. Phys.Chem. C 111, 10582 (2007).

132. K. Tanabe, Catal. Today 8, 1 (1990).

133. K. Tanabe, Catal. Today 78, 65 (2003).

134. I. Nowak and M. Ziolek, Chem. Rev. 99, 3603 (1999).

135. D.M. Antonelli, Microporous Mesoporous Mater. 33, 209 (1999).

136. O. Unal and M. Akinc, J. Am. Ceram. Soc. 78, 805 (1996).

137. J.H. Smatt, C. Weidenthaler, J.B. Rosenholm and M. Linden, Chem. Mater 18, 1443 (2006).

138. L. Vayssieres, K. Keis, A. Hagfeldt and S.E. Lindquis, Chem. Mater 13, 4395 (2001).

139. A. Wei, X.W. Sun, C.X. Xu, Z.L. Ding, Y. Yang, S.T. Tan and W. Huang, Nanotechnology 17, 1740 (2006).

140. J. Jiu, K.I. Kurumada and M. Tanigaki, Mater. Chem. Phys. 81, 93 (2003).

141. Z. Liu, Z. Jin, W. Li and X. Liu, J. Sol-Gel Sci. Technol. 40, 25 (2006).

142. F. Xu, P. Zhang, A. Navrotsky, Z.Y. Yuan, T.Z. Ren, M. Halasa and B.L. Su, Chem. Mater. (2007).

143. S.Y. Lai, W. Pan and C.F. Ng, Appl. Catal. B 24, 207 (2000).

144. M. Daturi, C. Binet, J.C. Lavalley, A. Galtayries and R. Sporken, Phys. Chem. Chem. Phys. 1, 5717 (1999).

145. J.B. Miller and E.I. Ko, Catal. Today 35, 269 (1997).

146. J.H. Schattka, D.G. Shchukin, J. Jia, M. Antonietti and R.A. Caruso, Chem. Mater. 14, 5103 (2002).

147. D.G. Shchukin, J.H. Schattka, M. Antonietti and R.A. Caruso, J. Phys. Chem B 107, 952 (2003).

148. N.Q. Minh, J. Am. Ceram. Soc. 76, 563 (1993).

149. M.C. Mamak, N. Coombs and G.A. Ozin, J. Am. Chem. Soc. 122, 8932 (2000).

150. D. Grosso, G.J.A.A. Soler-Illia, E.L. Crepaldi, B. Charleux and C. Sanchez, Adv. Funct. Mater. 13, 37 (2003).

151. E.L. Crepaldi, G.J.A.A. Soler-Illia, A. Bouchara, D. Grosso, D. Durand and C. Sanchez, Angew. Chem. Int. Ed. 42, 347 (2003).

152. G. Frenzer and W.H. Maier, Ann. Rev. Mater. Res. 36, 281 (2006).

153. C. Wang, A. Geng, Y. Guo, S. Jiang, X. Qu and L. Li, J. Coll. Interf. Sci. 301, 236 (2006).

154. J.H. Schattka, E.H.M. Wong, M. Antonietti and R.A Caruso, J. Mater. Chem. 16, 1414 (2006).

155. Z.Y. Yuan, T.Z. Ren, A. Vantomme and B.L. Su, Chem. Mater. 10, 5096 (2004).

156. T.Z. Ren, Z.Y. Yuan, A. Azioune, J.J. Piraux and B.L. Su, Langmuir 22, 3886 (2006).

157. H.C. Foley, Microporous Mater. 4, 407 (1995).

158. T. Kyotani, Carbon 38, 269 (2000).

159. S.H. Joo, S.J. Choi, I. Oh, J. Kwak, Z. Liu, O. Terasaki and R. Ryoo, Nature 412, 169 (2001).

160. M. Kruk, M. Jaroniec, R. Ryoo and S.H. Joo, J. Phys. Chem.. B 104, 7960 (2000).

161. R. Ryoo, S.H. Joo and S. Jun, J. Phys. Chem. B 103, 7743 (1999).

162. F. Schüth, Angew. Chem. Int. Ed. 42, 3604 (2003).

163. A. Vinu, T. Mori and K. Ariga, Sci. Technol. Adv. Mater. 7, 753 (2006).

164. Y. Meng, D. Gu, F.Q. Zhang, Y.F. Shi, L. Cheng, D. Feng, Z.X. Wu, Z.X. Chen, Y. Wan, A. Stein and D.Y. Zhao, Chem. Mater. 18, 4447 (2006).

165. Y. Huang, H.Q. Cai, T. Yu, F.Q. Zhang, F. Zhang, Y. Meng, D. Gu, Y. Wan, X.L. Sun, B. Tu and D.Y. Zhao, Angew. Chem. Int. ed. 46, 1089 (2007).

166. J. Lee, S. Han and T. Hyeon, J. Mater. Chem. 14, 478 (2004).

167. G.S. Chai, I.S. Shin and J.S. Yu, Adv. Mater. 16, 22 (2004).

168. A.H. Lu, J.H. Smatt, S. Backlund and M. Linden, Microporous Mesoporous Mater. 72, 59 (2004).

169. H. Minakuchi, N. Nakanishi, N. Soga, N. Ishizuka and N. Tanaka, Anal. Chem. 68, 3498 (1996)

170. N. Ishizuka, H. Minakuchi, K. Nakanishi, N. Soga and N. Tanaka, J. Chromatogr. A, 797, 133 (2001).

171. E. Prouzet and C. Boissière, C.R. Chim. 8, 579 (2005).

172. A. Ghanbari-Siahkali, A. Philippou, J. Dwyer and M.W. Anderson, Appl. Catal. A: Gen. 192, 57 (2000).

173. M.J. Verhoef, P.J. Kooyman, J.A. Peters and H. van Bekkum, Microporous Mesoporous Mater. 27, 365 (1999).

174. E. Riviera-Munoz, D. Lardizabal, G. Alonso, A. Aguilar, M.H. Siadati and R.R. Chianelli, Catal. Lett. 85, 147 (2003).

175. A. Corma, A. Martinez and V. Martinez-Soria, J. Catal. 169, 480 (1997).

176. C. Schüth, S. Disser, F. Schüth and M. Reinhard, Appl. Catal. B: Environ. 28, 147 (2000).

177. M. Okumura, S. Tsubota, M. Iwamoto and M. Haruta, Chem. Lett. 27, 315 (1998).

178. R. Burch, N. Cruise, D. Gleeson and S.C. Tsang, Chem. Commun. 951 (1996).

179. S. Krijnen, H.C.L. Abbenhuis, R.W.J.M. Hanssen, J.H.C. van Hooff and R.A. van Santen, Angew. Chem. Int. Ed. 37, 356 (1998).

180. J.M. Thomas, T. Maschmeyer, B.F.G. Johnson and D.S. Shephard, J. Mol. Catal. A: Chem. 141, 139 (1999).

181. C.E. Song and S.G. Lee, Chem. Rev. 102, 3495 (2002).

182. B.J.S. Johnson and A. Stein, Inorg. Chem. 40, 801 (2001).

183. S. Velu, M.P. Kapoor, S. Inagaki and K. Suzuki, Appl. Catal. A: Gen. 245, 317 (2003).

184. V. Idakiev, L. Ilieva, D. Andreeva, J.L. Blin, L. Gigot and B.L. Su, Appl. Catal. A: Gen. 243, 25 (2003).

185. M.R. Hoffmann, S.T. Martin, W. Choi and D.W. Bahnemann, Chem Rev. 95, 69 (1995).

186. A.L. Linsebigler, G. Lu and J.T. Yates, Chem. Rev. 95, 735 (1995).

187. L. Zhang and J.C. Yu, Chem. Comm. 2078 (2003).

188. H.L. Tidahy, S. Siffert, J.F. Lamonier, E.A. Zhilinskaya, A. Aboukais, Z.Y. Yuan, A. Vantomme, B.L. Su, X. Canet, G. De Weireld, M. Frère, T.B. N'Guyen, J.M. Giraudon and G. Leclercq, Appl. Catal. A : Gen. 310, 61 (2006).

189. W.S. Ahn, K.I. Min, Y.M. Chung, K.K. Rhee, S.H. Joo and R. Ryoo, Stud. Surf. Sci. Catal. 135, 313 (2001).

190. M. Choi and R. Ryoo, Nature Mater. 2, 473 (2003).

CHAPTER 10

ENVIRONMENTAL APPLICATION OF NANOTECHNOLOGY

G. Ali Mansoori

BioEngineering, Chemical Engineering and Physics Departments,
University of Illinois at Chicago (m/c 063), Chicago, IL 60607-7052 USA
Corresponding author — Email: mansoori@uic.edu

Tahereh Rohani Bastami

Department of Chemistry,
Ferdowsi University, Mashhad, P.O. Box 9177948944-1111, I.R. Iran
Email: tahereh.rohani@gmail.com

Ali Ahmadpour

Department of Chemical Engineering,
Ferdowsi University, Mashhad, P.O. Box 9177948944-1111, I.R. Iran
Email: ahmadpour_ir@yahoo.com

Zarrin Eshaghi

Payame Noor University of Mashhad, Mashhad, Iran
Email: zarrin_eshaghi@yahoo.com

Nanotechnology is an emerging field that covers a wide range of technologies which are presently under development in nanoscale. It plays a major role in the development of innovative methods to produce new products, to substitute existing production equipment and to reformulate new materials and chemicals with improved performance resulting in less consumption of energy and materials and reduced harm to the environment as well as environmental remediation. Although, reduced consumption of energy and materials benefits the environment, nanotechnology will give possibilities to remediate problems associated with the existing processes in a more sustainable way. Environmental applications of nanotechnology address the development of solutions to the existing environmental problems, preventive measures for future problems resulting from the interactions of energy and materials with the environment, and any possible risks that may be posed by nanotechnology itself.

This article gives a comprehensive review on the ongoing research and development activities on environmental remediation by nanotechnology. First, the essential aspects of environmental problems are reviewed and then the application of nanotechnology to the compounds, which can serve as environmental cleaning, is described. Various environmental treatments and remediations using different types of nanostructured materials from air, contaminated wastewater, groundwater, surface water and soil are discussed. The categories of nanoparticles studied include those which are based on titanium dioxide, iron, bimetallics, catalytic particles, clays, carbon nanotube, fullerenes, dendrimers and magnetic nanoparticles. Their advantages and limitations in the environmental applications are evaluated and compared with each other and with the existing techniques. The operating conditions such as pH, required doses, initial concentrations, and treatment performances are also presented and compared. The report covers the bulk of the published researches during the period of 1997 to 2007.

1. General Introduction

Nanotechnology is a field of applied science, focused on the design, synthesis, characterization and application of materials and devices on the nanoscale. This branch of knowledge is a sub-classification of technology in colloidal science, biology, physics, chemistry and other scientific fields and involves the study of phenomena and manipulation of materials in the nanoscale [1-3]. This results in materials and systems that often exhibit novel and significantly changing physical, chemical and biological properties due to their size and structure [3]. Also, a unique aspect of nanotechnology is the "vastly increased ratio of surface area to volume", present in many nanoscale materials, which opens new possibilities in surface-based sciences [4].

Similar to nanotechnology's success in consumer products and other sectors, nanoscale materials have the potential to improve the environment, both through direct applications of those materials to detect, prevent, and remove pollutants, as well as indirectly by using nanotechnology to design cleaner industrial processes and create environmentally responsible products. For example, iron nanoparticles can remove contaminants from soil and ground water; and nanosized sensors hold promise for improved detection and tracking of contaminants.

Behavior of materials at nanoscale is not necessarily predictable from what we know at macroscale. At the nanoscale, often highly desirable, properties are created due to size confinement, dominance of interfacial phenomena, and quantum effects. These new and unique properties of nanostructured materials, nanoparticles, and other related nanotechnologies lead to improved catalysts, tunable photoactivity, increased strength, and many other interesting characteristics [5-7].

As the exciting field of nanotechnology develops, the broader environmental impacts of nanotechnology will also need to be considered. Such considerations might include: the environmental implications of the cost, size and availability of advanced technological devices; models to determine potential benefits of reduction or prevention of pollutants from environmental sources; potential new directions in environmental science due to advanced sensors; effects of rapid advances in health care and health management as related to the environment; impact of artificial nanoparticles in the atmosphere; and impacts from the development of nanomachines [8].

Research is needed using nanoscale science and technology to identify opportunities and applications to environmental problems, and to evaluate the potential environmental impacts of nanotechnology. Also, approaches are needed to offer new capabilities for preventing or treating highly toxic or persistent pollutants, which would result in the more effective monitoring of pollutants or their impact in the ways not currently possible.

Early application of nanotechnology is remediation using nanoscale iron particles. Zero-valent iron nanoparticles are deployed *in situ* to remediate soil and water contaminated with chlorinated compounds and heavy metals.

Among the many applications of nanotechnology that have environmental implications, remediation of contaminated groundwater using nanoparticles containing zero-valent iron is one of the most prominent examples of a rapidly emerging technology with considerable potential benefits. There are, however, many uncertainties regarding the fundamental features of this technology, which have made it difficult to engineer applications for optimal performance or to assess the risk to human or ecological health. This important aspect of nanoparticles needs extensive considerations as well.

One of the main environmental applications of nanotechnology is in the water sector. As freshwater sources become increasingly scarce due to overconsumption and contamination, scientists have begun to consider seawater as another source for drinking water. The majority of the world's water supply has too much salt for human consumption and desalination is an option but expensive method for removing the salt to create new sources of drinking water. Carbon nanotube membranes have the potential to reduce desalination costs. Similarly, nanofilters could be used to remediate or clean up ground water or surface water contaminated with chemicals and hazardous substances. Finally, nanosensors could be developed to detect waterborne contaminants.

Air pollution is another potential area where nanotechnology has great promise. Filtration techniques similar to the water purification methods described above could be used in buildings to purify indoor air volumes. Nanofilters could be applied to automobile tailpipes and factory smokestacks to separate out contaminants and prevent them from entering the atmosphere. Finally, nanosensors could be developed to detect toxic gas leaks at extremely low concentrations. Overall, there is a multitude of promising environmental applications for nanotechnology. Much of the current research is focused on energy and water technologies.

Environmental remediation includes the degradation, sequestration, or other related approaches that result in reduced risks to human and environmental receptors posed by chemical and radiological contaminants. The benefits, which arise from the application of nanomaterials for remediation, would be more rapid or cost-effective cleanup of wastes.

Cost-effective remediation techniques pose a major challenge in the development of adequate remediation methods that protect the environment. Substances of significant concern in remediation of soils, sediment, and groundwater include heavy metals (e.g., mercury, lead, cadmium) and organic compounds (e.g., benzene, chlorinated solvents, creosote, and toluene). Specific control and design of materials at the molecular level may impart increased affinity, capacity, and selectivity for pollutants. Minimizing quantities and exposure of hazardous wastes to the air and water and providing safe drinking water are among the environmental protection agencies' goals. In this regards, nanotechnology could play a key role in pollution prevention technologies [9-11].

In the present article, we have reviewed processes used for the environmental treatment by nanotechnology. It should be mentioned that due to the high variety of techniques and conditions used by different researchers, the results are only summarized in the form of tables and various nanosystems are described in the text.

2. Nano-Materials and Their Environmental Applications

2.1. Titanium Dioxide (TiO₂) Based Nanoparticles

Titanium dioxide (TiO_2) is one of the popular materials used in various applications because of its semiconducting, photocatalytic, energy converting, electronic and gas sensing properties. Titanium dioxide crystals are present in three different polymorphs in nature that in the order of their abundance, are Rutile, Anatase and Brookite (See Figure 1) [12].

Many researchers are focused on TiO_2 nanoparticle and its application as a photocatalyst in water treatment. Nanoparticles that are activated by light, such as the large band-gap semiconductors titanium dioxide (TiO_2) and zinc oxide (ZnO), are frequently studied for their

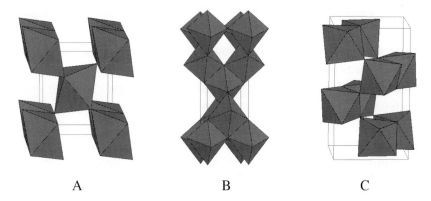

Figure 1. The crystal structures of A) rutile, B) anatase, C) brookite [Ref. 13], http://ruby.colorado.edu/~smyth/min/tio2.html.

ability to remove organic contaminants from various media. These nanoparticles have the advantages of readily available, inexpensive, and low toxicity. The semiconducting property of TiO_2 is necessary for the removal of different organic pollutants through excitation of TiO_2 semiconductor with a light energy greater than its band gap, which could generate electron hole pairs. These may be exploited in different reduction processes at the semiconductor/solution interface.

A semiconductor can adopt with donor atoms that provide electrons for the conduction band where they can carry a current. These materials can also adopt with acceptor atoms that take electrons from the valence band and leave behind some positive charges (holes). The energy levels of these donors and acceptors fall into the energy gap.

The most affecting properties of semiconducting nanoparticles are distinguished changes in their optical properties compared to those of bulk materials. In addition, there is a significant shift in optical absorption spectra toward the blue shift (shorter wavelengths) as the particle size is reduced [14].

Stathatos *et al.* [15] used reverse micelle technique to produce TiO_2 nanoparticles and deposited them as thin films. They deposited TiO_2 mesoporous films on glass slides by dip-coating in reverse micellar gels containing titanium isopropoxide. The films demonstrated high capacity in adsorption of several dyes from aqueous and alcoholic solutions. It had also a rapid degradation of the adsorbed dyes when the colored films were exposed to the visible light.

It is known that, the semiconducting properties of TiO_2 materials is responsible for the removal of various organic pollutants, but the rapid recombination of photo-generated electron hole pairs and the non-selectivity of the system are the main problems that limit the application of photocatalysis processes [16]. It was suggested that, replacing adsorbed solvent molecules and ions by chelating agents, i.e. surface modification, changes the energetic situation of surface states and considerably alters the chemistry, which is taking place at the surface of titanium dioxide (TiO_2).

Phenol is one of the toxic materials found in municipal and waste waters. Synthesized titanium dioxide nanoparticles of both Anatase and Rutile forms were used for wet oxidation of phenols by hydrothermal treatment of microemulsions and their photocatalytic activity [17]. Such treatment has the advantage that the size of particles is affected by the ratio of surfactant to water. Size of water droplets in the reverse microemulsions is found to be almost the same as that of formed particles. The main reactions proposed for phenol degradation are [17]:

$$TiO_2 + h\upsilon \rightarrow TiO_2(h)^+ + e^- \tag{1}$$

$$TiO_2(h)^+ + H_2O(ads) \rightarrow OH^\bullet + H^+ + TiO_2 \tag{2}$$

$$OH^\bullet + Phenol \rightarrow intermediate\ products\ (e.g.,\ benzoquinone) \tag{3}$$

$$TiO_2(h)^+ + Intermediate\ products \rightarrow CO_2 + H_2O + TiO_2 \tag{4}$$

In another study, a novel composite reactor with combination of photochemical and electrochemical system was used for the degradation of organic pollutants [18]. In this process, UV excited nanostructure TiO_2, was served as the photocatalyst. The reactor performance was evaluated by the degradation process of Rhodamine 6G (R-6G) (See Figure 2).

Fine TiO_2 particles have shown better efficiency than the immobilized catalysts, but complete separation and recycling of fine particles (less than $0.5\,\mu m$) from the treated water, are very expensive. Therefore, from the economics point of view, this method is not suitable for the industrial-scale. This problem was solved by fixing the carbon-black-modified nano-TiO_2 (CB-TiO_2) on aluminum sheet as a support [19]. The photocatalytic activity of CB-TiO_2 thin films was observed to be 1.5 times greater than that of TiO_2 thin films in the degradation of reactive Brilliant Red X-3B. Core $SrFe_{12}O_{19}$ nanoparticles and TiO_2 nanocrystals were also synthesized as the magnetic photocatalytic particles [20]. This system recovers the photocatalyst particles and protects them from the treated water stream by applying an external magnetic field (See Table 1). In the presence of natural organic matters in water, many problems may happen since they can occupy the catalytically active surface sites and lead to much lower decomposition efficiency. One useful method for overcoming the mentioned problem is the combination of adsorption and oxidative destruction technique.

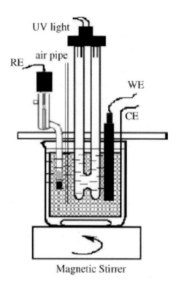

Figure 2. The diagram of the composite reactor with combination of photochemical and electrochemical system used for the degradation of organic pollutants (RE = reference electrode; WE = working electrode; and CE = counter electrode). From [Ref. 18], J. Chen *et al.* Water Research. 37, 3815 (2003).

Ilisz *et al.* [21] used a combination of TiO_2-based photocatalysis and adsorption for the decomposition of 2-chlorophenol (2-CP). The three combined systems that they studied and compared were:

1. TiO$_2$ intercalated into the interlamellar space of a hydrophilic montmorillonite by means of a heterocoagulation method (TiO$_2$ pillared montmorillonite, TPM);
2. TiO$_2$ hydrothermally crystallized on hexadecylpyridinium chloride-treated montmorillonite (HDPM-T);
3. Hexadecylpyridinium chloride-treated montmorillonite (HDPM) applied as an adsorbent and Degussa P25 TiO$_2$ as a photocatalyst (HDPM/TiO$_2$).

The latter was shown the highest rate for pollutant decomposition compared to the others and it could be re-used without further regeneration. In another application, the work was focused on crystalline Titania with ordered nanodimensional porous structures [22]. In this regard, the mesoporous spherical aggregates of Anatase nanocrystal were first fabricated and then cetyltrimethylammonium bromide was employed as the structure-directing agent. After that, the interaction between cyclohexane micro-droplets and the cetyltrimethylammonium bromide self-assemblies was applied to photo-degrade a variety of organic dye pollutants in aqueous media such as methyl orange (See Table 1).

In addition, in the study of Peng *et al.* [23], the mesoporous titanium dioxide nanosized powder was synthesized using hydrothermal process by applying cetyltrimethylammonium bromide (CTAB) as surfactant-directing and pore-forming agent. They synthesized and applied the nanoparticle products for the oxidation of Rhodamine B (RB) (See Table 1).

The mesoporous structures with high surface area are able to provide simple accessibility and more chances for guest molecules and light to receive by the active sites. In this regard, Paek *et al.* [24] fabricated mesoporous photocatalysts with delaminated structure. The exfoliated layered titanate in aqueous solution was reassembled in the presence of Anatase TiO$_2$ nanosol particles to make a large number of mesopores and eventually a large surface area TiO$_2$ photocatalysts (See Table 1).

P-25 TiO$_2$ is a highly photoactive form of TiO$_2$ composed of 20-30% Rutile and 70-80% Anatase TiO$_2$ with particle sizes in the range of 12 to 20 nm. Adams *et al.* [25] synthesized SBA-15 mesoporous silica thin films encapsulating Degussa P-25 TiO$_2$ particles via a block copolymer templating process. High calcination temperature (above 450°C) is usually required to form a regular crystal structure. However, in the meantime, high temperature treatment would reduce the surface area and loose some hydroxyl or alkoxide group on the surface of TiO$_2$, which prevent simple dispersion. This problem was solved by hydrothermal process to produce pure Anatase-TiO$_2$ nanoparticles at low temperature (200°C, 2 h). These TiO$_2$ nanoparticles have several advantages, such as fully pure Anatase crystalline form, fine particle size (8 nm) with more uniform distribution and high-dispersion in either polar or non-polar solvents, stronger interfacial adsorption and convenient coating on different supporting materials compared to the other TiO$_2$ powders.

Asilturk et al. [26] examined the behavior of Anatase nano-TiO$_2$ in catalytic decomposition of Rhodamine B (RB) dye. Rhodamine B was fully decomposed with the catalytic action of nano-TiO$_2$ in a short time of about 60 min's. It was found that, the nano-TiO$_2$ could be repeatedly used with increasing the photocatalytic activity.

In recent years, the technology of ultrasonic degradation has been studied and extensively used to treat some organic pollutants. The ultrasound with low power was employed as an irradiation source to make heat-treated TiO$_2$ powder. This method was used for decomposition of parathion with the nanometer Rutile titanium dioxide (TiO$_2$) powder as the sonocatalyst after treatment of high-temperature activation [27]. There is an appropriate method to increase the photocatalytic efficiency of TiO$_2$, which consists of adding a co-adsorbent such as activated carbon (AC) to it. The resulting synergy effect can be explained by the formation of a common contact interface between different solid phases. Activated carbon acts as an adsorption trap for the organic pollutant, and then immediately degraded through mass transfer of organic substances to the photoactivated TiO$_2$ on the surface. In the study of Li et al. [28], carbon grain coated with activated nano-TiO$_2$ (20-40 nm) (TiO$_2$/AC) was prepared and used for the photodegradation of methyl orange (MO) dyestuff in aqueous solution under UV irradiation (See Table 1). They have summarized the benefits arise from the applications of these activated carbons as:

- The adsorbent support provides a high concentration environment of target organic substances around the loaded TiO_2 particles by adsorption. Therefore, the rate of photo-oxidation is enhanced.
- The organic substances are oxidized on the photocatalyst surfaces via adsorption states. The resulting toxic intermediates are also adsorbed and oxidized and as a result, they are not released in the air atmosphere to cause secondary pollution.
- Since the adsorbed substances are finally oxidized to give CO_2, the high adsorbed ability of the hybrid photocatalysts for organic substances is maintained for a long time. The amount of TiO_2 in catalysts play significant role upon the photo-efficiency of hybrid catalysts.

In another investigation, Wu *et al.* [29] studied the dye decomposition kinetics in a batch photocatalytic reactor under various operational conditions including agitation speed, TiO_2 suspension concentration, initial dye concentration, temperature and UV illumination intensity in order to establish reaction kinetic models.

In general, it can be concluded that all the modified and thin film samples prevent rapid recombination, while CB-TiO_2 films and TiO_2/strontium ferrite samples have the advantage of easy separation because of their fixation on the support.

Mahmoodi *et al.* [30] studied the effect of immobilized titanium dioxide nanoparticles on the removal of Butachlor (N- butoxymethyl-2-chloro-2, 6-diethylacetanilide) which is one of the organic pollutants in agricultural soil and wastewater. Due to high preparation cost and toxicity of nanoparticles in the environment, they used the immobilized form of TiO_2, because of their easy recovery from aqueous media. The effective parameters investigated in this study were inorganic anions (NO_3^-, Cl^-, and SO_4^{2-}), H_2O_2 concentration and PH. In another work, Mahmoodi *et al.* [31] immobilized TiO_2 nanoparticles for the degradation and mineralization of two agricultural pollutants (Diazinon and Imidacloprid as *N*-heterocyclic aromatics). Dai *et al.* [32] examined the removal of methyl orange in aqueous suspension containing titania nanoparticles with meso-structure (m-TiO_2) under UV-irradiation. As mentioned above, immobilization of nanoparticle was useful for their easy recovery. The photocatalytic efficiency of immobilized TiO_2 nanoparticle with 6 nm diameter (supported by glass substrate) as well as conventional suspended catalysts is investigated recently by Mascolo *et al.* [33] for the degradation of methyl red dye. The results have shown that the conventional suspended TiO_2 degussa P25 is more effective than the supported nanoparticles. Although, they found the mechanism for dye degradation was the same for both cases, but lowering the photodegradation was due to reduction in active surface area for adsorption and subsequent catalyst action.

The effect of thermal treatment on the photodegradation of rhodamine B (RhB) in water with titania nanorod film was investigated by Wu [34]. In addition, Srinivasan and White [35] studied the photodegradation of methylene blue using three-dimensionally ordered macroporous (3DOM) titania. Titania (3DOM) was prepared by colloidal crystal templating against the polystyrene spheres. It was found that the interconnected framework structure of (3DOM) titania provide more active surface sites for the photodegradation through diffusion.

Sobana *et al.* [36] prepared silver nanoparticles doped TiO_2 and used them for the photodegradation of direct azo dyes (Direct red 23, and Direct blue 53). The noble metals such as Ag (or Pt, Au, and ...) could act as electron traps, since they facilitate electron-hole separation and, as mentioned earlier, prevention of electron-hole recombination is useful and gives higher efficiency of photodegradation. The optimum dosage of Ag, which doped on TiO_2 and enhanced the photodegradation of dyes was 1.5%. Chuang *et al.* [37] investigated the synergy effect of TiO_2 nanoparticle and carbonized bamboo for the enhancement of benzene and toluene removal. They prepared carbonized moso bamboo powder (CB), mixture of TiO_2 nanoparticles and carbonized bamboo powder (CBM), and composite of TiO_2 nanoparticle and carbonized bamboo powder (CBC) with two weight ratios of CB to TiO_2 i.e. 1:1 and 1:2. They also compared their performances for the removal of benzene and toluene and found that at the same ratio of TiO_2 to CB, the efficiency increased as follow: CBC > CBM > CB (See Table 1).

Degradation of nitrobenzene by using nano-TiO_2 and ozone were recently studied by Yang *et al.* [38]. They compared the effect of nano-TiO_2 catalyzed plus ozone and ozone only and found that the catalyzed

Table 1. Removal of pollutants using TiO_2 nanoparticles

Type of nanoparticle	Removal target	Initial concentration	Dose of nanoparticle	Irradiation time (min)	Removal efficiency (%)	Ref.
TiO_2 nanoparticle	Rhodamine 6G	125 mmol/L	0.1%(w/w)	12	90	18
$TiO_2/SrFe_{12}O_{19}$ composite	Procion Red MX-5B	10 mg/L	2.0mg/50 ml TiO_2, 30% $TiO_2/SrFe_{12}O_{19}$	300	98	20
Mesoporous Anantase nanocrystal	Methyl orange	30 mg/L	3 g/L	45	100	22
Mesoporous TiO_2 nanopowder[1]	Rhodamine B	1.0×10^{-5} M	50 mg/50ml	120	97	23
Mesoporous titania nanohybrid (naohybrid-I)[2]	4-chlorophenol	1.0×10^{-5} M	25 mg/100ml	240	99	24
Mesoporous titania nanohybrid (naohybrid-I)[2]	Methyl orange	1.0×10^{-5} M	25 mg/100ml	120	100	24
Rutile TiO_2 nanoparticle	Parathion	50 mg/L	1000 mg/L	120	>70	27
TiO_2/AC nanoparticle[3]	Methyl orange	1.0×10^{-3} mol/L[3]	0.5 g/200ml (47wt% TiO_2)	140	100	28
Pure TiO_2 nanoparticle	Methyl orange	1.0×10^{-3} mol/L[3]	0.5 g/200ml	200	80	28
CBC2[4]	Benzene	45 mg/L	5 g	180	72	37
CBC2[4]	Toluene	45 mg/L	5 g	180	71	37

[1] Calcinated at 400°C

[2] [Ti] nanoparticles/[Ti] layered titaate

[3] TiO_2 + activated carbon

[4] TiO_2 nanoparticle and carbonized bamboo composite (CB:TiO2, 1:2)

ozonation was more efficient than ozone alone. Titanate nanotubes/ anatase nanocomposite was also synthesized via hydrothermal method for the photocatalytic decolorization of rhodamine B under visible light [39]. Lee *et al.* [40] prepared titanate nanotubes (TNTs) by a hydrothermal process and then washed them by HCl solutions with different concentrations. They used TNT treated samples for the removal of basic dyes. In addition, they modified TNTs with surfactant hexa-decyltrimethyl ammonium chloride (HDTMA) via cation exchange process to remove acid dyes. The adsorption capacities for basic and acid dyes were 380 and 400 mg/g, respectively.

2.2. Iron Based Nanoparticles

Nanoparticles could provide very high flexibility for both *in situ* and *ex situ* remediations. For example, nanoparticles are easily deployed in *ex situ* slurry reactors for the treatment of contaminated soils, sediments, and solid wastes. Alternatively, they can be anchored onto a solid matrix such as carbon, zeolite, or membrane for enhanced treatment of water, wastewater, or gaseous process streams. Direct subsurface injection of nanoscale iron particles, whether under gravity-feed or pressurized conditions, has already been shown to effectively degrade chlorinated organics such as trichloroethylene, to environmentally benign compounds. The technology also holds great promise for immobilizing heavy metals and radionuclides.

The use of zero-valent iron (ZVI or Fe^0) for *in situ* remedial treatment has been expanded to include all different kinds of contaminants [41]. Zero-valent iron removes aqueous contaminants by reductive dechlorination, in the case of chlorinated solvents, or by reducing to an insoluble from, in the case of aqueous metal ions. Iron also undergoes "Redox" reactions with dissolved oxygen and water:

$$2Fe^o_{(s)} + O_{2(g)} + 2H_2O \rightarrow 2Fe^{2+}_{(aq)} + 4OH^-_{(aq)} \tag{5}$$

$$Fe^o_{(s)} + 2H_2O \rightarrow Fe^{2+}_{(aq)} + H_{2(g)} + 2OH^-_{(aq)} \tag{6}$$

Supported zero-valent iron nanoparticles with 10-30 nm in diameter were also prepared [41]. These nanoparticles were used for separation and immobilization of Cr (VI) and Pb (II) from aqueous solution by reduction of chromium to Cr (III) and Pb to Pb (0) [41].

In another research, nanopowder of zero-valent iron (<100 nm, with the specific surface area of 35 m^2/g) was used for the reduction and immobilization of Cr (VI) too [42]. Nitrogen oxidants also react with Fe^0, as illustrated by the de-nitrification of nitrate (NO_3^-).

$$5Fe^0 + 2NO_3^- + 6H_2O \leftrightarrow 5Fe^{2+} + N_2 + 12OH^- \tag{7}$$

$$Fe^0 + NO_3^- + 2H^+ \rightarrow Fe^{2+} + H_2O + NO_2^- \tag{8}$$

$$NO_3^- + 6H_2O + 8e^- \rightarrow NH_3 + 9OH^- \tag{9}$$

Nanopowder of zero-valent iron (ZVI or Fe^0) was used for the removal of nitrate in water. These nanoparticles have a large ratio of surface area to mass ($31.4 m^2/g$) [43]. Nanoscale ZVI was employed by Lowry *et al.* [44] for dechlorination of polychlorinated biphenyl (PCB) to lower-chlorinated products under ambient conditions.

More recently, it was demonstrated that nano-sized zero-valent iron (nZVI) oxidizes organic compounds in the presence of oxygen [45]. The high surface area of nano scale nZVI may allow for more efficient generation of oxidants. A decrease in reactivity is expected with the build-up of iron oxides on the surface, particularly at high pH. Feitz *et al.* [45] investigated the oxidization of herbicide molinate by nano scale zero-valent iron (nZVI), when it is used in the presence of oxygen.

The EZVI (emulsified zero-valent iron) technology with nanoscale or microscale iron was enhanced to address this limitation associated with the conventional use of ZVI [46]. Quinn *et al.* [46] evaluated the performance of nanoscale emulsified zero-valent iron (nEZVI) to improve in-situ de-halogenation of dense, nonaqueous phase liquids (DNAPLs) containing trichloroethene (TCE) from ground water and soil (See Figure 3).

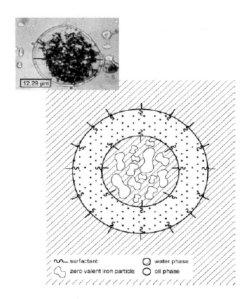

Figure 3. Schematic and photograph of EZVI (emulsified zero-valent iron) droplet showing the oil-liquid membrane surrounding particles of ZVI in water. From [Ref. 46], J. Quinn *et al.* Environ. Sci. Technol. 39, 1309 (2005).

Lindane (γ-hexachloroccyclohexane) is one of the persistent organic pollutants (POP) in the drinking water. FeS nanoparticle could degrade Lindane from water. These nanoparticles were synthesized by the wet chemical method and stabilized using a polymer from basidiomycetous [47]. One of the applications of ZVI is the removal and sorption of Arsenic contamination from water, ground water and soil [48].

Nanopowder of ZVI as a fine powder cannot be used in fixed-bed columns unless they have granular shape [49]. Cellulose beads are a promising adsorbent due to their special characteristics including hydrophilic, porous, high surface area, and excellent mechanical and hydraulic properties. Cellulose and its derivative in the form of beads are widely applied as ion exchangers, adsorbents for heavy metal ions and proteins, and as the carriers for immobilization of biocatalysts. Guo and Chen [49] prepared and used new adsorbent, bead cellulose loaded with iron oxyhydroxide (BCF), for the adsorption and removal of arsenate and arsenite from aqueous systems (See Figure 4).

Figure 4. ESEM micrograph of BCF (bead cellulose loaded with iron oxyhydroxide). From [Ref. 49], X. Guo, and F. Chen. Environ. Sci. Technol. 39, 6808 (2005).

It is recognized that oxides of poly-valent metals such as: Fe (III), Al (III), Ti (IV), and Zr (IV), show ligand sorption properties through formation of inner-sphere complexes. Furthermore, hydrated Fe (III) oxide (HFO) is inexpensive, readily available and chemically stable over a wide pH range. Iron (III) oxides have high sorption affinity toward both As (V) or arsenates and As (III) or arsenites, which are the Lewis bases [50]. In the study of Cumbal and Sengupta [50], sizes of the fresh precipitated amorphous HFO particles were found to vary from 20 to 100 nm. Despite their high arsenic removal capacity, such fine submicron particles and their aggregates are shown to be unusable in fixed beds or any flow through systems due to excessive pressure drops and poor mechanical strength. To overcome these problems, HFO nanoparticles were dispersed within a macro porous polymeric cation exchanger and the resulting hybrid material were then employed for arsenic removal.

Cation and anion exchangers were used as host materials for dispersing HFO nanoparticles within the polymer phase. The resulting polymeric/inorganic hybrid adsorbent, referred to as hybrid ion exchanger or HIX, combines excellent mechanical properties of spherical polymeric beads. HIX was amenable to efficient *in situ* regeneration with caustic soda and could subsequently be brought into service following a short rinse with carbon dioxide spiked water [51].

Xu and Zhao [52] used carboxy methyl cellulose (CMC) stabilized ZVI nanoparticles to reduce Cr (VI) in both aqueous and soil media through batch and continuous flow column study. They found that the stabilized ZVI nanoparticle is more effective than the non-stabilized one for the removal of Cr (VI). In the batch experiments, the reduction of Cr (VI) was improved from 24% to 90% as the dosage of ZVI increased from 0.04 to 0.12 g/L [52].

In another work, Xiong *et al.* [53] studied the degradation of perchlorate (ClO_4^-) in water and ion exchange brine. They used CMC-stabilized ZVI nanoparticles and compared CMC-stabilized Fe (0), non-stabilized Fe (0), and CMC-stabilized Fe-metal catalysts (such as Fe-Pd catalyst), for the reduction of perchlorate. The results showed that the stabilized ZVI nanoparticle is more efficient than the other nanoparticles for perchlorate reduction. The results also illustrated the stabilized ZVI nanoparticles could increase perchlorate reduction rate by 53% in saline water (with concentration of NaCl up to 6% w/w) [53].

Giasuddin *et al.* [54] investigated the removal of humic acid (HA) with ZVI nanoparticles (nZVI) and also their interaction with As (III) and As(V). The effect of competing anion was also studied and the results indicated the complete removal of HA in the presence of 10 mM NO_3^- and SO_4^{2-}, whereas HA removal were only 0%, 18%, and 22% in the presence of 10 mM $H_2PO_4^{2-}$, HCO_3^-, and $H_4SiO_4^0$, respectively. Li and Zhang [55] used core-shell structure of iron nanoparticle as a sorbent and reductant to remove of Ni (II) from aqueous solution. The results indicated that the sorption capacity for the removal of Ni (II) was 0.13 gNi/gFe or 4.43 meq Ni (II)/g. Cheng *et al.* [56] also applied ZVI-nanoparticle and commercial form of Fe^0 powder with different mesh sizes for the dechloronation of p-chlorophenol from water. Comparison between those particles indicated that the nanoscale Fe^0 was more effective for the reduction process. Celebi *et al.* [57] synthesized nanoparticles of zero-valent iron (nZVI) and used them to remove Ba^{2+} ion from aqueous solution. Hristovski *et al.* [58] prepared a hybrid ion-exchange (HIX) for the simultaneous removal of arsenate and perchlorate by impregnation of nano-crystaline iron hydroxide nanoparticle onto strong base ion-exchange (IX) resin.

2.3. Bimetallic Nanoparticles

Destruction of halogenated organic compound (HOCs) by zero-valent iron represents one of the latest innovative technologies for environmental remediation. Laboratory investigation in the past few years indicated that granular iron could degrade many HOCs, such as chlorinated aliphatics, chlorinated aromatics and polychlorinated biphenyls. Wang and Zhang [59] stated that the implementation of zero-valent iron technique would encounter challenges such as:

- Production and accumulation of chlorinated by-products due to the low reactivity of iron powders toward lightly chlorinated hydrocarbons. For example, reduction of tetrachloroethene (also known industrially as perchloroethylene, PERC or PCE) and trichloroethene (TCE) by zero-valent iron has been observed to produce cis-1, 2-dichloromethane (DCE) and vinylchloride (VC), both being of considerable toxicological concern.

- Decrease in iron reactivity over time, probably due to the formation of surface passive layers or the precipitation of metal hydroxides (e.g. $Fe(OH)_2$, $Fe(OH)_3$) and metal carbonates (e.g., $FeCO_3$) on the surface of iron.

Some other metals, especially zinc and tin, can transform HOCs quicker than iron. Palladium, with its superior catalytic ability produced spectacular results as well. For example, recent studies found out that palladized iron can completely dechlorinate many chlorinated aliphatic compounds to hydrocarbons [59]. In some researches, synthesized nanoscale iron and palladized iron particles are used for degradation of chlorinated compounds (See Table 3) [59-61].

Another metal acting as a catalyst is nickel, Ni(II). This metallic catalyst could prevent formation of toxic by-products by dehalogenation of chlorinated compounds via hydrogen reduction rather than electron transfer [63]. Many researchers have focused on the synthesis of Ni/Fe nanoparticles for the reduction of chlorinated compounds (See Table 3) [62, 63].

One of the major problems associated with ground- and surface-waters is nitrate contamination. The pH value is a means to control the reduction of nitrate by iron and, in effect, the formation of a passive oxide layer. Huang *et al.* [64] reported that iron powder can effectively reduce nitrate only at a pH$\leq$4. They suggested that the acidity is required to effectively remove surface passivation and to trigger nitrate reduction.

The deposition of small amounts of a second metal, such as Pd, Pt, Ag, Ni, and Cu, on iron has been shown to accelerate the reaction rate [65]. Whereas iron is deposited with the second metal, a relative potential difference drives the electron from iron to that metal [65].

Liou *et al.* [66] used uncatalyzed and catalyzed nanoscale Fe^0 systems for the denitrification of unbuffered 40 mg/L nitrate solutions at initial neutral pH. Compared to microscale Fe^0 ($<$100 mesh), the efficiency and rate of nitrate removal using uncatalyzed and catalyzed nano-Fe^0 were highly promoted. The maximum elevated rate was obtained using copper-catalyzed nano-Fe^0 (nano-Cu/Fe). Figure 5 shows the proposed scheme for reaction of nitrate reduction in the Cu/Fe system [66].

Another synthetic bimetallic nanoparticle is Pd/Au, which reduced the chlorinated compounds from water and ground water. Nutt *et al.* [67] synthesized Pd supported on gold nanoparticles (Au NPs). They found that these catalysts were considerably more active than Pd NPs. Joo and Zhao [68] prepared Fe-Pd bimetallic with 0.2% w/w of sodium carboxy methyl cellulose (CMC) as stabilizer and used them for the degradation of lindane and atrazine, the chlorinated herbicides.

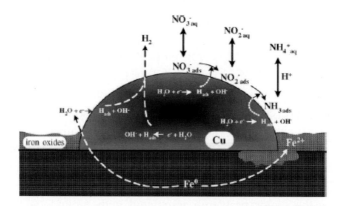

Figure 5. Proposed scheme of the nitrate reduction reaction at Cu/Fe system. From [Ref. 66], Y.H. Liou *et al.* J. Hazardous Materials B, 127, 102, (2005).

Table 2. Removal of pollutants using iron nanoparticles

Type of nanoparticle	Removal target	Initial concn. (mg/L)	Dose of nanoparticle	Contact time (min)	Removal efficiency (%)	pH	Adsorption capacity	Ref.
Iron sulfide nanoparticle[1]	Lindane	5.0	-	480	94	7.0	-	47
Zero-valent powder iron	Arsenic(V)	1.0	1.0 g/L	60	99.9	7.0	-	48
BCF[2]	Arsenate	-	-	-	-	7.0	33.2 (mg/g BCF)	49
BCF[2]	Arsenite	-	-	-	>99.9	7.0	99.6(mg/g BCF)	49
ZVI	Humic acid	20	1.0 g/L	5	>99.9	6	-	54
Iron nanoparticle	Ni (II)	100	5.0 g/L	< 180	>99.9	-	-	55

[1] FP1(polymer from the basidiomycetous fungus, *Itajahia* sp. -stabilized FeS nanoparticles
[2] Fe content of 220 mg/mL

Table 3. Removal of pollutants using bimetallic nanoparticles

Type of nanoparticle	Removal target	Initial concentration	Dose of nanoparticle	Contact time (min)	Removal efficiency (%)	Ref.
Pd/Fe nanoparticle	Trichloroethene	20 mg/L	2 g /100 ml	15	>99.9	59
Pd/Fe nanoparticle	Tetrachloroethene	20 mg/L	5.0 g /L	90	>99.9	60
Ni/Fe nanoparticle	Trichloroethene	23.4 mg/L	0.1g /40ml	120	>90	62
Cu/Fe nanoparticle	Nitrate	40 mg/L	0.5 g of 0.44% bimetallic particles /65 mL	60	100	65

2.4. Nanoparticulate Photocatalysts and Catalysts

Catalysis involves the modification of a chemical reaction rate, mostly speeding up or accelerating the reaction rate by a substance called catalyst that is not consumed throughout the reaction. Usually, the catalyst participates in the reaction by interacting with one or more of the reactants and at the end of process; it is regenerated without any changes.

There are two main kinds of catalysts, homogeneous and heterogeneous. The homogeneous type is dispersed in the same phase as the reactants. The dispersal is ordinarily in a gas or a liquid solution. Heterogeneous catalyst is in a different phase from the reactants and is separated by a phase boundary. Heterogeneous catalytic reactions typically take place on the surface of a solid support, e.g. silica or alumina. These solid materials have very high surface areas that usually arise from their impregnation with acids or coating with catalytically active material e.g. platinum-coated surfaces.

Catalysts usually have two principal roles in nanotechnology areas:

- In macro quantities, they can be involved in some processes for the preparation of a variety of other nanostructures like quantum dots, nanotubes, etc.
- Some nanostructures themselves can serve as catalysts for certain chemical reactions.

The chemical activity of a conventional heterogeneous catalyst is proportional to its overall specific surface area per unit volume, which is customarily reported in the unit of square meters per grams, with typical values for commercial catalysts in the range of 100 to 400 m^2/g. There are different procedures to enhance the surface area of the catalyst, which result in voids or empty spaces within the material. It is quite common for these materials to have pores with diameters in the nanometer range. The pore surface areas are usually determined by the Brunauer-Emmett-Teller (BET) method.

The active component of a heterogeneous catalyst can be a transition ion. Example of some metallic oxides that serve as catalysts, either by themselves or by distribution on a supporting material, are NiO, Cr_2O_3, Fe_2O_3, Fe_3O_4, Co_3O_4. For some reactions, the catalytic activity arises from the presence of acid sites on the surface. These sites correspond to either Bronsted acids, which are proton donors, or Lewis acids, which are electron pair acceptors [14].

Pajonk [69] prepared nanoparticles from the sol-gel chemistry combined with the supercritical drying method (aerogels) to enhance the catalyst properties such as textural and thermal qualities. Rare earth metal oxides modification of automotive catalysts (e.g. CeO_2, ZrO_2) for exhaust gas treatment has resulted in the structural stability, catalytic functions and resistance to sintering at high temperatures [70]. Owing to the low redox potential of non-stochiometrics CeO_2, oxygen could release with conversion between 3^+ and 4^+ oxidation states of the Ce ions. This is shown to be essential for effective catalytic functions under the dynamic air-to-fuel ratio cycling [70, 71] (See Figure 6).

Nanocrystal surfaces usually are coated with capping ligands. These ligands manage the surface character (both the chemistry and physical structures) which expressed greatly the photocatalytic properties, such as fluorescence lifetimes, quantum yields, surface charge and particle solubility [72].

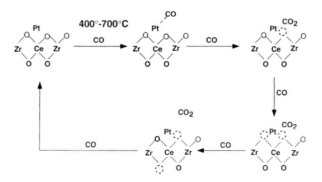

Figure 6. A schematic model for oxidation of CO by a $CeO_2.ZrO_2$ /Pt catalyst promoted by metal–oxide support interaction. The dashed circles represent oxygen vacancies. A key step is the release of oxygen atoms in conjunction with the $Ce^{4+} \rightarrow Ce^{3+}$ conversion. Oxygen vacancies are generated initially at the ceria (cerium(IV) oxide)/Pt interface and subsequently migrate into the interior of the lattice. From [Ref. 70], C.K. C.K. Loong, and M. Ozawa. J. Alloys and Compounds, 303, 60, (2000).

The metal complex porphyrins and other biological porphyrin-type molecules like vitamin B_{12}, which are referred to as metalloporphyrinogens, have several characteristics that make them very applicable for the treatment of persistent organic pollutants [73]. They act as redox catalysts for many reactions and known to be active over a long range of redox potentials, electrochemically active with almost any metal, function well in aqueous solutions in the groundwater environment, and highly stable, which enables reactions under severe conditions that prevent other treatment methods (e.g., bioremediation).

Metalloporphyrinogens are molecules with nanometer size and known to catalyze the decomposition of COC (chloro-organic compounds) by reduction reactions. Dror *et al.* [73] applied these catalysts immobilized in sol-gel matrix, for the reduction of COC. They performed experiments under conditions suitable for ground water systems with titanium citrate and zero-valent iron as electron donors. All the chloro-organic compounds used in these experiments were reduced in the presence of several sol-gel-metalloporphyrinogen hybrids (heterogeneous catalysts).

Wang *et al.* [74] investigated the effect of size, fabrication method, and morphology of ZnO nanoparticles as photocatalysts on the decomposition of methyl orange. They have used ZnO nanoparticles with different diameters of 10, 50, 200, and 1000 nm. ZnO particles were prepared by two methods of chemical deposition and thermal evaporation. It was found that the preparation method was the most important step and ZnO-nanoparticle, with 50 nm diameter synthesized via thermal evaporation method, provided the highest photocatalyst activity [74].

Huang *et al.* [75] used $ZnWO_4$ nanoparticle as photocatalyst for the degradation of rhodamine B in water and decomposition of formaldehyde in gas phase. The sample had the highest photocatalytic activity when prepared at 450°C for 1h. The temperature and time of annealing was observed to be effective for photoatalytic activity. In addition, Lin *et al.* [76] prepared $ZnWO_4$ nanoparticles and nanorods and used them as photocatalysts for the photodegradation of Rhodamine B and gaseous formaldehyde.

2.5. Nanoclays

Clays are layered minerals with space in between the layers where they can adsorb positive and negative ions and water molecules. Clays undergo exchange interactions of adsorbed ions with the outside too. Although clays are very useful for many applications, they have one main disadvantage i.e. lack of permanent porosity. To overcome this problem, researchers have been looking for a way to prop and support the clay layers with molecular pillars. Most of the clays can swell and thus increase the space in between their layers to accommodate the adsorbed water and ionic species. These clays were employed in the pillaring process.

As stated previously, ultra-fine TiO_2 powders have large specific surface areas, but due to their easy agglomeration, an adverse effect on their catalytic performance has been observed [77]. Ding *et al.* [77] experienced that the recovery of pure TiO_2 powders from water was very hard when they used them in aqueous systems. They dispersed TiO_2 particles in layered clays and it appeared to provide a feasible solution to such problems. The composite structures, known as pillared clay, could stabilize TiO_2 particles and give access of different molecules to the surface of TiO_2 crystals. In addition, the interlayer surface of pillared clays is generally hydrophobic, and this is an advantage in adsorption and enriching diluted hydrophobic organic compound in water.

Ooka *et al.* [78] prepared four kinds of TiO_2 pillared clays from different raw clays such as montmorillonite, saponite, fluorine hectorite and fluorine mica. They have tested the surface hydrophobicities and performances of clays in adsorption- photocatalytic decomposition of phthalate esters. It was found that surface hydrophobicity of pillared clays (especially TiO_2) largely varied with the host clay. Since the TiO_2 particles in the pillared clays are too small to form a crystal phase, they presented a poor photocatalytic activity. To overcome this problem, nanocomposite of titanium dioxide (TiO_2) and silicate nanoparticles were made by reaction between titanium hydrate sol of strong acidity and smectite clays in the presence of polyethylene oxide (PEO) surfactants [79]. It resulted in forming larger precursors of TiO_2 nanoparticles and condensing them on the fragmentized pieces of the silicate. Introducing

PEO surfactants into the synthesis process significantly enhanced the porosity and surface area of the composite solid.

In other works, nanocomposite of iron oxide and silicate was also synthesized for degradation of azo-dye orange (II) [80]. To improve the sorption capacity, clays were modified in different ways, such as treatment by inorganic and organic compounds, acids and bases. Organoclays have recently attracted lots of attention in a number of applications, such as dithiocarbamate-anchored polymer/organosmectite for the removal of heavy metal ions from aqueous media [81] (See Table 4).

A new class of nano-sized large porous titanium silicate (ETAS-10) and aluminum-substituted ETAS-10 with different Al_2O_3/TiO_2 ratios were successfully synthesized and applied to the removal of heavy metals, in particular Pb^{2+} and Cd^{2+} (See Table 4). Since tetra-valent Ti is coordinated by octahedral structure, it creates two negative charges that must be normally balanced by two mono-valent cations. This leads to a great interest in ion exchange or adsorption property of this material [82].

Wang and Wang [83] prepared a series of biopolymer chitosan/ montmorillonite (CTS/MMT) nanocomposites and used them as sorbents for the adsorption of Congo Red. They investigated the effect of pH and temperature and found that the sorption capacity was increased with increasing the CTS to MMT ratio.

Table 4. Removal of pollutants using nanoclays

Type of nanoparticle	Removal target	Adsorption capacity (q_m)	Ref.
Dithiocarbamate-anchored nanocomposite	Pb(II)	170.70 mg/g	81
Dithiocarbamate-anchored nanocomposite	Cd(II)	82.20 mg/g	81
Dithiocarbamate-anchored nanocomposite	Cr(III)	71.10 mg/g	81
ETAS-10 (A)[1]	Pb(II)	1.75 mmol/g	82
ETAS-10(A)	Cd(II)	1.24 mmol/g	82
ETAS-10(B)[2]	Pb(II)	1.68 mmol/g	82
ETAS-10(B)	Cd(II)	1.12 mmol/g	82
CTS/MMT nanocomposite	Congo Red	54.52 mg/g	83

[1] ETAS-10 (A). (Al_2O_3/TiO_2=0.1), T=25°C

[2] ETAS-10(B): (Al_2O_3/TiO_2=0.2), T=25°C

2.6. Nanotubes

The discovery of fullerenes and carbon nanotubes has opened a new chapter in carbon chemistry. Superconducting and magnetic fullerides, atoms trapped inside the fullerene cage, chemically bonded fullerene complexes, and nanometer-scale helical carbon nanotubes are some of the leading areas that have generated much excitement. The creation of the hollow carbon buckminsterfullerene molecule as well as methods to produce and purify bulk quantities of it has triggered an explosive growth of research in the field [84-89].

Carbon nanotubes, in particular, hold tremendous potential for applications because of their unique properties, such as high thermal and electrical conductivities, high strength, high stiffness, and special adsorption properties [90] (See Figures 7 and 8).

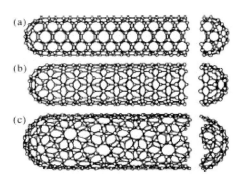

Figure 7. Some SWNTs (single-walled carbon nanotubes) with different chiralities. The difference in structure is easily shown at the open end of the tubes. a) armchair structure b) zigzag structure c) chiral structure. From [Ref. 91], students.chem.tue.nl/ifp03/default.htm.

Carbon nanotubes have cylindrical pores and adsorbent molecules interact with their carbon atoms on the surrounding walls. This interaction between molecules and solid surface depends on the pore size and geometry of pores. When a molecule is placed in between two flat surfaces, i.e., in a slit-shaped pore, it interacts with both surfaces, and the potentials on the two surfaces overlap. The extent of the overlap depends on the pore size. However, for cylindrical and spherical pores, the potentials are greater because more surface atoms interact with the adsorbed molecule [90]. In addition, carbon nanotubes are highly graphitic (much more than the activated carbons). Hence, the carbon nanotubes can adsorb molecules much stronger than activated carbons, which have slit-shaped or wedge-shaped pores [90].

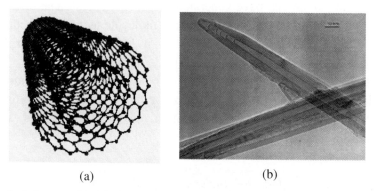

(a)   (b)

Figure 8. a) Structure Model of Multiwall Carbon Nanotube (2-layer) and b) TEM Image of MWNTs. From [Ref. 92], http://www.noritake-elec.com/itron/english/nano.

Carbon nanotubes (CNTs) show adsorption capability for removal of heavy metals such as lead [93, 94]. The results were shown that the as-grown CNTs have week affinity toward Lead. The adsorption capacity of CNTs was improved by oxidization with oxidized acid (HNO_3), since the acid can introduce many functional group such as hydroxyl (-OH), carboxyl (-COOH), and carbonyl (>C=O) on the surface of CNTs. (See Table 5) [93]. Li *et al.* [95] oxidized carbon nanotube (CNTs) with H_2O_2, $KMnO_4$, and HNO_3, and found that cadmium (II) adsorption capacities enhanced for three types of oxidized CNTs, due to the functional groups introduced by oxidation compared with the as-grown CNTs (See Table 5).

Lu and Chiu. [96] purified commercial single-walled carbon nanotubes (SWCNTs) and multi-walled carbon nanotubes (MWCNTs) by sodium hypochlorite solutions and used them as adsorbent for the removal of zinc from water. Likewise, fluoride is one of the pollutants in the drinking water and it has been adsorbed from water by amorphous Al_2O_3 supported on carbon nanotubes (Al_2O_3/CNTs) [97]. Also, aligned carbon nanotubes (ACNTs), a new kind of carbon material, were prepared by catalytic degradation of xylene via ferrocene as catalyst and used for the adsorption of fluoride from drinking water [98].

Carbon nanotube shows the adsorption capability for the removal of organic pollutants like 1, 2-dichlorobenzene, trihalomethanes, n-nonane, and CCl_4 with different modification and purification from water [99-101]. Agnihotri *et al.* [102] used gravimetric techniques to determine the adsorption capacities of commercially carbon nanotubes for organic compounds (toluene, methyl-ethyl-ketone, hexane and cyclo-hexane).

Stafiej and Pyrzynska [103] prepared purified carbon nanotubes via soaking them in HNO_3 for 12 h at room temperature and then washed them with deionized water until natural pH. They used the treated CNTs for the removal of heavy metals such as Cu, Co, Cd, Zn, Mn, and Pb. It was found that the affinity of heavy metals toward CNTs at pH of 9 were in the order of Cu (II) > Pb (II) > Co (II) > Zn (II) > Mn (II). Wang *et al.* [104] employed pristine MWCNTs, acidified MWCNTs (with different durations of soaking in nitric acid solution), and annealed MWCNTs for the removal of Pb (II). The results indicated that the maximum adsorption capacity of acidified MWCNTs and pridtine MWCNTs for Pb(II) were 91 and 7.2 mg/g, respectively. In addition, Wang *et al.* [105] prepared manganese oxide-coated carbon nanotubes (MnO_2/CNTs) as an adsorbent for the removal of lead (II) from aqueous solution. They found that the adsorption capacity of Pb (II) was 78.74 mg/g from the Langmuir isotherm model. Xu *et al.* [106] applied oxidized (MWCNTs) as adsorbent to remove Pb (II) from aqueous solution. Similar studies done by Kandah and Meunier [107] revealed that the oxidized MWCNTs have higher affinity for Ni (II) removal than the non-oxidized sample (See Table 5). In addition, Lu and Su [108] used thermally treated MWCNTs as sorbent for the adsorption of NOM (natural organic matters) from aqueous solution.

2.7. Dendrimer and Nanosponges

Another example of environmental treatment and remediation-related application of nanomaterials includes dendritic nanoscale chelating agents for polymer-supported ultrafiltration (PSUF). Dendrimers are

Table 5. Removal of pollutants using carbon nanotubes (CNTs)

Type of nanoparticle	Removal target	Adsorption capacity (mg/g)	pH	Ref.
CNTs[1]	Pb(II)	17.44	5	93
CNTs[2]	Cd(II)	11	5.5	95
SWCNTs	Zn(II)	43.66	-	96
MWCNTs	Zn(II)	32.68	-	96
ACNTs	Fluoride	4.5	7.0	98
MWCNTs[3]	Ni(II)	18.08	6	107
MWCNTs[4]	Ni(II)	49.26	6	107

[1] Acid-refluxed CNTs
[2] KMnO₄ oxidized
[3] As-produced CNTs
[4] Oxidized CNTs

highly branched polymers with controlled composition and an architecture that consists of nanoscale features. In other words, dendrimers or cascade molecules have branching construction similar to a tree, in which one trunk forms several large branches; each forming smaller branches, and so on. The roots of the tree also have the same branching mode of growth. This kind of architecture characterized by fractal geometry in which dimensions are not just integers such as 2 or 3, but also fractions (See Figure 9).

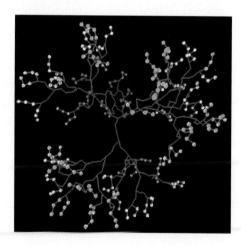

Figure 9. Schematic representation of a dendrimer. From [Ref. 109], rati.pse.umass.edu/usim/gallery.html.

These nanostructures can be designed to encapsulate metal ions and zero-valent metals, enabling them to dissolve in suitable media or bind to appropriate surfaces.

Modification of any components in these branched polymeric structures give variety of means for controlling critical macromolecular parameters such as internal and external rigidity, hydrophilicity and hydrophobicity, degrees of void, excluded volumes, and response to stimuli such as changes in solvent polarity and temperature [110].

The ability of synthesize water soluble dendrimers with metal ion chelating functional groups and also surface groups with weak binding affinity, provided opportunities for developing the treatment efficiency [110]. Some dendrimers can trap molecules such as radicals, charged moieties (part of molecules) and dyes. When molecules with different sizes are trapped inside the dendrimer, they can be selectively released by gradual hydrolysis (reaction with water) of the outer and middle layers.

Poly amidoamine (PAMAM) dendrimers are a new class of nanoscale materials that can be carried as water-soluble chelators. Usually, PAMAM macromolecules are synthesized by repeatedly attaching amidoamine monomers in their radial branched layers, termed "generations", to a starting ammonia core [111].

The environmental applications of dendrimers were first explored by Diallo *et al.* [110]. They have reported the removal of copper from water via different generations of PAMAM dendrimers. Later, Diallo *et al.* [111] studied the feasibility of using dendrimer-improved ultrafiltration to recover Cu(II) from aqueous solution. The dendrimer-Cu(II) complexes can be efficiently separated from aqueous solutions by ultrafiltration. The metal ion laden dendrimers can be regenerated by decreasing the solution pH to 4.0, thus enabling the recovery of the bound Cu (II) ions and recycling of the dendrimers

The soil treatment of PAMAM dendrimers was also tested. Different generation and terminal functional groups for removal of copper (II) and lead from a sandy soil were investigated [112, 113].

Rether and Schuster [114] made a water-soluble benzoylthiourea modified ethylenediamine core-polyamidoamine dendrimer for the selective removal and enrichment of toxicologically relevant heavy metal ions. They studied complexation of Co (II), Cu (II), Hg (II), Ni (II), Pb (II) and Zn (II) by the dendrimer ligand and using the polymer-supported ultrafiltration process. The interactions of the different heavy metal ions with the dendritic ligands were determined by measuring the metal ion retention, which was dependent upon the pH of the solution. The results indicate that all metal ions can be retained almost quantitatively at pH = 9. Cu (II) as well as Hg (II) formed the most stable complexes with the benzoylthiourea modified PAMAM derivatives and can be separated selectively from the other investigated heavy metal ions. The bound metal ions can typically be recovered by decreasing the pH of the solution.

Diaminobutane poly (propylene imine) dendrimers functionalized with long aliphatic chains were employed to remove organic impurities such as polycyclic aromatic hydrocarbons from water and produce ultra pure water. These types of dendrimers are completely insoluble in water. The encapsulating properties of these new dendrimeric derivatives for lipophilic molecules should not be hindered by the introduction of the alkyl chains [115].

One of the novel systems for encapsulating organic pollutants is cross linked dendritic derivatives. In the research carried out by Arkas *et al.* [116] for the preparation of ultra pure water, the amino groups of poly propyleneimine dendrimer and hyper branched polyethylene imine were interacted under extremely mild conditions with 3-(triethoxysilyl) propyl isocyanate. They produced porous ceramic filters and employed these dendritic systems for water purification. In this experimental work, the concentration of polycyclic aromatic compounds in water was reduced to few ppb's by continuous filtration of contaminated water through these filters. Then, the filters loaded with pollutants were effectively regenerated by treatment with acetonitrile.

In another work, Arkas *et al.* [117] developed a method that permits removal of organic pollutants with employing a simple filtration step, which can be easily scaled-up. They used the long-alkyl chain functionalized polypropylene imine dendrimers, polyethylene imine hyper branched polymers and β-cyclodextrin derivatives which are completely insoluble in water.

2.8. Self-Assemblies (In General)

Self-assembly is defined as a reversible process in which pre-existing parts, or disordered components of a pre-existing system, form structures of patterns. In other words, a self-assembly process is the spontaneous organization of small molecules into larger well-defined stable, ordered molecular complexes or aggregates, and spontaneous adsorption of atoms or molecules onto a substrate in a systematic ordered manner [14]. The most well-studied subfield of self-assembly is molecular self-assembly, but in recent years it has been demonstrated that self-assembly is possible with micro and millimeter scale structures lying in the interface between two liquids like micelles.

Molecular self-assembly is gathering of molecules without guidance or management from an outside source. There are two types of self-assembly i.e. intramolecular and intermolecular, although the term self-assembly itself usually refers to intermolecular one. Intramolecular self-assembling molecules are complex polymers with the ability to assemble from the random coil conformation into a well-defined stable structure (secondary and tertiary structure). An example of this type of self-assembly is protein folding. Intermolecular self-assembly is the ability of molecules to form supramolecular assemblies (quaternary structure). A simple example is the formation of a micelle by surfactant molecules in solution.

Attaching a monolayer of molecules to mesoporous ceramic supports gives materials known as Self-Assembled Monolayers on Mesoporous Supports (SAMMS). The highly ordered nanostructure of SAMMS is the result of three molecular-self-assembly stages. The first stage is the aggregation of the surfactant molecules to make the micelle template. The second generation is the aggregation of the silicate-coated micelles into the mesostructured body, and the third is the self assembly of the silane molecules into an ordered monolayer structure across the pore interface (see Figure 10).

The resulted functionalized hexagonal structure is a base to build an environmental sorbent material [118]. The rigid, open pore structure of the supports makes all of the interfacial binding sites accessible to

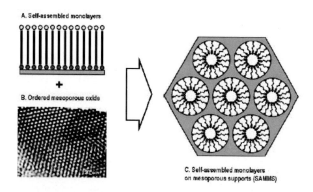

Figure 10. Schematic representation of SAMMS (Self-Assembled Monolayers on Mesoporous Supports). From [Ref. 121], G.E. Fryxell *et al.* Environ. Sci. Technol, 39, 1324 (2005).

solution species, and results in fast sorption kinetics. All the above advantages of SAMMS make it ready for various chemical bindings.

SAMMS of silica-based materials are highly efficient sorbents for target species, such as heavy metals, tetrahedral oxometalate anions, and radionuclides. The self-assembly technique is also used for preparation of catalysts in the form of thin films on a support material. Some of these thin film catalysts have great uniformities and high photocatalytic activities, but their thicknesses can hardly be controlled. Thus, preparing them in large areas is rather difficult.

A suitable technique of preparing ultra-thin films with precise thickness adjustment is the layer-by-layer self-assembly procedure [119]. The principle of multilayer assembly is quite simple; colloidal particles will self-assemble on the surface of a suitable solid substrate because of its surface forces. In most cases, electrostatic interaction provides the stability for the films.

Szabó *et al.* [119] provided a new process for synthesizing $Zn(OH)_2$ and ZnO nanoparticles. For this purpose, multilayer films of $Zn(OH)_2$ and ZnO nanoparticles were prepared by the layer-by-layer self-assembly technique on glass surface (See Figure 11). Photocatalytic measurements were made with model organic materials β-naphtol and industrial kerosene in a loop-type batch reactor.

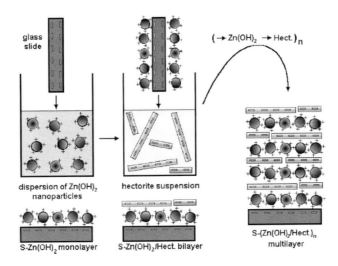

Figure 11. Side view schematics depicting of the self-assembly preparation procedure for the S–(Zn(OH)$_2$/Hect)$_{10}$ multilayer films. From [Ref. 122], T. Szabó *et al.* Colloids and Surfaces A: Physicochem. Eng. Aspects, 230, 23, (2004).

2.9. Micelles (Self-Assembled Surfactants)

Micelles are self-assembled surfactant materials in a bulk solution. Surfactants or "Surface active agents" are usually organic compounds that are amphipathic, meaning they contain both hydrophobic groups (tails) and hydrophilic groups (heads). Therefore, they are typically soluble in both organic solvents and water.

There are hundreds of compounds that can be used as surfactants and are usually classified by their ionic behavior in solutions; anionic, cationic, non-ionic or amphoteric (zwiterionic). Each surfactant class has its own specific properties. A surfactant can be classified by the presence of formally charged groups in its head. There are no charge groups in a head of nonionic surfactant. The head of an ionic surfactant carries a net charge. If the charge is negative, the surfactant is more specifically called anionic and if the charge is positive, it is called cationic. If a surfactant contains a head with two oppositely charged groups, it is termed zwitterionic.

The concentration at which surfactants begin to self-assemble and form micelles is known as critical micelle concentration or CMC. When micelles appear in the water, their tails form a core that is like an oil droplet, and their (ionic) heads form an outer shell that maintains favorable contact with water. The self-assembled surfactant is referred to as "reverse micelle" when surfactants assemble in the oil. In this case, the heads are in the core and the tails have favorable contact with oil [14]. Pacheco-Sanchez and Mansoori [120] and Priyanto *et al.* [121] focused on behavior of asphaltene micelles nano-structures that might be formed to serve as elements of nano-materials.

Surfactant-enhanced remediation techniques have shown significant potential in their application for the removal of polycyclic aromatic hydrocarbon (PAHs) pollutants in the soil. Increasing surfactant concentration in the solution has shown higher effectiveness in the extraction of NAPLs (non-aqueous phase liquids) and PAHs. At high concentrations, surfactant solutions improve the formation of pollutant emulsions that are hard to extract from the sample [122].

On the other hand, surfactant solutions with low concentrations are not very effective in solubilizing the pollutants. As a result, recent research has been directed towards the design of a surfactant that minimizes their losses and the development of surfactant recovery and recycling techniques [122-124]. To overcome these problems, Kim *et al.* [122] tested amphiphilic polyurethane (APU) nano-network polymer particles. They examined the APU efficiency to remove a model hydrophobic pollutant (phenantrene) from a contaminated sandy aquifer material. One of the advantages of the APU particle emulsion is the wide range of concentration that can be used in soil remediation. APU nano-network suspensions extracted up to 98% of the phenanthrene adsorbed on the aquifer material with extremely low loss of particles [124].

2.10. Magnetic Nanoparticles

When a material is placed within a magnetic field, the magnetic forces of the material's electrons will be affected. However, materials can react quite differently to the presence of an external magnetic field. Their reaction is dependent on a number of factors, such as the atomic and molecular structure of the material, and the net magnetic field associated with the atoms. The magnetic moments associated with atoms have three origins. These are the electron orbital motion, the change in orbital motion caused by an external magnetic field, and the spin of the electrons. In most atoms, electrons occur in pairs and they spin in opposite directions. So, their opposite spins cause their magnetic fields to cancel each other. Therefore, no net magnetic field exists. Alternately, materials with some unpaired electrons will have a net magnetic field and will react more to an external field. Most materials can be classified as diamagnetic, paramagnetic or ferromagnetic. Diamagnetic metals have a very weak and negative susceptibility to magnetic fields, while paramagnetic metals have small positive susceptibility to magnetic fields.

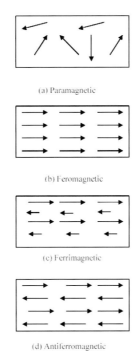

(a) Paramagnetic

(b) Feromagnetic

(c) Ferrimagnetic

(d) Antiferromagnetic

Figure 12. Illustration of various arrangements of individual atomic magnetic moments that constitute paramagnetic (a), ferromagnetic (b), ferromagnetic (c), and antiferromagnetic (d), material. From [Ref. 14] C.P. Poole Jr., and F.J. Owens, Introduction to nanotechnology, 2003, John Wiley & Sons Inc.

Ferromagnetic materials exhibit a strong attraction to magnetic fields and are able to retain their magnetic properties after the external field has been removed. These materials get their magnetic properties due to the presence of magnetic domains. In these domains, large numbers of atom's moments (10^{12} to 10^{15}) are aligned parallel so that the magnetic force within the domain is strong. Some transition ion atoms such as iron, manganese, nickel, and cobalt are examples of ferromagnetic materials (See Figure 12).

Depending on the size and subsequent change in magnetic property, the magnetic nanoparticles are used in different applications. Since the relaxation time of magnetic nanoparticles can be changed by changing the size of the nanoparticles or using different kinds of materials, magnetic nanoparticles have been a very useful tool in different kind of applications, from biomedical to data storage systems.

One of the major applications of magnetic particles is in the area of magnetic separation. In this case, it is possible to separate a specific substance from a mixture of different other substances. The separation time is one of the important parameters in the magnetic separation method. Separations using magnetic gradients, such as "High Magnetic Gradient Separation" (HGMS), are now widely used in the fields of medicine, diagnostics and catalysis to name a few. In HGMS, a liquid phase containing magnetic particles is passed though a matrix of wires that are magnetized by applying a magnetic field [125]. The particles are held onto the wires and at the conditions that the field is cut off, they can be released. If these particles are used in order to be fixed to specific molecules, the latter can be isolated from waste water or slurries. For such applications, the materials can be recycled and does not generate secondary waste. These processes sometime called "magnetically assisted chemical separation (MACS)".

In MACS processes, particles are typically micrometric and are made of magnetite nanoparticles embedded into a polymer microsphere with a diameter ranging between 0.1 and 25 µm. However, the magnetic particles have the disadvantage of small adsorption capacity and slow adsorption rates due to their small surface area or their porous properties [125].

Application of nanoparticles, with diameters ranging between 4 and 15 nm, either dispersed in an extracting solvent, and/or specifically coated by complex species has also been described for MACS [125]. The magnetic component of the micro and nanoparticles employed for MACS process is typically magnetite (Fe_3O_4) or its products of oxidation γ-Fe_2O_3.

Various studies by different research groups have been employed for treating contaminated water by magnetic nanoparticles some of which will be discussed here.

Takafuj *et al.* [126] prepared polymer poly (1-vinylimidazole)-grafted nanosized magnetic particles as an organic-inorganic hybrid magnetic materials for expanding the sorbent-based separation technology to a multiphase complex system (See Figure 13).

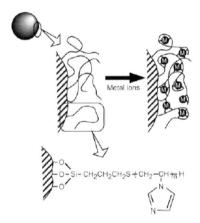

Figure 13. Schematic illustration of polymer-grafted magnetic particles. From [Ref. 126], Takafuj *et al.* Environ. Sci. Technol, Langmuir, 21, 11173 (2005).

It is well known that Cr(VI) is toxic to animals and plants, while Cr(III) is considered to be less harmful. Hu *et al.* [127] developed an innovative process combining nanoparticle adsorption and magnetic separation for the removal and recovery of Cr (VI) from wastewater. They produced ten nanometer modified $MnFe_2O_4$ nanoparticles as a new adsorbent using a co-precipitation way followed by a surface redox reaction. The results exhibited that surface-modified $MnFe_2O_4$ nanoparticles were efficient adsorbents for the rapid removal of Cr (VI) from aqueous solutions.

In another study, the nanoscale maghemite was synthesized, characterized, and evaluated as adsorbent of Cr (VI) [128]. It is found that some coexisting ions, such as Na^+, Ca^{2+}, Mg^{2+}, Cu^{2+}, Ni^{2+}, NO_3^-, and Cl^- had no significant effect on the process which illustrated the selective adsorption of Cr(VI) from wastewater.

Magnetic nano-carriers can be easily manipulated by an external magnetic field and therefore should be appropriate as the support for adsorbents. Chang *et al.* [129] prepared the magnetic chitosan nanoparticles with diameter of 13.5 nm as a magnetic nano-adsorbent. They have done this by the carboxymethylation of chitosan and followed with binding on the surface of Fe_3O_4 nanoparticles via carbodiimide activation. Magnetic chitosan nano-adsorbent was shown to be quite efficient for the fast removal of Co (II) ions at the pH range of 3–7 and the temperature range of 20–45°C.

Ngomsik *et al.* [130] have studied the removal of nickel ions from the aqueous solution using magnetic alginate microcapsules. They have found that the sorption capacity for nickel removal were increased by increasing the pH of the solution. Also, magnetic particles in the microcapsules allowed easy isolation of the microcapsule beads from aqueous solutions after the sorption process.

Different kinds of magnetic nanoparticles were also employed for the removal of organic pollutants, such as sorption of methylene blue on polyacryclic acid-bound iron oxide from an aqueous solution [131]. In this work, novel magnetic nanoparticle made up of iron oxide nanoparticles as cores which bounded by polyacryclic acid as ion exchange groups were applied for the removal of basic dye (methylene) blue. The results indicated that, these magnetic nanoparticles are efficient for the separation of bromelain [132] (See Table 6).

Cumbal and Sengupta [133] prepared a new class of hybrid (dual-zone) magnetic sorbents as shown in Figure 14 with the characteristics of magnetically active, selective for inorganic and organic environmental contaminants and involving efficient regeneration and reuse. Experimental results showed that the imparted magnetic activity, in terms of magnetic susceptibility within polymer beads, was dependent on the chemical nature of the functional group [133].

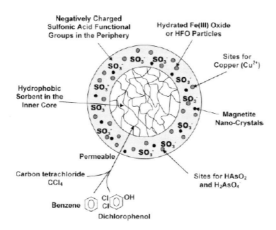

Figure 14. Illustration of a dual-zone magnetic sorbent allowing sorption of a wide array of target contaminants. From [Ref. 131], L.H. Cumbal and A.K. Sengupta, Ind. Eng. Chem. *Res*, 44, 600 (2005).

Recently, Hu *et al.* [134] synthesized several kinds of magnetic nanoparticles listed below:

$CoFe_2O_4$, $CuFe_2O_4$, $MgFe_2O_4$, $MnFe_2O_4$, $NiFe_2O_4$, $ZnFe_2O_4$.

They compared their performances in the removal of Cr (VI). They investigated many parameters such as contact time, pH, shaking rate, and magnetic properties. The results indicated their adsorption capacities were in the following order:

$MnFe_2O_4 > MgFe_2O_4 > ZnFe_2O_4 > CuFe_2O_4 > NiFe_2O_4 > CoFe_2O_4$.

Mayo *et al.* [135] also studied the effect of particle sizes in the adsorption and desorption of AS(III) and AS(VI). They found that as the particle size decreases from 300 to 12 nm, the adsorption capacity increases nearly 200 times.

Banerjee and Chen [136] studied removal of copper ions from aqueous solution with modified magnetic nanoparticles. They treated Fe_3O_4 with gum Arabic. Gum was attached to Fe_3O_4 via the interaction between carboxylic groups of gum Arabic and the surface hydroxyl groups of Fe_3O_4. The maximum capacity obtained were 17.6 and 38.5 mg/g for MNP and GA-MNP (gum Arabic- magnetic nanoparticle), respectively.

2.11. Nanomembrane and Nanosieve

A membrane is a semi-permeable and selective barrier between two phases (retentive and permeate) through which only selected chemical species may diffuse. Membrane filtration is frequently employed for the separation of dissolved solutes in a fluid or the separation of a gas mixture [137].

Historically, membrane technology has had wide application in wastewater treatment and desalination via reverse osmosis. In this method, a pressure difference across a membrane is employed to overcome the osmotic pressure gradient. The smaller water molecules are literally pushed through the membrane while the large solute species are retained behind [138].

Among different classes of membranes, reverse osmosis (RO) filtration is a well known process in the desalination of seawater and ultrafiltration (UF) is a well established process in the fractionation of Natural Organic Matter (NOM). Nanofiltration (NF) is a process with membrane permeability between RO and UF. Another membrane design is emulsion liquid membrane (ELM). An ELM is formed by first encapsulating an aqueous "receiving" or strip phase within a hydrophobic membrane liquid. This emulsion is then further dispersed within the continuous aqueous feed phase. This technology was used for the extraction of phenols, removal of heavy metal cations such as zinc, cadmium, chromium, copper, lead, palladium and mercury from wastewater and also removal of alkali metal cations such as Na^+, K^+, Li^+ and Cs^+, radioactive fission products, such as Cs-137, Sr-90, Ce-139 and Eu-152 and anions, such as chlorides, sulfate, phosphate and chromate [138].

The efficiency of nitrate removal by three commercial nanofiltration membranes, NF90, NF270 (Dow-FilmTec) and ESNA1-LF (Hydranautics) was studied by Santafe-Moros *et al.* [139]. The results

Table 6. Removal of pollutants using magnetic nanoparticles

Type of nanoparticle	Removal target	Initial concentration	Removal efficiency (%)	Adsorption capacity	Dose of nanoparticle	pH	Ref.
Modified jacobsite (MnFe$_2$O$_4$)	Cr (VI)	-	-	31.55 mg/g	-	2.0	127
Maghemite nanoparticle	Cr (VI)	50 mg/L	97.3	-	5 g/L	2.5	128
Magnetic chitosan nanoparticle	Co (II)	-	-	27.50 mg/g		5.5	129
Magnetic alginate microcapsules	Ni (II)	-	-	0.52 mmol/g	-	5.3	130
PAA-bound iron oxide	Methylene blue	-	-	0.199 mg/mg	-	9.0	131
PAA-bound iron oxide	Bromelain	-	-	0.476 mg/mg	-	4.0	132

indicated that NF270 gave very high flux but very low nitrate rejection, even at the lowest concentrations of the feed.

Natural organic matters (NOM) such as humic acid and fulvic acid are widely distributed in soil, natural water and sediments. These materials include mixture of the degradation products of plant and animal residues [140]. Membrane TiO_2-modified photocatalytic oxidation (PCO) has been focused on the NOM removal and/or destruction processes.

Fu *et al.* [140] designed a submerged membrane photocatalysis reactor (SMPR) for degradation of fulvic acid via novel nano-structured TiO_2/silica gel photocatalyst. The possibility of using novel TiO_2 for the prevention of microfiltration membrane fouling for water purification was tested as well.

Ultrafiltration has been applied in most membrane separation processes. The hydrophilicity of the membrane and its porous structure play important roles in these processes. An appropriate porous membrane must have high permeability, good hydrophilicity and excellent chemical resistance to the feed streams. In order to obtain high permeability, membranes should have high surface porosity, and good pore structure. Polyvinylidene fluoride (PVDF) is a material that can form such asymmetric membranes, since it is thermally stable and resistant to corrosion by most chemicals and organic compounds. PVDF-based membranes exhibit outstanding anti-oxidation activities, strong thermal and hydrolytic stabilities and good mechanical properties [141]. Yan *et al.* [141] studied the modification of polyvinylidene fluoride (PVDF) ultrafiltration membrane by dispersing nano-sized alumina (Al_2O_3) particles uniformly in a PVDF solution (19% polymer weight).

Using selective polymeric membranes for gas separation is also a promising process. Efficient gas separation membranes are required to have both high permeability and selectivity. Compared with flat membranes, hollow fiber membranes are more favored due to a larger membrane area per volume, good flexibility and easy handling in the module fabrication [142]. Mixed matrix composite membranes that are fabricated by encapsulating the molecular sieves into the polymer matrix have also been recognized. The hollow fibers are developed with a thin zeolite beta-polysulfone mixed matrix selective layer and improved selectivity for He/N_2 and O_2/N_2 separation [143].

3. Conclusions

With developing different aspects of nanotechnology, the broader environmental impacts of that will also need to be considered. Such considerations might include models to determine potential benefits of reduction or prevention of pollutants from industrial sources.

Nanoscience technology holds great potential for the continued improvement of technologies regarding environmental protection. The present review has given further evidence to this issue and it has tried to address what all the potential environmental impacts of the technology might be. For a quick review, the summary of practical aspects of nanotechnology applications for efficient removal of pollutants in the environment is briefly presented in Table 7.

Acknowledgements

The authors would like to thank Professor Christophe Darnault, Professor Amid Khodadoust, Dr. Shawn Niaki, Professor Krishna Reddy and the Fall Semester 2007 students of the course on "Atomic and molecular nanotechnology" at UIC for their comments and inputs to this review. In preparation of this report we have made every attempt to review all the relevant referred publications on environmental applications of nanotechnology. However, the limitation of our resources and the sheer number of publications in this field may have prevented the inclusion of all such publications in this review. Our sincere apologies are extended to any and all authors whose works are not included in this report.

Table 7. Summary of environmental treatments using nanoparticles

Type of nanoparticle	Type of treatment	Removal target	Advantage	Disadvantage
Nanoparticles based TiO$_2$	Photocatalyst oxidation	Organic pollutants	Non toxicity, Water insolubility under most conditions, photo-stability	High operation cost, Hard to recovery, sludge generation
Nanoparticles based iron	Reduction, adsorption	Heavy metals, anions, organic pollutants (dechlorination)	In situ remediation, soil and water treatment, Low cost, safe to handle	Hard to recovery, sludge generation, cost for sludge disposal, Health risk
Nanoparticles based Bimetalic	Reduction, adsorption	Dechlorination, denitrification	Higher reactivity than the iron nanoparticle	Hard to recovery, sludge generation,
Nanoclay	Adsorption	Heavy metals, organic pollutants	Low cost, Unique structures, Long-term stability, reuse, High sorption capacity, Easy recovery, large surface and pore volume	sludge generation
Nanotube and fullerene	Adsorption	Heavy metals, anions, organic pollutants	Treatment of pollution from air and water, exceptional mechanical properties, unique electrical properties, Highly chemical stability	High capital cost, low adsorption capacity, Hard to recovery, sludge generation, Health risk
Dendrimers	Encapsulation	Heavy metals, organic pollutants	Simple separation, renewable, large binding capacity, cost-effective, no sludge generation, reduce pollutant to the level of a few ppb, Treatment of pollution from soil and water	Costly
Micelles	Adsorption	Organic pollutants from soil	In situ treatment, high affinity for hydrophobic organic pollutant	Costly
Metal-sorbing vesicles	Adsorption	Heavy metals	Re-use, high selective uptake profile, high metal affinity	
Magnetite nanoparticles	Adsorption	Heavy metals, organic pollutants	Simple separation, no sludge generation	External magnetically field are required for separation, Costly
Nanofiltration and nanosieve membranes	Nanofiltration	Organic and inorganic compound	Low pressure than RO	Costly, prone to membrane fouling

Glossary

Adsorbate: Any substance that is or can be adsorbed [144].

Adsorbent: Any solid having the ability to concentrate significant quantities of other substance of other surface [144].

Adsorption: A process in which fluid molecules are concentrated on a surface by chemical or physical forces, or both [144].

Breakthrough: The first appearance in the effluent of an adsorbate of interest under specified condition [144].

Cavitation: Normally, cavitation is a nucleated process; that is, it occurs at pre-existing weak points in the liquid, such as gas-filled crevices in suspended particulate matter or transient microbubbles from prior cavitation events [145].

Desorption: The separation of an adsorbate as such from a sorbent [144].

Dosage: The quantity of substance applied per unit weight or volume of the fluid being treated [144].

Encapsulation: Encapsulation is the confinement of a guest molecule inside the cavity of a supramolecular host molecule (molecular capsule, molecular container or cage compounds) [146].

Equilibrium adsorption capacity: The quantity of a given component adsorbed per unit of adsorbent from a fluid or fluid mixture at equilibrium temperature and concentration, or pressure [144].

Expanded bed: A bed of granular particles through which a fluid flows upward at a rate sufficient to slightly elevate and separate the particles without changing their relative positions [144].

Ex situ: *Ex situ* means the contaminants are removed from the ground, either by digging up contaminated soil or pumping contaminated ground water and treating the contamination in treatment facilities built at the site. The remediated soils or ground water are then either placed back into the ground or disposed off-site [147].

Fines: Particles smaller than the smallest nominal specification conditions [144].

Fixed bed: A bed of granular particles through which the fluid flows without causing substantial movement of the bed [144].

Fluidized bed: A bed of granular particles in which the fluid flows upward at a rate sufficient to suspend the particles completely and randomly in the fluid phase [144].

Freundlich isotherm: A logarithmic plot of quantity of component adsorbed per unit of adsorbent versus concentration of that component at equilibrium and at constant temperature, which approximates the straight line postulate by Freundlich adsorption equation:

$$X\!\!\!\big/_M = kC^n$$

X = quantity adsorbed, M = quantity of adsorbent, C = concentration, k and n = constant [144].

In situ: In situ means the technology is delivered directly to the subsurface soils or ground water to treat the contaminants where they are located [147].

Isotherm: A plot of quantity adsorbed per unit of adsorbent against equilibrium concentration, or pressure, when temperature is held constant [144].

Langmuir isotherm: A plot of isotherm adsorption data which to a reasonable degree fit the Langmuir adsorption equation [144].

Macropore: Pores with width exceeding 50 nanometers (500 angstrom units) [144].

Mesopore: Pores of width between 2 and 50 nanometers (20 and 500 angstrom units) [144].

Micropore: Pores of width between no exceeding 2 nanometers (20 angstrom units) [144].

Nanocrystal: Most solids are crystalline with their atoms arranged in a regular manner. They have what is called long-range order because the regularity can extend throughout the crystal. When the size of the crystal approaches the order of the de Broglie wavelength of the conduction electrons, the metal clusters may exhibit novel electronic properties [14].

Nanoparticles: Nanoparticles are generally considered to be a number of atoms or molecules bonded together with a radius of < 200 nm [14].

Nanotechnology: Nanotechnology means a billionth (1×10^{-9}) [14].

Pollution prevention: Pollution prevention refers to "source reduction" and other practices that efficiently use raw materials, energy, water, or other resources to reduce or eliminate creation of waste. This strategy also includes using less toxic and renewable reagents and processing materials, where possible, and the production of more environmentally benign manufactured products [148].

Pores: The complex network of channels in the interior of a particle of a sorbent [144].

Pore diameters: The diameter of a pore in a model in which the pores in a sorbent are assumed to be cylindrical in shape and which calculated from data obtained by a specified procedure [145].

Pore volume: volume of the pores in a unit weight of a sorbent [144].

Quantum yield: The number of defined events which occur per *photon* absorbed by the system. The integral quantum yield is:

ϕ = (number of events)/(number of photons absorbed)

For a photochemical reaction:

ϕ = (amount of reactant consumed or product formed)/ (amount of photons absorbed)

The differential quantum yield is: $\phi = \{(d[x]/dt)/n\}$

where $d[x]/dt$ is the rate of change of a measurable quantity, an d n the amount of photons (mol or its equivalent Einstein) absorbed per unit time. ϕ can be used for photophysical processes or photochemical reactions [149].

Regeneration: Distillation or elution –type process for restoring the adsorptive properties of a spent sorbent [144].

Sorption: A process in which fluid molecules are taken up by absorption and adsorption [144].

Sonochemistry: The chemical applications of ultrasound, "sonochemistry" [145].

Surface area (B.E.T): The total surface area of a solid calculated by the B.E.T. (Brunauer, Emmett, Teller) equation, from nitrogen adsorption or desorption data obtained under specified conditions [145].

Surfactant: The name "surfactant" refers to molecules that are surface active, usually in aqueous solutions [150].

Surface oxide: Oxygen containing compounds and complexes formed at the surface of an adsorbent [144].

Ultrasound: Sound is nothing more than waves of compression and expansion passing through gases, liquids or solids. We can sense these waves directly through our ears if they have frequencies from about

Hertz to 16 kHz (the Hertz unit is cycles of compression or expansion per second; kiloHertz, abbreviated kHz, is thousands of cycles per second). Ultrasound has frequencies pitched above human hearing (above roughly 16 kHz) [145].

References

1. G.A. Mansoori, and T.A.F. Soelaiman, J. ASTM International **2**, 21 pages (2005).
2. G.A. Mansoori, United Nations Tech. Monitor, Special Issue, **53** (2002).
3. http://www.nano.gov/nsetrpts.htm. *National Nanotechnology Initiative: The Initiative and Its Implementation Plan;* NSTC/NSET report, Washington D.C., March (2001).
4. W.X. Zhang, and T. Masciangioli, Environ. Sci. Technol. **37**, 102A (2003).
5. G.A. Mansoori, G.R. Vakili-Nezhaad, and A.R. Ashrafi, Int'l J. Pure & Applied Math. Sci. 2, 58 (2005).
6. G.A. Mansoori, "Phase Transitions in Small Systems", Proceed. NanoSci. Tech. Workshop, Kashan Univ., May (2003).
7. G.A. Mansoori, "Nanothermodynamics & Phase Transitions in Nanosystems", The 4th Int'l Conf. Fluids & Thermal Energy Conversion, 7, (2003).
8. Hutchison and E. James, Third Green Chemistry Conference, Barcelona, Spain, Nov. (2001).
9. C. Darnault, K. Rockne, A. Stevens, G.A. Mansoori, and N. Sturchio, Water. Environ. Res, **177**, 2576, (2005).
10. A. Ahmadpour, A. Shahsavand, and M.R. Shahverdi, "Current Application of Nanotechnology in Environment", Proceedings of the 4th Biennial Conference of Environmental Specialists Association, Tehran, February (2003).
11. A. Shahsavand, and A. Ahmadpour, "The Role of Nanotechnology in Environmental Culture Development", In Proceedings of the First International Seminar on the Methods for Environmental Culture Development, Tehran, June (2004).
12. U. Diebold, Surf. Sci. Rep. **48**, 53 (2003).
13. http://ruby.colorado.edu/~smyth/min/tio2.html.
14. C.P. Poole Jr., and F.J. Owens, "Introduction to nanotechnology" John Wiley & Sons Inc., Hoboken, New Jersey (2003).
15. E. Stathatos, D. Tsiourvas, and P. Lianos, Colloids and Surfaces A: Physicochemical and Engineering Aspects **149**, 49, (1999).
16. O.V. Makarova, T. Rajh, M.C. Thurnauer, A. Martin, P.A. Kemme, and D.Cropek, Environ. Sci. Technol. **34**, 4797, (2000).
17. M. Andersson, L. Osterlund, S. Ljungstrom and A. Palmqvist, J. Phys. Chem. B **106**, 10674 (2002).

18. J. Chen, M. Liu, L. Zhang, J. Zhang and L. Jin, Water Research **37**, 3815 (2003).

19. L. Li, W. Zhu, P. Zhang, Z. Chen, and W. Han, Water Research **37**, 3646 (2003).

20. W. Fu, H. Yang, L. Chang, H. Bala, M. Li, and G. Zou, Colloids and Surfaces A: Physicochem. Eng. Aspects **289**, 47 (2006).

21. I. Ilisz, A. Dombi, K. Mogyorósi and I. Dékány, Colloids and Surfaces A: Physicochem. Eng. Aspects **230**, 89 (2004).

22. H. Wang, J.J. Miao, J.M. Zhu, H.M. Ma, J.J. Zhu, and H.Y. Chen, Langmuir **20**, 11738 (2004).

23. T. Peng, D. Zhao, K.D.W. Shi, and K. Hirao, J. Phys. Chem. B **109**, 4947 (2005).

24. S.M. Paek, H. Jung, Y.J. Lee, M. Park, S.J. Hwang, and J.H. Choy, Chem. Mater. **18**, 1134 (2006).

25. W.A. Adams, M.G. Bakker, and T.I. Quickenden, J. Photochemistry and Photobiology A: Chemistry **81**, 166 (2006).

26. M. Asilturk, F. Sayýlkan, S. Erdemoglu, M. Akarsu, H. Sayýlkan, M. Erdemoglu, and E. Arpac, J. Hazard. Mater. B **129**,164 (2006).

27. J. Wang, T. Ma, Z. Zhang, X. Zhang, Y. Jiang, D. Dong, P. Zhang, and Ying. Li, J. Hazard. Mater. **137**, 972 (2006).

28. Y. Li, X. Li, J. Li, and J. Yin, Water research **40**, 1119 (2006).

29. C.H. Wu, H.W. Chang, and J.M. Chern, J. Hazard. Mater. **137**, 336 (2006).

30. N.M. Mahmoodi, M. Arami, N. Yousefi Limaee, K. Gharanjig, and F. Nourmohammadian, Mater. Research Bulletin **42**, 797 (2007).

31. N.M. Mahmoodi, M. Arami, N. Yousefi Limaee, and K. Gharanjig, J. Hazard. Mater. **145**, 65 (2007).

32. K. Dai, H. Chen, T. Peng, D. Ke, and H. Yi, Chemosphere **69**, 1361 (2007).

33. G. Mascolo, R. Comparelli, M.L. Curri, G. Lovecchio, A. Lopez, and A. Agostiano, J. Hazard. Mater. **142**, 130 (2007).

34. J.M. Wu, Environ. Sci. Technol. **41**, 1723 (2007).

35. M. Srinivasan, and T. White, Environ. Sci. Technol. **41**, 4405 (2007).

36. N. Sobana, M. Muruganadham, and M. Swaminathan, J. Molecular Catalysis A: Chemical **258**, 124 (2006).

37. C.S. Chuang, M.K.Wang, C.H. Ko, C.C. Ou, and C.H. Wu, Bioresource Technol., (2007).

38. Y. Yang, J. Ma, Q. Qin, and X. Zhai, J. Molec. Cataly. A: Chem. **267**, 41 (2007).

39. Y. Yan, X. Qiu, H.Wang, L. Li, X. Fu, L. Wu, and G. Li, J. Alloys Comp. (2007).

40. C-K. Lee, S-S Liu, L-C. Juang, C-C. Wang, M.D. Lyu, and S.H. Hung, J. Hazard. Mater. **148**, 756 (2007).

41. S. M. Ponder, J.G. Darab, and T.E. Mallouk, Environ. Sci. Technol. **34**, 2564 (2000).

42. J. Cao, and W.X. Zhang, J. Hazard. Mater. B **132**, 213 (2006).

43. S. Choe, Y.Y. Chang, K.Y. Hwang, J. Khim, Chemosphere **41**, 1307 (2000).

44. G.V. Lowry. K and M. Johnson, Environ. Sci. Technol. **38**, 5208 (2004).

45. A.J. Feitz, S.H. Joo, J. Guana, Q. Suna, D.L. Sedlak, and T. D. Waite, Colloids Surf. A: Physicochem. Eng. Aspects **265**, 88 (2005).

46. J. Quinn, C. Geiger, C. Clausen, K. Brooks, C. Coon, S. O'hara, T. Krug, D. Major, W.S. Yoon, A. Gavaskar, and T. Holdswoth, Environ. Sci. Technol. **39**, 1309 (2005).

47. K.M. Paknikar, V. Nagpal, A.V. Pethkar, and J.M. Rajwade, Sci. Technol. Adv. Mater. **6**, 370 (2005).

48. S.R. Kanel, J.M. Greneche, and H. Choi, Environ. Sci. Technol. **40**, 2045 (2006).

49. X. Guo, and F. Chen, Environ. Sci. Technol. **39**, 6808 (2005).

50. L. Cumbal, and A.K. Sengupta, Environ. Sci. Technol. **39**, 6508 (2005).

51. M.J. DeMarco, A.K. SenGupta, and J.E. Greenleaf, Water Research **37**, 164 (2003).

52. Y. Xu, and D. Zhao, Water research **41**, 2101 (2007).

53. Z. Xiong, D. Zhao, and G. Pan, Water research **41**, 3497 (2007).

54. A.B.M. Giasuddin, S.R.Kanel, and H. Choi, Environ. Sci. Technol. **41**, 2022 (2007).

55. X.Q. Li, and W.X. Zhang, Langmuir **22**, 4638 (2006).

56. R. Cheng, J.L. Wang, and W.X. Zhang, J. Hazard Mater **144**, 334 (2007).

57. O. Celebi, C. Uzum, T. Shahwan, and H. N. Erten, J. Hazard. Mater. **148**, 761 (2007).

58. K. Hristovski, P. Westerhoff, T. Moller, P. Sylvester, W. Condit, and Heath Mashe, J. Hazard. Mater. (2007).

59. C.B. Wang, and W.X. Zhang, Environ. Sci. Technol. **31**, 2154 (1997).

60. H.L. Lien, and W.X. Zhang, Colloids and Surfaces A: Physicochem. Eng. Aspects **191**, 97 (2001).

61. D.W. Elliott, and W.X. Zhang, Environ. Sci. Technol. **35**, 4922 (2001).

62. B. Schrick, J.L. Blough, A.D. Jones, and T.E. Mallouk, Chem. Mater **14**, 5140 (2002).

63. J. Feng, and T.T. Lim, Chemosphere **59**, 1267 (2005).

64. C.P. Huang, H.W. Wang, and P.C. Chiu, Water Research **32**, 2257 (1998).

65. Y.H. Liou, S.L. Lo, C.J. Lin, C.Y. Hu, W.H. Kuan, and S.C. Weng, Environ. Sci. Technol. **39**, 9643 (2005).

66. Y.H. Liou, S.L. Lo, C.J. Lin, W.H. Kuan, and S.C. Weng, J. Hazard. Mater. B **127**, 102 (2005).

67. M.O. Nutt, J.B. Hughes and M.S. Wong, Environ. Sci. Technol. **39**, 1346 (2005).

68. S H. Joo, and D. Zhao, Chemosphere, (2007).

69. G.M. Pajonk, Catalysis Today **52**, 3 (1999).

70. C.K. Loong, and M. Ozawa, J. Alloys Comp. **303**, 60 (2000).

71. A. Bumajdad, M.I. Zaki, J. Eastoe, and L. Pasupulety, Langmuir **20**, 11223 (2004).

72. A. Korgel, and H. G. Monbouquette, J. Phys. Chem. B. **101**, 5010 (1997).

73. I. Dror, D. Baram, and B. Berkowitz, Environ. Sci. Technol. **39**, 1283 (2005).

74. H. Wang, C. Xie, W. Zhang, S. Cai, Z. Yang, and Y. Gui, J. Hazard. Mater. **141**, 645 (2007).

75. G. Huang, C. Zhang, and Y. Zhu, J. Alloys Comp. **432**, 269 (2007).

76. J. Lin, and Y. Zhu, Inorg. Chem. **46**, 8372 (2007).
77. Z. Ding, H.Y. Zhu, G.Q. Lu, and P.F. Greenfield, J. Colloid Inter. Sci. **209**,193 (1999).
78. C. Ooka, H. Yoshida, K. Suzuki and T. Hattori, Micro. Meso. Mater. **67**, 143 (2004).
79. H.Y. Zhu, J.Y. Li, J.C. Zhaob, and G.J. Churchman, Applied Clay Sci. **28**, 79 (2005).
80. J. Feng, X. Hu, P.L.Yue, H. Y. Zhu, and G.Q. Lu, Ind. Eng. Chem. Res. **42**, 2058 (2003).
81. R. Say, E. Birlik, A. Denizli, and A. Ersöz, Applied Clay Sci. **31**,298 (2006).
82. J.H. Choi, S.D. Kim, S.H. Noh, S.J. Oh and W.J. Kim, Micro. Meso. Mater. **87**,163 (2006).
83. L. Wang, and A. Wang, J. Hazard. Mater. **147**, 979 (2007).
84. A. Eliassi, M.H. Eikani and G.A. Mansoori, *"Production of Single-Walled Carbon Nanotubes - A Review"*, Proceed. of the 1st Conference on Nanotechnology - The next industrial revolution, 2, 160-178, March (2002).
85. A.M. Rashidi, M.M. Akbarnejad, A. Ahmadpour, A.A.Khodadadi, Y. Mortazavi, M. Mahdiarfar, and H.R. Ghafarian,, *"Preparation of SWNTs by CVD method over Co-Mo/MgO Catalyst using different Additives"*, International Conference on Carbon "CARBON 2003", Oveido, Spain, July (2003).
86. A.M. Rashidi, M.M. Akbarnejad, A. Ahmadpour, A.A. Khodadadi, Y. Mortazavi, H.R. Ghafarian, and F. Tayari, *"Synthesis of CNTs by Catalytic Vapor Deposition of Methane"*, The 8th National Iranian Chemical Engineering Conference, Mashhad, Iran, October (2003).
87. A.M. Rashidi, M.M. Akbarnejad, Y. Mortazavi, A.A. Khodadadi, M. Attarnejad, and A. Ahmadpour, *"The Preparation of Bamboo Structured Carbon Nanotube by CVD of Acetylene on Co-Mo/MCM-41 with the Controlled Porosity"*, 1st Iran-Russia Joint Seminar & Workshop on Nanotechnology (IRN 2005), Tehran, (2005).
88. A. Shahsavand, and A. Ahmadpour, Compu. and Chem. Eng. **29**, 2134 (2005).
89. A.M. Rashidi, M.M. Akbarnejad, A.A. Khodadadi, Y. Mortazavi, and A. Ahmadpour, Nanotechnology **18**, 1 (2007).
90. Ralph. T. Yang, *"Adsorbents: Fundamentals and applications"*, John Wiley & Sons, Inc, (2003).
91. students.chem.tue.nl/ ifp03/default.htm
92. http://www.noritake-elec.com/itron/english/nano/
93. Y.H. Li, S. Wang, J. Wei, X. Zhang, C. Xu, Z. Luan, D. Wu and B. Wei, Chem. Phy. Lett. **357**, 263 (2002).
94. Y.H. Li, Z. Di, J. Ding, D. Wu, Z. Luan and Y. Zhu, Water Research **39**, 605 (2005).
95. Y.H. Li, S. Wang, Z. Luan, J. Ding, C. Xu and D. Wu, Carbon **41**,1057 (2003).
96. C. Lu, and H. Chiu, Chem. Eng. Sci. **61**, 1138 (2006).

97. Y.H. Li, S. Wang, A. Cao, D. Zhao, X. Zhang, C. Xu, Z. Luan, D. Ruan, J. Liang, D. Wu, and B. Wei, Chem. Phy. Lett. **350**, 412 (2001).

98. Y.H. Li, S. Wang, X. Zhang, J. Wei, C. Xu, Z. Luan, and D. Wu, Mater. Research Bulletin 38, 469 (2003).

99. X. Peng, Y. Li, Z. Luan, Z. Di, Hu. Wang, B. Tian, and Z. Jia, Chem. Phy. Lett. 376, 154 (2003).

100. C. Lu, Y.L. Chung, and K.F. Chang, Water Research 39, 1183 (2005).

101. P. Kondratyuk, and J.T. Yates Jr, Chem. Phy. Lett. **410**, 324 (2005).

102. S. Agnihotri, M.J. Rood, and M. Rostam-Abadi, Carbon **43**, 2379 (2005).

103. A. Stafiej, and K. Pyrzynska, Sep. Purif. Technol., (2007).

104. H. Wang, A. Zhou, F. Peng, H. Yu, and J. Yang, J. colloid. Inter. Sci., (2007).

105. S-G. Wang, W-X. Gong, X-W. Liu, Y-W. Yao, B-Y. Gao, and Q-Y. Yue, Sep. Purif. Technol., (2007).

106. D. Xu, X. Tan, C. Chen, and X. Wang, J. Hazard. Mater. (2007).

107. M. I. Kandah, and J-L. Meunier, J. Hazard. Mater. **146**, 283 (2007).

108. C. Lu, and F. Su, Sep. Purif. Technol. (2007).

109. rati.pse.umass.edu/ usim/gallery.html

110. M.S. Diallo, L. Balogh, A. Shafagati, J.H. Johnson Jr, W.A. Goddard, and D.A. Tomalia, Environ. Sci. Technol. **33**, 820 (1999).

111. M.S. Diallo, S. Christie, P. Swaminathan, J.H. Johnson Jr, and W.A. Goddard, Environ. Sci. Technol. **39**, 1366 (2005).

112. Y. Xu, and D. Zhao, Environ. Sci. Technol. **39**, 2369 (2005).

113. Y. Xu, and D. Zhao, Ind. Eng. Chem. Res. 45, 1758 (2006).

114. A. Rether, and M. Schuster, Reactive & Func. Poly. **57**, 13 (2003).

115. M. Arkas, D. Tsiourvas, and C.M. Paleos, Chem. Mater. **15**, 2844 (2003).

116. M. Arkas, D. Tsiourvas, and M. Paleos, Chem. Mater. 17, 3439 (2005).

117. M. Arkas, R. Allabashi, D. Tsiourvas, E.M. Mattausch, and R. Perfler, Environ. Sci. Technol. 40, 2771 (2006).

118. G.E. Fryxell, Y. Lin, S. Fiskum, J.C. Birnbaum, H. Wu, K. Kemner, and S. Kelly, Environ. Sci. Technol. 39, 1324 (2005).

119. T. Szabó, J. Németh and I. Dékány, Colloids and Surfaces A: Physicochem. Eng. Aspects 230, 23 (2004).

120. J.H. Pacheco-Sanchez and G.A. Mansoori, Petrol. Sci. & Technol. **16**, 377 (1998).

121. S. Priyanto, G.A. Mansoori and A. Suwono, Chem. Eng. Sci. 56, 6933 (2001).

122. J.Y. Kim, C. Cohen, M.L. Shuler, and L.W. Lion, Environ. Sci. Technol. **34**, 4133 (2000).

123. J.Y. Kim, S.B. Shim, and J.K. Shim, J. Hazard. Mater. B **98**, 145 (2003).

124. J.Y. Kim, S.B. Shim, and J.K. Shim, J. Hazard. Mater. B **116**, 205 (2004).

125. A.F. Ngomsik, A. Bee, M. Draye, G. Cote, V. Cabuil, and C.R. Chimie, Water Research **8**, 963 (2005).

126. M. Takafuj, S. Ide, H. Ihara, and Z. Xu, Chem. Mater. **16**, 1977 (2004).

127. J. Hu, I.M.C. Lo, and G. Chen, Langmuir **21**, 11173 (2005).

128. J. Hu, G. Chen, and I.M.C. Lo, Water Research **39**, 4528 (2005).
129. Y.C. Chang, S.W. Chang, and D.H. Chen, Reactive & Func. Poly. **66**, 335 (2006).
130. A.F. Ngomsik, A. Bee, J.M. Siaugue, V. Cabuil and G. Cote, Water Research **40**, 1848 (2006).
131. S.Y. Mak, and D.H. Chen, Dyes and Pigments **61**, 93 (2004).
132. D.H. Chen, and S.H. Huang, Process Biochemistry **39**, 2207 (2004).
133. L.H. Cumbal, and A.K. Sengupta, Ind. Eng. Chem. Res. **44**, 600 (2005).
134. J.G. Hu, I.M.C. Lo, and G. Chen, Sep. Purif. Technol. **56**, 249 (2007).
135. J.T. Mayo, C. Yavuz, S. Yean, L. Cong, H. Shipley, W. Yu, J. Falkner, A. Kan, M. Tomson, and V.L. Colvin, Sci. Technol. Adv. Mater. **8**, 71 (2007).
136. S.S. Banerjee, and D.H. Chen, J. Hazard. Mater. **147**, 792 (2007).
137. M. Alborzfar, G. Jonsson, and C. Gron, Water Research **32**, 2983 (1998).
138. S.E. Kentish, and G.W. Stevens, Chem. Eng. J. **84**, 149 (2001).
139. A. Santafe-Moros, J.M. Gozalvez-Zafrilla and J. Lora-Garca, Desalination **185**, 281 (2005).
140. J. Fu, M. Ji, Z. Wang, L. Jin, and D. An, J. Hazard. Materi. B **131**, 238 (2006).
141. L. Yan, Y. S. Li, C.B. Xiang, and S.Xianda, J. Mem. Sci. **276**, 162 (2006).
142. L.Y. Jiang, T. S. Chung, C. Cao, Z.Huang, and S. Kulprathipanja, J. Mem. Sci. **252**, 89 (2005).
143. L.Y. Jiang, T.S. Chung, and S. Kulprathipanj, J. Mem. Sci. **276**, 113 (2006).
144. ASTM D2652-93, Standard Terminology relating to activated carbon.
145. http://www.scs.uiuc.edu/suslick/britannica.html
146. http://en.wikipedia.org/wiki/Molecular_encapsulation
147. http://www.chem.ox.ac.uk/nanotemp/info.html, 8-9, May, (2003).
148. T.M, Masciangioli. Nora, Savage. Barbara P, Karn, EPA,2001.
149. http://www.iupac.org/goldbook/Q04991.pdf
150. R.M. Pashley, M.E. Karaman, *"Applied Colloid and Surface Chemistry"*, John Wiley & Son, Ltd., (2004).

CHAPTER 11

NANOSTRUCTURED IONIC AND MIXED CONDUCTING OXIDES

Xin Guo

Institute of Solid State Research and Center of Nanoelectronic Systems for Information Technology, Research Center Jülich, 52425 Jülich, Germany
E-mails: x.guo@fz-juelich.de; guo@iwe.rwth-aachen.de

Sangtae Kim

Department of Chemical Engineering and Materials Science, University of California at Davis, One Shields Avenue, Davis, CA 95616, USA
E-mail: chmkim@ucdavis.edu

The nano-effects in three groups of ionic and mixed conducting oxides are summarized. Compared with microstructured counterparts, the total conductivity of nanostructured oxygen ion conductors (*e.g.* doped-ZrO_2 or CeO_2) is lower, even though the grain-boundary conductivity is actually higher. The grain-boundary conductivity of nanostructured ZrO_2 or CeO_2 is always more than one order of magnitude lower than the bulk conductivity. This fact, together with the high-density of the grain boundaries in nanostructured ZrO_2 or CeO_2, results in a lower total conductivity. Electrons are accumulated in the space-charge layer of mixed conductors of oxygen ions and electrons (*e.g.* slightly doped and undoped CeO_2). With the comparatively high electronic bulk contribution and high density of grain boundaries, the grain boundaries in nanocrystalline CeO_2 become electronically conducting and dominate the overall behavior. Therefore, the *n*-type conductivity of nanocrystalline CeO_2 is enhanced by four orders of magnitude. When the grain size decreases to the nanometer scale, the *p*-type conductivity of mixed conductors of oxygen ions and holes (*e.g.* nanocrystalline

BaTiO$_3$) is enhanced by one to two orders of magnitude, which is due to a significantly reduced oxidation enthalpy. The defect thermo-dynamics on the nanometer scale is different.

1. Introduction

In ionic conductors, ions are the predominant charge carriers; while ions and electrons or holes are responsible for the electrical conduction in mixed conductors. In technologically important electrochemical devices, such as lithium-ion batteries,[1] fuel cells,[2,3] chemically sensors,[4] and hydrogen permeation membranes,[5] ionic conduction is of prime importance. Owing to the relatively low mobility of ionic charge carriers, the ionic conductivity of most ionic solids is usually small at low to intermediate temperatures; therefore, there is always significant interest in enhancing the ionic conductivity. Unlike the electronic conductivity that can be enhanced by orders of magnitude simply by increasing the charge carrier (electron or hole) concentration, there exists a maximum ionic conductivity corresponding to an optimal ionic charge carrier concentration,[6] such that increasing the ionic charge carrier concentration above the optimal value results in a decrease in ionic conductivity, which is due to, for example, the association of ionic defects.

Nanostructured materials, in general, include bulk materials with nanometer-sized grains, nanocrystalline composites, microcrystalline materials with nanostructures, and thin films or multilayers with nanoscale thickness, *etc.* In comparison with their microstructured counterparts, the most remarkable feature of nanostructured materials is the high interfacial density. This feature leads to two nano-effects:[7-11] *trivial size effect* and *true size effect*. The *trivial size effect* is defined as the increased contribution of the interfacial properties to the overall materials properties, due to the drastically increased interface to bulk fraction. For example, the ionic conduction in nanocrystalline materials is dominated by the grain-boundary conductivity.[12] When the spacing of the interface becomes comparable with the Debye length, local properties change as a function of distance and the *true size effect* occurs. The CaF$_2$/BaF$_2$ hetero-layer best embodies the *true size effect*.[13] The ionic conductivity of the CaF$_2$/BaF$_2$ hetero-layer increases almost two orders of magnitude when the hetero-layer period (the thickness of CaF_2 plus BaF_2 unit layer) deceases from 500 to 16 nm, because the neighboring space-charge layers overlap and the individual layers lose their bulk properties when the period decreases to 16 nm. An important implication of this finding is that the materials properties can be tuned by varying the spacing of interfaces. In this context an interesting question emerges: Is it always possible to achieve notably higher ionic conductivity in nanostructured materials?

In this article, we confine ourselves to the electrical properties of oxygen ion and mixed conductors, in light of the fact that oxygen ion and mixed conductors are both scientifically and technologically important. This article is structured as follows: at first, pure oxygen ion conductors are discussed. The situation of pure oxygen ion conductors is comparatively simple, because only one type of charge carriers is involved. Basic theory, for example, the Schottky barrier model, is introduced in this section, which is also the basis for the next sections. In the second section, oxygen ion-electron conductors are discussed, and the last section is dedicated to oxygen ion-hole conductors. In each section, nanostructured materials are compared with microcrystalline counterparts, and the *trivial size effect* and/or the *true size effect* are highlighted.

2. Oxygen Ion Conductors

The main oxygen ion conductors known to date belong to five distinct groups: fluorite (*e.g.* doped-ZrO_2 and CeO_2),[6] perovskite (*e.g.* $LaGaO_3$),[14] intergrowth perovskite/Bi_2O_2 layer (*e.g.* BIMEVOX),[15] pyrochlore (*e.g.* $Gd_2Ti_2O_7$),[16] and $La_2Mo_2O_9$ compound.[17] Among all these oxygen ion conductors, the group with the fluorite structure, *e.g.* Y_2O_3, Gd_2O_3 or CaO-doped ZrO_2 and CeO_2 find the broadest applications in technologically important devices, *e.g.* solid-oxide fuel cells,[2] oxygen sensors,[18] and oxygen pumps.[19] In most of these applications, doped-ZrO_2 and CeO_2 are present in the form of poly-crystals; consequently, the grain boundaries are a crucial part of the microstructure.

Over wide temperature and oxygen partial pressure ranges, doped-ZrO_2 and CeO_2 are pure oxygen ion conductors, with oxygen vacancies being the predominant charge carriers. The grain boundaries of doped-ZrO_2 and CeO_2 present a blocking effect to the ionic transport across them, *i.e.* the specific grain-boundary conductivity of doped-ZrO_2 or CeO_2 is usually at least two orders of magnitude lower than that of the bulk (see *e.g.* Refs.20-29 for ZrO_2, and Refs.30-39 for CeO_2), depending on temperature and dopant level. This blocking effect was previously attributed to an intergranular siliceous phase (*i.e.* an amorphous phase containing high SiO_2 concentration),[22-26,30,31] but it has been gradually realized that the oxygen-vacancy depletion in the space-charge layer at the grain boundaries is the decisive cause of the low grain-boundary conductivity (see *e.g.* Refs. 40-48 for ZrO_2, and Refs. 49-52 for CeO_2). In addition to the lower grain-boundary conductivity, the activation energy for the grain-boundary conductivity is higher than that for the bulk conductivity.[20,23,24,30,32]

The electrical properties of the grain boundaries of oxygen ion conductors have been extensively reviewed in Ref. 53. However, the new features of nanostructured materials are emphasized in this article.

2.1. Grain-Boundary Core and Space-Charge Layer

From a structural point of view, a grain boundary is a crystallographic mismatch zone (*i.e.* grain-boundary plane or grain-boundary core), observable by means of transmission electron microscopy (TEM). At thermodynamic equilibrium, the grain-boundary core of an ionic crystal carries an electrical charge due to the presence of excess ions of a given sign. This charge is compensated by adjacent space charges of an opposite sign. Owing to the charged grain-boundary core, the concentrations of charged point defects in the space-charge layer deviate from their bulk values; the accumulation or depletion of charge carriers in the space-charge layer significantly influences the electrical properties of polycrystalline ionic and mixed conductors.

Structurally, a space-charge layer is part of the bulk, but electrically, the space-charge layer is part of the grain boundary. In this sense an "electrical grain boundary" consists of a grain-boundary core and two adjacent space-charge layers. Thus the thickness of an "electrical grain boundary" $\delta_{gb} = 2\lambda^* + b$, here λ^* is the width of the space-charge layer, and b that of the grain-boundary core. Since a "crystallographic grain boundary" is typically around 1 nm thick, the thickness of an "electrical grain boundary" $\delta_{gb} \approx 2\lambda^*$. The electrical contribution of a space-charge layer can be taken into account of by introducing effective charge carrier concentrations and an effective width, and the conduction mechanism in the space-charge layer is usually bulk-like.

In 2 mol% Y_2O_3-doped (or stabilized) ZrO_2, the enrichment of additionally added divalent and trivalent minority solutes (with effective negative charge(s) in the ZrO_2 lattice, *e.g.* Y'_{Zr} and Ca''_{Zr}) at the grain boundaries was found to be significant, whereas the enrichment of pentavalent minority solutes (with an effective positive charge in the ZrO_2 lattice, *e.g.* $Nb^{\bullet}_{Zr}$ and $Ta^{\bullet}_{Zr}$) was not observed, pointing to a positive potential in the ZrO_2 grain-boundary core.[54]

Gadolinium has an almost perfect match of the ionic radius in the CeO_2 lattice, therefore, the elastic strain resulting from the ion size mismatch is too small to be an effective segregation driving force; any Gd (*i.e.* Gd'_{Ce}) segregation at the CeO_2 grain boundaries is mostly driven by the Coulomb interaction with the positive boundary charge. The grain-boundary segregation of Gd in Gd_2O_3-doped CeO_2 was observed,[55] being in accordance with the expected positive core potential.

Table 1. Atomic ratios in 10 mol% Y_2O_3-doped ZrO_2 bicrystal with a symmetric 24° [001] tilt grain boundary[57]

	Y/Zr	O/Zr	O/Y
Grain bulk	0.25 ± 0.04	2.12 ± 0.13	8.48 ± 1.45
Grain-boundary core	0.50 ± 0.07	1.65 ± 0.23	3.30 ± 0.65

Molecular dynamics simulations[56] of a $\Sigma 5$ symmetrical tilt grain boundary $((310)/[001]$ misorientation $\theta = 36.9°)$ in 8 mol% Y_2O_3-doped ZrO_2 shows that the structure relaxation can produce intrinsic oxygen vacancies in the grain-boundary core. The electron energy-loss spectroscopy (EELS) investigations[57] (summarized in Table 1) of a symmetric tilt grain boundary ((310)/[001] misorientation $\theta = 24°$) in 10 mol% Y_2O_3-doped ZrO_2 show an increase in the Y/Zr ratio, and a decrease in the O/Zr and O/Y ratios, indicating an enhanced yttrium and oxygen-vacancy concentration in the grain-boundary core. However, the yttrium segregation is insufficient to charge balance the oxygen-vacancy enrichment. Studies[57] of the grain boundaries in Gd_2O_3-doped CeO_2 ceramic samples reveal similar changes in the O/Ce ratio, indicating that these effects may be generic to the grain boundaries in fluorite-structured materials. The high concentration of oxygen vacancies in the grain-boundary cores of doped-ZrO_2 and CeO_2 may attribute to the positive core charge. By assuming a proper grain-boundary conduction model, the core potential can be calculated from experimentally determined parameters. The positively charged grain-boundary cores of doped-ZrO_2 and CeO_2 lead to the depletion of oxygen vacancies in the space-charge layer.

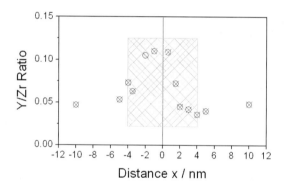

Figure 1. Y/Zr ratio profile across a grain boundary in 2.5 mol% Y_2O_3-doped ZrO_2 ceramic as determined by energy-dispersive X-ray spectroscopy (after Ikuhara *et al.*[60]). The yttrium accumulation mainly occurs in the shaded area.

At high temperatures, *e.g.* at sintering temperatures, acceptor cations are sufficiently mobile to segregate to the grain-boundary core, and accumulate in the space-charge layer, as a result of elastic strain and Coulomb interactions with the positively charged grain-boundary core. By means of various techniques, *e.g.* Auger electron spectroscopy (AES),

EELS, energy-dispersive X-ray spectroscopy (EDXS) and X-ray photoelectron spectroscopy (XPS), different researchers independently determined that the yttrium accumulation at the ZrO_2 grain boundaries occurs mostly within a distance of 2-4 nm from the grain-boundary core,[58-63] as demonstrated by the yttrium concentration profile given in Figure 1. The situation in doped-CeO_2 is very similar.[57] The 2-4 nm accumulation width represents the effective widths of the space-charge layer. Owing to the effective negative charge of acceptors in the lattice, the acceptor accumulation in the space-charge layer partly compensates the positive charge of the grain-boundary core. Since the conventional composition measurement techniques cannot distinguish between free and associated defects, the seemingly low yttrium or gadolinium accumulation factor must actually correspond to a much higher space-charge effect, because in the bulk most of the cations are essentially associated.

2.2. Grain-Boundary Electrical Properties

2.2.1. Schottky Barrier Model

SiO_2 is one of the major impurities present in ZrO_2 and CeO_2 ceramics, along with alkali and some transition metal oxides. During sintering, these impurities accumulate/disperse at grain boundaries, react with each other and the bulk components (Zr, Ce, Y, Gd *etc.*) to form an intergranular siliceous phase. It is commonly accepted that the intergranular siliceous phase significantly affects the grain-boundary electrical properties of doped-ZrO_2 and CeO_2.[22-26,30,31] The intergranular siliceous phase is poorly conductive or even insulating. In view of this fact, the ionic conduction across the grain boundaries occurs solely through the grain-to-grain contacts. The presence of the siliceous phase only determines the fraction of the grain-to-grain contacts, and constricts the ionic current across the grain boundaries.

However, in materials of high purity in which the siliceous phase is not observed, the specific grain-boundary conductivity is still at least 2 orders of magnitude lower than that of the grain bulk.[26,32,41-43,46-52] It is thus evident that, although the presence of a siliceous phase undoubtedly results in a grain-boundary blocking effect, the presence of the siliceous phase is definitely not a pre-requisite for the blocking effect. The grain-to-grain contacts are themselves blocking in nature, being the intrinsic cause of the grain-boundary blocking effect.

The equilibrium concentration of charge carrier species j with an effective charge of z at a locus x is

$$\frac{c_j(x)}{c_j(\infty)} = \exp\left(-\frac{ze\Delta\varphi(x)}{k_BT}\right). \tag{1}$$

In Eq. (1), $c_j(x)$ is the defect concentration at the locus x, $\Delta\varphi(x)$ the electrostatic potential in relation to the bulk, e the elementary charge, k_B the Boltzmann constant and T the absolute temperature. The depletion of charge carriers, *e.g.* oxygen vacancies, gives rise to a back-to-back double Schottky barrier. The grain-boundary conductivity, σ_{gb}, resulting from the charge-carrier depletion in the space-charge layer, can be derived by integrating Eq. (1). The integration gives[41]

$$\frac{\sigma_{bulk}}{\sigma_{gb}} = \frac{\exp\left(ze\Delta\varphi(0)/k_BT\right)}{2ze\Delta\varphi(0)/k_BT}. \tag{2}$$

In Eq. (2), $\Delta\varphi(0) = \varphi(0) - \varphi(\infty)$, it is the space-charge potential. $\Delta\varphi(0)$ is also the Schottky barrier height when the Schottky barrier model is assumed. It has to be pointed out that Eq. (2) is only valid for the charge-carrier depletion situation.[40,41,64] For oxygen vacancies, $z = 2$; therefore, the grain-boundary conductivity of doped-ZrO_2 or CeO_2 is

$$\frac{\sigma_{bulk}}{\sigma_{gb}} = \frac{\exp\left(2e\Delta\varphi(0)/k_BT\right)}{4e\Delta\varphi(0)/k_BT}. \tag{3}$$

The activation energy, E_a, is defined as $E_a = -d \ln \sigma / d(1/k_BT)$. Accordingly, differentiating Eq. (3) yields

$$E_a^{gb} - E_a^{bulk} = (2e\Delta\varphi(0) - k_BT)\left[1 + \frac{1}{T\Delta\varphi(0)}\frac{d\Delta\varphi(0)}{d(1/T)}\right]. \tag{4}$$

As E_a^{bulk} is independent of $\Delta\varphi(0)$, the activation energy for the grain-boundary conductivity (E_a^{gb}) is thus determined by the Schottky barrier height. In the temperature range of 250-500 °C, the average value of $E_a^{gb} - E_a^{bulk}$ calculated from Eq. (4) for 8 mol% Y_2O_3-doped ZrO_2 is about 0.12 ± 0.01 eV, agreeing well with the experimentally determined result (~0.11 eV[42]). Similar agreement has also been proved for Y_2O_3-doped CeO_2 ceramics.[52]

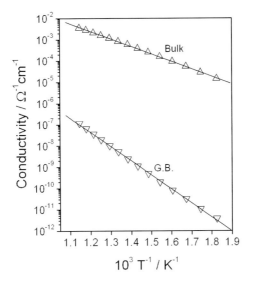

Figure 2. Bulk and grain-boundary conductivities of 8 mol% Y_2O_3-doped ZrO_2 of high purity as a function of temperature (after Guo and Maier[41]). The activation energy for the bulk conductivity is 1.05 eV, and that for the grain-boundary conductivity is 1.16 eV.

Assuming cubic grains of the same size and homogeneous grain boundaries, the grain-boundary thickness, δ_{gb}, can be calculated from[41]

$$\delta_{gb} = \frac{\varepsilon_{gb}}{\varepsilon_{bulk}} \frac{C_{bulk}}{C_{gb}} d_g \approx \frac{C_{bulk}}{C_{gb}} d_g, \tag{5}$$

if the dielectric constant of the space-charge layer, ε_{gb}, is approximated to that of the grain bulk ε_{bulk}. This approximation is not unreasonable considering the fact that the dielectric constant of ZrO_2 or CeO_2 is insensitive to concentration.[42] In Eq. (5), C_{bulk} and C_{gb} are the capacitances of the grain bulk and the grain boundaries, and d_g the grain size. The grain-boundary thickness, δ_{gb}, determined from Eq. (5) is ~4.8 nm for 8 mol% Y_2O_3-doped ZrO_2,[42] ~5.4 nm for 8.2 mol% Y_2O_3-doped ZrO_2,[20] ~5.0 nm for 2 mol% Y_2O_3-doped ZrO_2,[21] ~4.0 nm for 10 mol% Y_2O_3-doped CeO_2,[52] and ~6.0 nm for 1.0 mol% Y_2O_3-doped CeO_2,[52] being independent of grain size. Typical width of a space-charge layer (roughly one half of the grain-boundary thickness) is thus ~2.5 nm. This space-charge layer width value agrees very well with the width value determined from segregation (2-4 nm[58-63]).

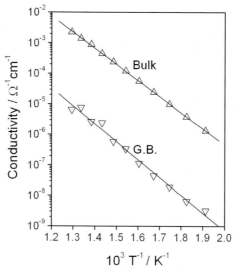

Figure 3. Bulk and grain-boundary conductivities of 1.0 mol% Y_2O_3-doped CeO_2 under $pO_2 = 10^5$ Pa as a function of temperature (after Guo *et al.*[52]). The activation energy for the bulk conductivity is 0.70 eV, and that for the grain-boundary conductivity is 1.36 eV.

When the grain-boundary thickness is known, the specific grain-boundary conductivity, σ_{gb}^{sp}, is simply

$$\sigma_{gb}^{sp} = \frac{L}{R_{gb}A} \frac{\delta_{gb}}{d_g}. \tag{6}$$

In Eq. (6), R_{gb} is the grain-boundary resistance, L the sample thickness, and A the cross-section area. Typical specific grain-boundary conductivity values for 8 mol% Y_2O_3-doped ZrO_2 ceramic of high purity are plotted in Figure 2 as a function of temperature; the bulk conductivities are also plotted for comparison. The specific grain-boundary conductivity of 8 mol% Y_2O_3-doped ZrO_2 ceramic is about 2 orders of magnitude lower than the bulk conductivity.

Typical specific grain-boundary conductivity values for 1.0 mol% Y_2O_3-doped CeO_2 ceramic of high purity under an oxygen partial pressure of 10^5 Pa are plotted in Figure 3 as a function of temperature; the bulk conductivities are also plotted for comparison. The specific grain-boundary conductivity of 1.0 mol% Y_2O_3-doped CeO_2 ceramic is about 5 to 7 orders of magnitude lower than the bulk conductivity. But when 10 mol% Y_2O_3 is doped to CeO_2, the specific grain-boundary conductivity is about 3 orders of magnitude lower than the bulk conductivity.[52]

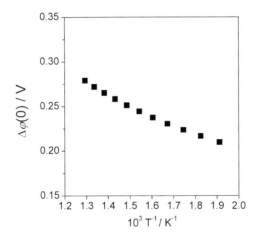

Figure 4. Schottky barrier heights, $\Delta\varphi(0)$, of 8 mol% Y_2O_3-doped ZrO_2 as a function of inverse temperature (after Guo *et al.*[42]).

With the specific grain-boundary conductivity and the bulk conductivity, the Schottky barrier height, $\Delta\varphi(0)$, can be calculated from Eq. (3) for materials of high purity. Results for 8 mol% Y_2O_3-doped ZrO_2 are given in Figure 4, and the results for 10 and 1.0 mol% Y_2O_3-doped CeO_2 under $pO_2 = 10^5$ Pa in Figure 5. The Schottky barrier height all increases with increasing temperature. Depending on temperature, the Schottky barrier height of 8 mol% Y_2O_3-doped ZrO_2 is in the range of

0.20-0.28 V, the Schottky barrier height of 10 mol% Y_2O_3-doped CeO_2 is in the range of 0.30-0.40 V, and that of 1.0 mol% Y_2O_3-doped CeO_2 around 0.50 V.

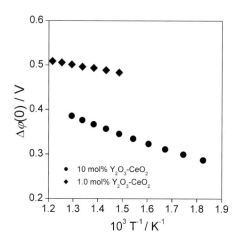

Figure 5. Schottky barrier heights, $\Delta\varphi(0)$, of 10 and 1.0 mol% Y_2O_3-doped CeO_2 under $pO_2 = 10^5$ Pa as a function of inverse temperature (after Guo *et al.*[52]).

2.2.2. Oxygen-Vacancy Concentration Profile

It has been proven that the acceptor, *e.g.* yttrium, accumulation profile formed at a high temperature is frozen at temperatures below 1000 °C, and an yttrium grain-boundary accumulation factor of about 2 was determined for Y_2O_3-doped ZrO_2 by AES.[58,59] As a first approximation, one can assume that the acceptor concentration is constant up to the grain-boundary core. In view of the acceptor profiles given in Figure 1, such an assumption is not unreasonable.

The defect profiles in the space-charge layer is governed by the Poisson equation

$$\frac{d^2\Delta\varphi(x)}{dx^2} = -\frac{1}{\varepsilon}Q(x).$$

(7)

In Eq. (7), $Q(x)$ is the charge density, and ε the dielectric constant. Under the approximation of constant acceptor concentration, the analytical solution of the Poisson equation for Y_2O_3-doped ZrO_2 gives[41]

$$\Delta\varphi(x) = \frac{ec_{Y'_{Zr}}(\infty)}{2\varepsilon}(x-\lambda^*)^2 . \tag{8}$$

Substituting Eq. (8) into Eq. (1) and taking $z = 2$, one gets the oxygen-vacancy concentration in the space-charge layer, which is[41]

$$\frac{c_{V_O^{\bullet\bullet}}(x)}{c_{V_O^{\bullet\bullet}}(\infty)} = \exp\left[-\frac{1}{2}\left(\frac{x-\lambda^*}{L_D}\right)^2\right], \tag{9}$$

but for $x \geq \lambda^*$, $c_{V_O^{\bullet\bullet}}(x)/c_{V_O^{\bullet\bullet}}(\infty) = 1$. In Eqs. (8) and (9), λ^* is the width of the space-charge layer, expressed by

$$\lambda^* = 2L_D\left(\frac{e\Delta\varphi(0)}{k_BT}\right)^{1/2}, \tag{10}$$

and L_D is the Debye length, given by

$$L_D = \left(\frac{k_BT\varepsilon}{2e^2c_{Y'_{Zr}}(\infty)}\right)^{1/2}. \tag{11}$$

$c_{V_O^{\bullet\bullet}}$ and $c_{Y'_{Zr}}$ are the concentrations of oxygen vacancies and yttrium ions. Normalized oxygen-vacancy concentration, $c_{V_O^{\bullet\bullet}}(x)/c_{V_O^{\bullet\bullet}}(\infty)$, in the space-charge layer of 8 mol% Y_2O_3-doped ZrO_2 calculated from Eq. (9) at 500 °C is shown in Figure 6. As shown in this figure, oxygen vacancies are depleted by more than 3 orders of magnitude in the space-charge layer. The ratio of the grain-boundary conductivity to the bulk conductivity at 500 °C calculated from such an oxygen-vacancy profile is in the range of 0.002 to 0.01 for 8 mol% Y_2O_3-doped ZrO_2, which is consistent with the experimental value (~0.004).[41] The vacancy depletion can thus account for the low grain-boundary conductivity in ZrO_2 materials of high purity. And normalized oxygen-vacancy concentrations, $c_{V_O^{\bullet\bullet}}(x)/c_{V_O^{\bullet\bullet}}(\infty)$, in the space-charge layers of 10 and 1.0 mol% Y_2O_3-doped CeO_2 under $pO_2 = 10^5$ Pa at 500 °C are shown in Figure 7. As shown in this figure, oxygen vacancies are depleted by 7 to 10 orders of magnitude in the space-charge layers.

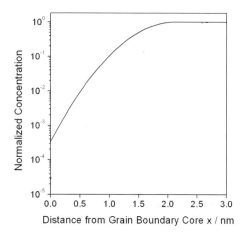

Figure 6. Normalized oxygen-vacancy concentration, $c_{v_o^{\cdot\cdot}}(x)/c_{v_o^{\cdot\cdot}}(\infty)$, in the space-charge layer of 8 mol% Y_2O_3-doped ZrO_2 at 500 °C (after Guo and Maier[41]).

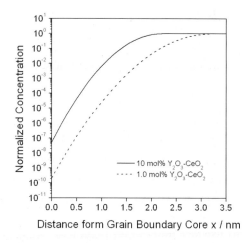

Figure 7. Normalized oxygen-vacancy concentrations, $c_{v_o^{\cdot\cdot}}(x)/c_{v_o^{\cdot\cdot}}(\infty)$, in the space-charge layers of 10 and 1.0 mol% Y_2O_3-doped CeO_2 under $pO_2 = 10^5$ Pa and at 500 °C (after Guo et al. [52]).

2.3. Grain Size Dependent Grain-Boundary Conductivity

When the grain size decreases from the micrometer to the nanometer scale, the nano-effects appears. The works on ZrO_2 bulk ceramics by Guo and Zhang[46] and Hahn *et al.*[65] cover a grain size range of 41-1330 nm. Guo and Zhang[46] prepared 3 mol% Y_2O_3-doped ZrO_2 samples with average grain sizes of 120 to 1330 nm from high-purity powder (quoted impurity contents by weight: SiO_2 ~20 ppm, Al_2O_3 ~50 ppm, Fe_2O_3 < 20 ppm, Na_2O ~190 ppm). After sintering, the relative densities of the samples were all above 94%, and the phase was confirmed by X-ray diffraction to be tetragonal. Figure 8 shows a typical high-resolution TEM (HRTEM) micrograph of 3 mol% Y_2O_3-doped ZrO_2: the microstructure is well developed, only atomic level disorder at the grain boundaries is observed, the absence of a second phase at the grain boundaries, even at the triple grain junction, is obvious. The microstructural observation is consistent with the very low impurity (especially silicon) content in the samples.

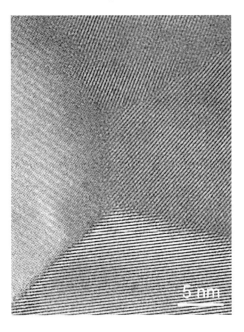

Figure 8. HRTEM micrograph of grain boundaries in the 3 mol% Y_2O_3-doped ZrO_2 sample (after Guo and Waser[53]).

To illustrate the effect of grain size, the bulk conductivity, σ_{bulk}, and the specific grain-boundary conductivity, σ_{gb}^{sp}, at 550 °C are plotted against average grain size in Figure 9. Data for 2.9 mol% Y$_2$O$_3$-doped ZrO$_2$ with an average grain size of about 41 nm are also plotted for comparison. As shown in this figure, σ_{bulk} decreases and σ_{gb}^{sp} increases with decreasing grain size. The decreasing σ_{bulk} is probably due to the bulk "de-doping" resulting from the grain size dependent grain-boundary segregation:[66] segregation occurs over a much larger grain-boundary area as the grain size decreases, therefore, more solute within grains is "drained" to the grain boundaries.

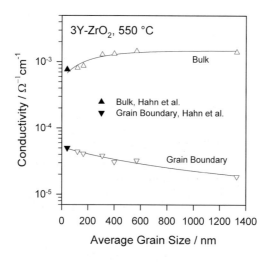

Figure 9. Bulk conductivities and specific grain-boundary conductivities for 3 mol% Y$_2$O$_3$-doped ZrO$_2$ at 550 °C as a function of average grain size (after Guo and Zhang[46]). Data for 2.9 mol% Y$_2$O$_3$-doped ZrO$_2$ with an average grain size of about 41 nm (Hahn *et al.*[65]) are also plotted for comparison.

$\Delta\varphi(0)$ values calculated from the σ_{bulk} and σ_{gb}^{sp} are presented in Figure 10: the Schottky barrier height of 3 mol% Y$_2$O$_3$-doped ZrO$_2$ at 550 °C decrease with decreasing grain size. The increasing specific grain-boundary conductivity is due to the decreasing Schottky barrier height. Substituting L_D and λ^* in Eq. (9), one gets the oxygen-vacancy profile in the space-charge layer; such profiles for 3 mol% Y$_2$O$_3$-doped

ZrO$_2$ at 550 °C are shown in Figure 11: the concentration of oxygen vacancies in the space-charge layer increases with decreasing grain size.

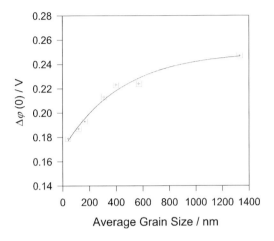

Figure 10. Schottky barrier height, $\Delta\varphi(0)$, as a function of average grain size for 3 mol% Y$_2$O$_3$-doped ZrO$_2$ at 550 °C (after Guo and Zhang[46]).

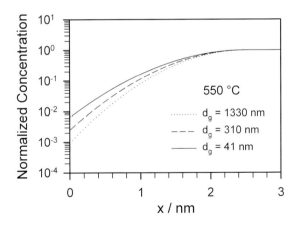

Figure 11. Normalized oxygen-vacancy concentrations, $c_{V_O^{\bullet\bullet}}(x)/c_{V_O^{\bullet\bullet}}(\infty)$, in the space-charge layers of 3 mol% Y$_2$O$_3$-doped ZrO$_2$ of different grain sizes at 550 °C (after Guo and Zhang[46]).

Although the specific grain-boundary conductivity increases with decreasing grain size, even for the finest grain size (41 nm) the specific grain-boundary conductivity is still more than one order of magnitude lower than the bulk conductivity.

Chiang *et al.*[66] achieved a grain size of about 10 nm for 26 mol% Gd_2O_3-doped CeO_2; but even for such a small grain size, the specific grain-boundary conductivity is still about two orders of magnitude lower than the bulk conductivity. Thus the refinement of grain size actually increases the total sample resistance.

2.4. Zirconia Films with Nanometer Thickness

It is difficult to prepare bulk ceramics with nanometer-sized grains, but it is not so difficult to prepare films with nanometer thickness. Guo *et al.*[67] prepared 8 mol% Y_2O_3-doped ZrO_2 films with thicknesses of 12 and 25 nm on (100) MgO substrates by pulsed laser deposition from a stoichiometric ceramic target. The in-plane electrical conductivities, measured parallel to the film plane, are presented in Figure 12. For comparison, the electrical conductivity of 8 mol% Y_2O_3-doped ZrO_2 bulk ceramic (with an average grain size > 15 µm) is also plotted in Figure 12.

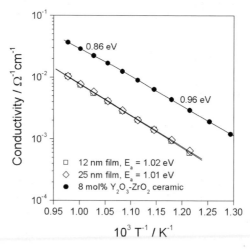

Figure 12. Comparison of electrical conductivities of nanostructured films and bulk ceramic (after Guo *et al.*[67]).

From this Figure, one can see that the ionic conductivity of the nanostructured films is lower than that of the bulk ceramic by about a factor of 4, and the activation energy for the conductivity is higher for the nanostructured films.

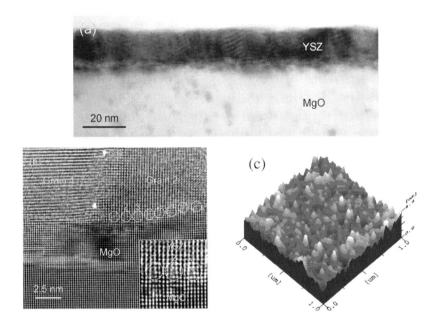

Figure 13. (a) TEM, (b) HRTEM and (c) atomic force microscope (AFM) micrographs of film with a thickness of ~12 nm. The misfit dislocations at the ZrO$_2$/MgO interface are highlighted by circles (after Guo *et al.*[67]).

Figure 13 (a) shows the TEM micrograph of the thin film with a thickness of ~12 nm. The film is continuous and homogeneous in thickness. It is quite obvious that the film is polycrystalline, consisting of columnar grains perpendicular to the substrate. The HRTEM micrograph of the film (Figure 13 (b)) clearly displays the ZrO$_2$ columnar grains, and it should be noted that the grain boundaries are free of any second phases. The ZrO$_2$/MgO interface is more interesting: occasionally the orientation $\langle 001 \rangle_{ZrO2}$ //$\langle 001 \rangle_{MgO}$ is fulfilled, and misfit dislocations are clearly observed. As demonstrated in the Figure 13 (b) inset, three MgO planes bond with two ZrO$_2$ planes at the interface, giving rise to edge dislocations at the interface that relieve the compressive strain within the YSZ film. The lattice mismatch between ZrO$_2$ and MgO is quite large (~ -17.6%); such a large lattice mismatch impedes the epitaxial growth of ZrO$_2$ films on MgO substrates, even though the crystal structures of ZrO$_2$ and MgO are both cubic. The interface is atomically nonflat, but free of any second phases. According to Figure 13 (c), the average grain size (d_g) of this film is ~76 nm. Another film, the thickness of which is ~25 nm, is also polycrystalline, consisting of columnar grains perpendicular to the substrate, and the average grain size is ~88 nm. The ZrO$_2$/ZrO$_2$ grain boundaries and the ZrO$_2$/MgO interface are free of any second phases as well.

The nanostructured ZrO$_2$ films consist of columnar grains, and the nanostructure can be schematically represented by Figure 14 (a). In the current direction, charge carriers can diffuse either parallel or perpendicular to the grain boundaries, so there are two kinds of grain-boundary resistances: the perpendicular resistance $R_{gb}^{\perp}$, which is

$$R_{gb}^{\perp} = \frac{1}{\sigma_{gb}^{\perp}} \frac{L}{A} \frac{\delta_{gb}}{d_g}, \tag{12}$$

and the parallel resistance $R_{gb}^{//}$, given by

$$R_{gb}^{//} = \frac{1}{\sigma_{gb}^{//}} \frac{L}{A} \frac{d_g}{\delta_{gb}}. \tag{13}$$

Usually the grain-boundary conductivity perpendicular to and that parallel to the current direction are different. For microstructured materials, $d_g/\delta_{gb}\sim10^3$, it is obvious that $R_{gb}^{//} \gg R_{gb}^{\perp}$; the parallel resistance $R_{gb}^{//}$ is therefore neglected, and the so-called grain-boundary resistance R_{gb} is simply $R_{gb}^{\perp}$. However, the parallel resistance $R_{gb}^{//}$ in nanostructured materials cannot be neglected, because the grain size and the grain-boundary thickness are comparable. When the parallel grain-boundary resistance $R_{gb}^{//}$ is taken into consideration, the equivalent circuit becomes Figure 14 (b). The total dc resistance of a film sample, R_{tot}, in this case is

$$R_{tot} = \frac{(R_{bulk} + R_{gb}^{\perp})R_{gb}^{//}}{R_{bulk} + R_{gb}^{\perp} + R_{gb}^{//}}.$$ (14)

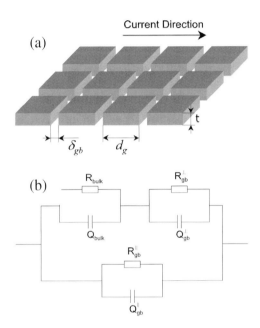

Figure 14. (a) Simplified structure of nanostructured film and (b) corresponding equivalent circuit (after Guo *et al.*[67]). Q$_{bulk}$ and Q$_{gb}$ are the constant phase elements.

Owing to the high proportion of grain-boundary regions in the nano-structured films, the grain-boundary resistances ($R_{gb}^{\perp}$ and $R_{gb}^{//}$) constitute a high proportion of the film resistance. This is consistent with the higher activation energy obtained for the film conductivity (Figure 12). In doped-ZrO$_2$, the activation energy for the grain-boundary conductivity is higher, because the ionic transport across the grain boundaries should overcome a Schottky barrier.[41]

As a result of the bulk "de-doping", the bulk conductivity of the nanostructured ZrO$_2$ films is expected to be lower than that of micro-crystalline ceramics. And owing to the oxygen-vacancy depletion in the space-charge layer, the grain-boundary conductivity remains low even when the film thickness and/or the grain size decreases to the nanometer scale. Therefore, the ionic conductivity of the nanostructured ZrO_2 films is lower than that of microcrystalline ceramics by about a factor of ~4. If nanostructured ZrO_2 films are coarsened (with larger thickness and grain size), the conductivity of the coarsened films is expected to be comparable with that of bulk materials. Kosacki *et al.*[68] demonstrated that the ionic conductivity of ZrO_2 films is the same as that of bulk ZrO_2 when the grain size is $\geq$ 60 nm.

12 nm is by far the smallest ZrO_2 film thickness reported in literature. More literature results are available for thicker films. For example, the electrical conductivity of 8 mol% Y_2O_3-doped ZrO_2 films, with thicknesses of 0.6-1.5 μm and grain sizes of 60-100 nm, was also found to be slightly smaller than that of bulk ceramics.[69] However, enhanced conductivity has been reported for nanometer thick ZrO_2 films as well.[68,70,71] Kosacki *et al.*[68,71] deposited epitaxial 9.5 mol% Y_2O_3-doped ZrO_2 films on MgO substrates. Therefore, the grain-boundary blocking effect is avoided. They found that the film conductivity increases with decreasing film thickness when the film thickness is smaller than 60 nm. The conductivity of 15-nm-thick film is about one to two orders of magnitude higher than that of films with thicknesses of 58-2000 nm. The ZrO_2/MgO interface is supposed to be responsible for the conductivity enhancement.

The nano-effects reported in Secs. 2.3 and 2.4 are just *trivial size effect*. Since a typical space-charge layer width in doped-ZrO_2 is ~2.5 nm,[41] the *true size effect* can only be expected when the grain size decreases down to a few nanometers, *e.g.* $\leq$ 5 nm. No *true size effect* has been reported for doped-ZrO_2 and CeO_2 yet.

2.5. Conclusions

(1) The potential of the grain-boundary core of doped-ZrO_2 or CeO_2 is positive; the enrichment of oxygen vacancies in the grain-boundary core is most probably responsible for the positive potential. The positively charged grain-boundary core expels oxygen vacancies, while it attracts acceptor cations, thus causing the oxygen-vacancy depletion and the acceptor accumulation in the space-charge layer.

(2) The grain-to-grain contacts are electrically resistive in nature, which is due to the oxygen-vacancy depletion in the space-charge layer; this is the decisive cause of the grain-boundary blocking effect. A Schottky barrier model explains the grain-boundary electrical properties. The activation energy for the grain-boundary conductivity is determined by the Schottky barrier height.

(3) Only *trivial size effect* has been observed for doped-ZrO_2 or CeO_2. The grain-boundary conductivity increases with decreasing grain size; this is due to the decreasing Schottky barrier height. As a result, the concentration of oxygen vacancies in the space-charge layer increases with decreasing grain size. But the grain-boundary conductivity is still about 2 orders of magnitude lower than the bulk conductivity, even when the grain size is on the nanoscale, *e.g.* 41 nm for doped-ZrO_2 and 10 nm for doped-CeO_2. The conductivity of 12-nm-thick ZrO_2 film is also lower than that of bulk ceramics.

3. Mixed Conductors of Oxygen Ions and Electrons

Owing to the positively charged grain-boundary core, electrons, as a result of the negative charge, should accumulate in the space-charge layer when present. The accumulation of electrons in the space-charge layer can be of importance in situations of low oxygen partial pressures, high temperatures and high space-charge potentials.

Doped and undoped CeO_2 can be readily reduced to introduce electrons into the lattice, according to

$$O_O^\times \leftrightarrow 2e' + V_O^{\bullet\bullet} + 1/2 O_2 ,$$ (15a)

$$K_{Re}(T) = c_{V_O^{\bullet\bullet}} n^2 pO_2^{1/2} = K_{Re}^0 \exp\left(-\frac{\Delta H_{Re}}{k_B T} \right) ,$$ (15b)

where K_{Re}^0 is the pre-exponential constant, ΔH_{Re} the enthalpy of reduction. Slightly doped and undoped CeO_2 materials are mixed conductors of oxygen vacancies and excess electrons, with the oxygen vacancies being doubly ionized over a wide range of temperature and oxygen partial pressure. Oxygen vacancies are mainly generated by introducing acceptor dopants (even in undoped CeO_2 of high purity, background impurities are still considerable[52]). It is generally agreed that the *n*-type (electron) conductivity takes place by small polaron transport.[72,73] The mobility of electrons is[73]

$$\mu_{eon}(T) = \frac{3.9 \times 10^2 \, cm^2 KV^{-1} s^{-1}}{T} \exp\left(-\frac{0.4 eV}{k_B T}\right). \tag{16}$$

The mobility of oxygen vacancies $\mu_{ion}(T)$ is roughly one order of magnitude lower than $\mu_{eon}(T)$.[74]

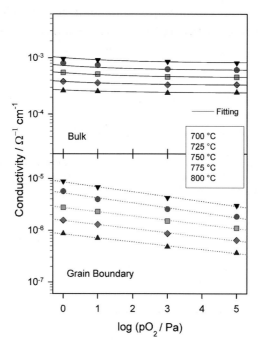

Figure 15. Oxygen partial pressure dependences of the specific grain-boundary conductivity and the bulk conductivity of 0.1 mol% Y_2O_3-doped CeO_2. The bulk conductivity line curves are the fitting results according to Eq. (17) (after Guo *et al.*[52]).

3.1. Microcrystalline CeO₂

Guo *et al.*[52] studied 0.1 mol% Y_2O_3-doped CeO_2 of high purity, in which the effect of the siliceous phase is negligible. 0.1 mol%

Y_2O_3-doped CeO_2 samples show remarkable *n*-type conductivity at low oxygen partial pressures and/or high temperatures, and the expected electron accumulation in the space-charge layer was demonstrated.

The oxygen partial pressure dependences of 0.1 mol% Y_2O_3-doped CeO_2 are presented in Figure 15. At temperatures higher than 700 °C, the bulk conductivities slightly increase with decreasing oxygen partial pressure, while the grain-boundary conductivities significantly increase. To give an example, at 800 °C the bulk conductivity increases by about 25 % when the oxygen partial pressure decreases from 10^5 to 2 Pa, whereas the grain-boundary conductivity increases by a factor of 3. This significant increase in the grain-boundary conductivity indicates that the electronic partial conductivity is more pronounced at the grain boundaries, being consistent with the expected accumulation of electrons in the space-charge layer.

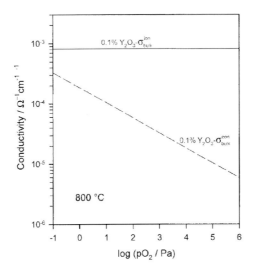

Figure 16. Ionic and electronic partial conductivities in the bulk of 0.1 mol% Y_2O_3-doped CeO_2 at 800 °C as obtained from fitting according to Eq. (17) (after Guo *et al.*[52]).

The conductivities given in Figure 15 are all total (ionic plus electronic) conductivities. The oxygen partial pressure dependence of the bulk total conductivity (σ_{bulk}) can be expressed by

$$\sigma_{bulk} = \sigma_{bulk}^{ion} + \sigma_{bulk}^{eon} = \alpha + \beta \, pO_2^{-1/m} \qquad (17)$$

with $\alpha = \sigma_{bulk}^{ion}$ and $m = 4$ in the extrinsic region, while $\alpha = 0$ and $m = 6$ in the intrinsic region. By fitting experimental σ_{bulk} vs. pO_2 relations according to Eq. (17), one can separate the ionic and electronic partial conductivities (σ_{bulk}^{ion} and σ_{bulk}^{eon}) in the bulk. The fitting results of the bulk total conductivities are given in Figure 15, and the ionic and electronic partial conductivities in the bulk at 800 °C thus determined for 0.1 mol% Y_2O_3-doped CeO_2 are presented in Figure 16. As shown in Figure 16, in the oxygen partial pressure range of 2 to 10^5 Pa, the bulk is predominantly ionically conductive, *i.e.* $\sigma_{bulk}^{ion} > \sigma_{bulk}^{eon}$ at 800 °C. The bulk concentrations, $c_{V_O^{\bullet\bullet}}(\infty)$ and $n(\infty)$, can be calculated from the bulk conductivities according to $c_k = \sigma_k/z_k\mu_k F$, where c_k denotes the concentration of the species k, z_k denotes the charge number ($z_k = 2$ for oxygen vacancies and $z_k = 1$ for electrons), μ_k the mobility, F the Faraday constant.

The effective grain-boundary thickness, δ_{gb}, can be calculated from $d_g C_{bulk}/C_{gb}$. The effective grain boundary thickness of 0.1 mol% Y_2O_3-doped CeO_2 is around 12 nm, and it is temperature and oxygen partial pressure dependent. For $pO_2 = 10^5$ Pa and at low temperatures, the electron concentration in 0.1 mol% Y_2O_3-doped CeO_2 is still much smaller than the dopant concentrations, the Schottky barrier model then still holds. Therefore, as an approximation, Eq. (9) may still be used to calculate the concentration of oxygen vacancies, $c_{V_O^{\bullet\bullet}}(x)$, in the space-charge layer. Then the electron concentration, $n(x)$, can be calculated from

$$\frac{n(x)}{n(\infty)} = \left(\frac{c_{V_O^{\bullet\bullet}}(x)}{c_{V_O^{\bullet\bullet}}(\infty)} \right)^{-1/2}. \qquad (18)$$

The profiles of oxygen vacancies and electrons thus obtained for 0.1 mol% Y_2O_3-doped CeO_2 at 500 °C are presented in Figure 17. It must be emphasized here that an inversion layer is formed even at relatively low temperature (*e.g.* 500 °C) and high oxygen partial pressure (10^5 Pa) in

0.1 mol% Y_2O_3-doped CeO_2. Within the inversion layer, the electron concentration is higher than the oxygen-vacancy concentration.

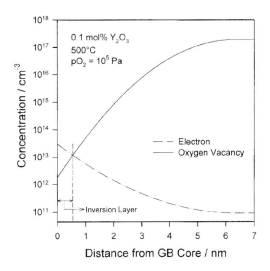

Figure 17. Profiles of oxygen vacancies and electrons in space-charge layer of 0.1 mol% Y_2O_3-doped CeO_2 at 500 °C and under $pO_2 = 10^5$ Pa (after Guo *et al.*[52]).

With increasing temperature and/or decreasing oxygen partial pressure, the electron concentration in the space-charge layer becomes higher, and the inversion layer becomes thicker. After annealing in a reducing atmosphere, for example, 2% H_2-Ar mixture, the electronic conductivity dominates in 0.1 mol% Y_2O_3-doped CeO_2. The inversion layer should be dominant in the space-charge layer as well; the space-charge layer thus becomes even more conductive than the bulk. The electrical conductivity of 0.1 mol% Y_2O_3-doped CeO_2 after annealing in the mixture of 2% H_2-Ar as a function of inverse temperature is presented in Figure 18. Please note that the activation energy is very close to 0.4 eV, which is the migration enthalpy of electrons in CeO_2,[72] suggesting that the conduction is purely electronic.

Electrons can also be introduced into ZrO_2 by, for example, doping with TiO_2 [75] or annealing under extremely reducing condition.[76] The accumulation of electrons in the ZrO_2 space-charge layer has also been demonstrated.

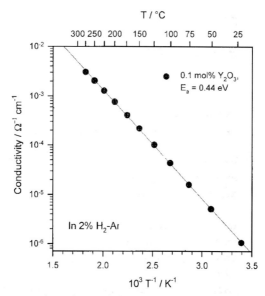

Figure 18. Temperature dependence of the conductivity of 0.1 mol% Y_2O_3-doped CeO_2 after annealing in 2 % H_2-Ar mixture (after Guo et al.[52]).

3.2. Nanocrystalline CeO₂

With the comparatively high electronic bulk contribution and high-density of grain boundaries, the grain boundaries in nanocrystalline CeO_2 can become electronically conducting and dominate the overall behavior. The electrical conductivity of nanocrystalline CeO_2 is usually several orders of magnitude higher than that of microcrystalline counterparts.[49-51,55,77] Such an observation was initially attributed to the enhanced concentration of electrons in the grain-boundary core in nanocrystalline CeO_2.[77] The lower activation energy of the conductivity, thus the defect formation enthalpy, measured for nanocrystalline CeO_2 supports this argument, and it is also consistent with a theoretical calculation.[78] Note that in the *neutral layer model* the electroneutrality condition in the grain-boundary core is assumed.

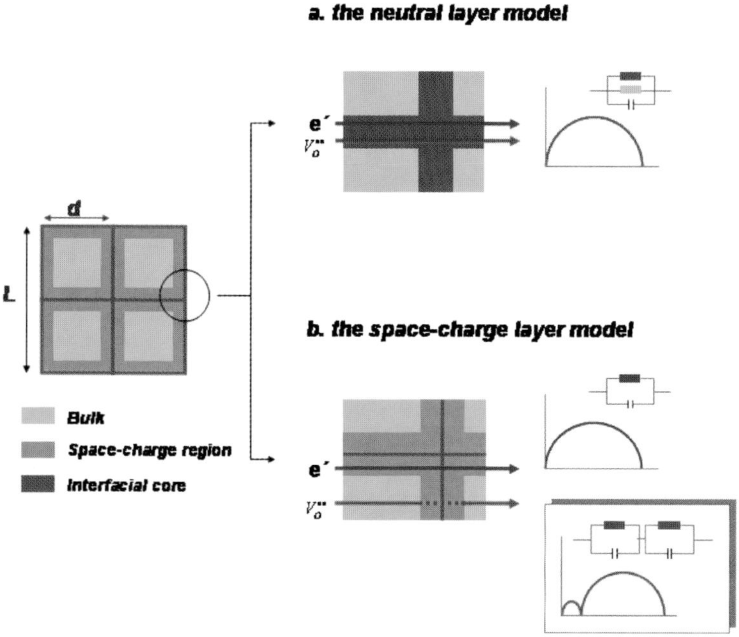

a. the neutral layer model

b. the space-charge layer model

Figure 19. Schematic comparison of the neutral layer and the space-charge models. The impedance spectra anticipated from both models are also shown (after Kim and Maier[51]).

The *a priori* significance of the space-charge concept lies in the fact that positive space-charge potentials of ~0.3-0.5 V have been obtained for CeO_2,[52] as discussed in the previous section. The magnitude of ~0.3 V for CeO_2 is sufficient to cause an electronic conductivity enhancement of the observed orders of magnitude for nanocrystalline CcO_2. It becomes clear that a reliable answer as to which explanation is correct requires a direct separate determination of ionic and electronic conductivities and a quantitative analysis of the models. While in the neutral layer model, the ion conductivity should also — as long as the mobility of the vacancies is not significantly lowered — be enhanced in the grain-boundary core, owing to Eq. 15(a). A severe depletion of $V_O^{\bullet\bullet}$ and hence a severe depression of the ionic conductivity should occur in the space-charge region.

By separately determining the electronic and ionic contributions in the nominally pure and 0.15 mol % Gd-doped nanocrystalline CeO_2 (the

grain size is about 30 nm) employing ion blocking cells, Kim and
Maier[51] proved that the much higher conductivity of nanocrystalline
CeO_2 is due to the electron accumulation in the space-charge layer.
Under these conditions, electrons are expected to flow along the grain
boundaries parallel to the current direction, while the migration of
oxygen vacancies is expected to occur in the bulk, interrupted by the
grain boundaries perpendicular to the current direction. Figure 19
schematically distinguishes between the neutral layer model and the
space-charge model. The impedance spectra for electronic and ionic
conduction in nanocrystalline CeO_2 predicted by both models are also
shown in Figure 19.

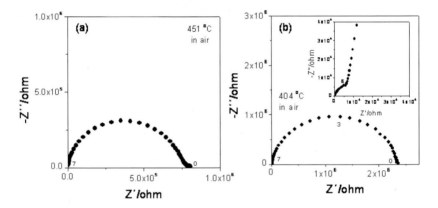

Figure 20. The impedance spectra measured from (a) nominally pure and (b) 0.15 mol %
Gd-doped nanocrystalline CeO_2. The inset shown in (b) presents the spectrum at high
frequencies (after Kim and Maier[51]).

Figure 20 shows the measured impedance spectra. Indeed, Figure
20(b) shows two separated semicircular arcs, while Figure 20(a) shows
only one semicircular arc. Note that such spectra are expected only based
on the space-charge model as shown in Figure 19.

Furthermore, the oxygen partial pressure and the temperature
dependences of the electronic and ionic conductivities of nanocrystalline
CeO_2 were precisely reproduced by the Schottky barrier model. The
space-charge potential estimated using Eq. (3) is 0.3 V, and the width of
the space-charge layer is 2-3 nm depending upon temperature, consistent

with the value measured for microcrystalline CeO_2. Figure 21 shows the oxygen-vacancy concentration profile in the space-charge layer in 0.15 mol % Gd-doped CeO_2 calculated using Eq. (9). The width of the space-charge layer shown in Figure 21 is consistent with the measured value.

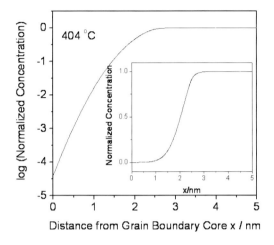

Figure 21. Normalized oxygen-vacancy concentration, $c_{v_o^{..}}(x)/c_{v_o^{..}}(\infty)$, in the space-charge layer at 404 °C when the space-charge potential is 0.3 V. Insert is the linear scale (after Kim and Maier[51]).

3.3. Conclusion

While present, electrons are accumulated in the space-charge layer; such a phenomenon has been demonstrated in micro- and nanocrystalline CeO_2 ceramics. With the comparatively high electronic bulk contribution and high density of grain boundaries, the grain boundaries in nanocrystalline CeO_2 becomes electronically conducting and dominate the overall behavior. Under these conditions, electrons are expected to flow along the grain boundaries parallel to the current direction, while the migration of oxygen vacancies is expected to occur in the grain bulk, interrupted by the grain boundaries perpendicular to the current direction.

4. Mixed Conductors of Oxygen Ions and Holes

In oxidizing to moderately reducing atmosphere, typically 10^{-5} to 10^5 Pa at 600 K, acceptor-doped $SrTiO_3$ or $BaTiO_3$ is a mixed conductor of oxygen vacancies and holes. Electrons are also present, however, with a bulk concentration several orders of magnitude lower than those of oxygen vacancies and holes.[79-82]

A positive potential in the grain-boundary core has also been found in nominally undoped and acceptor-doped $SrTiO_3$, which is ascribed to an enhanced concentration of oxygen vacancies in the grain-boundary core, and the excess of oxygen vacancies (relative to the bulk) in the core is theoretically demonstrated to be energetically favorable.[83-88] An atomic resolution analysis[84] shows that the incomplete oxygen octahedra in the grain-boundary core act as immobile effective oxygen vacancies and lead to a fixed, positive core charge. The positive core charge leads to the depletion of oxygen vacancies and holes in the space-charge layer, giving rise to a back-to-back double Schottky barrier situation. As a result, the specific conductivity of the grain boundary is several orders of magnitude lower than the bulk conductivity.[89-97]

The constancy of the electrochemical potential of oxygen vacancies and holes demands, for a dilute solution, that

$$\frac{p^2(x)}{p^2(bulk)} = \frac{c_{V_O^{\bullet\bullet}}(x)}{c_{V_O^{\bullet\bullet}}(bulk)} = \exp\left(-\frac{2e\Delta\varphi(x)}{kT}\right), \tag{19}$$

the depletion of oxygen vacancies is thus quadratic, compared with that of holes. The comparison of the concentrations of various charge carriers in the bulk and at the grain boundary ought to deliver significant information on the validity of this space-charge picture. In order to enable this, ionic and electronic contributions to the grain-boundary conductivity have to be separated.

4.1. Electronic and Ionic Contributions to the Grain-Boundary Conductivity

Here we demonstrate how the ionic and electronic partial conductivities of a blocking grain boundary perpendicular to the current direction can be determined, and applied it to a Fe-doped $SrTiO_3$ bicrystal. More detailed description can be found in Ref. 98.

As illustrated in Figure 22, if a single crystal with identical geometry and composition is used to "imitate" the grain bulk of a bicrystal, the electronic and ionic contributions to the grain-boundary conductivity of the bicrystal can be separated by means of the Hebb-Wagner polarization. The ionic and electronic partial conductivities of the grain bulk can be obtained from the single crystal, and the ionic and electronic partial conductivities of the single grain boundary can be obtained from the bicrystal.

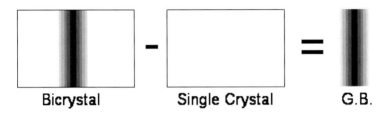

Bicrystal　　　　**Single Crystal**　　　　**G.B.**

Figure 22. Imitation of the grain bulk of a bicrystal by a single crystal with identical geometry and composition (G.B.: grain boundary).

Free polished surfaces as well as Au coated surfaces exhibit such a sluggish oxygen exchange rate for $T < 750$ K that neither a special blocking electrode nor a glass-sealing of the non-contacted surfaces is necessary to perform such a stoichiometric polarization on $SrTiO_3$.[89,90] This inhibition is efficiently suspended at $YBa_2Cu_3O_{6+x}$ (YBCO) electrodes.[89,90] Hence, a reversible YBCO electrode and an ionically blocking Au electrode were used to determine the electronic partial conductivity of $SrTiO_3$. All experiments were carried out on a $SrTiO_3$ single crystal and a $SrTiO_3$ bicrystal with a $\Sigma 5$ grain boundary perpendicular to the current direction. The single crystal and the bicrystal were both prepared from a $SrTiO_3$ single crystal boule doped with 0.016 wt.% Fe. The single crystal was used for the purpose of comparison, and as already pointed out, to "imitate" the grain bulk of the bicrystal. The $\Sigma 5$ grain boundary is a symmetric boundary with a tilt angle of 36.8° with respect to the grain-boundary plane (310).

Experimentally obtained electronic and ionic partial conductivities of the grain boundary of the $SrTiO_3$ bicrystal are given in Figure 23, with corresponding bulk values. The atmosphere used during measurements was argon with an oxygen partial pressure of about 2.0 Pa. As shown in this figure, the electronic and ionic partial conductivities in the bulk are comparable, and the ionic partial conductivity becomes increasingly important with decreasing temperature; contrary to the bulk, the electronic partial conductivity is always dominant at the grain boundary. This is qualitatively exactly what we expect from the space-charge picture.

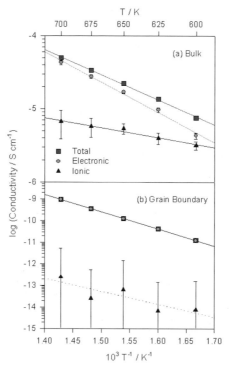

Figure 23. Temperature dependence of total and partial conductivities of (a) bulk and (b) grain boundary (after Guo *et al.*[98]).

If estimated from the defect chemistry,[90,99] the electron concentration in the bulk of the $SrTiO_3$ bicrystal is 7 to 9 orders of magnitude lower than the hole concentration in the temperature range of 700 to 600 K.

Therefore, the electronic partial conductivities in the bulk shown in Figure 23(a) can be almost solely attributed to holes. The concentrations of charge carriers in the bulk can be calculated from the bulk conductivities; calculated concentrations at 700 and 600 K are shown in Figure 24. At both temperatures, the concentrations of oxygen vacancies are about 3 to 4 orders of magnitude higher than those of holes; but due to the much higher mobility of holes, the electronic partial conductivities are always higher.

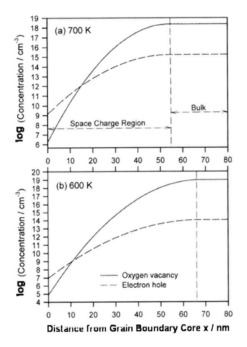

Figure 24. The concentration profiles of oxygen vacancies and electron holes in the bulk and in the space-charge region at (a) 700 K and (b) 600 K (after Guo *et al.*[99]).

Since the hole conduction is dominant both in the grain bulk and at the grain boundary, and $z = 1$ for holes, thus

$$\frac{\sigma_{bulk}}{\sigma_{gb}} = \frac{\exp\left(e\Delta\varphi(0)/k_{B}T\right)}{2e\Delta\varphi(0)/k_{B}T}. \tag{20}$$

The Schottky barrier height, $\Delta\varphi(0)$, determined from Eq. (20) is ~0.84 V. And the grain-boundary thickness, δ_{gb}, calculated from the grain-boundary capacitance (Eq. (5)), is in the range of about 110 to 130 nm when the temperature decreases from 700 K to 600 K. Calculated from the Schottky barrier height, the concentration of oxygen vacancies at $x = 0$ is about 12 orders of magnitude lower than the bulk value, and the concentration of holes about 6 orders of magnitude lower at 700 K. The oxygen-vacancy concentration profile is given by Eq. (9) (see Sec. 2.2.2), and the hole concentration profile by Eq. (19). The calculated profiles of oxygen vacancies and holes in the space-charge region at 700 and 600 K are also plotted in Figure 24. As shown in this figure, (*i*) both charge carriers are depleted in the space-charge layer, and depletion of oxygen vacancies is more severe; (*ii*) the space-charge layer becomes thicker at lower temperature.

4.2. Thickness Dependent p-Type Conductivity of Epitaxial SrTiO₃ Thin Films

Unlike ZrO_2 and CeO_2, the space-charge layer width of $SrTiO_3$ can be as large as tens of nanometers; one good example is Figure 24. This is due to the relatively large dielectric constant and low dopant concentration. Therefore, the effects of the space-charge layer are very pronounced in thin films and ceramics with fine grain size. The effects of the space-charge layer in epitaxial $SrTiO_3$ thin films[100] are analyzed in the following.

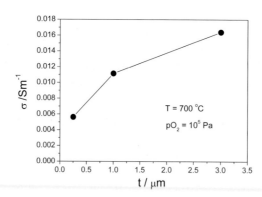

Figure 25. Thickness dependence of the *p*-type conductivities of the epitaxial SrTiO₃ thin films at 700 °C (after Guo *et al.*[100]).

The epitaxial (100) $SrTiO_3$ thin films with thicknesses of 3 µm, 1 µm and 250 nm were prepared by pulsed laser deposition (PLD) on (100) MgO substrates. The complication of grain boundaries is avoided in epitaxial thin films. MgO is highly insulating, and the lattice mismatch between MgO and the perovskite thin film is sufficiently low to allow for the growth of single crystalline titanate thin films. Furthermore, the thermal expansion coefficients are favorable to the stability of the films. Though nominally undoped, the chemical analysis of impurity concentrations revealed a slight presence of acceptor type elements, mainly Fe <150 ppm, Mg <40 ppm and Sc <193 ppm.

The existence of the space-charge layer at the titanate surfaces has been experimentally demonstrated by the ^{18}O tracer diffusion experiments.[101,102] In bulk materials, the effect of the surface space-charge layer is negligible in most cases. However, the surface space-charge layer plays an important role in thin films due to the very large surface-to-volume ratio. Epitaxial thin films of different thicknesses allow studying the extent to which the distance between surfaces as well as the surface itself affects the conduction characteristics. The thickness dependence of the p-type conductivities of the epitaxial $SrTiO_3$ thin films, measured at 700 °C in pure oxygen, is displayed in Figure 25. In the p-type region, the absolute conductivity values decrease with decreasing film thickness.

In the measurement current direction (y axis), the resistance, ΔR_{sc}, of a slice in the space-charge layer with a thickness of Δx is (see Figure 26):

$$\Delta R_{sc} = \frac{1}{\sigma(x)} \frac{L}{w\Delta x}. \tag{21}$$

The $SrTiO_3$/MgO interface is different from the $SrTiO_3$/atmosphere interface, but it is not possible to separate the $SrTiO_3$/MgO interface and the $SrTiO_3$/atmosphere interface with present electrical characterization techniques. Therefore, as an approximation, both interfaces are assumed to be identical. Then the resistance, R_{sc}, of two space-charge layers is:

$$\frac{1}{R_{sc}} = 2\sum \frac{1}{\Delta R_{sc}} = 2\sum \frac{w}{L}\sigma(x)\Delta x = \frac{2w}{L} \int_0^{\lambda^*} \sigma(x)dx. \tag{22}$$

In Eqs. (21) and (22), $\sigma(x)$ is the conductivity at locus x. The resistance, R_{in}, of the inner portion (excluding the space-charge layers) is

$$R_{in} = \frac{1}{\sigma(in)} \frac{L}{w(t - 2\lambda^*)} .$$ (23)

Therefore, the conductivity, σ, of the film is

$$\sigma = \sigma(in)\left[\left(1 - \frac{2\lambda^*}{t}\right) + \frac{2}{t}\int_0^{\lambda^*} \frac{\sigma(x)}{\sigma(in)} dx\right] .$$ (24)

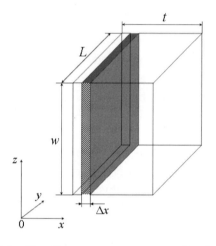

Figure 26. Sketch of the film. The measurement current direction is parallel to axis y (after Guo et al.[100]).

When $t \gg 2\lambda^*$, a more detailed analysis is possible. When $t \gg 2\lambda^*$, the Poisson equation can be analytically solved for semi-infinite boundary conditions. For acceptor-doped $SrTiO_3$, $\Delta\varphi(0)$ was determined to be mostly in the range of 0.5 to 0.8 V.[89,99,103] Taking $\Delta\varphi(0) = 0.5$ V, λ^* of the epitaxial $SrTiO_3$ thin films is estimated from the impurity level to be ~15 nm at 700 °C. Even for the 250 nm thick film, the film thickness is still much larger than the combination of two space-charge layers. Therefore, we have $t \gg 2\lambda^*$ for the films. According to the Schottky barrier model, the electrostatic potential, $\Delta\varphi(x)$, of the space-charge layer is[89]

$$\Delta\varphi(x) = \frac{ec_A}{2\varepsilon}(x - \lambda^*)^2 ,$$ (25)

and the space-charge layer width, λ^*, is[89]

$$\lambda^* = \left(\frac{2\varepsilon}{ec_A} \Delta\varphi(0) \right)^{1/2} . \tag{26}$$

In Eqs. (25) and (26), c_A is the acceptor concentration.

In the p-type region, acceptor-doped SrTiO$_3$ is a mixed conductor of oxygen vacancies and electron holes, however, with the hole conduction being dominant. Oxygen vacancies and holes are depleted in the space-charge layers at the surfaces. Such a situation for the epitaxial SrTiO$_3$ thin films is depicted in Figure 27.

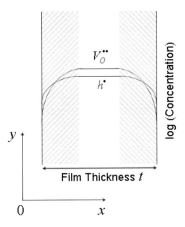

Figure 27. Oxygen-vacancy and hole concentration profiles across epitaxial SrTiO$_3$ thin film in the p-type region. The space-charge layers are the shaded regions (after Guo et al.[100]).

At equilibrium, the p-type conductivity is given by

$$\frac{\sigma_p(x)}{\sigma_p(in)} = \exp\left(-\frac{e\Delta\varphi(x)}{k_B T} \right), \tag{27}$$

where $\sigma_p(x)$ and $\sigma_p(in)$ are the p-type conductivity at locus x and the inner portion of the film, respectively. Accordingly, Eq. (24) becomes

$$\sigma_p = \sigma_p(in)\left[\left(1 - \frac{2\lambda^*}{t} \right) + \frac{2}{t}\int_0^{\lambda^*} \exp\left(-\frac{e\Delta\varphi(x)}{k_B T} \right) dx \right]. \tag{28}$$

The integration of Eq. (28) yields[100]

$$\sigma_p = \sigma_p(in)\left[1-\left(2-\frac{k_BT}{e\Delta\varphi(0)}\right)\frac{\lambda^*}{t}\right].\tag{29}$$

Since the value of λ^*/t increases with decreasing film thickness, then according to Eq. (29), the p-type conductivity decreases with decreasing film thickness.

In summary, the surface space-charge layers play an increasingly important role in the charge carrier transport with decreasing film thickness. In the p-type region, oxygen vacancies and holes are depleted in the space-charge layers, and the proportion of the depletion layers increases with decreasing film thickness, which causes the decreasing p-type conductivity with decreasing film thickness. The film thickness dependent conductivity is due to the *trivial size effect*.

4.3. Overlapping of Neighboring Space-Charge Layers in Nanocrystalline SrTiO₃

Because of the relatively large space-charge layer width, there is a good chance to achieve the overlapping of neighboring space-charge layers in nanostructured SrTiO₃. Balaya *et al.*[104,105] prepared nanocrystalline SrTiO₃ ceramics with a grain size of 80 nm. The acceptor impurity level was determined to be about 100 ppm, and the oxygen partial pressure dependence of the electrical conductivity is described by a power law with an exponent +0.21, indicating p-type conductivity. The space-charge potential was estimated to be about 0.20 V. The space-charge layer width calculated from the impurity level and the space-charge potential is then 48 nm, which is more or less half the grain size. Therefore, the neighboring space-charge layers are expected to overlap. The following observations support this argument:

1. The impedance spectrum of nanocrystalline SrTiO₃ shows only one semicircle, whereas the grain bulk and the grain boundary impedances are well separated for microcrystalline SrTiO₃.

2. The dielectric constant of nanocrystalline SrTiO₃ is very close to that of the grain bulk of microcrystalline SrTiO₃.

In addition, a simulation based on the space-charge layer overlapping reproduces the major features of the measured impedance spectra.

When the neighboring space-charge layers overlap, oxygen vacancies and holes are depleted over entire grains; therefore, even the grain bulk is negatively charged. However, the material remains electrically neutral, because the negative charge of the grain bulk is compensated by the positive charge of the grain-boundary core.

4.4. Enhancement of p-Type Conductivity in Nanocrystalline BaTiO₃

BaTiO₃ ceramic samples, nominally undoped but actually doped with acceptor impurities (mainly 82 ppm Mn, the others, *e.g.* Al, Sc, Fe, *etc.*, are below the detection limit), with an average grain size of ~35 nm were prepared and the electrical properties investigated by Guo *et al.*[106] For comparison, microcrystalline samples with an average grain size of ~5.6 μm were prepared by firing the nanocrystalline samples at 1100 °C for 2 hours. Figure 28 shows the TEM micrograph of the nanocrystalline BaTiO₃ ceramic, indicating an average grain size of ~35 nm. The HRTEM micrograph shown in the inset demonstrates that the grain boundaries were free of impurity phases.

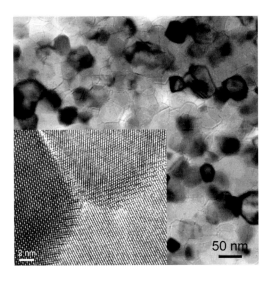

Figure 28. TEM micrograph of BaTiO₃ ceramic with an average grain size of ~35 nm. The HRTEM micrograph in the inset demonstrates atomically abrupt grain boundaries with no intergranular impurity phase (after Guo *et al.*[106]).

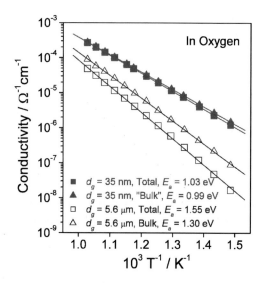

Figure 29. Temperature dependence of the conductivities of nano- and microcrystalline BaTiO₃ samples measured in pure oxygen (after Guo *et al.*[106]).

The electrical conductivities of the nano- and microcrystalline BaTiO₃ ceramics measured in pure oxygen ($pO_2 = 10^5$ Pa) are plotted in Figure 29 as a function of temperature. Within the temperature range of 400 to 700 °C (well above the ferroelectric transformation temperature (~120 °C[107])), the conductivity of the nanocrystalline sample is about 1 to 2 orders of magnitude higher than that of the microcrystalline sample, and the activation energy is remarkably lower for the nanocrystalline sample.

As the samples are doped with acceptor impurities, the principle defect reaction in an oxidizing atmosphere is[108,109]

$$V_O^{\bullet\bullet} + 1/2 O_2 \leftrightarrow O_O^x + 2h^\bullet, \tag{30a}$$

$$K_{Ox}(T) = \frac{p^2}{[V_O^{\bullet\bullet}]pO_2^{1/2}} = K_{Ox}^0 \exp\left(-\frac{\Delta H_{Ox}}{k_B T}\right). \tag{30b}$$

In Eq. (30b), ΔH_{Ox} is the oxidation enthalpy, and K_{Ox}^0 the pre-exponential constant. Oxygen vacancies are generated by the compensation of acceptor impurities. Owing to the higher mobility of holes, the conductivity of the nanocrystalline sample is of *p*-type, as demonstrated by the oxygen partial pressure dependence (Figure 30). Although an electron accumulation at the grain boundaries similar to nanocrystalline CeO_2 is expected for acceptor-doped $BaTiO_3$, the higher conductivity of nanocrystalline $BaTiO_3$ shown in Figure 29 cannot be due to the electron accumulation at the grain boundaries; this is because of the *p*-type conductivity of nanocrystalline $BaTiO_3$.

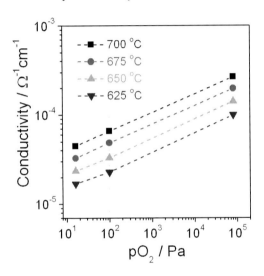

Figure 30. Oxygen partial pressure dependence of the bulk conductivity of nano-crystalline $BaTiO_3$ (after Guo *et al.*[106]).

At 700 °C, both the nano- and microcrystalline samples were annealed under pO_2 = 2 Pa for 30 hours, afterwards, pure oxygen as introduced and both samples were annealed in oxygen for another 30 hours. This treatment caused a weight gain of ~0.023 wt.% in the microcrystalline sample, whereas ~0.098 wt.% in the nanocrystalline sample. This phenomenon suggests that the $K_{Ox}(T)$ value of Eq. (30b) is higher for nanocrystalline $BaTiO_3$, which may be due to a smaller oxidation enthalpy for nanocrystalline $BaTiO_3$. From Eq. (30b), one readily gets $p \sim \exp(-\Delta H_{Ox}/2k_B T)$. In addition, the mobility $\mu \sim \exp(-\Delta H_m / k_B T)$, so that the conductivity $\sigma \sim \exp\left(-\dfrac{\Delta H_{Ox}/2 + \Delta H_m}{k_B T}\right)$ as $\sigma = ep\mu$; therefore, the activation energy for the p-type conductivity $E_a = \Delta H_{Ox}/2 + \Delta H_m$. In the above equations, ΔH_m is the migration enthalpy of holes. For microcrystalline BaTiO$_3$, ΔH_{Ox} was determined to be ~0.92 eV.[108] Assuming that ΔH_m is similar for micro- and nanocrystalline BaTiO$_3$, one estimates ΔH_{Ox} to be ~0.3 eV for nanocrystalline BaTiO$_3$ from the activation energies for the bulk conductivities given in Figure 29. This indicates an amazing reduction of the oxidation enthalpy. A smaller ΔH_{Ox} for nanocrystalline BaTiO$_3$ naturally results in a higher hole concentration and higher p-type conductivity.

At nanometer scale, the defect thermodynamics of BaTiO$_3$ is most probably dominated by the grain boundaries. This is a *true size effect*.

4.5. Conclusion

Oxygen vacancies and holes are depleted in the space-charge layer of acceptor-doped SrTiO$_3$ and BaTiO$_3$, resulting in a grain-boundary conductivity orders of magnitude lower than the bulk conductivity. The grain-boundary electrical properties can be described by a Schottky barrier model. The space-charge layer width of acceptor-doped SrTiO$_3$ and BaTiO$_3$ is comparatively large, therefore, both *trivial size effect* and *true size effect* have been observed. The surface space-charge layers play an increasingly important role in the charge carrier transport with decreasing film thickness. This is a *trivial size effect*, as long as the film thickness is considerably larger than the space-charge layer width. The work on nanocrystalline BaTiO$_3$ demonstrates that the defect thermodynamics is different when the grain size decreases to the nanometer scale. This is a *true size effect*.

5. Concluding Remarks

The *trivial size effect* only modifies the electrical properties of nanostructured oxides. With continuously decreasing feature size (*e.g.*

grain size or film thickness), one can observe the continuous variation of the electrical properties. This is because the impact of the interfacial properties on the overall materials properties increases continuously with decreasing feature size. Doped-ZrO_2 or CeO_2 ceramics best embody the *trivial size effect*. The grain-boundary conductivity of nanocrystalline ZrO_2 and CeO_2 increases with decreasing grain size; however, it still remains more than one order of magnitude lower than the bulk conductivity, even when grain size is only 10 nm. Owing to the high-density of grain boundaries, the total conductivity of nanocrystalline ZrO_2 or CeO_2 is lower than that of microcrystalline ZrO_2 or CeO_2.

The nanostructured oxides exhibit distinctively different electrical properties when the *true size effect* appears. The grain-boundary conductivity of $SrTiO_3$ and $BaTiO_3$ is many orders of magnitude lower than the bulk conductivity. Therefore, the *p*-type conductivity decreases with decreasing feature size. Such a phenomenon has been observed for $SrTiO_3$ epitaxial thin films. However, comparing with microcrystalline $BaTiO_3$, the *p*-type conductivity of nanocrystalline $BaTiO_3$ is about one to two orders of magnitude higher and the activation energy remarkably lower. This phenomenon is ascribed to a greatly reduced oxidation enthalpy in nanocrystalline $BaTiO_3$ (~0.3 eV vs. ~0.92 eV for micro-crystalline $BaTiO_3$). The greatly reduced oxidation enthalpy is due to the *true size effect*.

References

1. C. A. Angell, C. Liu and E. Sanchez, *Nature*, 362, 137 (1993).
2. N. Q. Minh and T. Takahashi, *Science and Technology of Ceramic Fuel Cells* (Elsevier, Amsterdam, 1995).
3. K. D. Kreuer, S. J. Paddison, E. Spohr and M. Schuster, *Chem. Rev.*, 104, 4637 (2004).
4. C. O. Park, S. A. Akbar and W. Weppner, *J. Mater. Sci.*, 38, 4639 (2003).
5. T. Norby and Y. Larring, *Solid State Ionics*, 136-137, 139 (2000).
6. T. H. Etsell and S. N. Flengas, *Chem. Rev.*, 70, 339 (1970).
7. J. Maier, *Nature Mater.*, 4, 805 (2005).
8. J. Maier, *Solid State Ionics*, 148, 367 (2002).
9. J. Maier, *Solid State Ionics*, 154, 291 (2002).
10. J. Maier, *Solid State Ionics*, 157, 327 (2003).
11. J. Maier, *Solid State Ionics*, 175, 7 (2004).

12. H. L. Tuller, *Solid State Ionics*, 131, 143 (2000).
13. N. Sata, K. Eberman, K. Eberl and J. Maier, *Nature*, 408, 946 (2000).
14. T. Ishihara, H. Matsuda and Y. Takita, *J. Am. Chem. Soc.*, 116, 3801 (1994).
15. F. Abraham, J. C. Boivin, G. Mairesse and G. Nowogrocki, *Solid State Ionics*, 40-41, 934 (1990).
16. S. A. Kramer and H. L. Tuller, *Solid State Ionics*, 82, 15 (1995).
17. P. Lacorre, F. Goutenoire, O. Bohnke, R. Retoux and Y. Laligant, *Nature*, 404, 856 (2000).
18. W. J. Fleming, *J. Electrochem. Soc.*, 124, 21 (1977).
19. A. Q. Pham and R. S.Glass, *Electrochim. Acta*, 43, 2699 (1998).
20. M. J. Verkerk, B. J. Middelhuis and A. J.Burggraaf, *Solid State Ionics*, 6, 159 (1982).
21. J. S. Lee and D. Y.Kim, *J. Mater. Res.*, 16, 2739 (2001).
22. S. P. S. Badwal and A. E.Hughes, *J. Eur. Ceram. Soc.*, 10, 115 (1992).
23. M. Gödickemeier, B. Michel, A. Orliukas, P. Bohac, K. Sasaki, L. Gauckler, H. Heinrich, P. Schwander, G. Kostorz, H. Hofmann and O. Frei, *J. Mater. Res.*, 9, 1228 (1994).
24. S. P. S. Badwal, *Solid State Ionics*, 76, 67 (1995).
25. M. Kleitz, L. Dessemond and M. C. Steil, *Solid State Ionics*, 75, 107 (1995).
26. M. Aoki, Y. M. Chiang, I. Kosacki, L. J. R. Lee, H. L. Tuller and Y. P. Liu, *J. Am. Ceram. Soc.*, 79, 1169 (1996).
27. F. Boulc'h, L. Dessemond and E. Djurado, *Solid State Ionics*, 154, 143 (2002).
28. J. H. Lee, J. H. Lee, Y. S. Jung and D. Y. Kim, *J. Am. Ceram. Soc.*, 86, 1518 (2003).
29. M. C. Martin and M. L. Mecartney, *Solid State Ionics*, 161, 67 (2003).
30. D. Y. Wang and A. S. Nowick, *J. Solid State Chem.*, 35, 325 (1980).
31. R. Gerhardt and A. S. Nowick, *J. Am. Ceram. Soc.*, 69, 641 (1986).
32. G. M. Christie and F. P. F. van Berkel, *Solid State Ionics*, 83, 17 (1996).
33. S. J. Hong, K. Mehta and A. V. Virkar, *J. Electrochem. Soc.*, 145, 638 (1998).
34. J. H. Hwang, D. S. Mclachlan and T. O. Mason, *J. Electroceram.*, 3, 7 (1999).
35. C. Tian and S. W. Chan, *Solid State Ionics*, 134, 89 (2000).
36. C. M. Kleinlogel and L. J. Gauckler, *J. Electroceram.*, 5, 231 (2000).
37. X. D. Zhou, W. Huebner, I. Kosacki and H. U. Anderson, *J. Am. Ceram. Soc.*, 85, 1757 (2002).
38. T. S. Zhang, P. Hing, H. T. Huang and J. A. Kilner, *Solid State Ionics*, 148, 567 (2002).
39. D. Perez-Coll, P. Nunez, J. R. Frade and J. C. C. Abrantes, *Electrochem. Acta*, 48, 1551 (2003).
40. D. Bingham, P. W. Tasker and A N. Cormack, *Phil. Mag.* A60, 1 (1989).
41. X. Guo and J. Maier, *J. Electrochem. Soc.*, 148, E121 (2001).
42. X. Guo, W. Sigle, J. Fleig and J. Maier, *Solid State Ionics*, 154-155, 555 (2002).

43. J. C. M'Peko, D. L. Spavieri and M. F. De Souza, *Appl. Phys. Lett.*, 81, 2827 (2002).
44. J. C. M'Peko and M. F. De Souza, *Appl. Phys. Lett.*, 83, 737 (2003).
45. X. Guo, *J. Am. Ceram. Soc.*, 86, 1867 (2003).
46. X. Guo and Z. Zhang, *Acta Mater.*, 51, 2539 (2003).
47. X. Guo and Y. Ding, *J. Electrochem. Soc.*, 151, J1 (2004).
48. X. Guo, S. Mi and R.Waser, *Electrochem. Solid-State Lett.*, 8, J1 (2005).
49. A. Tschöpe, E. Sommer and R. Birringer, *Solid State Ionics*, 139, 255 (2001).
50. A. Tschöpe, *Solid State Ionics*, 139, 267 (2001).
51. S. Kim and J. Maier, *J. Electrochem. Soc.*, 149, J73 (2002).
52. X. Guo, W. Sigle and J. Maier, *J. Am. Ceram. Soc.*, 86, 77 (2003).
53. X. Guo and R.. Waser, *Prog. Mater. Sci.*, 51, 151 (2006).
54. S. L. Hwang and I. W. Chen, *J. Am. Ceram. Soc.*, 73, 3269 (1990).
55. D. A. Blom and Y. M. Chiang, *Mat. Res. Soc. Symp. Proc.*, 458, 127 (1997).
56. C. A. J. Fisher and H. Matsubara, *Comput. Mater. Sci.*, 14, 177 (1999).
57. Y. Lei, Y. Ito, N. D. Browning and T. J. Mazanec, *J. Am. Ceram. Soc.*, 85, 2359 (2002).
58. A. J. A. Winnubst, P. J. M. Kroot and A. J. Burggraaf, *J. Phys. Chem. Solids*, 44, 955 (1983).
59. G. S. A. M. Theunissen, A. J. A. Winnubst and A. J. Burggraaf, *J. Mater. Sci.*, 27, 5057 (1992).
60. Y. Ikuhara, P. Thavorniti, T. Sakuma, *Acta Mater.*, 45,5275 (1997).
61. S. Stemmer, J. Vleugels and O. Van Der Biest, *J. Eur. Ceram. Soc.*, 18, 1565 (1998).
62. L. Gremillard, T. Epicier, J. Chevalier and G. Fantozzi, *Acta Mater.*, 48, 4647 (2000).
63. E. C. Dickey, X. Fan and S. J. Pennycook, *J. Am. Ceram. Soc.*, 84, 1361 (2001).
64. J. Fleig, S. Rodewald and J. Maier, *J. Appl. Phys.*, 87, 2372 (2000).
65. P. Mondal, A. Klein, W. Jaegermann and H. Hahn, *Solid State Ionics.*, 118, 331 (1999).
66. Y. M. Chiang, E. B. Lavik and D. A. Blom, *NanoStruct. Mater.*, 9, 633 (1997).
67. X. Guo, E. Vasco, S. Mi, K. Szot, E. Wachsman and R. Waser, *Acta Mater.*, 53, 5161 (2005).
68. I. Kosacki, C. M. Rouleau, P. F. Becher, J. Bentley and D. H. Lowndes, *Electrochem. Solid-State Lett.*, 7, A459 (2004).
69. J. H. Joo and G. M. Choi, *Solid State Ionics*, 177, 1053 (2006).
70. I. Kosacki, T. Suzuki, V. Petrovsky, H. U. Anderson, *Solid State Ionics*, 136-137, 1225 (2000).
71. I. Kosacki, C. M. Rouleau, P. F. Becher, J. Bentley and D. H. Lowndes, *Solid State Ionics*, 176, 1319 (2005).
72. H. L. Tuller and A. S. Nowick, *J. Phys. Chem. Solids*, 38, 859 (1977).
73. I. K. Naik and T. Y. Tien, *J. Phys. Chem. Solids*, 39, 311 (1978).

74. S. Wang, T. Kobayashi, M. Dokiya and T. Hashimoto, *J. Electrochem. Soc.,* 147, 3606 (2000).
75. X. Guo and Y. Ding, *J. Electrochem. Soc.*, 151, J1 (2004).
76. J. S. Lee, U. Anselmi-Tamburini, Z. A. Munir and S. Kim, *Electrochem. Solid State Lett.*, 9, J34 (2006).
77. Y.-M. Chiang, E. B. Lavik, I. Kosacki, H. L. Tuller and J. Y. Ying, *Appl. Phys. Lett.*, 69, 185 (1996).
78. T. X. T. Sayle, S. C. Parker and C. R. A. Catlow, *Surf. Sci.*, 316, 329 (1994).
79. S. Steinsvik, R. Bugge, J. GjØnnes, J. TaftØ and T. Norby, *J. Phys. Chem. Solids*, 58, 969 (1997).
80. R. Brydson, H. Sauer, W. Engel and F. Hofer, *J. Phys.: Condens. Matter*, 4, 3429 (1992).
81. F. M. De Groot, J. Faber, J. J. M. Michiels, M. T. Czyzyk, M. Abbate and J. C. Fuggle, *Phys. Rev.* B 48, 2074 (1993).
82. R. Moos and K. H. Härdtl, *J. Appl. Phys.*, 80, 393 (1996).
83. M. M. McGibbon, N. D. Browning, M. F. Chisholm, A. J. McGibbon, S. J. Pennycook, V. Ravikumar and V. P. Dravid, *Science*, 266, 102 (1994).
84. N. D. Browning, J. P. Buban, H. O. Moltaji, S. J. Pennycook, G. Duscher, K. D. Johnson, R. P. Rodrigues and V. P. Dravid, *Appl. Phys. Lett.*, 74, 2638 (1999).
85. R. F. Klie and N. D. Browning, *Appl. Phys. Lett.*, 77, 3737 (2000).
86. M. Kim, G. Duscher, N. D. Browning, K. Sohlberg, S. T. Pantelides and S. J. Pennycook, *Phys. Rev. Lett.*, 86, 4056 (2001).
87. H. Chang, Y. Choi, J. D. Lee and H. Yi, *Appl. Phys. Lett.*, 81, 3564 (2002).
88. Z. Zhang, W. Sigle, F. Phillipp and M. Rühle, *Science*, 846, 302 (2003).
89. I. Denk, J. Claus and J. Maier, *J. Electrochem. Soc.*, 144, 3526 (1997).
90. I. Denk, F. Noll and J. Maier, *J. Am. Ceram. Soc.*, 80, 279 (1997).
91. J. Maier, *J. Eur. Ceram. Soc.*, 19, 675 (1999).
92. J. Maier, *Solid State Phenomena*, 67-68, 45 (1999).
93. M. Leonhardt, J. Jamnik, and J. Maier, *Electrochem. Solid-State Lett.*, 2, 333 (1999).
94. J. Fleig, S. Rodewald and J. Maier, *J. Appl. Phys.*, 87, 2372 (2000).
95. M. Vollman and R. Waser, *J. Am. Ceram. Soc.*, 77, 235 (1994).
96. R. Waser, *Solid State Ionics*, 75, 89 (1995).
97. R. Hagenbeck and R. Waser, *J. Appl. Phys.*, 83, 2083 (1998).
98. X. Guo, J. Fleig and J. Maier, *J. Electrochem. Soc.*, 148, J50 (2001).
99. X. Guo, J. Fleig and J. Maier, *Solid State Ionics*, 154-155, 563 (2002).
100. C. Ohly, S. Hoffmann-Eifert, X. Guo, J. Schubert and R. Waser, *J. Am. Ceram. Soc.*, 89, 2845 (2006).
101. R. V. Wang and P. C. McIntyre, *J. Appl. Phys.*, 97, 023508 (2005).
102. R. A. De Souza, J. Zehnpfenning, M. Martin and J. Maier, *Solid State Ionics*, 176, 1465 (2005).
103. R. A. De Souza, J. Fleig, R. Merkle and J. Maier, *Z. Metallkd.*, 94, 218 (2003).

104. P. Balaya, J. Jamnik, J. Fleig and J. Maier, *Appl. Phys. Lett.*, 88, 062109 (2006).
105. P. Balaya, M. Ahrens, L. Kienle, J. Maier, B. Rahmati, S. B. Lee, W. Sigle, A. Pashkin, C. Kuntscher and M. Dressel, *J. Am. Ceram. Soc.*, 89, 2804 (2006).
106. X. Guo, C. Pithan, C. Ohly, C.-L. Jia, J. Dornseiffer, F.-H. Haegel and R. Waser, *Appl. Phys. Lett.*, 86, 082110 (2005).
107. M. H. Frey, Z. Xu, P. Han and D. A. Payne, *Ferroelectrics*, 206-207, 337 (1998).
108. N.-H. Chan, R. K. Sharma and D. M. Smyth, *J. Am. Ceram. Soc.*, 64, 556 (1981).
109. N.-H. Chan, R. K. Sharma and D. M. Smyth, *J. Am. Ceram. Soc.*, 65, 167 (1982).

CHAPTER 12

NANOSTRUCTURED CATHODE MATERIALS FOR ADVANCED Li-ION BATTERIES

Ying Wang[1,†] and Guozhong Cao*

*Department of Materials Science and Engineering, University of Washington, Seattle, WA 98195, USA; *Email: gzcao@u.washington.edu; [1]Current address: Materials Research Institute and Department of Materials Science and Engineering, Northwestern University, Evanston, IL 60208, USA [†]Email: ying-wang@northwestern.edu*

Nanostructured materials lie at the heart of the fundamental advances in efficient energy storage/conversion in which surface process and transport kinetics place determining roles. This review describes some recent developments in the synthesis and characterizations of nanostructured cathode materials, including lithium transition metal oxides, vanadium oxides, manganese oxides, lithium phosphates, and various nanostructured composites. The major topic of this article is to highlight some new progress in using these nanostructured materials as cathodes to develop lithium batteries with high energy density, high rate capability and excellent cycling stability.

1. Introduction

1.1 General Background

Recent increases in demand for oil, associated price increases, and environmental issues are continuing to exert pressure on an already stretched world energy infrastructure. Significant progress has been made in the development of renewable energy technologies such as solar cells, fuel cells, and bio-fuels. In the past, these types of energy sources have been marginalized, but as new technology makes alternative energy more practical and price competitive with fossil fuels, it is expected that the coming decades will usher in a long expected transition away from oil and gasoline as our primary fuel. Although a variety of renewable energy technologies have been developed, they have not reached wide spread use. High performance of such technologies is mainly achieved through designed sophisticated device structures with multiple materials, for example tandem cells in photovoltaic devices. Almost all the alternative energy technologies are limited by the materials properties. For example, poor charge carrier mobilities in organic/polymer semiconductors limit the energy conversion efficiency of organic photovoltaic cells less than 6%. Thermoelectrics typically possess a figure of merit less than 2.5. Portable electric power sources have lower energy and power density due largely to poor charge and mass transport properties. New materials that are chemically modified through molecular or atomic engineering and/or possess unique microstructures would offer significantly enhanced properties for more efficient energy conversion devices and high density energy/power storage.

One alternative energy/power source under serious consideration is electrochemical energy production, as long as this energy consumption is designed to be more sustainable and more environmentally benign. The lithium-ion battery is the representative system for such electrochemical energy storage and conversion. At present lithium-ion batteries are efficient, light-weight and rechargeable power sources for consumer electronics, such as laptop computers, digital cameras and cellular phones. Moreover, it has been intensively studied for use as power supplies of electric vehicles (EVs) and hybrid electric vehicles (HEVs). High energy and high power densities are required for such devices. Lithium-ion batteries are attractive power-storage devices owning to their high energy density [1]. However, their power density is relatively low because of a large polarization at high charging-discharging rates. This polarization is caused by slow lithium diffusion in the active material and increases in the resistance of the electrolyte when the charging-discharging rate is increased. To overcome these problems, it is important to design and fabricate nanostructured electrode materials that provide high surface area and short diffusion paths for ionic transport and electronic conduction.

Nanomaterials offer the unusual mechanical, electrical and optical properties endowed by confining the dimensions of such materials and the overall behavior of nanomaterials exhibit combinations of bulk and surface properties [2]. Thus, nanostructured materials are drawing a tremendous amount of attention because of their novel properties, and because of their potential applications in a variety of nanodevices, such as field-effect transistors (FETs) [3,4,5,6], chemical and biological sensors[7,8,9,10], nanoprobes [11], and nanocables [12]. The reports on the processing, properties and applications of nanomaterials are rapidly appearing on daily basis. Many synthesis methods have been reported for the synthesis of nanostructured electrode materials. Among them, solution-based methods are well known for their advantages in tailoring the size and morphology of the nanostructures. It is the uncomplicated sol-gel processing (soft chemistry) method in combination with template synthesis or hydrothermal treatment that produces the most desirable nanostructures with remarkable reliability, efficiency, selectivity, and variety. Template sysnthesis is a general method for preparing ordered arrays of nanostructures with nanorods/nanotubes/nanocables protruding from the underlying current collector [13]. Hydrothermal synthesis is another powerful tool to transform transition metal oxides into high-quality nanostructures. Other fabrication methods of nanostructures include reverse micelle technique and the size of nanostructures can be tuned easily by keeping the freshly made nanorods in the micellar solution [14]. This article aims to give a concise and useful survey of recent progress on synthesis and characterizations of nanostructured cathode materials for lithium-ion batteries, starting with a brief overview on lithium-ion batteries and cathode materials as follows.

1.2 Lithium Batteries and Cathode Materials

A battery consists of three basic components: an anode, a cathode, and an electrolyte; and is a device that converts chemical potential to electric energy through faradaic reactions, that include heterogeneous charge transfer occurring at the surface of an electrode [15]. Batteries are broadly grouped as primary and secondary batteries. Primary batteries are single-use devices and can not be recharged; secondary batteries are also called rechargeable batteries and can be recharged for many times. In a typical secondary battery, energy storage involves faradaic reactions occurring at the surface of an electrode, and mass and charge transfer through the electrode; therefore, the surface area and the transport distance play important roles in determining the performance of the battery in question. Chemical composition, crystal structure, and microstructure will have significant impacts on the surface reaction and transfer processes, as well as, its cyclic stability.

Intercalation electrodes in batteries are electroactive materials and serve as a host solid into which guest species are reversibly intercalated from an electrolyte. Intercalation compounds are a special family of materials. The intercalation refers to the reversible intercalation of mobile guest species (atoms, molecules or ions) into a crystalline host lattice that contains an interconnected system of empty lattice site of appropriate size, while the structural integrity of the host lattice is formally conserved [16]. The intercalation reactions typically occur around room temperature. A variety of host lattice structures have been found to undergo such low temperature reactions [17]. However, the intercalation reactions involving layered host lattices have been most extensively studied, partly due to the structural flexibility, and the ability to adapt to the geometry of the intercalated guest species by free adjustment of the interlayer separation. The readers are referred to a comprehensive and excellent article on inorganic intercalation compounds [16]. Despite the differences in chemical composition and lattice structure of the host sheets, all the layer hosts are characterized by strong interlayer covalent bonding and weak interlayer intercalations. The weak interlayer intercalations include van der Waals force or electrostatic attraction through oppositely charged species between two layers. Various host lattices are metal dichalcogenides, metal oxyhalides, metal phosphorous trisulphides, metal oxides, metal phosphates, hydrogen phosphates, and phosphonates, graphite and layered clay minerals. Guest materials include metal ions, organic molecules and organometallic molecules. When guest species are incorporated into host lattices, various structural changes will take place. The principle geometrical transitions of layered host lattice matrices upon intercalation of guest species include: (1) change in interlayer spacing, (2) change in stacking mode of the layers and (3) formation of intermediate phases at low guest concentrations may exhibit staging [18]. There are various synthesis methods for the formation of intercalation compounds [16,19]. The most commonly used and simplest method is the direct reaction of the guest species with the host lattice [20]. For direction reactions, intercalation reagent must be good reducing agents of the host crystals. Ion exchange is a method to replace the guest ion in an intercalation compound with another guest ion, which offers a useful route for intercalating large ions that do not directly intercalate [21]. Appropriate chosen solvents or electrolytes may assist the ion exchange reactions by flocculating and reflocculating the host structure [22]. Electrointercalation is yet another method, in which the host lattice serves as the cathode of an electrochemical cell [23]. Electrochemical lithium intercalation occurs together with compensating electrons leading to the formation of vanadium bronzes as follows:

$$V_2O_5 + xLi^+ + xe^- \Leftrightarrow Li_xV_2O_5 \qquad (1)$$

The principal concept of lithium-ion batteries is illustrated in Figure 1. A combination of a negative lithium intercalation material (anode) with another lithium intercalation material (cathode) having a more positive redox potential gives a Li-ion transfer cell. Anode and cathode are separated by the electrolyte which is an electronic insulator but a Li-ion conductor. Upon charging, lithium ions are released by the cathode and intercalated at the anode. When the cell is discharged, lithium ions are extracted by the cathode and inserted into the anode. Early batteries used metallic lithium as anode which combines a very negative redox potential with a low equivalent weight. It was later replaced by carbon because of safety concerns. Replacement of the metallic lithium or carbon by lithium intercalation compounds improves both cell life and safety but at the expense of cell voltage, specific charge, and rate capability. Electrode materials must fulfill three fundamental requirements to reach the goal of a high specific energy and energy density: (1) a high specific charge and charge density, i.e., a high number of available charge carriers per mass and volume unit of the material; (2) a high cell voltage, resulting from a high (cathode) and low (anode) standard redox potential of the respective electrode redox reaction; and (3) a high reversibility of electrochemical reactions at both cathodes and anodes to maintain the specific charge for hundreds of charge/discharge cycles.

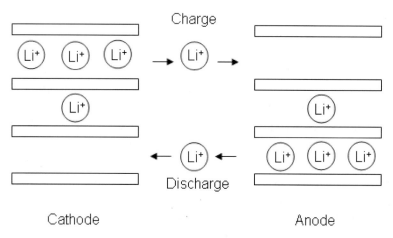

Figure 1. Schematic illustration of a lithium-ion battery.

Ever since the idea of a rechargeable lithium cell based on Li intercalation reactions was initiated in the early 1970s, numerous lithium intercalation electrodes have been proposed to date. The area of cathodes is much less developed than anodes [24], and we will focus on the cathode materials in this article. Details on lithium-ion battery cathode materials can be found in recent reviews by M. Whittingham et al [25,26]. There are two categories of cathode materials. One is layered compounds with anion close-packed lattice; transition metal cations occupy alternate layers between the anion sheets and lithium ions are intercalated into remaining empty layers. $LiTiS_2$, $LiCoO_2$, $LiNi_{1-x}Co_xO_2$, and $LiNi_x Mn_x Co_{1-2x}O_2$ are all belonged to this group. The spinels with the transition metal cations ordered in all the layers can be considered to be in this group as well. This class of materials have the inherent advantage of higher energy density (energy per unit of volume) owning to their more compact lattices. The other group of cathode materials has more open structures, such as vanadium oxides, the tunnel compounds of manganese oxides, and transition metal phosphates (e.g., the olivine $LiFePO_4$). These materials generally provide the advantages of better safety and lower cost compared to the first group. This article will review methods to synthesize nanostructures of these two groups of cathode materials for use as high-performance electrodes in lithium-ion batteries, choosing $LiCoO_2$, $LiNiO_2$, $LiMn_2O_4$, substituted lithium metal oxides, solid solutions of lithium metal oxides, V_2O_5, MnO_2, and $LiFePO_4$ as representatives. Section 2 covers lithium transition metal oxides and section 3 reviews vanadium oxides and manganese oxides. Lithium phosphates are discussed in section 4. In each section, we look firstly at structural and electrochemical properties of the materials, then at synthesis and characterization of nanostructures of these materials, how synthesis methods and parameters affect properties, and how to improve electrochemical performance even further by incorporating other nanomaterials such as nanosized oxide coatings on lithium metal oxides and nanosized carbon coatings on lithium phosphates. The last point leads to an overview on nanostructured composites in section 5, including discussions about nanostructured composites of carbon-oxide, polymer-oxide, metal-oxide, carbon-polymer, oxide-oxide, and metal-polymer-oxide. To the best of our knowledge, there is no comprehensive review to cover a large variety of nanostructured cathode materials to date. Herein we highlight some recent progress in using various nanostructured cathode materials to develop high-performance lithium-ion batteries.

2. Nanostructured Lithium Transition Metal Oxides and Nanosized Coatings on Lithium Transition Metal Oxides

Currently there are three intercalation materials that are used commercially as cathode materials for rechargeable lithium batteries: $LiCoO_2$, $LiNiO_2$ and $LiMn_2O_4$. $LiCoO_2$ is the most popular among the possible cathode materials due to the convenience and simplicity of preparation. This material can be easily synthesized using both solid state and chemical approaches [27,28]. The Li_xCoO_2 system has been studied extensively thus far [29,30]. The Li_xCoO_2 exhibits excellent cyclability at room temperature for $1 > x > 0.5$. Therefore, the specific capacity of the material is limited to the range of 137 to 140 mAh/g, although the theoretical capacity of $LiCoO_2$ is 273 mAh/g [31]. On the other hand, Li_xCoO_2 is very expensive and highly toxic, in spite of its good electrochemical property and easy synthesis. The reversible capacity of Li_xNiO_2 is higher than that of Li_xCoO_2, since the amount of lithium that can be extracted/intercalated during redox cycles is around 0.55 in comparison with 0.5 for $LiCoO_2$, allowing the specific capacity to be more than 150 mAh/g with appropriate cyclability [32]. Although $LiNiO_2$ is isostructural to $LiCoO_2$, preparation of $LiNiO_2$ is more complicated. Since there are additional nickel ions on the lithium sites, and vice versa in the crystal structure of $LiNiO_2$, the Li-Ni-O system is represented by $Li_{1-y}Ni_{1+y}O_2$ with a deviation from the normal stoichiometry [33,34]. This special structure makes it very difficult to synthesize the stoichiometric oxide with all the lithium sites completely filled by lithium. $LiMn_2O_4$ is the third most popular cathode material for lithium-ion batteries. In comparison with $LiCoO_2$ and $LiNiO_2$, $LiMn_2O_4$ possesses essential advantages of less toxicity and having an abundant materials source. In principle, $Li_xMn_2O_4$ permits the intercalation/extraction of lithium ions in the range of $0 < x < 2$ [35]. For intermediate values of x between 1 and 2 the material consists of two different phases—cubic in bulk and tetragonal at the surface. Simultaneously the intercalation of lithium ions effectively decreases the average valence of manganese ions and leads to a pronounced cooperative Jahn-Teller effect, in which the cubic spinel crystal becomes distorted tetragonal with a c/a ≈ 1.16 and the volume of the unit cell increases by 6.5%. This high c/a ratio causes a low capacity restricted to 120~125 mAh/g and significant capacity degradation at moderate temperatures in the range of 50 to 70°C [36]. To enhance the poor rate capability of lithium metal oxides often owning to the structural unstability of some lithium metal oxides such as $LiNiO_2$ or $LiMnO_2$, light substitutions or preparing solid solutions of several lithium metal oxides have been explored and the new compounds have shown promising electrochemical characteristics [37]. However, lithium metal oxides still suffer from some intrinsic limitations. For example, $LiCoO_2$ has decent lithium diffusion coefficient, 5×10^{-9} cm^2/s [38], whereas the conductivity of this material is as low as 10^{-3} S/cm [39]. To improve the intercalation/deintercalation kinetics of the material, it is necessary to downsize the material to achieve short diffusion distance and large surface area.

2.1 Nanostructured Lithium Transition Metal Oxides

There have been several recent reports on the synthesis and electrochemical properties of nanostructured lithium metal oxides. Chen and coworkers synthesized nanotubes of $LiCoO_2$, $LiNi_{0.8}Co_{0.2}O_2$ and $LiMn_2O_4$ using the template-based method and proposed an "in situ reacted nanoparticle nanotube" formation mechanism [40]. Figure 2a and b show TEM images of a bundle of $LiCoO_2$ nanotubes and the magnified view of a single $LiCoO_2$ nanotube, respectively. A layer separation of 0.466 nm, corresponding to (003) planes, is observed in the HRTEM image (Figure 2c). HRTEM images and FFT analysis show polycrystalline nature of these nanotubes. The nanotube electrodes show high intercalation capacities and better cyclability than their nanocrystalline counterparts because of the high surface area and short Li diffusion distance of nanotubes. Thin film of $LiCoO_2$ could be deposited at room temperature in a nanocrystalline state using planar magnetron rf sputtering [41]. Subsequent heating the films at 300°C causes the average grain size to increase but still within the nanosized dimensions, while the lattice distortion is reduced by the heating. Such nanocrystalline film of $LiCoO_2$ annealed at low temperature demonstrates improved electrochemical performance. As for the spinel $LiMn_2O_4$, synthesis of $LiMn_2O_4$ nanoparticles could be carried out by a sol-gel method combined with post-calcination and the particle size is affected by the calcination temperature [42]. The nanoparticles have a size of 10 nm at the low temperature of 350°C, whereas submicron-sized particles are obtained at the high temperature of 550°C. The $LiMn_2O_4$ nanoparticles were found to behave differently in different voltage ranges. In comparison with large nonporous cathode, the nanoparticle cathode shows improved capacity and cycleability in the 3 V discharge range, while in the 4 V discharge region, it exhibits decreased capacity and improved cycleability. The enhancement in capacity and cycleability is due to the reduced charge-transfer resistance of nanoparticle cathode in comparison with large cathode material. Nanostructured solid solutions of lithium metal oxides such as $LiNi_{0.5}Mn_{1.5}O_4$ can be synthesized using solution methods as well. Kunduraci and Amatucci employed a modified Pechini method to obtain nanostructured $LiNi_{0.5}Mn_{1.5}O_4$ by adding aqueous solutions of $Li(NO_3)$, $Mn(NO_3)_2 \cdot 4H_2O$ and $Ni(NO_3)_2 \cdot 6H_2O$ into mixture of citric acid and ethylene glycol [43]. The as-prepared samples were sintered at different temperatures and resulted in ordered $P4_332$ (P) spinel or disordered $Fd3m$(F) spinel. The disordered spinel contains a small amount of Mn^{3+} and has higher electronic conductivity than the ordered sample by two orders of magnitude. Therefore, the disordered spinel shows higher rate capability than the ordered sample and exhibits capacity retention of 80% at 6C. Furthermore, shape of nanostructured $LiNi_{0.5}Mn_{1.5}O_4$ can be tailored by a polymer-assisted method [44]. First nanocrystalline oxalates were obtained by grinding hydrated salts and oxalic acid in the presence of polytheyleneglycol 400. Then nanorodlike $LiNi_{0.5}Mn_{1.5}O_4$ was prepared by thermal decomposition of mixed nanocrystalline oxalates at 400°C. Heating the nanorodlike $LiNi_{0.5}Mn_{1.5}O_4$ up to 800°C breaks the nanorods into nanoparticulate $LiNi_{0.5}Mn_{1.5}O_4$ with size in the range of 70-80 nm. Such nanoparticulate $LiNi_{0.5}Mn_{1.5}O_4$ cathode shows good electrochemical characteristics at a wide range of rates (from C/4 to 15C) when cycled between 3.5 and 5 V.

2.2 Nanosized Coatings on Lithium Transition Metal Oxides

It should be noted that the nanoparticulate forms of lithium transition metal oxides such as $LiCoO_2$, $LiNiO_2$, or their solid solutions, can react with the electrolyte and lead to safety problems. In the case of $LiMn_2O_4$, the use of nanoparticles causes undesirable dissolution of Mn. Significant efforts have been made to increase the stability of these nanocrystalline lithium metal oxides. Better stability can be achieved by coating the electrode materials with a nanosized stabilizing surface layer that alleviates these problems. As for $LiCoO_2$, coatings of various phosphates and oxides have been studied and significant improvements in capacity retention have been demonstrated. Kim et al. made an extensive study on the effect of the MPO_4 (M = Al, Fe, SrH and Ce)

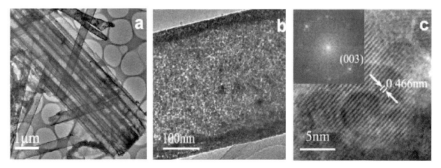

Figure 2. TEM (a,b) and HRTEM (c) images of the LiCoO$_2$ nanotubes. The inset of panel c is the corresponding FFT analysis. Reprinted with permission from Ref. 40, X. Li, et al., J. Phys. Chem. B, **109**, 14017 (2005), Copyright @ American Chemical Society.

nanoparticle coatings on LiCoO$_2$ cathode material [45]. They found that the extent of the coating coverage is affected by the nanoparticle size and morphology despite the same coating concentration and annealing temperature. Smaller nanoparticles of AlPO$_4$ or FePO$_4$ with a size less than 20 nm fully encapsulate LiCoO$_2$, whereas CePO$_4$ particles with a size larger than 150 nm or whisker-shaped SrHPO$_4$ only partially cover LiCoO$_2$. Not surprisingly, the LiCoO$_2$ fully covered by AlPO$_4$ or FePO$_4$ exhibits the highest intercalation capacity of 230 mAh/g in a voltage range of 4.8 and 3 V at a rate of 0.1C. The AlPO$_4$-coated LiCoO$_2$ also shows the best capacity retention. Nevertheless, the CePO$_4$- and SrHPO$_4$-coated cathodes shows better capacity retention than the FePO$_4$-coated cathode at 90°C, which is attributed to the continuous Fe metal ion dissolution at this temperature. The improvement in the electrochemical performance in the coated cathode is ascribed to the suppression of cobalt dissolution and the non-uniform distribution of local strain by the coating layer. In a further investigation of AlPO$_4$-coated LiCoO$_2$, electrochemical properties of AlPO$_4$-nanoparticle-coated LiCoO$_2$ at various cutoff-voltages were found to depend on the annealing temperature [46]. The AlPO$_4$-coated cathodes exhibit excellent electrochemical performance with high cutoff voltages larger than 4.6 V when annealed at 600 and 700 °C, while such cathodes annealed at 400°C show a lower capacity and poorer rate capability. However, the AlPO$_4$-coated LiCoO$_2$ annealed at 400°C showed optimal capacity retention [47].

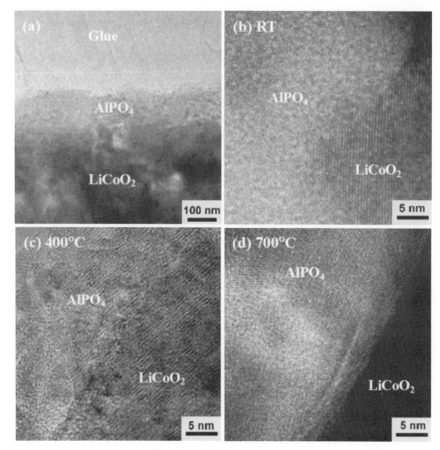

Figure 3. (a) Cross-sectional TEM images of AlPO$_4$-coated LiCoO$_2$. A ~100 nm thick AlPO$_4$ continuous layer is coated on LiCoO$_2$. High resolution images of the AlPO4-coated LiCoO$_2$ at (b) room temperature, (c) 400°C, and (d) 700°C. Reprinted with permission from Ref. 47, B. Kim et al., J. Electrochem. Soc. **153(9)**, A1773, (2006), Copyright @ The Electrochemical Society.

Figure 3 shows typical TEM images of AlPO$_4$-coated LiCoO$_2$ deposited at room temperature, 400°C and 700°C. A continuous layer of AlPO$_4$ with thickness of about 100 nm is coated on the surface of LiCoO$_2$, as shown in Figure 3a. The coating layer deposited at room temperature is amorphous (Figure 3b). The coating deposited at 400°C is composed of nanocrystals with size in the range of 3-5 nm (Figure 3c), and the coating deposited at 700°C consists of ~20-30 nm sized nanocrystals (Figure 3d). The dependence of electrochemical properties on annealing temperature can be explained by the effect of temperature on the nanostructures of the coating layer and the interdiffusion at the interface between the coating layer and the $LiCoO_2$ cathode. In addition to coatings of phosphates, surface modification of $LiCoO_2$ by coating various oxides such as ZrO_2 [48], Al_2O_3 [49], SnO_2 [50], MgO [51] or ZnO [52] has been widely investigated. In the case of ZnO-coated $LiCoO_2$, the ZnO coating reduces the cobalt dissolution and prevents the inorganic surface films such as LiF from covering the $LiCoO_2$ particles.[52] Moreover, the ZnO coating alleviates the cycle-life degradation caused by inappropriate conductive carbon. Based on the impedance spectra, the charge-transfer resistance of ZnO-coated-$LiCoO_2$ is much smaller than the uncoated cathode, although the ZnO coating layer is more resistant than the $LiCoO_2$ surfaces. It can be concluded that surface modification with ZnO improves the high-voltage cycleability of the $LiCoO_2$ cathodes. In a similar manner, ZrO_2 coating protects the $LiCoO_2$ cathode surface and reduces the electrolyte decomposition at high voltages [53,54]. The ZrO_2-coated $LiCoO_2$ shows much better structural change behaviors than the bare $LiCoO_2$, as evidenced by in situ XRD data. The battery cells discussed above all employ liquid organic electrolytes which are flammable and cause safety concerns. Replacing the liquid electrolyte with nonflammable solid electrolyte such as sulfide electrolyte is a solution to the safety problems, however, the energy densities and power densities of solid-state lithium batteries are relatively low for practical applications. One way to improve the rate capability of solid-state batteries is to add a buffer film with a thickness in nanometer scale between the electrode and electrolyte materials. A thin layer of $Li_4Ti_5O_{12}$ with thickness of a few nanometers was chosen to be coated on the $LiCoO_2$ cathode [55]. The $Li_4Ti_5O_{12}$ is also a Li intercalation material which ensures the electronic conduction, however, this material intercalates lithium ions at voltages lower than 1.5 V and thus does not act as intercalation material in the voltage range of $LiCoO_2$. The power densities of the solid-state batteries with the thin $Li_4Ti_5O_{12}$ layer between the $LiCoO_2$ cathode and sulfide electrolyte are greatly increased and comparable to those of commercial lithium batteries, which is attributed to the suppression of the lithium-ion transfer and subsequent prevention of the formation of the spacing charge layer by $Li_4Ti_5O_{12}$ coating.

$LiMn_2O_4$ or substituted $LiMn_2O_4$ is very attractive as a cathode material because it is safer and cheaper than $LiCoO_2$. However, this material suffers from capacity fading especially at elevated temperatures. Coating of nanosized oxides on $LiMn_2O_4$ will help to improve its cycling performance. The electrochemical behavior of nanosized ZnO-coated $LiMn_2O_4$ was examined at 55°C [56]. After 50 cycles at 55°C, the coated $LiMn_2O_4$ shows capacity retention of 97%, much higher than the capacity retention (58%) of the bare cathode. ZnO coating collects HF from the electrolyte and thus decreases the Mn dissolution in the electrolyte then subsequently reduces the interfacial resistance. For the same reason, nanosized ZnO homogenously coated on the $Li_{1.05}Al_{0.1}Mn_{1.85}O_{3.95}F_{0.05}$ by a hydrothermal process was found to significantly improve cycling performance of the cathode at 55°C [57]. The coated $Li_{1.05}Al_{0.1}Mn_{1.85}O_{3.95}F_{0.05}$ shows high capacity retention of 98.5% after 50 cycles. Similarly, coating of amorphous ZrO_2 on $LiMn_2O_4$ can improve the high-temperature cycleability by picking up acidic species from electrolyte [58]. Moreover, the ZrO_2-coated $LiMn_2O_4$ exhibits tremendously improved cycling stability at high rates up to $10C$ due to the following mechanisms. First, ZrO_2 can form a few stable phases with Li and thus amorphous ZrO_2 matrix possibly possesses high solubility of Li. Therefore, the ZrO_2 coating can act as a highly-Li conducting solid electrolyte interface which reduces the interfacial resistance. Second, the rigid oxide coating strongly bonds to $LiMn_2O_4$ which tolerates the lattice stress resulted from volume expansion during lithium intercalation. Lastly, ZrO_2 can collect HF from electrolyte to reduce Mn dissolution like ZnO does. The electrochemical behavior of ZrO_2-coated stoichiometric $LiMn_2O_4$ and substituted $Li_{1.05}M_{0.05}Mn_{1.9}O_4$ (M = Al, Ni) cathodes were further compared with those of cathodes coated with Al_2O_3 and SiO_2. ZrO_2-coated $Li_{1.05}M_{0.05}Mn_{1.9}O_4$ (M = Al, Ni) shows the best cycling stability at 50°C [59]. The ZrO_2 coating, deposited from colloidal suspensions, is porous network connected by ZrO_2 nanoparticles with dimensions less than 4 nm. This ZrO_2 network effectively scavenges HF from the electrolyte and allows the access of the electrolyte to the cathode, and thus improves the high-temperature cycleability of the cathode.

3. Nanostructured Metal Oxides

3.1 Nanostructured Vanadium Oxides

Vanadium oxide is a typical intercalation compound as a result of its layered structure. For Li-ion intercalation applications, vanadium oxide offers the essential advantages of low cost, abundant source, easy synthesis, and high energy densities. Orthorhombic crystalline V_2O_5 consists of layers of VO_5 square pyramids that share edges and corners [60,61]. The reversible electrochemical lithium intercalation into V_2O_5 at room temperature was first reported by Whittingham in 1975 [62]. In addition to crystalline V_2O_5, high Li intercalation capacity has been reported for hydrated vanadium pentoxide ($V_2O_5 \cdot nH_2O$), such as $V_2O_5 \cdot nH_2O$ glasses with P_2O_5 or other network formers [63], $V_2O_5 \cdot nH_2O$ xerogels [64,65], and $V_2O_5 \cdot nH_2O$ aerogels [66]. Specific energies of over 700 WAh/kg were measured for lithium cells with a xerogel cathode [65]. $V_2O_5 \cdot nH_2O$ xerogels are composed of ribbonlike particles and display lamellar ordering, with water molecules intercalated between the layers [67]. These water molecules expand the distance between the layers, and the intercalation capacities of $V_2O_5 \cdot nH_2O$ xerogels are enhanced as a result [65]. The structure of the $V_2O_5 \cdot nH_2O$ xerogel can be illustrated as an assembly of well-defined bilayers of single V_2O_5 layers made of square pyramidal VO_5 units with water molecules residing between them [67]. This structure possesses enough atomic ordering perhaps to be characterized as nanocrystalline.

To date there are a large number of publications on nanostructures of vanadium oxides. Pioneering work on the synthesis and electrochemical properties of vanadium oxide nanorolls was carried out by Spahr et al. [68]. In their synthesis, a combination of sol-gel reaction and hydrothermal treatment of vanadium oxide precursor is conducted in the presence of an amine that acts as structure directing template [68,69,70,71,72]. The resultant nanoroll is either constructed in closed concentric cylinders (nanotubes) or formed by scrolling one or more layers (nanoscrolls). If amine is replaced by ammonia during the hydrolysis step, a new type of vanadium oxide nanoroll (nanotube) with alternating interlayer distances is yielded [73]. Such a unique structure is first observed in a tubular phase. Compared to other tubular systems, the vanadium oxide nanorolls are especially interesting because they possess four different contact regions, that is, tube opening, outer surface, inner surface, and interstitial region. VO_x nanorolls can intercalate a variety of molecules and ions reversibly without change in the crystalline structure. The Li intercalation capacities have been found up to 200 mAh/g, however, there is structural breakdown during redox cycles and degradation in cycling performance due to the morphological flexibility. The cyclic voltammetry measurements show that the well-ordered nanorolls behave closely to classic crystalline vanadium pentoxide, while the defect-rich nanorolls have electrochemical behavior similar to that of sol-gel-prepared hydrated vanadium pentoxide materials. The specific capacity of defect-rich nanorolls (340 mAh/g) is higher than that of the well-ordered nanorolls (240 mAh/g) under comparable conditions.

Martin and coworkers have reported a series of studies on polycrystalline V_2O_5 nanorod arrays. They used a template-based method by depositing triisopropoxyvanadium(V) oxide (TIVO) into the pores of polycarbonate filtration membranes followed by removal of membranes at high temperature [74]. The V_2O_5 nanorod arrays deliver three times the capacity of the thin film electrode at a high rate of $200C$ and four times the capacity of the thin-film control electrode above $500C$. After that, Li and Martin achieved improved volumetric energy densities of V_2O_5 nanorod arrays by chemically etching the polycarbonate membrane to increase its porosity prior to template synthesis [75]. In the latest work of Sides and Martin, V_2O_5 nanorods of different diameters were prepared and their electrochemical properties at low temperature were compared [76]. V_2O_5 nanorods with nanometersized diameters (e.g. 70 nm) deliver dramatically higher specific discharge capacities at low temperature than V_2O_5 nanorods with micrometersized diameters. Thus Li-ion battery electrodes composed of nanosized material meet the low-temperature performance challenge, because nanomaterials palliate the

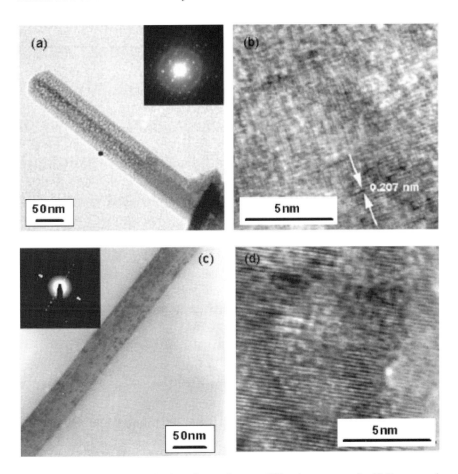

Figure 4. (a) TEM image and selected area electron diffraction pattern of a V_2O_5 nanorod prepared from template-based electrochemical deposition from $VOSO_4$ solution. (b) High-resolution TEM image of the V_2O_5 nanorod in (a), showing lattice fringes. The spacing of the fringes was measured to be 0.207 nm. (c) TEM image and selected area electron diffraction pattern of a V_2O_5 nanorod prepared from template-based electrophoretic deposition from V_2O_5 sol. (d) High-resolution TEM image of the V_2O_5 nanorod in (c). The spacing of the fringes was measured to be 0.208 nm. Reprinted with permission from Ref. 78, K. Takahashi et al. Jpn. J. Appl. Phys. **44**, 662 (2005), Copyright @ The Japan Society of Applied Physics.

problems of slow electrochemical kinetics and the slow diffusion by offering high surface area and short diffusion distance.

Synthesis and electrochemical properties of single-crystal V_2O_5 nanorod arrays were first reported by Cao's group [77,78,79]. They utilized a template-based electrodeposition method by depositing V_2O_5 into pores of polycarbonate templates with the assistance of electric field from three different types of solutions or sol, i.e., VO^{2+} solution, VO_2^+ solution and V_2O_5 sol. Figure 4a and c show TEM images of a V_2O_5 nanorod and selected area electron diffraction pattern, which clearly demonstrated the single-crystalline nature or, at least, well textured nature of the grown nanorods with a [010] growth direction for nanorods grown from both routes. Figure 4b and d also show high-resolution TEM images of a single V_2O_5 nanorod, in which lattice fringes are clearly visible. The spacing of the fringes was measured to be 0.207 nm for nanorod grown from route A, and 0.208 nm for nanorod made from V_2O_5 sol. These values are similar for different synthesis route and correspond well with the spacing of (202) planes at 0.204 nm. These fringes make an angle of 88.9° with the long axis of the nanorod, which is consistent with a growth direction of [010]. Similar measurements made on high-resolution images of other nanorods also yield results consistent with a [010] growth direction. The formation of single-crystal nanorods from solutions by electrochemical deposition is attributed to evolution selection growth (Figure 5a). The initial heterogeneous nucleation or deposition on the substrate surface results in the formation of nuclei with random orientation. The subsequent growth of various facets of a nucleus is dependent on the surface energy, and varies significantly from one facet to another [80]. In the case of nanorods made from the V_2O_5 sol by electrophoretic deposition, the formation of single-crystal nanorods is explained by homoepitaxial aggregation of crystalline nanoparticles (Figure 5b). Thermodynamically it is favorable for the crystalline nanoparticles to aggregate epitaxially; such growth behavior and mechanism have been well reported in literature [81,82]. As a result, V_2O_5 nanorods grown by electrochemical deposition from solutions are dense single crystals, while the nanorods grown from sol electrophoresis are also single-crystalline but have many defects inside the crystal. Such difference in nanostructure determines the different electrochemical behavior of nanorods grown from different solutions or sol. The nanorods grown from V_2O_5 sol by electrophoresis show the best kinetic

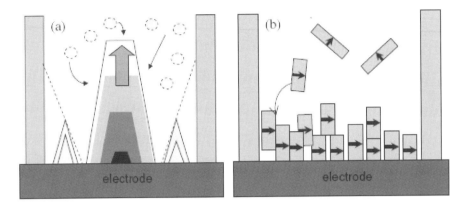

Figure 5. Schematic illustration of growth mechanisms of single crystalline nanorods: (a) evolution selection growth and (b) homoepitaxial aggregation. Reprinted with permission from Ref. 78, K. Takahashi et al. Jpn. J. Appl. Phys. 44, 662 (2005), Copyright @ The Japan Society of Applied Physics.

property for Li-ion intercalation. All the V_2O_5 nanorod arrays show higher capacity and enhanced rate capability in comparison with the sol-gel derived polycrystalline V_2O_5 film. For example, the V_2O_5 nanorod arrays grown from VO^{2+} solution deliver 5 times the capacity of the film at a current density of 0.7 A/g. For the single-crystal nanorod arrays, the long axis (growth direction) is parallel to the interlayers of V_2O_5, thus the nanorods provide shorter and simpler diffusion path for lithium ions and allow the most freedom for dimension change. Using the similar template-based electrodeposition method but with different growth conditions, Wang et al. prepared nanotube arrays of $V_2O_5 \cdot nH_2O$ [83]. The authors found that nanotubes were resulted when using lower voltage and shorter deposition time compared to the conditions for preparing nanorods. The $V_2O_5 \cdot nH_2O$ nanotube arrays demonstrate an initial high capacity of 300 mAh/g, about twice the initial capacity of 140 mAh/g from the $V_2O_5 \cdot nH_2O$ film. Such enhancement of capacity is due to the large surface area and short diffusion distances offered by the nanotube array. Subsequently, the authors used a two-step electrodeposition method to prepare $Ni-V_2O_5 \cdot nH_2O$ core-shell nanocable arrays [84]. Ni nanorod arrays were first grown by the template-based electrochemical deposition. In the second step, the hydrated vanadium

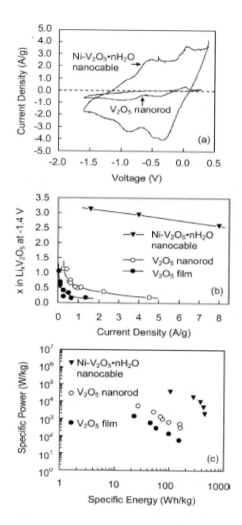

Figure 6. (a) Cyclic voltammograms of Ni-V$_2$O$_5$·nH$_2$O nanocable array and V$_2$O$_5$ nanorod array using a scan rate of 10 mV/s. (b) Relationship between current density and Li-ion intercalation capacity of and (c) Ragone plot for Ni-V$_2$O$_5$·nH$_2$O nanocable array, V$_2$O$_5$ nanorod array and sol-gel film. Reprinted with permission from Ref. 84, K. Takahashi et al. J. Phys. Chem. B **109**, 48 (2005), Copyright @ American Chemical Society.

pentoxide shell was deposited onto the surface of nickel nanorods through sol electrophoretic deposition. Figure 6 compares the electrochemical performance of Ni-V$_2$O$_5$·nH$_2$O nanocable arrays, single-crystal V_2O_5 nanorod arrays and sol-gel derived V_2O_5 films. Obviously Ni-$V_2O_5 \cdot nH_2O$ nanocable arrays demonstrate remarkably improved capacity and rate capability in comparison with the other two. The intercalation capacities of both nanorod arrays and sol-gel films decrease rapidly as the current density increases, while nanocable arrays are able to retain the high capacity at high current density (discharge rate), indicating the excellent high-rate performance of nanocable arrays. As shown in Figure 6c, Ni-$V_2O_5 \cdot nH_2O$ nanocable array has significantly higher energy density and power density than those of the nanorod array and sol-gel film by at least one order of magnitude, which is ascribed to the enhanced surface area and the reduced internal resistance.

Following the systematic studies on ordered arrays of V_2O_5 nanorods, nanotubes and nanocables, Lee and Cao reported the synthesis and electrochemical properties of V_2O_5 films with nanosized features [85]. Typically, platelet and fibrillar structured V_2O_5 films were prepared by solution methods, and the discharge capacities and cyclic performance of these films were compared with those of the conventional plain structured film. The platelet film consists of 20-30 nm sized standing platelets perpendicular to the substrate with random orientation, whereas fibrillar film is comprised of randomly oriented fibers though most of them protrude from the substrate surface. The initial discharge capacities of platelet and fibrillar structured V_2O_5 films are 1240 and 720 mAh/g, respectively, which are far larger than the initial discharge value (260 mAh/g) of the plain structure film. Such large discharge capacity values are ascribed to the combined effects of the reduced Li^+ diffusion distance, which prevents concentration polarization of Li^+ in the V_2O_5 electrode and poor interlayered cross-linking offering more Li^+ intercalation. However, platelet and fibrillar structured V_2O_5 films were easily degraded during electrochemical cyclic tests. Similarly platelet structured V_2O_5 films are also obtained by DC sputtering, but shows good cycling performance [86]. The capacity only changes from 80 to 73 $\mu Ah/cm^2$ after 100 cycles and to 70 $\mu Ah/cm^2$ after 200 cycles at a current density of 100 $\mu A/cm^2$. These results can be explained by the *h00* preferred orientation of the film which ensures a good homogeneity for Li intercalation/deintercalation and thus a good cycleability.

In addition to V_2O_5 thin films with structural features on the nanoscale, three-dimensional ordered macroporous (3DOM) V_2O_5 electrode materials with nanometer-sized features were synthesized using a colloidal-crystal-templated method [87]. The method is based on soaking the poly(methyl methacrylate) (PMMA) colloidal crystals in the $NaVO_3$ solution so that the interstitial spaces are infiltrated with precursor solution, followed by chemical conversion, drying and sintering to remove polymer spheres. The resultant material possesses photonic-crystal structures composed of interconnected open pores with nanometer-sized thin walls. Such three-dimensional ordered structure provides several advantageous features for Li-ion intercalation/deintercalation process: the continuous network ensures the electrical conductivity; the large open pores facilitate the transport of electrolyte; and the thin walls shorten the Li diffusion distances. Such photonic structures were later utilized for a real electrochemical cell system, in which the anode is 3DOM carbon and the pores are filled with polymer electrolyte [88]. The top surface of the 3DOM carbon is removed so that only electrolyte is in contact with the V_2O_5 gel cathode and the bottom carbon is adhered to the current collector. This work clearly demonstrates the feasibility of constructing three-dimensional electrochemical cells based on nanostructured materials. Other highly porous materials include mesoporous vanadium oxide with nanometer-sized pores that permit the easy diffusion of lithium ions. Liu et al. synthesized mesoporous vanadium oxide with pore sizes ranging from 3 to 4 nm by electrodepositing from a $VOSO_4$ solution in the presence of a block polyalkylene oxide polymer (P123) [89]. This polymer surfactant plays the key role in the formation of mesoporous structure. The authors specifically investigated the rate performance of the mesoporous vanadium oxide electrode and found that the material delivered a capacity of 125 mAh/g at a high rate of $50C$, corresponding to a capacitance of 450 F/g which is comparable to that of porous carbon capacitors. Therefore, the mesoporous vanadium oxide is very promising as cathode material for high-power lithium-ion batteries and fills in the gap between batteries and capacitors. Moreover, the mesoporous structure provides elasticity that allows for dimensional change during Li-ion intercalation/deintercalation, and thus offers good cycleability.

Hydrothermal synthesis is a powerful tool to transform transition metal oxides into high-quality nanostructures and nanostructured vanadium oxides in different morphologies can be produced via this procedure. Examples include long belt like nanowires, which are several tens of micrometers long and a few tens of nanometers wide, and are crystallized well growing along the [010] direction [90], and new types of vanadium oxide belts exhibiting a boomerang shape [91]. The structure of these nanobelts is unique in that it originates from twinning along the [130] direction, which is the first observation of twins within individual nanosized crystals. Liu et al. synthesized vanadium pentoxide nanobelts for highly selective and stable ethanol sensor materials by acidifying ammonium metavanadate followed by hydrothermal treatment [92]. In a separate report, $V_2O_5 \cdot nH_2O$ crystalline sheets, the intermediate products between nanobelts and nanowires, are fabricated hydrothermally using V_2O_5, H_2O_2 and HCl [93]. Nevertheless, the intercalation properties of these vanadium oxide nanobelts or nanosheets are not further investigated. More recently, Li et al. have studied the synthesis and electrochemical behavior of orthorhombic single-crystalline V_2O_5 nanobelts [94]. The V_2O_5 nanobelts with widths of 100-300 nm, thicknesses of 30-40 nm and lengths up to tens of micrometers are obtained by hydrothermal treatment of aqueous solutions of V_2O_5 and H_2O_2. The authors proposed a dehydration-recrystallization-cleavage mechanism for the formation of V_2O_5 nanobelts. A high initial discharge capacity of 288 mAh/g is found for the V_2O_5 nanobelts in a voltage range of 4.0 - 1.5 V; subsequently, the capacity decreases to 191 mAh/g for the second cycle then remains steady for the next four cycles. Apart from anhydrous crystalline V_2O_5 nanobelts, $V_2O_5 \cdot 0.9H_2O$ nanobelts and $V_2O_5 \cdot 0.6H_2O$ nanorolls are synthesized with hydrothermal treatment of NH_4VO_3 in the presence of difference acids [95]. The $V_2O_5 \cdot 0.9H_2O$ nanobelts are tens of micrometers long, 100-150nm wide and 20-30 thick. The $V_2O_5 \cdot 0.6H_2O$ nanorolls are half-tube nanostructured as a result of incomplete scrolling. It is interesting to note that $V_2O_5 \cdot 0.6H_2O$ nanorolls show higher intercalation capacity (253.6 mAh/g) than $V_2O_5 \cdot 0.9H_2O$ nanobelts (223.9 mAh/g) under a current density of 0.6 mA/g, which can be ascribed to the higher surface area and lower water content of nanorolls. Furthermore, the capacities of nanorolls and nanobelts increases to 287.8 mAh/g and 307.5 mAh/g, respectively, after annealing and dehydration of these nanostructures, which suggests the significant effect of water content on the electrochemical behavior. Single-crystalline nanowires can also be achieved by a combination of hydrothermal process of polycrystalline V_2O_5 and post-calcination treatment [96]. The nanowires have the diameter of 50-200 nm and the length up to 100 μm and deliver high initial capacity of 351 mAh/g. A combination of hydrothermal method and post-annealing process was also used by Lutta et al. to obtain vanadium oxide nanofibers [97]. In their method, polylactide fibers were hydrothermally treated in the mix of ammonium vanadate and acetic acid, followed by annealing in oxygen, resulting in vanadium oxide nanofibers, 60-140 nm in width and several microns in length. The V_2O_5 nanofibers deliver capacities exceeding 100 mAh/g which remain stable over 10 cycles. Obviously, morphology and water content have significant effect on the electrochemical performance of nanostructured vanadium oxides. For certain morphology, size is another factor affecting the electrochemical property. In this regard, Cui and coworkers prepared V_2O_5 nanoribbons and investigated the dependence of the electrochemical property on the width and thickness of nanoribbons by studying the chemical, structural and electrical transformations of V_2O_5 nanoribbons at the nanostructured level [98]. They found that transformation of V_2O_5 into the ω-$Li_3V_2O_5$ phase takes place within 10 s in thin nanoribbons and the efficient electronic transport can be maintained to charge ω-$Li_3V_2O_5$ nanoribbon within less than 5 s. Therefore, it is suggested that Li diffusion constant in nanoribbons is faster than that in bulk materials by three orders of magnitude, leading to a remarkable enhancement in power density (360C). It can be concluded that lithium-ion batteries based on nanostructured vanadium oxides have not only higher energy density but also higher power density and thus will find applications in electric and hybrid electric vehicles.

3.2 Nanostructured Manganese Oxides

Apart from traditional nanostructured layered materials that intercalate guest species between the interlayers, there are other inorganic compounds demonstrating high lithium storage capacity by electrochemically reacting with lithium ions. For example, the lithium intercalation of nanostructured manganese oxide involves formation/decomposition of lithium oxide, which is facilitated by formation of metallic manganese. Interestingly, nanostructured manganese oxide can act as either cathode or anode materials by controlling the working voltage range. Arrays of amorphous MnO_2 nanowires have been prepared by anodic electrodeposition into alumina templates [99]. These MnO_2 nanowires function as rechargeable cathodes for lithium-ion battery cells and deliver a capacity of 300 mAh/g when cycled between 3.5 and 2 V *vs.* Li/Li$^+$. More recently, Wu's group have reported the electrochemical synthesis of interconnected MnO_2 nanowires without using any template or catalyst [100]. These interconnected MnO_2 nanowires act as cathode materials when cycled to a middle voltage (1.5 V *vs.* Li/Li$^+$). When further cycled to a low voltage (0 V *vs.* Li/Li$^+$), such nanostructures exhibit a high capacity of over 1000 mAh/g, higher than that of commercially used carbon families as anode materials.

In addition to MnO_2 nanowires, nanostructured MnO_2 of different crystallographic types and morphologies have been synthesized through solution route and investigated as Li-ion battery cathode materials. Chen's group have selectively synthesized α-, β-, and γ-MnO_2 by using simple hydrothermal decomposition of a $Mn(NO_3)_2$ solution [101]. Typically, β-MnO_2 crystals are produced with a variety of novel shapes, including 1-D nanowires, 2-D hexagonal starlike structures and dentritelike hierarchical nanostructures. However, β-MnO_2 nanostures show low capacity and poor cycling stability, while α- and γ-MnO_2 1-D nanostructures demonstrate favorable electrochemical performance. α-MnO_2 nanowires deliver a capacity of 204 mAh/g when discharged to 1.5 V *vs.* Li/L$^+$ and retain a capacity of 112 mAh/g after 20 cycles at the current rate of 50 mA/g. γ-MnO_2 nanorods deliver a capacity of more than 210 mAh/g and retain a capacity of 148 mAh/g after 20 cycles at the

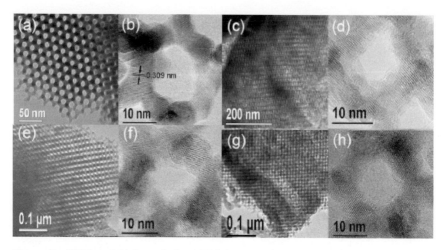

Figure 7. TEM and high-resolution TEM images of mesoporous β-MnO₂: (a, b) as-prepared; (c, d) after discharge; (e, f) end of discharge after 30 cycles; and (g, h) end of charge after 30 cycles. Reprinted with permission from Ref. 103, F. Jiao and P. T. Bruce, Adv. Mater. **19**, 657 (2007), Copyright @ Wiley-VCH.

current rate of 50 mA/g. In another report, Ho and Yen prepared α/γ-MnO₂ mixed-phase coating on Pt through cathodic deposition from Mn(NO₃)₂ aqueous solution [102]. The morphology of such α/γ-MnO₂ coating resembles a honeycomb consisting of flake structures in the nanometer scale. Remarkably, the α/γ-MnO₂ coating shows a gradual increase in capacity and crystalline stability after cyclic test. Its Li-intercalation capacity increases from 182 mAh/g for the first cycle to 209 mAh/g for the tenth cycle between 4.0 and 2.0 V *vs.* Li/Li⁺. Such enhancement in capacity and crystallization after cycling is ascribed to the mixed α/γ-MnO₂ phases and the nanosized structure. As mentioned above, bulk β-MnO₂ or nanostructured β-MnO₂ rapidly converts to LiMn₂O₄ spinel upon Li intercalation, resulting in unfavorable electrochemical performance. However, mesoporous β-MnO₂ demonstrates a remarkably high Li intercalation capacity of 284 mAh/g, corresponding to a composition of Li₀.₉₂MnO₂ [103]. Bruce's group reported the first synthesis of mesoporous β-MnO₂ with a highly ordered pore structure and highly crystalline walls [103]. Figure 7a and b show typical TEM images of mesoporous β-MnO₂, clearly demonstrating the highly ordered pore structure with a wall thickness of 7.5 nm. Figure 7c-7h show TEM images of mesoporous β-MnO$_2$ after first charge, end of discharge after 30 cycles and end of charge after 30 cycles. Both the β-MnO$_2$ crystal structure and the mesoporous structure are preserved upon cycling. The thin walls of the mesoporous β-MnO$_2$ allow volume changes during Li intercalation/deintercalation and 81% capacity is retained after 50 cycles. High capacities of more than 230 mAh/g are also reported for layered MnO$_2$ nanobelts, synthesized by Ma et al. using the hydrothermal treatment of Mn$_2$O$_3$ powders in an aqueous solution of NaOH [104]. The nanobelts are self-assembled into bundles with narrow size dispersion of 5-15 nm width and demonstrate a high capacity of 230 mAh/g up to 30 cycles.

Amorphous manganese oxides have also received increasing attention as cathode materials used in lithium-ion batteries [105,106]. Yang and Xu prepared nanostructured amorphous MnO$_2$ cryogels using two different sol-gel routes and investigated the influence of synthesis conditions on their electrochemical properties [107]. The cryogels are obtained by freeze drying MnO$_2$ hydrogels and the hydrogels are synthesized by reacting sodium permanganate with disodium fumarate (route 1) or with fumaric acid (route 2), respectively. Cryogels obtained from hydrogels synthesized via route 2 deliver much higher Li intercalation capacities than those obtained from hydrogels synthesized via route 1. For both routes, cryogels obtained from hydrogels using higher precursor concentration exhibit higher capacities. The capacity of the cryogel with the best performance can reach 289 mAh/g at a C/100 rate.

4. Nanostructured Lithium Phosphates and Nanostructured Carbon-Lithium Phosphate Composites

Lithium phosphate is presently the center of much interest as the cathode for lithium-ion batteries, because it is inexpensive, abundantly available, environmentally friendly, thermally stable in the fully charged state and has a large theoretical capacity of 170 mAh/g. The results on the diffusion coefficient of LiFePO$_4$ are controversial, because there is no compositional variation and what is measured is the movement of the LiFePO$_4$/FePO$_4$ interface.   A diffusion coefficient around 10^{-13}-10^{-14} cm^2/s over a whole range of composition was reported by Franger et al for LiFePO$_4$.[108] Another experimental work reported a value of 2×10^{-14} cm^2/s [109]. Most recently, a systematic study of LiFePO$_4$ with cyclic voltammetry (CV) has been presented [110]. In this study, the lithium diffusion coefficients were determined by CV to be 2.2×10^{-14} and 1.4×10^{-14} cm^2/s for charging and discharging LiFePO$_4$ electrodes in 1 M LiPF$_6$ ethylene carbonate/diethyl carbonate, respectively. There are essentially no electronically conducting species in pure LiFePO$_4$. Therefore, the conductivity of the material is only 10^{-11} S/cm partially due to the motion of lithium ions [111]. Carbon containing precursors (e.g. carbonates, acetates and oxalates) are used to prepare LiFePO$_4$ so that some residual carbon will prevent the formation of ferric ions. The as-prepared samples show higher conductivities, in the range of 10^{-5}-10^{-6} S/cm, however, it is not yet high enough for high power lithium-ion batteries [112].

To increase the conductivity, the material could be doped as suggested by Chiang and coworkers.[111] However, doping may have deleterious impact if it occurs on the lithium sites. Conductive coatings deposited on the surface of LiFePO$_4$ are usually employed to solve the conductivity issue. Most coatings are carbonaceous and deposited during the synthesis process. Pioneering work on carbon coated LiFePO$_4$ was carried out by Ravet et al. [113,114] Sucrose was used as one carbon source [114] and was added on the initial hydrothermal samples [115] or during pyrolysis [116]. Other methods include thermal decomposition of pyrene [117] or citric acid based sol-gel processing [118]. It should be noted that the electrochemical properties of LiFePO$_4$ are influenced by the quality of carbon coatings. Wilcox et al. found that the conductivity and rate behavior of LiFePO$_4$ are strongly affected by carbon structural factors such as sp^2/sp^3 and disordered/grapheme (D/G), as determined by Raman spectroscopy, and H/C ratios determined from elemental analysis [119]. The structure of carbon can be controlled by the use of additives during LiFePO$_4$ synthesis. LiFePO$_4$ coated with the more graphitic carbon has higher conductivity and shows better electrochemical performance.   Another factor that influences the electrochemical performance of LiFePO$_4$/C composites is the porosity. Gaberscek et al.

nanoribbons in the aerogel have similar morphology and dimensional scale, and thus have intimate contact with each other in the nanoscale. Moreover, the porous structure of carbon nanotubes and V_2O_5 aerogel permits electrolyte access throughout the composite material. As a result, such nanocomposite electrode shows high capacities exceeding 400 mAh/g at high rates. Apart from vanadium oxides, some nanostructured lithium vanadium oxides have also been reported to form nanocomposite with carbon which exhibits excellent electrochemical characteristics. It was reported that mixing the precursor of $Li_{1+a}V_3O_8$ with a suspension of carbon black resulted in nanocomposites of $Li_{1+a+x}V_3O_8/\beta$-Li $_{1/3}V_2O_5/C$ [132]. β-$Li_{1/3}V_2O_5$ was a by-product formed when the initial $Li_{1+a}V_3O_8$ was reduced by carbon. Here carbon particles play critical roles as a reducing agent, a growth-limiting agent to restrict the electroactive material within the nanoscale, and as an electronically conducting agent. The $Li_{1+a+x}V_3O_8/\beta$-$Li_{1/3}V_2O_5/C$ nanocomposite shows significantly better electrochemical performance in comparison with the standard $Li_{1+a}V_3O_8$. Similarly, acetylene black was used to prompt the reduction of potassium permanganate, yielding amorphous manganese oxide/carbon composites [133]. The as-prepared composite delivers a high capacity of 231 mAh/g at a current density of 40 mA/g, showing good electrochemical performance at high rates. The energy density of MnO_2/C nanocomposite can be further increased by optimization of the synthesis conditions. Hibino and coworkers used a sonochemical synthesis method to prepare MnO_2/C nanocomposite with acetylene black and sodium permanganate and optimized synthesis conditions such as the reaction temperature and specific surface area of the carbon to achieve the best electrochemical performance of the nanocomposite [134]. The active material content increases by increasing the reaction temperature. It is interesting to note that the capacity increases with the increasing amount of active material then decreases, because the excessive formation of active material increases the electrochemicaly effective volume, leading to capacity drop. On the other hand, using carbon with higher surface area results in higher capacity; the highest capacities are 126 and 99.9 mAh/g at current densities of 1 and 10 A/g, respectively. A number of lithium phosphates/cabon composites have also been studied as cathode materials for lithium batteries, including those of general

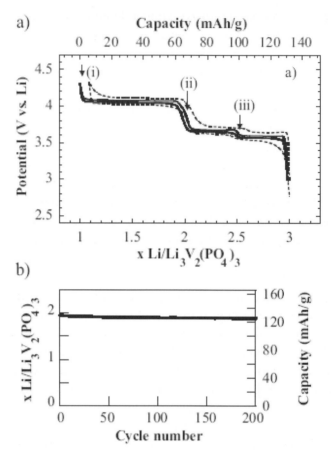

Figure 10. (a) Voltage-composition plot for C/Li$_3$V$_2$(PO$_4$)$_3$ composites at rates of C/5 (solid line) and 5C (dotted line) in the potential window 3.0 – 4.3 V; single phase compositions are indicated: x = 2.5 (i); 2.0 (ii); and 1.0 (iii). (b) Cycling stability at a rate of 1C. Reprinted with permission from Ref. 136, H. Huang et al., Adv. Mater. **14**, 1525 (2002), Copyright @ Wiley-VCH.

formula LiMPO$_4$ (M = Fe, Mn, Co, Ni) [135] and Li$_3$V$_2$(PO$_4$)$_3$ [136]. Yang and Xu have reported the synthesis and characterization of carbon-coated lithium metal phosphates LiMPO$_4$ (M = Fe, Mn, Co, Ni) [135]. The authors developed an organic sol-gel method using ethylene glycol as the solvent and synthesized well-dispersed submicron-sized particles with uniform size distribution.   Among the carbon-coated LiMPO$_4$

(M = Fe, Mn, Co, Ni) composites, the LiFePO$_4$/C with surface carbon coating of 1.8 wt% achieves an electronic conductivity of 10^{-2} S/cm and shows the best electrochemical performance. Compared to LiFePO$_4$ that attracts a lot of attention, Li$_3$V$_2$(PO$_4$)$_3$ is relatively unexplored. Like LiFePO$_4$, this material also suffers from low electronic conductivity. To solve this issue, Li$_3$V$_2$(PO$_4$)$_3$ crystallites were wrapped within a conductive carbon network to form a nanocomposite which delivers almost full capacity at high rates [136]. The potential curves in Figure 10(a) reveal that two lithium ions per formula unit are completely extracted in three steps to give a theoretical capacity (100%) of 132 mAh/g at a rate of $C/5$. 95% theoretical capacity is still achieved at a high rate of 5C. The flat plateaus in the curve correspond to Li$_x$V$_2$(PO$_4$)$_3$, where x = 2.5 (i); 2.0 (ii); and 1.0 (iii). Such a sequence of phase transitions between two single phases shows the very low degree of polarization in the discharge curve owning to the facile ion and electron transport. Excellent cycling stability is also demonstrated by this material, as shown in Figure 10(b). When cycled between 3.0 V and 4.8 V, the Li$_3$V$_2$(PO$_4$)$_3$/C composite delivers a specific energy density of 2330 mWh/cm^{-3} comparable to LiCoO$_2$ (2750 mWh/cm^3) or LiFePO$_4$ (2065 mWh/cm^3).

5.2 Nanostructured Polymer-Oxide Composites

Over the past two decades much interest has been placed on the conductive polymer/transition metal oxide nanocomposite. The hybrid material consists of conductive organic polymers (e.g. polyacetylene, polyaniline and polypyrrole (PPy)) interleaved between the layers of an oxide lattice such as V$_2$O$_5$. Both oxide and polymer are electrochemically active and this feature makes the polymer/oxide nanocomposite very attractive as the cathode material for lithium-ion batteries. The layer-by-layer (LbL) technique, based on physical adsorption of oppositely charged layers, has been widely used to prepare V$_2$O$_5$ nanocomposites alternating with polymer layers. One popular example is V$_2$O$_5$/polyaniline nanocomposite film fabricated by the LbL technique and the intimate contact between the oxide and polymer within nanoscale results in an improved intercalation capacity [137]. Later,

Huguenin et al. prepared V_2O_5 nanocomposite alternating with blends of chitosan and poly(ethylene oxide) (PEO) using the LbL technique and investigated the charge storage capability in such nanoarchitectures [138]. A small amount of chitosan (1%) is added to blend with PEO because the adsorption of alternate layers of PEO and V_2O_5 is not efficient. The V_2O_5/blend shows higher capacity and intercalates 1.77 moles of lithium per mole of V_2O_5. The enhanced electrochemical performance of V_2O_5/blend in comparison with V_2O_5/chitosan is due to a larger number of electrochemically active sites and faster lithium diffusion within the host material. At 20 mV/s, the charges injected were 3.29 mC/cm^2 and 8.02 mC/cm^2 for V_2O_5/chitosan and V_2O_5/blend, respectively. In a more recent report, polyaniline homogeneously distributed into V_2O_5/polyaniline nanocomposite was found to stabilize the capacity [139]. In this study, a reverse micelle method was used to prepare V_2O_5/polyaniline nanofibers which exhibit improved cycling performance compared to the V_2O_5 nanofibers [139]. The V_2O_5/polyaniline nanofibers containing 30 mol% polyaniline delivers a steady capacity of about 300 mAh/g without morphology change over 10 cycles, whereas the V_2O_5 nanofibers do not retain the morphology after cycling. Some V_2O_5/polymer nanocomposite shows lower storage capacity but better cycling stability compared to pure nanostructured V_2O_5 [140]. As reported by Reddy et al., $PVP_xV_2O_5$ ($x = 0.5$, 1) nanobelts synthesized by a hydrothermal method exhibit lower capacity but better cycleability compared with V_2O_5 nanobelts. The authors studied the interaction between the oxide and polymer with Fourier transformation infrared spectroscopy (FTIR) and found that the hydrogen atoms in PVP are hydrogen-bonded with the oxygen atoms of the V=O bonds of V_2O_5 nanobelts, which effectively shields the electrostatic interaction between V_2O_5 interlayer and lithium ions. As discussed above, polymers can be intercalated between the interlayers of V_2O_5, on the other hand, V_2O_5 can be interleaved within a block polymer matrix as well [141]. Mayes and coworkers used a sol-gel method to prepare continuous and amorphous V_2O_5 phase within the poly(oligooxythylene methacrylate) (POEM) domains of a poly(oligooxythylene methacrylate)-block-poly(butyl methacrylate) (POEM-*b*-PBMA) copolymer (70 wt% POEM) up to weight ratios of 34% V_2O_5 [141]. The resulted nanocomposite film is flexible and semi-transparent and the redox activity of V_2O_5 is preserved in such nanocomposite.

Cathode materials other than V_2O_5 can form nanocomposites with conductive polymer as well. Poly(ethylene oxide) (PEO) was used as an electroactive polymeric binder to mix with carbon containing $Li_{1.1}V_3O_8$ [142]. The resulted composite electrode shows a capacity of 270 mAh/g at a rate of 5/C, higher than the capacity (180 mAh/g at C/5 rate) of the standard electrode without PEO. Such improved electrode performance is attributed to the more efficient charge-carrier collection within the composite electrode. Among all known cathode materials, elemental sulfur is the cheapest and has the highest theoretical capacity density of 1672 mAh/g assuming a complete reaction to yield Li_2S [143]. However, Li/S cells suffer from low utilization of active material, because electrochemical reaction with the interior active materials is hindered by the insulated reaction products covering the sulfur particles. Moreover, the dissolved polysulfides transfer onto the surface of the Li anode, causing lithium corrosion and poor rechargeability of Li/S cells. To overcome these two problems, nanodispersed composites with sulfur embedded in a conductive polymer matrix were designed and prepared by heating the mixture of polyacrylonitrile (PAN) and sublimed sulfur [144,145]. The composite also show excellent cycling life due to the suppressed dissolution of polysulfides into the electrolyte and thus demonstrates a great potential as cathode material for lithium batteries. Conductive polymers themselves can act as cathode material, however, they suffer from low capacities and display sloping charge-discharge curves. For example, polyryrrole (PPy) is one of the most popular conductive polymers and has a specific energy ranging from 80 to 390Wh/kg [146]. To improve its capacity, a Fe^{III}/Fe^{II} redox couple is physically or chemically attached to the PPy polymer backbone [147]. The examination of the $PPy/LiFePO_4$ composite electrode shows that the composite has higher specific capacity and rate capability.

5.3 Nanostructured Metal-Oxide Composites and Other Composites

The third most popular composite electrode is metal based cathode material, exemplified by the $Ni-V_2O_5 \cdot nH_2O$ core-shell structure discussed earlier in section 3.1. Accordingly, Wang et al. synthesized $Ag\text{-}Ag_{0.08}V_2O_5\cdot nH_2O$ composite films by dispersing silver nanowires into $V_2O_5\cdot nH_2O$ matrix [148]. The composite film is found to deliver twice the capacity of the $V_2O_5\cdot nH_2O$ xerogel film, due to further amorphization of $V_2O_5\cdot nH_2O$, the increased porosity and the enhanced electronic conductivity. In a similar concept, $LiCoO_2/Ag$ multilayer film was fabricated by magnetron sputtering and showed enhanced rate capability in comparison with $LiCoO_2$ film of the same thickness [149]. Thickness of Ag layer is restricted within nanoscale and the rate capability of the multilayer film improves with the increased thickness of Ag layer as a result of the enhanced electronic conductivity.

More recently, oxide/metal/polymer composites have been obtained and been shown to have very good electrochemical performance. Li et al. prepared freestanding $V_2O_5/Pt/PVA$ multilayer films and the thicknesses of the V_2O_5, Pt, and PVA are 22, 57, and 704 nm [150]. Other types of composite structures include oxide/oxide composite and carbon/polymer composite. Imachi et al. designed and synthesized a double-layer cathode composed of a $LiCoO_2$ main layer with a $LiFePO_4$ sublayer on top of Al current collector which showed better tolerance against overcharging than other electrodes including ($LiCoO_2\text{-}LiFePO_4$ mixture)/Al single layer and $LiFePO_4/LiCoO_2/Al$ double layer [151]. The authors attributed such enhanced electrochemical performance to a large increase in the ohmic resistance of the delithiated Li_xFePO_4 layer which shuts the charging current down during overcharging without shut-down of the separator. In the case of polymer/carbon nanocomposite, Sivakkumar et al. synthesized a polyaniline (PANI)/multiwalled carbon nanotube (CNT) composite by in situ chemical polymerization and utilized the nanocomposite as a cathode material in a lithium metal-polymer cell assembled with ionic liquid electrolyte [152]. Such cell demonstrates a maximum discharge capacity of 139 mAh/g with good cycleability and shows decent high rate performance (111 mAh/g at the 2.0*C* rate).

6. Concluding Remarks

This review clearly reveals how moving from bulk materials to the nanoscale can significantly change device performance for energy storage and conversion. The development of high-performance lithium-ion batteries can benefit from the distinct properties of nanomaterials, such as high surface areas, short diffusion paths and a large quantity of active sites. Among a wide range of synthetic methods to prepare nanomaterials, simple and elegant are soft chemistry routes that involve sol-gel reactions and that frequently use organic molecules as structure-directing templates.

As discussed in this review, there are two groups of Li-ion battery cathode materials in general: the one with more compact lattices such as $LiCoO_2$, $LiNiO_2$, $LiMn_2O_4$, substituted lithium transition metal oxides, or solid solutions of lithium transition metal oxides, and the other group of cathode materials with more open structure including V_2O_5, MnO_2 and $LiFePO_4$. Nanoparticulate forms and one-dimensional nanostructures of lithium transition metal oxides are fabricated with solid state approaches or solution chemistry methods. To increase the stability of these nanocrystalline lithium transition metal oxides, it is necessary to coat these materials with nanosized thick layers to suppress metal dissolution. In the case of $LiCoO_2$, coatings of various phosphates (e.g. $AlPO_4$) and oxides (e.g. ZnO or ZrO_2) have been studied and significant improvements in capacity retention have been demonstrated. Nanosized ZnO or ZrO_2 coatings on $LiMn_2O_4$ and substituted $LiMn_2O_4$ also help to improve the cycling performance of the cathodes by collecting acidic species from electrolyte to reduce Mn dissolution. Vanadium oxide is one of the earliest studied oxides as cathode materials. There are many reports on synthesis and electrochemical properties of nanostructured vanadium oxides. Sol-gel processing and hydrothermal treatment are usually employed to prepared a large variety of nanostructured vanadium oxides, including nanorolls, nanobelts, nanowires, mesoporous structures, two-dimensional thin films with nanosized features and three-dimensional ordered photonic crystal structures with nanosized features. The template-based solution methods are utilized to prepare ordered arrays of nanostructures, such as polycrystalline or single-crystalline V_2O_5 nanorod arrays, $V_2O_5 \cdot nH_2O$ nanotube arrays and $Ni-V_2O_5 \cdot nH_2O$

core-shell nanocable arrays. In analogy to vanadium oxides, nanostructured manganese oxides are synthesized with soft chemistry methods and different morphologies are produced including nanowires, nanotubes, nanobelts, mesoprous structures and honeycomb-structured thin films. Morphology, structure, growth mechanisms and electrochemical properties of these nanostructures have been discussed in this article. All nanostructured electrodes exhibit significantly improved storage capacity and rate performance than thin film electrodes. There are only a few studies on LiFePO$_4$ nanoparticles, and submicron-sized or micron-sized LiFePO$_4$ are more commonly reported. However, to increase the conductivity of this material, carbon or metallic coatings with thickness on the order of a few nanometers are deposited on the surface of LiFePO$_4$, mostly during synthesis process. Such novel designs of nanostructured composites are generalized and applied to other oxides and conductive materials, including composites of carbon-oxide, polymer-oxide, metal-oxide, carbon-polymer, oxide-oxide or even metal-oxide-polymer.

Applications of nanotechnology in energy storage are in the stage of research and development. For realization of wide industrial applications, further work is required to achieve controlled and large-scale synthesis of nanostructures, to understand mechanisms of lithium storage in nanomaterials and kinetic transport on the interface between electrode and electrolyte. The effects of nanostructures in battery performance are not only simple consequences of a reduction in size. Interfacial properties are subtle and critical, considering space-charge effects at the interface between nanosized electrode materials and charge transport between electrode and electrolyte. This challenges researchers worldwide to carry out systematic experimental studies and to develop predictive theoretical tools for better fundamental understanding of relationships between nanostructures and electrochemical characteristics of electrode materials.

Acknowledgements

This work has been supported in part by National Science Foundation (DMI-0455994) and Air Office of Scientific Research

(AFOSR-MURI, FA9550-06-1-032). This work has also been supported by the Center for Nanotechnology at UW, Pacific Northwest National Laboratories (PNNL), Joint Institute of Nanoscience and Nanotechnology (JIN, UW and PNNL), Washington Technology Center (WTC), and JFE Steel Corporation, Japan. Y. W would like to acknowledge the Ford, Nanotechnology, and JIN graduate fellowships. A portion of the research (TEM study) described in this paper was performed in the Environmental Molecular Sciences Laboratory, a national scientific user facility sponsored by the Department of Energy's Office of Biological and Environmental Research and located at PNNL.

References

1. J. M. Tarascon and M. Armand, Nature **414**, 359 (2001).
2. G. Z. Cao, Nanostructures and Nanomaterials, Synthesis, Properties and Applications, Imperial College Press, London (2004).
3. A. Bachtold, P. Hadley, T. Nakanishi and C. Dekker, Science **294**, 1317 (2001).
4. R. Martel, T. Schmidt, H. R. Shea, T. Hertel and P. Avouris, Appl. Phys. Lett. **73**, 2447 (1998).
5. G. Zheng, W. Lu, S. Jin and C. M. Lieber, Adv. Mater. **16**, 1890 (2004).
6. Y. W. Heo, L. C. Tien, Y. Kwon, D. P. Norton, S. J. Pearton, B. S. Kang and F. Ren, Appl. Phys. Lett. **85**, 2274 (2004).
7. Y. Cui, Q. Wei, H. Park and C. M. Lieber, Science **293**, 1289 (2001).
8. A. Star, T. Han, V. Joshi, J. P. Gabriel and G. Grüner, Adv. Mater. **16**, 2049 (2004).
9. E. Comini, G. Faglia, G. Sberveglieri, Z. W. Pan and Z. L. Wang, Appl. Phys. Lett. **81**, 1869 (2002).
10. E. S. Snow, F. K. Perkins, E. J. Houser, S. C. Badescu and T. L. Reinecke, Science **307**, 1942 (2005).
11. H. J. Dai, J. H. Hafner, A. G. Rinzler, D. T. Golbert and R. E. Smalley, Nature **384**, 147 (1996).
12. Y. Zhang, K. Suenaga, C. Colliex and S. Iijima, Science **281**, 973 (1998).
13. J. C. Hulteen and C. R. Martin, J. Mater. Chem. **7**, 1075 (1997).
14. N. Pinna, U. Wild, J. Urban and R. Schlögl, Adv. Mater. **15**, 329 (2003).
15. M. Winter and R.J. Brodd, Chem. Rev. **104**, 4245 (2004).
16. D. O'Hare, in Inorganic Materials, eds. D. W. Bruce and D. O'Hare, John Wiley & Sons, New York (1991), p. 165.
17. R. Schöllhorn, Angew Chem. Int. Ed. Engl. **19**, 983 (1980).
18. R. Schöllhorn, Chemical Physics of Intercalation, A. P. Legrand, S. Flandrois, Plenum, New York, NATO Ser. B **172**, 149 (1987).

19. D. W. Murphy, S. A. Sunshine and S. M. Zahurak, in Chemical Physics of Intercalation, eds. A. P. Legrand and S. Flandrois, Plenum: New York, NATO Ser. B **172**, 173 (1987).

20. D. W. Murphy, P. A. Christian, F. J. Disalvo and J. V. Waszczak, Inorg. Chem. **24**, 1782 (1985).

21. R. Clement, J. Am. Chem. Soc. **103**, 6998 (1981).

22. L. F. Nazar and A. J. Jacobson. J. Chem. Soc. Chem. Commun. (1986), p570.

23. R. Schöllhorn, Physics of Intercalation Compounds; Springer-Verlag: Berlin, 1981.

24. A. S. Aricò, P. Bruce, B. Scrosati, J-M. Tarascon and W. V. Schalkwijk, Nat. Mater. **4**, 366 (2005).

25. M. S. Whittingham. Chem. Rev. **104**, 4271 (2004).

26. M. S. Whittingham, Y. Song, S. Lutta, P. Y. Zavalij and N. A. Chernova, J. Mater. Chem. **15**, 3362 (2005).

27. J. N. Reimers and J. R. Dahn, J. Electrochem. Soc. **139**, 2091 (1992).

28. P. N. Kumta, D.Gallet, A. Waghray, G. E.Blomgren and M. P. Setter, J. Power Sources **72**, 91 (1998).

29. T. Ohzuku, A. Ueda, N. Nagayama, Y. Iwakoshi and H. Komori, Electrochim. Acta, **38**, 1159 (1993).

30. G. G. Amatucci, J. M. Tarascon and L. C. Klein, J. Electrochem. Soc. **143**, 1114 (1996).

31. T. Ohzuku and A. Ueda, Solid State Ionics **69**, 201 (1994).

32. T. Ohzuku, A. Ueda and N. Nagayama, J. Electrochem. Soc. **140**, 1862 (1993).

33. J. R. Dahn, U. von Sacken and C. A. Michal, Solid State Ionics **44**, 87 (1990).

34. A. Rougier, P. Gravereau and C. Delmas, J. Electrochem. Soc. **143**, 1168 (1996).

35. L. Abello, E. Husson, Y. R.Repelin and G. Lucazeau, Spectrochim. Acta, **39A**, 641 (1983).

36. M. M. Thackeray, J. Electrochem. Soc. **142**, 2558 (1995).

37. B. Ammundesen and J. Paulsen, Adv. Mater. **13**, 943 (2001).

38. J. B. Goodenough and K. Mizuchima, U.S. Patent 4,302,518, (1981).

39. P. S. Herle, B. Ellis, N. Coombs and L. F. Nazar, Nat. Mater. **1**, 123 (2002).

40. X. Li, F. Cheng, B. Guo and J. Chen, J. Phys. Chem. B, **109**, 14017 (2005).

41. J. F. Whitacre, W. C. West, E. Brandon and B. V. Ratnakumar, J. Electrochem. Soc. **148**, A1078 (2001).

42. C. J. Curtis, J. Wang and D. L. Schulz, J. Electrochem. Soc. **151**, A590 (2004).

43. M. Kunduraci and G. G. Amatucci, J. Electrochem. Soc. **153**, A1345 (2006).

44. J. C. Arrebola, A. Caballero, M. Cruz, L. Hernán, J. Morales and E. R. Castellón, Adv. Funct. Mater. **16**, 1904 (2006).

45. J. Kim, M. Noh, J. Cho, H. Kim and K. Kim, J. Electrochem. Soc. **152(6)**, A1142 (2005).

46. J. Cho, B. Kim, J. Lee, Y. Kim and B. Park, J. Electrochem. Soc. **152(1)**, A32 (2005).

47. B. Kim, C. Kim, T. Kim, D. Ahn and B. Park, J. Electrochem. Soc. **153**(9), A1773, (2006).

48. Y. J. Kim, J. Cho, T.-J. Kim and B. Park, J. Electrochem. Soc. **150**, A1723 (2003).

49. L. Lui, Z. Wang, H. Li, L. Chen and X. Huang, Solid State Ionics **152-153**, 341 (2002).

50. J. Cho, C.-S. Kim and S.-I. Yoo, Electrochem. Solid-State Lett. **3**, 362 (2000).

51. Z. Wang, X. Huang and L. Chen, J. Electrochem. Soc. **150**, A199 (2003).

52. T. Fang, J. Duh and S. Sheen, J. Electrochem. Soc. **152**, A1701 (2005).

53. K. Y. Chung, W. Yoon, J. McBreen, X. Yang, S. H. Oh, H. C. Shin, W. I. Cho and B. W. Cho, J. Electrochem. Soc. **153**, A2152 (2006).

54. H. Miyashiro, A. Yamanaka, M. Tabuchi, S. Seki, M. Nakayama, Y. Ohno, Y. Kobayashi, Y. Mita, A. Usami and M. Wakihara, J. Electrochem. Soc. **153**, A348 (2006).

55. N. Ohta, K. Takada, L. Zhang, R. Ma, M. Osada and T. Sasaki, Adv. Mater. **18**, 2226 (2006).

56. Y. Sun, K. Hong and J. Prakash, J. Electrochem. Soc. **150**, A970 (2003).

57. J. Han, S. Myung and Y. Sun, J. Electrochem. Soc. **153**, A1290 (2006).

58. M. M. Thackeray, C. S. Johnson, J. S. Kim, K. C. Lauzze, J. T. Vaughey, N. Dietz, D. Abraham, S. A. Hackney, W. Zeltner and M. A. Anderson, Electrochem. Commun. **5**, 752 (2003).

59. J. S. Kim, C. S. Johnson, J. T. Vaughey, S. A. Hackney, K. A. Walz, W. A. Zeltner, M. A. Anderson and M. M. Thackeray, J. Electrochem. Soc. **151**, A1755 (2004).

60. D. W. Murphy, P. A. Christian, F. J. DiSalvo and J. V. Waszczak, Inorg. Chem. **18**, 2800 (1979).

61. L. Abello, E. Husson, Y. R. Repelin and G. Lucazeau, Spectrochim. Acta **39A**, 641 (1983).

62. M. S. Whittingham, J. Electrochem. Soc. **123**, 315 (1975).

63. Y. Sakurai, S. Okada, J. Yamaki and T. Okada, J. Power Sources **20**, 173 (1987).

64. K. West, B. Zachau-Christiansen, M. J. L. Ostergard and T. Jacobsen, J. Power Sources **20**, 165 (1987).

65. K. West, B. Zachau-Christiansen, T. Jacobsen and S. Skaarup, Electrochim. Acta **38**, 1215 (1993).

66. K. Salloux, F. Chaput, H. P. Wong, B. Dunn and M. W. Breiter, J. Electrochem. Soc. **142**, L191 (1995).

67. V. Petkov, P. N. Trikalitis, E. S. Bozin, S. J. L. Billinge, T. Vogt and M. G. Kanatzidis, J. Am. Chem. Soc. **124**, 10157 (2002).

68. R. Nesper, M. E. Spahr, M. Niederberger and P. Bitterli, Int. Patent Appl. PCT/CH97/00470 (1997).

69. M. E. Spahr, P. Bitterli, R. Nesper, M. Müller, F. Krumeich and H.-U. Nissen, Angew. Chem. **110**, 1339 (1998).

70. M. E. Spahr, P. Bitterli, R. Nesper, M. Müller, F. Krumeich and Nissen, H.-U. Angew. Chem., Int. Ed. **37**, 1263 (1998).

71. F. Krumeich, H.-J. Muhr, M. Niederberger, F. Bieri, B. Schnyder and R. Nesper, J. Am. Chem. Soc. **121**, 8324 (1999).

72. H.-J. Muhr, F. Krumeich, U. P. Schòholzer, F. Bieri, M. Niederberger, L. J. Gauckler and R. Nesper, Adv. Mater. **12**, 231 (2000).

73. K. S. Pillai, F. Krumeich, H.-J. Muhr, M. Niederberger and R. Nesper, Solid State Ionics **141-142**, 185 (2001).

74. C. J. Patrissi and C. R. Martin, J. Electrochem. Soc. **146**, 3176 (1999).

75. N. Li, C. J. Patrissi and C. R. Martin, J. Electrochem. Soc. **147**, 2044 (2000).

76. C. R. Sides and C. R. Martin, Adv. Mater. **17**, 125 (2005).

77. K. Takahashi, S. J. Limmer, Y. Wang and G. Z. Cao, J. Phys. Chem. B **108**, 9795 (2004).

78. K. Takahashi, S. J. Limmer, Y. Wang and G. Z. Cao, Jpn. J. Appl. Phys. **44**, 662 (2005).

79. K. Takahashi, Y. Wang and G. Z. Cao, Appl. Phys. Lett. **86**, 053102 (2005).

80. A. van der Drift, Philips Res. Rep. **22**, 267 (1968).

81. R. L. Penn and J. F. Banfield, Geochimica et Cosmochimica Acta **63**, 1549 (1999).

82. C. M. Chun, A. Navrotsky and I. A. Aksay, in Proc. Microscopy and Microanalysis (1995), p188.

83. Y. Wang, K. Takahashi, H. Shang and G. Z. Cao, J. Phys. Chem. B **109**, 3085 (2005).

84. K. Takahashi, Y. Wang and G. Z. Cao, J. Phys. Chem. B **109**, 48 (2005).

85. K. Lee, Y. Wang and G. Z. Cao, J. Phys. Chem. B **109**, 16700 (2005).

86. C. Navone, R. Baddour-Hadjean, J. P. Pereira-Ramos and R. Salot, J. Electrochem. Soc. **152(9)**, A1790 (2005).

87. H. Yan, S. Sokolov, J. C. Lytle, A. Stein, F. Zhang and W. H. Smyrl, J. Electrochem. Soc. **150(8)**, A1102 (2003).

88. N. S. Ergang, J. C. Lytle, K. T. Lee, S. M. Oh, W. H. Smyrl and A. Stein, Adv. Mater. **18**, 1750 (2006).

89. P. Liu, S. Lee, E. Tracy, Y. Yan and J. A. Turner, Adv. Mater. **14**, 27 (2002).

90. D. Pan, S. Zhang, Y. Chen and J. G. Hou, J. Mater. Res. **17**, 1981 (2002).

91. U. Schlecht, M. Knez, V. Duppel, L. Kienle and M. Burghard, Appl. Phys. A **78**, 527 (2004).

92. J. Liu, X. Wang, Q. Peng and Li, Y. Adv. Mater. **17**, 764 (2005).

93. X. K. Hu, D. K. Ma, J. B. Liang, S. L. Xiong, J. Y. Li and Y. T. Qian, Chem. Lett. **36(4)**, 560 (2007).

94. G. Li, S. Pang, L. Jiang, Z. Guo and Z. Zhang, J. Phys. Chem. B **110**, 9383 (2006).

95. B. Li, Y. Xu, G. Rong, M. Jing and Y. Xie, Nanotechnology, **17**, 2560 (2006).

96. X. Li, W. Li, H. Ma and J. Chen, J. Electrochem. Soc. **154(1)**, A39-A42 (2007).

97. S. T. Lutta, H. Dong, P. Y. Zavalij and M. S. Whittingham, Mater. Res. Bull. **40**, 383 (2005).

98. C. K. Chan, H. Peng, R. D. Tweten, K. Jarausch, X. F. Zhang and Y. Cui, Nano Lett. **7(2)**, 490 (2007).

99. W. C. West, N. V. Myung, J. F. Whitacre and B.V. Ratnakumar, J. Power Sources **126**, 203 (2004).

100. M. S. Wu, P. C. J. Chiang, J. T. Lee and J. C. Lin, J. Phys. Chem. B **109**, 23279 (2005).

101. F. Y. Cheng, J. Z. Zhao, W. E. Song, C. S. Li, H. Ma, J. Chen and P. W. Shen, Inorg. Chem. **45**, 2038 (2006).

102. W. H. Ho and S. K. Yen, J. Electrochem. Soc. **152(3)**, A506 (2005).

103. F. Jiao and P. T. Bruce, Adv. Mater. **19**, 657 (2007).

104. R. Ma, Y. Bando, L. Zhang and T. Sasaki, Adv. Mater. **16**, 918 (2004).

105. J. Kim and A. Manthiram, Nature **390**, 265 (1997).

106. D. Im and A. Manthiram, J. Electrochem. Soc. **149**, A1001 (2002).

107. J. Yang and J. J. Xu, J. Power Sources **122**, 181 (2003).

108. S. Franger, F. L. Cras, C. Bourbon and H. Rouault, Electrochem. Solid-State Lett. **5**, A231 (2002).

109. P. P. Prosini, M. Lisi, D. Zane and M. Pasquali, Solid State Ionics, **148**, 45 (2002).

110. D. Y. W. Yu, C. Fietzek, W. Weydanz, K. Donoue, T. Inoue, H. Kurokawa and S. Fujitani, J. Electrochem. Soc. **154**, A253 (2007).

111. S.-Y. Chung, J. T. Bloking and Y.-M. Chiang, Nat. Mater. **1**, 123 (2002).

112. S. Yang, Y. Song, K. Ngala, P. Y. Zavalij and M. S. Whittingham, J. Power Sources **119**, 239 (2003).

113. N. Ravet, J. B. Goodenough, S. Besner, M. Simoneau, P. Hovington and M. Armand, Electrochem. Soc. Abstracts, **99–2**, 127 (1999).

114. N. Ravet, S. Besner, M. Simoneau, A. Valle´e, M. Armand and J.-F. Magnan, Materiaux d'e´lectrode pre´sentant une conductivite de surface e´leve´e. 2000, Hydro-Quebec: European Patent 1049182A2.

115. S. Yang, P. Y. Zavalij and M. S. Whittingham, Electrochem. Commun. **3**, 505 (2001).

116. A. D. Spong, G. Vitins and J. R. Owen, J. Electrochem. Soc. **152**, A2376 (2005).

117. T. Nakamura, Y. Miwa, M. Tabuchi and Y. Yamada, J. Electrochem. Soc. **153**, A1108 (2006).

118. M. Gaberscek, R. Dominko, M. Bele, M. Remskar, D. Hanzel and J. Jamnik, Solid State Ionics **176**, 1801 (2005).

119. J. D. Wilcox, M. M. Doeff, M. Marcinek and R. Kostecki, J. Electrochem. Soc. **154**, A389 (2007).

120. R. Dominko, J. M. Goupil, M. Bele, M. Gaberscek, M. Remskar and D. Hanzel, J. Jamnik, J. Electrochem. Soc. **152**, A858 (2005).

121. F. Croce, A. D. Epifanio, J. Hassoun, A. Deptula, T. Olczac and B. Scrosati, Electrochem. Solid-State Lett. **5**, A47 (2002).

122. Y-H. Rho, L. F. Nazar, L. Perry and D. Ryan, J. Electrochem. Soc. **154**, A283 (2007).

123. H-M. Xie, R-S. Wang, J-R. Ying, L-Y. Zhang, A. F. Jalbout, H-Y. Yu, G-L. Yang, X-M. Pan and Z-M. Su, Adv. Mater. **18**, 2609 (2006).

124. Y-H. Huang, K-S. Park and J. B. Goodenough, J. Electrochem. Soc. **153**, A2282 (2006).

125. Y-S. Hu, Y-G. Guo, R. Dominko, M. Gaberscek and J. Jamnik, Adv. Mater. **19**, 1963 (2007).

126. H. Huang, S.-C. Yin and L. F. Nazar, Electrochem. Solid-State Lett. **4**, A170 (2001).

127. K-F. Hsu, S-Y. Tsay and B-J. Hwang, J. Mater. Chem. **14**, 2690 (2004).

128. K. Padhi, K. S. Nanjundaswamy and J. B. Goodenough, J. Electrochem. Soc. **144**, 1188 (1997).

129. J. M. Tarascon, C. Delacourt, A. S. Prakash, M. Morcrette, M. S. Hegde, C. Wurm and C. Masquelier, Dalton Trans. **19**, 2988 (2004).

130. M. Koltypin, V. Pol, A. Gedanken, D. Aurbach, J. Electrochem. Soc. **154**, A605 (2007).

131. J. S. Sakamoto, B. Dunn, J. Electrochem. Soc. **149**, A26 (2002).

132. D. Dubarry, J. Gaubicher, P. Moreau and D. Guyomard, J. Electrochem. Soc. **153**, A295 (2006).

133. X. Huang, H. Yue, A. Attia and Y. Yang, J. Electrochem. Soc. **154**, A26 (2007).

134. H. Kawaoka, M. Hibino, H. Zhou and I. Honma, J. Electrochem. Soc. **152**, A1217 (2005).

135. J. Yang and J. J. Xu, J. Electrochem. Soc. **153**, A716 (2006).

136. H. Huang, S-C. Yin, T. Kerr, N. Taylor and L. F. Nazar, Adv. Mater. **14**, 1525 (2002).

137. M. Ferreira, V. Zucolotto, F. Huguenin, R M. Torresi and O. N. Oliveira, Jr., J. Nanosci. Nanotechnol. **2**, 29 (2002).

138. F. Huguenin, D. S. dos Santos, Jr., A. Bassi, F. C. Nart and O. N. Oliveira, Jr., Adv. Funct. Mater. **14**, 985 (2004).

139. E. A. Ponzio, T. M. Benedetti and R. M. Torresi, Electrochim. Acta **52**, 4419 (2007).

140. C. V. S. Reddy, W. Jin, Q-Y. Zhu, W. Chen and R. R. Kalluru, Eur. Phys. J. Appl. Phys. **38**, 31 (2007).

141. E. A. Olivetti, J. H. Kim, D. R. Sadoway, A. Asatekin and A. M. Mayes, Chem. Mater. **18**, 2828 (2006).

142. D. Guy, B. Lestriez and D. Guyomard, Adv. Mater. **16**, 553 (2004).

143. N. Peter, M. Klaus, K. S. V. Santhannam and H. Otto, Chem. Rev. **97**, 207 (1997).

144. J. Wang, J. Yang, J. Xie and N. Xu, Adv. Mater. **14**, 963 (2002).

145. J. Wang, J. Yang, C. Wan, K. De, J. Xie and N. Xu, Adv. Funct. Mater. **13**, 487 (2003).

146. P. Novák, K. Müller, K. S. V. Santhanan and O. Haas, Chem. Rev. **97**, 207 (1999).

147. K-S. Park, S. B. Schougaard and J. B. Goodenough, Adv. Mater. **19**, 848 (2007).

148. Y. Wang, K. Lee, H. Shang, B. Wiley, Y. Xia and G. Cao, Phys. Stat. Sol. (a), **202**, R79 (2005).

149. J. S. Wook and S-M. Lee, J. Electrochem. Soc. **154**, A22 (2007).

150. Y. Li, T. Kunitake and Y. Aoki, Chem. Mater. **19**, 575 (2007).

151. N. Imachi, Y. Takano, H. Fujimoto, Y. Kida and S. Fujitani, J. Electrochem. Soc. **154**, A412 (2007).

152. S. R. Sivakkumar, D. R. MacFarlane, M. Forsyth and D-W. Kim, J. Electrochem. Soc. **154**, A834 (2007).

CHAPTER 13

NANOSTRUCTURED MATERIALS FOR SOLAR CELLS

Tingying Zeng,[1,*] Qifeng Zhang,[2] Jordan Norris,[1] Guozhong Cao[2]

[1]*Labratory of Nanostructures, Department of Chemistry, Western Kentucky University, 1906 College Heights Blvd #11079, Bowling Green, KY 42101-1079, U.S.A.,* [2]*Materials Science and Engineering, University of Washington, 302M Roberts Hall, Box 352120, Seattle, WA 98195-2120, U.S.A., *Corresponding author: tingying.zeng@wku.edu, Tel: 1-270-745-8980, fax: 1-270-745-5361*

Novel nanostructured materials synthesized in recent years are significantly attracting scientists and engineers for the development of new generation nanoscale solar cells. Major nanostructured materials include: semiconducting nanostructured porous materials, nanotubes, nanowires, different types of quantum dots, metal nanoparticles, carbon nanotubes and C_{60} families. All of these nanoscale materials have been found to have quantization size effects and unique optoelectrical properties, and therefore it is feasible to use them in photovoltaics. Extensive investigation of the possibility and feasibility of these nanostructured materials for high performance photovoltaic devices are concentrated in two areas: dye-sensitized solar cells or Grätzel solar cells and organic/inorganic nanocomposite photovoltaic devices. Why are nanoarchitectures so important to the two major nanoscale solar cells? This article highlights the most recent state-of-the art design and synthesis, as well as the characterization and applications of the novel structured materials for the two major types of solar cells. Identification of gaps in our current knowledge of these materials and discussion on the subject of achieving high overall power conversion light to electricity efficiency is included.

1. General Introduction

1.1. Photovoltaics and Conventional Inorganic Semiconductor Solar Cells

Solar cells or photovoltaics (PV) are devices that convert sunlight into electricity. Principally, electron and hole pairs are generated and separated once upon a light illuminates a semiconductor material, which has photon energy larger the semiconductor's bandgap [1, 2]. Conventional inorganic solid state p-n junction silicon solar cell (Figure 1) is a semiconductor diode, which is operated through a built-in-electrical field to drive the photon-induced electron and hole pairs separated. One layer is an "n-type" semiconductor with an abundance of electrons, which have a negative electrical charge, and the other layer is a "p-type" semiconductor with an abundance of "holes," which have a positive electrical charge. Sandwiching these two layers together creates a p/n junction at their interface, thereby creating an electric field. When n- and p-type silicon layers come into contact, excess electrons move from the n-type side to the p-type side. The result is a buildup of positive charge along the n-type side of the interface and a buildup of negative charge along the p-type side, forming the built-in-electrical-field to drive electrons drifting to one electrode and holes drifting to another electrode. Figure 2 gives the evaluation principles by measuring the photocurrent and photovoltage, called I~V curve under a light illumination. A Typical I~V characteristics of a solar cell presents three characteristic parameters: short-circuit current I_{sc}, open-circuit voltage V_{OC} and fill factor $ff = P_{max}/(V_{oc} \times I_{sc})$; P_{max} is the electrical power delivered by the cell at the maximum power point MPP [2]. Additionally, overall power conversion efficiency will be calculated based on the definition of

$$\eta_{global} = {i_{ph} V_{OC} (ff)} \Big/ {I_s}$$

where η_{global} is the overall efficiency of the photovoltaic cell, i_{ph} is the integral photocurrent density, V_{OC} is the open-circuit photovoltage, ff is the fill factor of the cell, and the I_s is the intensity of the incident solar radiation $I_s = 1000$ (W/m^2).

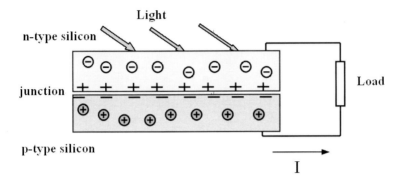

Figure 1. A single p-n junction in conventional silicon solar cell.

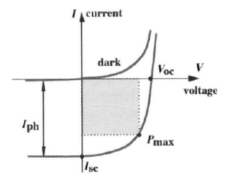

Figure 2. Typical I~V characteristics of a solar cell with three characteristic parameters: short-circuit current I_{sc}, open-circuit voltage Voc, and fill factor $FF = P_{max}/(V_{oc} \times I_{sc})$; P_{max} is the electrical power delivered by the cell at the maximum power point MPP. Reprinted with permission from Science, Ref. 2. Copyright 1999 Science Publishing Group.

Extensive interdisplinary research in material sciences, chemistry, physics, and engineering finally drove the solid-state inorganic p-n junction silicon semiconductor solar cell to the market in 1975 after it was invented in 1950s [1, 2]. At present, the best single-junction solar cells have efficiencies of 20-25%. Global energy risk imperatively requested a revolutionary progress in solar energy conversion efficiency since the late 1990s, and it became the research goals of solar energy utilization in this 21st century. Devices that operate about the existing performance limit of energy conversion efficiency of 32% calculated for single-junction cells will enable solar electricity from photovoltaics to be competitive with or cheaper than present fossil fuel electricity costs. Based on this promotion, multiple junction semiconductor thin film solar cells (tandems) with efficiency over 50%, optical frequency shifting (up/down conversion or thermophotonics) solar cells, multiple exciton generation from single photon, multiple energy level solar cells (such as intermediate band photovoltaics) and hot carrier solar cells have been emerged in recent years [3, 4].

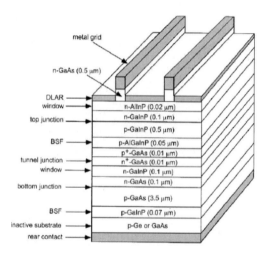

Figure 3. Depiction of an inorganic solar cell device with two-junctions, consisting of gallium-indium-phosphide (GaInP), gallium-arsenide (GaAs), indium-gallium-phosphide (InGaP), and aluminum-indium-phosphide (AlInP). Shown here is a monolithic tandem space cell using two stacked p–n junctions connected by a tunnelling junction. [Ref. 5]

Figure 3 depicts a solar cell with gallium-indium-phosphide (GaInP) in tandem with gallium-arsenide (GaAs) with the addition of an indium-gallium-phosphide (InGaP) tunnel junction layer and an aluminum-indium-phosphide (AlInP) barrier layer [5]. Although the two-junction solar cell device has obtained the high efficiency, the fabrication process is too complicated and the production cost is too high for commercial use. In addition, in order to widely utilize these solar cell devices, materials cost must be reduced.

1.2. Key Problems in Conventional Semiconductor Solar Cells

However, the high-temperature fabrication routes to single-crystal and polycrystalline silicon are very energy intensive and expensive [1]. Apart from the problems that they are heavy, fragile, and high cost, impurities usually cause significant efficiency decrease. The inorganic semiconductor thin film photovoltaics even though turns to use less expensive amorphous silicon and to develop compound semiconductor heterojunction cells (such as cadimium telluride and copper indium diselenide) [6], is still facing a grand challenge to develop high-efficiency, low-cost solar cells that can reach the ultimate thermodynamic efficiency limits. The key problem in optimizing the cost/efficiency ratio of such devices is that relative pure materials are needed to ensure that the photo-excited carriers are efficiently collected in conventional planar solar cell device designed and manufactured [7]. Since 2005, the overextended demand for raw silicon has been limited the conventional polysilicon solar photovoltaic market growth world-widely and it was estimated that it will keep more of the same in 2007 through 2008 [8]. Thus, search for alternative solar cells has become overwhelming research goals currently.

1.3. Nanostructured Solar Cells

In recent years, nanomaterial science and technology brought about fast growth of a new generation solar cells-nanosturctured solar cells, which consist of nanostructures using nanoscale materials and fabricated by nanotechnologies [9-13]. Nanosize materials have peculiar properties that are not expected in the bulk phase, and elucidation of there properties has already let to breakthroughs in various fields. The electrical and optical properties of nanoparticles are size and shape dependent. Hence, a proper organized nanostructure will bring about unique performances for optoelectrical devices. Therefore, the use of nanostructures offers an opportunity to circumvent the key limitation and therefore introduce a paradigm shift in the fabrication and design of solar cells to produce either electricity or non-carbon fuel [7]. At the current state-of-the-art, there are two major types of nanostructured solar cells: dye or quantum dot sensitized Grätzel solar cells [14-17] and organic-inorganic hybrid nanocomposite solar cells [18-21] or organic photovoltaics (OPVs). Inorganic semiconductor quantum dot multilayer solar cells may be ascribed to nanoscaled thin film multi-junction solar cells (Figure 4) [6]. It is not in our discussion coverage in this review since the solid-state inorganic thin film solar cells have different rationales from the nanostructured solar cells, and they are complicated in design and expensive in manufacture. The following sections may involve a little discussion in comparison to nanostructured photovoltaic devices, but we will not emphasize on them. More interested readings can be found from references [22-26].

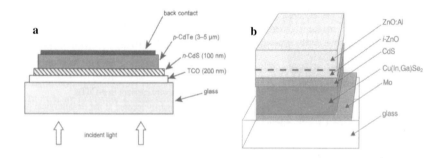

Figure 4. Schematic of the inorganic solar cell devices consisting of a) cadmium-telluride (CdTe), and b) copper-indium-gallium-selenide [Cu(In,Ga)Se$_2$] materials. [Ref. 6]

1.3.1 Grätzel Solar Cell and its Nanostructure

Based on photoelectrochemical cell operation principle, using dye or other proper sensitizers such as semiconductor quantum dots to sensitize a stable, non-toxic, wide bandgap oxide semiconductor film, such as TiO$_2$ or ZnO to convert light to electricity made the dye or quantum dot sensitized solar cells developed [27]. Dr. Grätzel and his colleagues have been pioneered in the dye-sensitized solar cells field for over ten years, this type of solar cells alternatively are called Grätzel solar cells.

Figure 5. Mesoporous Nanostructures of semiconductor nanocrystals used for Grätzel solar cell. (Ref. 16)

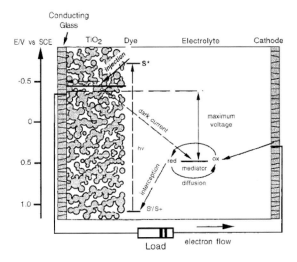

Figure 6. Operation principle of Grätzel solar cell: S is the photosensitizer. It may be a dye molecule, or a semiconductor quantum dot; the Redox mediate (electrolyte) may be a liquid redox couple such as iodine/iodide (I^-/I^{3-}), or a solid polymer hole transport material (HTM) which satisfy the energy level potentials. All the potentials here are referenced to the normal hydrogen electrode (NHE). The open-circuit voltage of the solar cell is dependent on the difference between the redox potential of the mediator and the Fermi level of the nanocrystalline film indicated with a dash line in this diagram. (Ref. 30)

Mesoporous nanostructures (Figure 5) of the wide bandgap semiconductor film that allows photosensitizer loading and absorbing light to generate electron and hole pairs are the core of this type of solar cells. Figure 6 schematically shows the operation principle which uses TiO_2 as the wide bandgap semiconductor functioning as electron collector (anode), and a redox mediate as the hole conducting material (HTM), regenerating the photosensitizer while itself is reproduced at the counter electrode by electrons passed through the external circuit load (cathode).

The sensitizer-coated mesoporous TiO_2 layer and the HTM penetrated in the porous photoactive film formed the heterojunction nanostructures. O'Regan and Grätzel reported their breakthrough work in 1991 using ruthenium complex dye to sensitized the mesoporous TiO_2 film and liquid electrolyte redox couple I^-/I^{3-} to regenerate a photosensitizer in the heterojunction nanostructures, which has been achieved overall PCE of 7.1 to 7.9%, and it has been well-known as Grätzel liquid solar cells (GLSCs) (Figure 7) [28, 29, 30]. Another significant contribution by Dr. Grätzel and his colleagues was the development of Grätzel solid solar cells (GSSCs) in 1998, in which they replaced the liquid electrolyte I^-/I^{3-} with a p-type semiconducting organic spiro-compound (Figure 8), which achieved an overall PCE of 0.74% [31].

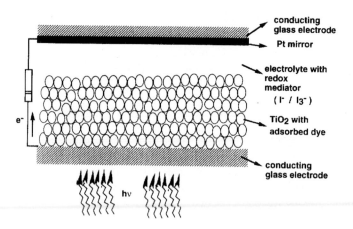

Figure 7. A Catoon of Grätzel liquid solar cells. (Ref. 29)

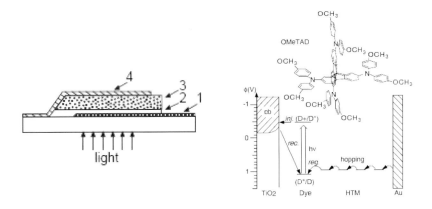

Figure 8. Solid Grätzel solar cell and the Spirobiflurene compound as the HTM (Ref. 31) Shown here is the spiro-MeOTAD [2, 2', 7, 7'-tetrakis (N, N-di-p-methoxyphenyl-amine)-9, 9'-spirobifluorene]. 1, conducting F-doped SnO2 coated glass; 2, compact TiO$_2$ layer; 3, dye-sensitized mesoporous TiO$_2$ layer and the HTM formed the heterojunction; 4, counter electrode Au.

The nanostructure of the wide bandgap semiconductor oxide material in Grätzal solar cells is the heterojunction heart. As a result, the development of mesoscopic semiconductor material that obtains high internal surface area and control of the thickness are technically critical. Up to today's arts, the GLSCs have been achieved overall PCE of 11% using ruthenium dyes [30], while GSSCs have been only reported to achieve about 4% overall PCE [32]. Overcoming interfacial charge recombination is a technical challenge for GSSC. It is an engineering challenge for GLSCs to overcome liquid electrolyte packing problem. Thus, development of GSSCs is preferable in a long term view. Our review will mostly focus on the state-of-the-arts in construction of this nanostructured heterojunction for high performance GSSCs.

1.3.2 Organic Polymer Solar Cell and its Nanostructure

Polymer-based organic/inorganic nanocomposite solar cells are another type of nanostructured photovoltaics [7]. The original organic solar cells are designed based on the photogeneration of excitons (bonded electron-hole pairs) and their effective separation, rather than the direct formation of charge carriers. Polymeric materials are versatile for this type of solar cell devices in that they can be synthesized and tailored to function at a specific region of the solar spectrum [33]. Many of the materials used in organic solar cells exhibit high coloration to absorb strongly in the visible region of the spectrum and good stability under illumination in air and moisture. In addition, these materials can be prepared in nanometer scale thin film to reduce the bulk resistance, can be chemically "doped" to enhance the conductivity, and can exhibit photoconducting behavior. Flat-junction organic solar cells, consisting of interpenetrating polymer nanostructured networks [34, 35], polymer/fullerene blends nanocomposite [36], and halogen-doped organic nanocrystals [37] have been studied. Newer devices with bulk donor-acceptor nanostructure heterojunctions formed by blending two organic materials [38], where one material serves as the electron donor (p-type conductor) and one material serves as the electron acceptor (n-type conductor) emerged to reduce the probability of surface charge recombination at the interface of the two materials. Figure 9 depicts a p-n organic solar cell showing the movement of electrons and holes after exciton separation.

In this type of nanocomposite structure, the electron-hole pair produced by the absorption of sunlight can reach the junction and dissociate into two free charge carriers. Typically, electron-hole pairs diffuse only a few nanometers before recombining. While the distance the electron-hole pair has to travel at most a few nanometers before reaching the interface in the nanostructure. Studied demonstrated that composite nanostructures have efficient photo-induced charge separation [39]. However, solar cells formed by only organic materials has very low electron mobility associated with conjugated polymers due to the presence of electron trapping species, specifically oxygen 40]. The presence of two different organic materials provides an interface for charge transfer via percolation pathways, but the efficiency is limited due to inefficient transport by charge hopping and the presence of structural traps associated with an incomplete network of percolation pathways [41].

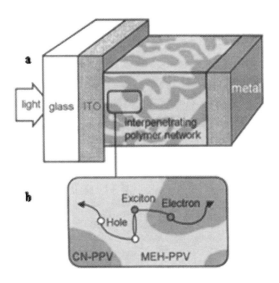

Figure 9. Schematic of a) an organic solar cell consisting of an n-type polymer and a p-type polymer interconnected to form a heterojunction, and b) the electron and hole migration through the polymer network after exciton generation. [Ref. 6]

To overcome low electron mobility problem, polymer-based inorganic nanoparticles or nanocrystals-doped OPVs have been explored in recent years [42, 43]. Thus, incorporating solid-state inorganic electron conducting nanocrystals into hole conducting organic materials to form hybrid nanostructures [44] generating a new generation OPVs. Dr. Alivisatos and his colleagues have been pioneered in this field, in which they reported to use semiconducting nanorods of cadmium selenide (CdSe) into a hole-conducting conjugated polymer, poly-3(hexylthiophene) or P3HT, to fabricate an OPV and achieved overall PCE of 1.7% [41, 45]. Using a sandwiched active layer of nanocomposite P3HT and [6,]-phenyl-C61-butyric acid methyl ester (PCBM), Li et al. achieved overall PCE of 4.4% [43]. The combination of inorganic and organic materials to fabricate hybrid solar cell devices is to correlate the advantages of both types of materials. The presence of inorganic semiconductor materials utilizes the high intrinsic carrier mobility to reduce current loss from recombination by quicker charge transport, and the advantage of using polymeric materials is the ease in incorporating organics by solution methods to provide an interface for charge transfer, which is typically favored between high electron affinity inorganics and relatively low ionization potential organics [41, 46]. It is found that charge transfer rates in organics that are chemically bound to nanocrystalline and bulk inorganic semiconductors with a high density of electron states can be very fast [45, 47].

1.3.3 Typical Characteristics of Nanostructured Solar Cells

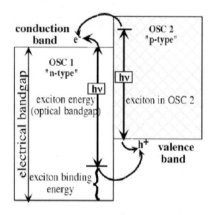

Figure 10. Energy level diagram for an excitonic heterojunction solar cell. Excitons created by light absorption in organic semiconductors 1 (OSC1) and 2 (OSC2) do not possess enough energy to dissociate in the bulk (except at trap sites), but the band offset at the interface between OSC1 and OSC2 provides an exothermic pathway for dissociation of excitons in both phases, producing electrons in OSC1 and holes in OSC2. The band offset must be greater than the exciton binding energy for dissociation to occur. [Ref. 40]

According to B. Gregg, the existing types of nanostructured solar cells, both dye-sensitized Grätzel solar cells and OPVs, can be categorized by their photoconversion mechanism as excitonic solar cells (XSCs) (Figure 10) [40]. The distinguishing characteristic is that charge generation and separation are simultaneous and this occurs via exciton dissociation at a heterointerface. Electrons are photogenerated on one side of the interface and holes on the other. This results in fundamental differences between XSCs and conventional PV cells. The open circuit photovoltage V_{oc} in conventional PVs is limited to less than the magnitude of the band bending (ϕ_{bi}); while the V_{oc} in XSCs is commonly greater than ϕ_{bi}, and it does not dependent on ϕ_{bi}, but is a function of both the built-in-electrical and the photo-generated chemical potential energy differences across the cells, since almost all of the carries are photogenerated in a narrow region near the interface, leading to a high photoinduced carrier concentration gradient [40]. Therefore, it is important for the development of nanostructured solar cells to construct and to optimize nanoarchitectures of heterointerfaces that benefit exciton or multiexcition created and separated, which is a technical challenge for both Gräztel solar cells and for the hybrid inorganic/organic polymer PVs.

Compared to conventional inorganic semiconductor thin film solar ells, nanostructured solar cells are very cheap in materials costs and easy to be fabricated. This review will only focus on the nanostructured materials for excitonic nanoscale solar cells. As lacking enough silicon materials for conventional solar cell industry in recent years, and as the crisis of the energy requests, we may expect that the extensive research and development for those new generation nanoscale solar cells will keep explosively growing in next a few years. Thus, we take this opportunity to highlight the most recent progresses in nanostructured material syntheses and characterizations for the use of nanoscale solar cells, and hope this will help the happen of high performance, low cost, nanoscale solar cells developed to reach a much higher competent level with conventional solar cells near future.

2. Nanostructured Materials for Grätzel Solar Cells

2.1. Materials Choice

Since Grätzel solar cell operates by sensitizing dyes or quantum dots bound to semiconductor nanocrystals participating in interfacial charge transfer under photoexcitation (Figure 11) [48], high internal surface area of the semiconductor mesoporous film and the high quality connection of

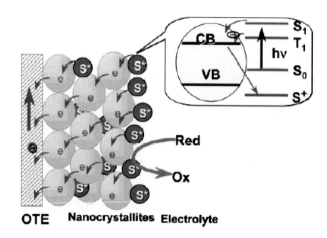

Figure 11. Dye or quantum dot sensitization of semiconductor nanostructures is the primary photochemical event in a Grätzel solar cell. (Ref. 48)

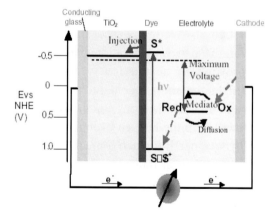

Figure 12. A schematic diagram of components used to build up a nanostructured Grätzel solar cells. S is the photosensitizer. It may be a dye molecule, or a semiconductor quantum dot; the Redox mediate (electrolyte) may be a liquid redox couple such as iodine/iodide (I^-/I^{3-}), or a solid polymer hole transport material (HTM) which satisfy the energy level potentials. All the potentials here are referenced to the normal hydrogen electrode (NHE). The open-circuit voltage of the solar cell is dependent on the difference between the redox potential of the mediator and the Fermi level of the nanocrystalline film indicated with a dash line in this diagram. (Ref. 16)

the semiconductor nanocrystals are basically requested to realize maximum photosensitizer mass loading and to establish electronic conduction, so as to effectively collect electrons injected by excited photosensitizer. A high efficient heterojunction in this nanostructure also allows hole-transport material, either liquid electrolyte or organic semiconduting materials effectively penetrating into the mesoporous film to contact photosensitizer, so as to efficiently transport holes to the counter electrode after the excitons separated.

Materials choice to build up this nanostructure needs to satisfy the energy levels for each component. Figure 12 and Figures 8, and 6 schematically present the basic relative requirements of the potentials for each component forming the Grätzel solar cells [16].

2.1.1 Wide Bandgap Semiconductor Materials

Wide bandgap semiconductor TiO_2 anatase has been widely used.

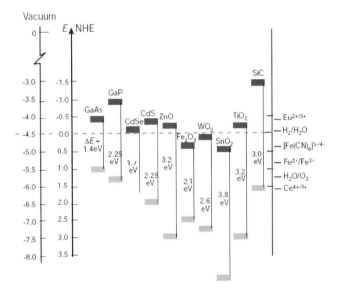

Figure 12. The band gap of various semiconducting materials used in various solid-state and dye-sensitized solar cell devices. It can be seen that ZnO and TiO_2 have the same band gap. [Ref. 13]

Figure 12 gives the referred bandgap energies of common semiconductor materials, indicating that there are alternative wide bandgap oxides such as ZnO, SnO$_2$ to be investigated. Nb$_2$O$_5$ has also been considered as one candidate [13].

2.1.2 Photosensitizers

The photosensitizer dye or semiconductor quantum dots as well as the hole transport material (HTM) that should be a p-type material, are required to have special properties: 1) the p-type material HTM must be transparent to the spectrum where the photosensitizer absorbs light; 2) the photosensitizer such as dye or quantum dot must be such that its LUMO level (or the QD's conduction band) is located above the bottom of the conduction band of TiO$_2$ and its HOMO level (QD's valence band) is located below the upper edge of the valence band of the p-type HTM. 3) photosensitizer dye or QDs usually are assembled as monolayer on the surface of the TiO$_2$ nanoparticles which are sintered to link together forming a high quality networks. The deposition of HTM layer should not damage the networks. A method to deposit the HTM without dissolving the photosensitizer monolayer is preferable [49, 50].

Based on the energy level requirement, ruthenium bipyridyle complex dyes are extensively used as the photosensitizers. Up to current state of the art, there are many kinds of dyes that have been synthesized and engineered to tail functional groups such as carboxylic group to favorite the electronic coupling with TiO$_2$ nanoparticle surface and to benefit the electron charge injection. Generally, the optical transition of Ru complex dyes has metal-to-ligand-charge-transfer (MLCT) character. Excitation of the dye involves transfer of an electron from metal to the π^* orbital of the ligand. N$_3$ has been recognized as an excellent dye for Gräztel solar cells [30]. It has two such MLCT transitions in the visible region. The absorption maxima in ethanolic solution are located at 518nm and 380nm, the extinction coefficients being 1.33×10^4 and 1.3×10^4 M^{-1}cm^{-1}, respectively [30]. Sensitization of the mesoporous TiO$_2$ nanostructure is realized by the molecular level thick monolayer dye molecules (usually only a few nanometers thick) through its low MLCT absorption and the interfacial redox reaction for the ultrafast electron injection is in the femtosecond time regime. The level of the vibronic state produced by 530-nm light excitation of N3 is about 0.25eV above the conduction band of TiO_2, and the 0-0 transition of N3 has a gap 1.65eV, indicating the excited-state level matches the lower edge of the TiO_2 conduction band [30]. Further research showed that the N3 emits at 750nm, and the excited-state lifetime is 60ns [30].

To compare the efficiency of different dyes in the full-sun spectrum range, an external quantum efficiency (EQE) is usually used to characterize the corresponding solar cell performance. Indeed, the incident photon to current conversion efficiency (IPCE), referred it as "external quantum efficiency" (EQE), corresponds to the number of electrons measured as photocurrent in the external circuit divided by the monochromic photon flux that illuminate the cells. It is defined as IPCE $(\lambda)=LHE(\lambda)\phi_{inj}\,\eta_{coll,}$ where $LHE(\lambda)$ is the light-harvesting efficiency for photons of wavelength λ, ϕ_{inj} is the quantum yield for electron injection from excited sensitizer in the conduction band of the semiconductor oxide, and $\eta_{coll,}$ is the electron collection efficiency [30]. N3 showed an IPCE of about 68% at 700nm, while black-dye, developed to shift the optical absorption to near IR range, presented an IPCE of about 77%, and its PV response extended to over 800nm and demonstrated an IPCE about 60% at this wavelength (Figure 13) [30].

Figure 13 summarizes a few popular dyes reported so far used as photosensitizers for Grätzel solar cells [51-54]. Those dyes were designed and molecularly engineered to tune either the redox potential [51], or the molar optical absorption extinction coefficient [52], or the optical absorption shifting from visible to near IR range spectrum [53], or to improve the thermal stability under a higher temperature operation environment [54]. Both ruthenium complex dyes and metal-free indoline dyes have been led to the GLSCs achieving overall PCE above 10%. More information about molecular engineering to design and synthesize dyes can be found from references [55-58].

Apart from dyes discussed above, quantum dots (QDs) are new generation and excellent photosensitizer for Grätzel solar cells. Based on the energy level of construction of Grätzel solar cells, semiconductor compound QDs of III-V and II-VI, mostly having the bandgaps in a

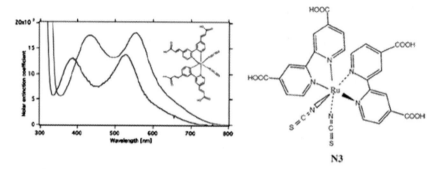

Ligand L = 4,4'-bis(carboxyvinyl)-2,2'-bipyridine (L) and its [Ru(II)L₂(NCS)₂] (K8) [51]

Cis-dithiocyanatobis(4,4'-dicarboxylic ruthenium(II) complex acid-2,2'-bipyridine)ruthenium(II) (N3) [30]

The red cure is the optical absorption of K8 dye, and the blue one is the N3 optical absorption [51].

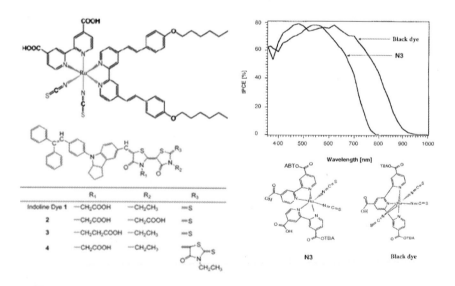

Metal-free Indoline, efficiency competing with N3, and thermally stable at 300°C [54].

Black-dye shifted the optical absorption to above 800nm [30].

Figure 13. A few popular dyes developed for Grätzel solar cells.

range of 1.0eV to 2.7eV based on their particle sizes and components, are good candidates as photosensitizers to sensitize wide bandgap oxide nanocrystals [59-62]. QDs have quantum confinement effects [63, 64]. Its optical absorption dependence to its particle size provides us the feasibility to flexibly control and tune the light harvesting spectrum region. Thus, near IR and IR range photosensitization to the wide bandgap oxide nanocrystals can be realized. Therefore, the QD-sensitized Grätzel solar cells can efficiently use sunlight, especially in the IR range [59]. In addition, an exciting discovery is that multiple excitions can be generated from the absorption of a single photon by a QD via impact ionization if the photon energy is three times higher than its band gap, as shown in Figure 14 [65].

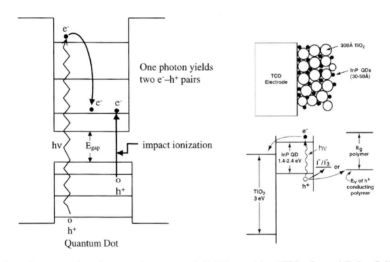

One photon produced two excitons   InP QD-sensitized TiO₂ Gratzel Solar Cell

Figure 14. Enhanced photovoltaic efficiency in QD solar cells by impact ionization (inverse Auger effect). (Ref. 65)

Efficiently rapid injection has been tested in many cases using InP QDs [59], CdSe QDs [60], PbSe QDs [66], which was in the time regime of femtosecond. Once upon light absorption, excitons are produced in the QD. The photon-induced electrons are subsequently injected into the semiconductor oxide conduction band, and the holes are transported to the hole conductor or an redox electrolyte within the mesoporous film (Figure 14) [65]. Research has demonstrated that electron transfer from thermally relaxed state occurs over a wide range of rate constant values. The injected charge carriers in a CdSe-modified TiO_2 film can be collected at a conducting electrode [60]. But at current state of the art, the reported QD-Grätzel solar cells have only achieved an IPCE of 45% and overall PCE 0.49% using PbS QDs [66], and a 12% IPCE at light response less than 380nm using CdSe to sensitize TiO_2 film [60]. Significant loss of electrons occurs due to scattering as well as charge recombination at the QDs/TiO_2 interfaces and internal TiO_2 grain boundaries has been observed [60]. QDs have advantages of high thermal stability, and much higher optical cross sections compared with dyes. Also, one high energy photon may generate multiple excitons. All those indicates that they are very promising photosensitizers for Grätzel solar cells. It may lead to an IPCE 100% achieved. Therefore, much effort need to be done to study the fundamental loss mechanisms within the QD-sensitized semiconductor mesoporous nanoarchitectures, and to synthesize high qualified QDs.

2.1.3 Hole-Transport Material (HTM)

Choosing HTM needs to concern the photosensitizer's HOMO or valance band level of the QD in the diagram of Figure12. Thus, dye's LUMO level is located above the bottom of the conduction band of TiO_2, and its HOMO level is located below the upper edge of the valence band of p-type HTM. Usually, an effective HTM is a p-type wide bandgap semiconductor materials or redox couple. Typical examples are the liquid electrolyte I^-/I^{3-} couple for GLSCs [28] and a p-type semiconducting organic spiro-MeOTAD [2, 2', 7, 7'-tetrakis (N, N-di-p-methoxyphenyl-amine)-9, 9'-spirobifluorene] for GSSCs [31]. CuI is another good candidate for GSSCs, which has a wide bandgap (3.1eV). The valence band edge of CuI is -5.3eV vs. the vacuum level that matches the HOMO level of the ruthenium bipyridyle dye used in GSSCs. CuSCN, another stable Cu(I) p-type semiconductor, having a bandgap of 3.6eV and a valence band edge of -5.1eV with respect to the vacuum scale, also satisfies the requirements of GSSCs [50].

P-type polythiophenes, exhibiting bandgaps in the range 1.9 ~ 2.0eV, and HOMO energy level of approximately 5 eV (vs. vacuum) have been tried as the HTM to target plastic solar cells and to be a cheaper alternative for the replacement of spiro-MeOTAD [67]. Recent study has been shown they are promising new HTMs for Grätzel solar cells [68]. Further research needs to focus on the mechanism of HTM in such type of heterojunction nanostructures for high performance of hole-transportation.

Up to today's art, using CuI and spiro-MeoTAD, for GSSCs, an overall PCE of 3% (less than 4%) in the full-sun spectrum were reported [50, 69]; while using liquid electrolyte I^-/I^{3-} couple for GLSCs, overall PCE above about 10~11% has been demonstrated for many cases of GLSCs [50, 51, 52]. Since the GLSCs have problems such as leakage, packaging and carrion due to the use of liquid electrolyte I^-/I^{3-}, GSSCs should be extensively explored and developed in research direction. There are two main obstacles for GSSCs to achieve higher overall PCE: insufficient light absorption and large internal interfacial recombination loss. Fundamental research needs to focus on those problems and explore new approach to overcome those obstacles.

2.2. Porous TiO₂ Nanostructures for Grätzel Solar Cells

For Grätzel solar cells, building-up a high quality mesoporous wide bandgap semiconductor film is first important step to construct the solar cell, since the heterojunction heart comes from the sensitization of dye or QD to mesoporous wide bandgap semiconductor nanocrystal film, which incorporates the HTM within the nanostructures. TiO₂ has been widely used so far because it has a bandgap of 3.2eV and nontoxic and chemically stable properties, and it is very cheap and easy to obtain.

2.2.1 Formation of the TiO₂ Mesoporus Films

Much of the research in Grätzel solar cells has surrounded the use of porous nanocrystalline titania (TiO₂) film in conjunction with an efficient light-absorbing dye, and have shown an impressive energy conversion efficiency of > 10% at lower production costs [50]. The TiO₂ semiconducting oxide functions as a suitable electron-capturing and electron-transporting material with a conduction band at 4.2eV and an energy bandgap of 3.2eV, corresponding to an absorption wavelength of ~ 387nm. Typical TiO$_2$ film thickness for solar cells with the highest light conversion efficiency ranges from 8µm to 12µm with a porosity of about 50%. This mesoscopic TiO$_2$ nanostructure provides enough surface area for monolayer dye chemisorption, allowing for enough dye adsorption on the surface at a given area so as to absorb almost of the incident light through scattering. Figure 5 is an SEM image of the typical TiO$_2$ mesoporous structure associated with 10% efficiency [16]. It has been shown that the grain size of the TiO$_2$ film can range from about 10nm up to 80nm depending on the processing technique. It has been shown that the structures of the anatase TiO$_2$ nanoparticles are square-bipyramidal, pseudocubic, and stab-like. The TiO$_2$ crystals are faceted with the (101) face mostly exposed, followed by the (100) face and the (001) face [30].

There are two general synthesis techniques to form nanocrystalline porous nanostructure film that provides a pathway for electrical conduction between particles. One approach applies a suspension of particles to a conducting substrate and then requires sintering at above 350°C to form sufficient contact between particles for charge transport to the underlying substrate [30, 70]. Another approach utilizes direct film formation onto a substrate by way of electrochemical or chemical deposition of nanocrystalline particles [71, 72, 73]. The first approach is the most common synthesis process to obtain TiO$_2$ film with high porosity and high surface-to-volume ratio. The preparation of crack-free mesoporous TiO$_2$ thick film for use as suitable electron-transporting electrodes involves the preparation of TiO$_2$ paste by way of sol-gel processing of commercially-available TiO$_2$ colloidal precursors containing an amount of organic additives. This conventional method requires the deposition of the prepared paste by either doctor-blading, spin-coating, or screen-printing on a transparent conducting substrate. High temperature sintering is utilized to remove the organic species and to connect the colloidal particles for electrical contact between particles. The pores between colloidal particles are also interconnected and can be filled with electrolyte. Typical thickness of mesoporous TiO$_2$ film using this method ranges from 2μm to 20μm, depending on the colloidal particle size and the processing conditions, and the maximum porosity obtained by this technique has reported to be around 50% with an average particle size approximately 20nm [27].

It has been shown that the film thickness is an important factor in the synthesis of this nanostructure film that is highly efficient for photosensitizer mass-loading and for electrons collection and transportation. Studies have shown that an increased probability of charge recombination with increasing film thickness occurs since electrons have to be transported across an increasing number of particles and grain boundaries [74]. In addition, a thicker film results in a resistance loss that can lead to a decrease in photovoltage and fill factor. Therefore, an optimal film thickness is necessary to obtain a maximum photocurrent. As a result, many other techniques have recently been investigated to synthesize TiO_2 electrodes with improved structure and film thickness for more efficient electron transport and good stability. Chemical vapor deposition (CVD) of Ti_3O_5 has been utilized to deposit layered crystalline anatase TiO_2 thin films that are optically responsive and stable [75]. Gas-phase hydrothermal crystallization of $TiCl_4$ in aqueous mixed paste has been done to obtain crack-free porous nanocrystalline TiO_2 thick film through low-temperature processing [76, 77, 78]. Compression techniques of TiO_2 powder have also been used to form porous and stable films. Recently, electrospinning [79] and electrodeposition [80] techniques have been used to deposit TiO_2 film on flexible substrates.

More recently, electrostatic layer-by-layer (ELBL) self-assembly has been used to build the TiO_2 nanostructure films for Grätzel solar cells [81]. Thus, negatively charge TiO_2 colloidal nanoparticles were self-assembled using a cation polymer moiety (polydiallyldimethyl ammonium chloride (PDAC) at a favorite pH larger than 7 through the ELBL processing with a deposition cycles over 50 bilayers. To create scattering nanostructures to efficiently incident light for the desired Grätzel solar cell, large particle size-TiO_2 particles with its particle size range in 250 nm to 400 nm were used to build the top film on small particle sized film on a conductive F-doped glass slide. The film thickness was precisely controlled through the deposited bilayer numbers.

Figure 15 shows the high resolution SEM images of the ELBL porous TiO$_2$ films obtained after sintered using furnace at 450°C. A high porosity of above 60% was obtained after removed the polymer molecules. A dye named N719 having a maximum absorption at 540nm was used to sensitize this active nanostructured film and has achieved IPCE at the day maximum absorption wavelength of 64%, and presented an overall PCE of 5% using a I⁻/I₃⁻ electrolyte to regenerate the dye N719 [81].

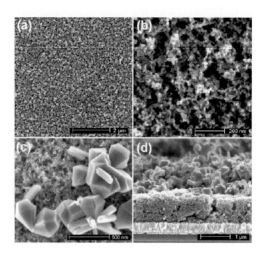

Figure 15. TiO$_2$ Nanostructures formed by ELBL processing (Ref. 81). HR-SEM images of (a, b) top view of TiO2 nanoparticulate films and (c) top view of scattering particles over nanoparticulate film, (d) cross-section of a nanoparticulate film topped with a scattering layer.

2.2.2 Photosensitization within the TiO$_2$ Nanostructures in Grätzel Cells

Studies on the charge transport of photoinjected electrons showed that electrons migrating through all the particles and grain boundaries in nanocrystalline TiO$_2$ films can be efficient enough for generating photocurrent [74, 82]. The common operation of nanocrystalline TiO$_2$ electrodes consist of the filling of trap states and the separation of charges controlled by kinetics [83-85] at the semiconductor-electrolyte interface [86-88]. The energy levels of a sensitized nanocrystalline TiO$_2$

film in contact with an electrolyte, and the process of light conversion is shown in Figure 6, 8, and 12.

The light conversion process of a dye-sensitized solar cell consists of TiO_2 as the semiconducting oxide material and an iodine-based redox system such as the liquid electrolyte. In the heterojunciton nanostructures, the dye adsorbed to TiO_2 is exposed to a light source, absorbs photons upon exposure, and injects electrons into the conduction band of the TiO_2 electrode. Regeneration of the dye is initiated by subsequent hole-transfer to the electrolyte and electron capture after the completion of the I^-/I_3^- redox couple at the solid electrode-liquid electrolyte interface. The photovoltage is also shown, which is the difference between the Fermi level of TiO_2 under illumination and the redox potential of the redox liquid electrolyte [74].

A typical ruthenium (Ru) complex dye sensitizer molecule is adsorbed to the surface of TiO_2, where the carboxylate groups serve to attach the Ru complex to the surface of TiO_2 and establish good electronic coupling. Figure 16 shows the desired pathway for a photoexcited electron, showing MLCT [74]. At the point of light absorption of the dye sensitizer, charge transfer occurs from the metal to the ligand. The excitation energy is channeled into the ligand where electron injection into the conduction band takes place.

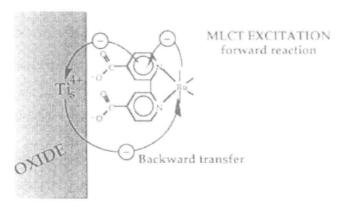

Figure 16. Schematic of the metal-to-ligand charge transfer in a ruthenium-based dye sensitizer anchored to the TiO_2 surface. [Ref. 74]

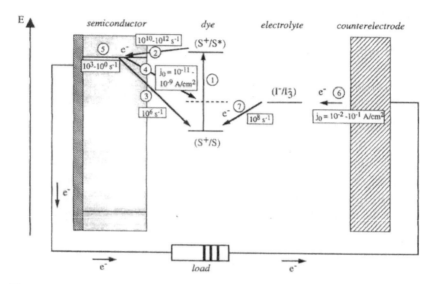

Figure 17. Schematic of the rate constants associated with each step of the light conversion process, showing the rate of (1) electron excitation, (2) electron injection, (3) electron and hole recombination, (4) reduction of the electrolyte by conduction band electrons, (5) electron migration, (6) reduction of the electrolyte at the counter electrode, and (7) reduction of the oxidized dye. [Ref. 74]

The TiO_2 material not only functions as the sensitizer support but also functions as the electron acceptor and electronic conductor. The electrons injected into the conduction band of TiO_2 migrate through the nanocrystalline film to the underlying conducting substrate, which acts as the current collector. The circuit is complete with the regeneration of the dye by electron transfer from the redox species in solution, which is then reduced at the counter electrode. During this process, it is possible that electron-hole recombination may occur at the interface where injected electrons can recombine with oxidized dye molecules or with oxidized species in the electrolyte. However, the chance of recombination is negligible if the rate of electron injection at the sensitizer-semiconductor interface is much higher than the rate of recombination at the semiconductor-electrolyte interface [74]. Figure 17 depicts the rate constants of the various steps associated with the light conversion process. Electron injection from the point of light absorption by the dye sensitizer into the conduction band has been measured to be in the picosecond range, whereas, the back reaction or recombination of electrons and holes has been measured in the microsecond range [74].

In nanocrystalline film, it has been shown that conduction band electrons preferentially become trapped at grain boundaries, and that charge carriers can be trapped in localized energy levels in the band gap region, which can be the limiting factors in obtaining higher efficiencies [89, 90]. Electron trapping in the bulk of the TiO_2 particles leads to a slow time response of the photocurrent but not to recombination losses, which can reduce the photovoltage [91, 92]. Electrons trapped at the surface of TiO_2 may lead to a recombination pathway, which can reduce the photocurrent and some decrease in photovoltage. Traps are likely to include Ti^{3+} states that result from nonstoichiometry, oxygen deficiency and ion intercalation, surface adsorbed species, or other surface or interface states [93]. It has been shown that the filling of trap sites increases the ratio of mobile electrons to trapped electrons which can lead to increased sensitivity of current to voltage [94]. The photocurrent generation process can be shown in Figure 18 [74].

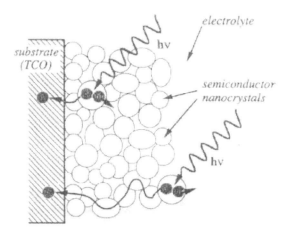

Figure 18. Schematic of the interconnected nanoparticle network showing photocurrent generation and electron migration through electron percolation pathways. [Ref. 74]

It can be seen that the particle network allows for the electrolyte to penetrate the entire colloidal film all the way through to the surface of the conducting substrate, forming a semiconductor-electrolyte junction at each nanoparticle. Each nanoparticle with a sensitizer layer will then generate an electron-hole pair after light absorption. It is assumed that the charge transfer of holes to the electrolyte is much faster than the recombination process, resulting in the electrons creating a gradient in the electrochemical potential between the particles and the conducting substrate. This gradient as described by excitonic solar cell driving force, thus allows for the transport of electrons through the interconnected network of particles to the conducting substrate, which produces current.

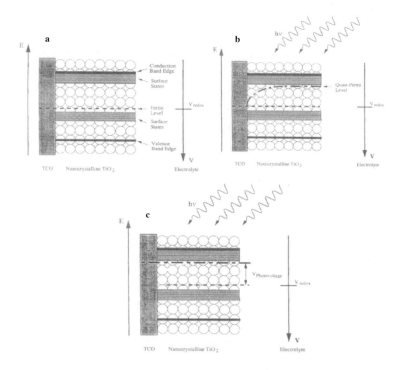

Figure 19. Schematic of the band edge positions associated with the energy levels a) under equilibrium in the dark, b) under illumination at short-circuit, and c) under illumination at open-circuit. [Ref. 74]

On the basis of a few studies [95, 96, 97], a schematic of the energy level diagrams of a system consisting of a mesoporous TiO_2 film on a conducting substrate (TCO) in an electrolyte can be assumed. Figure 19 shows the energy level diagram of the TCO-TiO_2-electrolyte system in various conditions.

In the first case, the system is under equilibrium in the dark. The penetration of the electrolyte through the entire film to the underlying conducting substrate results in a fixed band edge position through the entire TiO_2 film. In the second case, the system is under illumination with short-circuit conditions. A bent quasi-Fermi level of electrons is shown, where a gradient in the electron concentration from the outer layer to the conducting substrate is present and a current is drawn through the system. In the third case, the open-circuit condition results in a buildup of photovoltage. However, more exploration in the energetics [88, 98] and chemical nature of the energy levels in the band gap region and trapping states are required. The exact mechanism of charge transport through the colloidal particle network is not real clear. Some studies suggested a hopping mechanism of transport [90, 99] and other studies have suggested a tunneling mechanism through a potential barrier between particles [85, 87]. More work needs to be done focusing on sensitized single nanoparticle photovoltaic principles, and nanoscale charge migration in the network, for example, using ultrafast spectroscopy to track the excitons generated by the dye or QD, and probe the charge separation and injection, as well as their transportation in the nanoarchitectured networks. Those studies will help to understand the real charge loss reasons, so as to allow a better design of the fabrication process for the high efficiently photoactive TiO_2 porous nanostructures fro Grätzel solar cells.

2.2.3 Heterojunction of Nanostructured TiO₂ Film in Grätzel Solid Solar Cells

Attention should be paid to solid Grätzel solar cells [16, 17]. In GSSCs, liquid electrolyte needs to be replaced by solid HTM moiety in the TiO_2 porous networks to regenerate photosensitizer dye or QDs. Counterelectrode of Pt coated ITO glass as reflection coating or Au should be changed to a direct top electrode Pt or Au film, such as evaporated Au or Pt film. Usually, a 50 ~ 100nm thick compact TiO_2 thin film layer is necessary for GSSCs to avoid the direct contact between HTM and the conducting substrate. Then, a mesoporous TiO_2 film was deposited to enhance the dye mass loading and the light absorption of the photosensitizer monolayer. The GSSC is driven by majority carriers, and electrons flow in n-type TiO2 while holes flow in p-type materials. Apart from the optimization of HTM that needs to direct, research found that large internal charge recombination loss and high resistance [17, 59], low mass photosensitizer loading are the key problems for GSSCs [17]. Compared to GLSCs, it is difficulty to deposit a HTM to achieve intimate contact with the dye monolayer covering the mesoporous TiO_2 film. Such contact is so important for the efficient regeneration of dye molecules or QDs and for effective charge separation. Meanwhile, the heterojunction has an extremely large interface and very weak interfacial field. After the initial interfacial charge generation, the interfacial recombination between the electrons in the TiO_2 phase and the holes in the hole-conductor layer is unavoidable. It was found that the recombination in the GSSC using CuSCN as the HTM was 10 times faster than in the GLSC at the open circuit potential V_{oc} ($t_{1/2}$ ~ 150μs) and 100 times at short circuit ($t_{1/2}$ ~ 450μs), although both kinds of cells exhibited a similar charge transport rate ($t_{1/2}$ ~ 200μs) [69].

To suppress the interfacial charge recombination and avoid the contact between the HTM layer and the TiO_2 porous layer, an interfacial blocking layer was introduced into the CuI-based GSSC [17]. Thus, an Al_2O_3 insulating layer was inserted at theTiO$_2$/CuI interface to function as a physical blocking layer to avoid the direct contact between TiO_2 and CuI. The Al_2O_3 insulating thin layer thickness was precisely controlled to less than 1nm to keep tunneling efficiency of electrons from dye molecules to TiO_2, allowing the electrons being collected by the TiO_2 networks [100]. Figure 20 gives the illustration of the interfacial charge-transfer processes occurring in the TiO_2/dye/CuI GSSC, and Figure 21 presents two examples for the nanoarchitecture constructions of the ultrathin insulating layer which modified TiO_2 nanostructures for such GSSCs [17]. Atomic layer deposition (ALD) introducing surface reaction to create the ultrathin insulator layer is very effective [101]. Since the layer is so thin less than 1nm that this deposition will not increase the inner resistance of the film for electron transportation. Experiments have demonstrated that it is a good approach to suppress the interfacial recombination since the insulator layer treated TiO_2/dye/CuI presented almost 1% overall PCE higher than that of the GSSC without being treated.

In addition, when QDs used as photosensitizer, poor coverage for the QDs to TiO_2 nanocrystals was often observed [59, 60]. Significant interfacial recombination takes place between QDs and TiO_2 and internal TiO_2 grain boundaries (Figure 22). Organic monolayer-capped QDs and the linker between QD and TiO_2 may cause additional trapping of electrons on both kinds of nanoparticle surfaces [60]. Using different bifunctional molecules as the linkers may bring about different coverage to TiO_2, but also may lead to different rate interfacial charge injection and recombination. One can imagine the complicated differences of the photogenerated chemical potential energy across the cell at different nanoscale locations. Actually, shown in Figure 22 is a CdSe QD sensitized -TiO_2 GLSC. If assume to use this type of QD-TiO_2 nanostructures for GSSCs, it is important and a challenge to find a proper HTM to interpenetrate into the nanoscale networks well.

Compared to GLSCs, much less effort was done to increase its overall PCE at today's art. For the GLSCs, efforts have been made to develop new HTMs to replace I^-/I_3^- redox couple to try to solve the problems in sealing of volatile electrolytes in large scale modules, or to make quasi-solid state Grätzel solar cells using a sol-gel nanocomposite electrolyte containing I^-/I_3^- redox couple. This quasi-GSSC has achieved overall PCE 5.4% [102]. Ionic liquid 1-ethyl-3-methylimidazolium selenocyanate (EMISeCN) based on $SeCN^-/(SeCN)_3^-$ has been found a competed mediator with I^-/I_3^- redox couple for Gräztel solar cells, which have achieved overall PCE 7.5-8.3% under AM 1.5 sunlight illumination [103]. Those research results and ideas may be used to develop HTM for GSSCs. Optimizing and developing high efficiency HTMs while finding a good approach to make it interpenetrate the TiO_2 nanoscale networks, meanwhile keeping it working at a higher temperature environment are critical challenge in the high affiance GSSC development in next five to ten years.

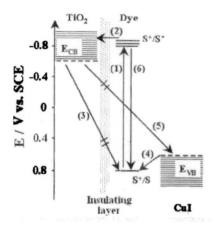

Figure 20. Illustration of the interfacial charge-transfer processes occurring in the TiO₂/dye/CuI GSSC. The blocking function of the insulating interlayer on interfacial recombination is shown as dash lines. [17]

Figure 21. Two configurations for the insulating layer-coated TiO₂ porous film electrode. In both cases, the insulating thin layer thickness was precisely controlled to less than 1nm to keep tunneling efficiency of electrons from dye molecules to TiO₂, allowing the electrons being collected by the TiO₂ networks. [17]

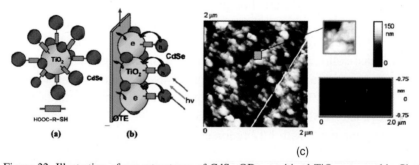

Figure 22. Illustration of nanostructures of CdSe QDs sensitized TiO₂ porous thin film: (a) linking CdSe QDs to TiO₂ particles with bifuncitonal surface modifier; (b) light harvesting assembly composed of TiO₂ film functionalized with CdSe QDs on optically transparent electrode; (c) the AFM images of an OTE/TiO₂/mercaptopropionic acid (MPA)/CdSe nanostructures. (Ref. 60)

2.3. Alternative Oxide Nanostructures for Grätzel Solar Cells

Although various techniques have been utilized and explored to synthesize a more efficient structure of TiO_2 nanocrystalline film to enhance the electrical and photovoltaic properties of dye-sensitized solar cell devices, the capability of these devices to surpass the 10% light conversion efficiency has been hindered. Efforts to find other dye-sensitized solar cell devices with various broad-band semiconducting oxide materials, including ZnO [104, 105] and SnO_2 [106, 107, 108] films, have been made for possible improvement of the current state of TiO_2-based devices. Composite structures consisting of a combination of TiO_2 and SnO_2, ZnO, or Nb_2O_5 materials, or a combination of other oxides, have also been examined in an attempt to enhance the overall light conversion efficiency [109-112]. In addition, hybrid structures comprised of a blend of semiconducting oxide film and polymeric layers for solid-state dye-sensitized solar cell devices have been explored in an effort to eliminate the liquid electrolyte completely for increased electron transfer and electron regeneration in hopes of increasing the overall efficiency [41, 45]. So far, these devices have achieved an overall light conversion efficiency of up to 5% for ZnO devices, up to 1% for SnO_2 devices, up to 6% for composite devices, and up to 2% for hybrid devices, all of which are still less efficient than solar cell devices based on dye-sensitized TiO_2 nanocrystalline film. Other methods incorporating insulating [113, 114, 115] or conducting oxides [116] to reduce electron-electron hole-recombination and enhance electron conduction have also been explored to improve the efficiency.

Among the alternative materials, zinc oxide (ZnO) has recently been explored more extensively. Since ZnO has a similar band gap to that for TiO_2 at 3.2eV, but has a much higher electron mobility of ~ 115-155cm^2/Vs than that for anatase TiO_2 at ~ 10^{-5}cm^2/Vs [117, 118], it has the greatest potential as an alternative material for improving the solar cell performance in Grätzel solar cells. Since ZnO has the same band gap as TiO_2, it has the same stability to photocorrosion as TiO_2. The highest overall efficiency obtained for ZnO nanoparticle film has been ~ 5% with an open-circuit voltage of ~ 560mV, a short-circuit current density of ~ 1.3mA/cm^2, and a fill factor of ~ 68% under 100mW/cm^2

illumination [104]. The solar cell performance of ZnO is still not as high as that of TiO$_2$, but since the use of ZnO is still new, as compared to TiO$_2$, it is still in the process of being optimized for further enhancement of photoresponse properties.

Another factor for using ZnO is the simple processing of ZnO through solution methods to tailor the nanostructure [119]. The simple process for tailoring the nanostructure is essential with the emergence of nanoscale materials, or nanowires, to enhance the solar cell performance by utilizing an ordered arrangement of nanowire arrays with simpler electron percolation pathways. These aligned nanowires are thought to provide a more ordered structure for dye adsorption and electron transport, as well as, provide a higher surface area for more light absorption, depending on the dimensions of the nanowires. Law et al used aqueous chemistry and a seeded growth process to synthesize a dense array of oriented, crystalline ZnO nanowires and obtained an efficiency of ~ 1.5% with nanowires ~ 16-17µm in length and 130-200nm in diameter [115]. The growth of single-crystalline ZnO nanowires is also essential to eliminate any barriers to electron transport typically found in polycrystalline material.

2.4. Ordered Semicondutor Nanoarchitectures for Gräztel Solar Cells

A desired morphology of the sensitized nanostructure films should have the mesorporous channels or nanorods aligned in parallel to each other and vertically with respect to the two electrodes. This would increase the electron diffusion length in the anode, facilitate pore diffusion, give easier access to the film surface avid grain boundaries and allow the junction to be formed under better control [27, 119]. Based on this assumption, much effort within recent three years have been made in constructing TiO$_2$ [120] or ZnO nanowires [119], nanotube [121], and nanosheets [122] for Gräztel solar cells.

2.4.1 Random TiO$_2$ Nanowires

Using an "orientated attachment mechanism" at low temperature, single-crystal-like anatase TiO$_2$ nanowires were formed in a network nanostructure. Figure 23 gives the organized TiO$_2$ nanowire images that grown through surfactant-assisted self-assembling processes (called as orientated attachment mechanism) at low temperature of 353K [120]. The direction of crystal growth could be controlled by changing the adsorption of surfactant molecules on the TiO$_2$ surface due to the reaction rate and the surface energy. Experiments demonstrated that an overall PCE above 9.3% was obtained using this TiO$_2$ nanowires powder to form the mesoporous TiO$_2$ nanostructures for a GLSC using N3 and I$^-$/I$_3^-$. Short-circuit photocurrent density, open-circuit voltage, and fill factor were 19.2mA/cm^2, 0.72V, and 0.675, respectively. One can imagine that the nanowires were not aligned vertically towards the substrate in this case.

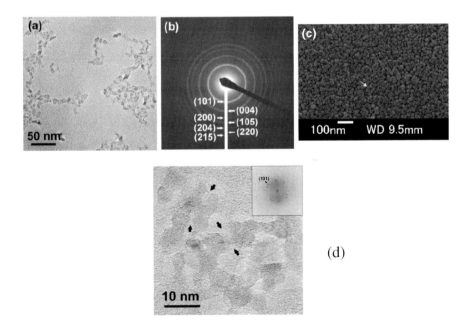

Figure 23. TEM images of TiO$_2$ Nanowire network structure (a) Nanowire formed by the connection of anatase nanoparticles, (b) pattern of titania nanowires, (c) SEM image of TiO$_2$ nanowire film, (d) HRTEM image of several titania nanowire with single anatase structure formed by oriented attachment. Shown is {101} spacing of the anatase phase. [Ref. 120]

2.4.2 Highly Organized ZnO Nanowires

Vertically aligned ZnO nanowires on a conductive substrate were fabricated via a seeded growth process [119], and they were used to build up a desired array GLSC. Figure 24 shows a schematic diagram of nanowire GSC model (a) and the prepared ZnO nanowires (b) and its nanostructure of diameter to length ratio (c ~ e). Using this-processed ZnO nanowire to create the sensitized nanostructures, a dye N719-sensitized GLSC using classic liquid I⁻/I₃⁻ redox couple as the mediator presented an overall full-sun PCE 1.5% [119]. A promising result was observed through transient mid-IR absorption experiments (Figure 25). The traces of the N719 dye-sensitized ZnO nanowire nanoarchitecture did show an ultrafast charge injection (<250fs) compared with ZnO nanoparticle nanostructure which was completed after pumped 5ps [119].

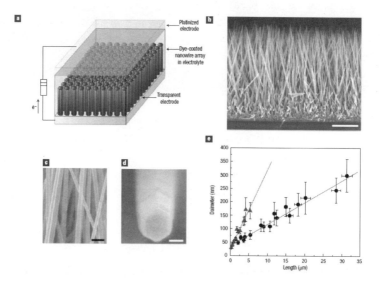

Figure 24. A schematic diagram of an ideal nanowire Grätzel solar cell (a) and ZnO nanowires prepared by a seeded growth process (b, c, d) and its morphology ratios of diameter to length under different growth conditions. [Ref. 119]

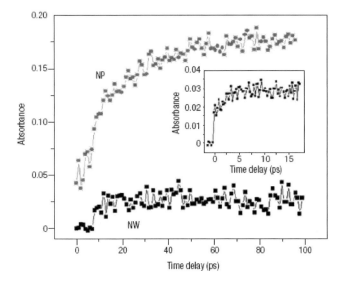

Figure 25. Transient mid-IR absorption traces of dye-sensitized ZnO nanowires (NW) and ZnO nanoparticle (NP) films: both films were sensitized by (Bu$_4$N)$_2$Ru(dcbpyH)$_2$(NCS)$_2$ (N719 dye) and were pumped at 400nm and the measured was made using a Ti:sapphire oscillator (30fs, 88MHz). [Ref. 119]

2.4.3 Highly Organized TiO$_2$ Nanotubes

TiO$_2$ nanotubes have been demonstrated to effectively improve electron lifetimes and to provide excellent pathways for electron percolation [Figure 26(a)] [121]. A 360-nm-thick, highly ordered nanotube arrays-based GLSC gave an overall PCE 2.9% using commercialized dye as photosensitizer and I$^-$/I$_3^-$ to regenerate the dye under AM 1.5 illumination [Figure 26 (b)] [121]. The TiO$_2$ nanotube has 46-nm pore diameter, and 17-nm wall thickness, and 360-nm length. They were perpendicularly grown on a fluorine-doped tin oxide-coated glass substrate by anodic oxidation of a titanium thin film. After crystallization by an oxygen anneal, the nanotube arrays were treated with TiCl$_4$ which enhanced the photogenerated current compared to the sample without being treated [121]. These results indicate that remarkable photoconversion efficiencies may be obtained with an increase of the nanotube-array and by an optimized post treatment.

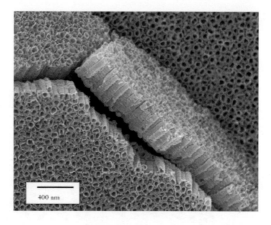

Figure 26. (a) TiO$_2$ Nanotubes obtained by anodic oxidation of a titanium thin film which was sputtered on ITO glass slide. (Ref. 121)

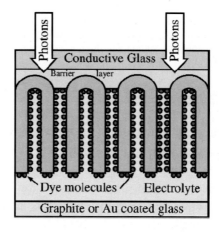

Figure 26. (b) Diagram of the Nanotube GLSC using I$^-$/I$_3^-$ as the electrolyte. (Ref. 121)

2.5. Discussion

However, in those highly organized TiO$_2$ nanowires or ZnO nanowires-based GLSCs, the overall PCE reported so far were still less than 4%. In addition, the ZnO nanowires-based GLSC only presented an overall PCE 1.5%, while the TiO$_2$ nanowire-based GLSC shown in Figure 24 gave an overall PCE above 9.5% using N$_3$ and the nanowires were not vertically organized. ZnO has a band gap similar to that of TiO$_2$ and has a much higher electron mobility ~115-155cm^2/Vs than that for anatase TiO$_2$, which is ~ 10^{-5}cm^2/Vs. This indicates that apart from the electron mobility, additional important factors need to be concerned. First, highly organized nanowires or nanotubes may have much less surface area to allow enough mass loaded for the photosensitizers. Thus inefficient light harvesting exists in the nanowires-based GLSCs. Second, photoinduced carrier concentration gradients always exist in the nanowired-based nanostructures across the cell under illumination, and the cell photovoltage is a function of both the built-in-electrical and the photo-generated chemical potential energy differences across the cells [40]. Thus, photodynamics in a narrow region near the interface where the carries are photogenerated to induce chemical potential gradients within the nanostructures have to be concerned. The reported results from different groups have indicated that there would much charge loss in ZnO nanowires-based GLSC compared with TiO$_2$ nanowires-based GLSCs. This may result in higher charge injection efficiency from dye to TiO$_2$ nanowires than that of ZnO nanowires. Further research could address this reason.

The greater potential for increasing the surface area of ZnO through surface structure modification, in conjunction with a higher electron mobility associated with the ZnO material, could provide a promising means for improving the solar cell performance of Grätzel solar cells. An excellent design was demonstrated very recently, which coated TiO$_2$ nanoparticles on indium-tin oxide nanorods in effort to improve the GLSC efficiency [123]. This study may generate a new type of nanostructures-nanocrystal coated nanowires. Thus, nanoassembly of TiO$_2$ or ZnO nanoparticles on ZnO nanowires or other promising nanowires or nanotubes may bring about significant efficiency improvement for the Grätzel solar cells.

It should be pointed out that, at the current state of the art, all the research are focusing on liquid quasi-solid state Grätzel solar cells using those organized TiO$_2$ or ZnO nanostructures. This includes a recent reported CdSe QDs sensitized ZnO nanowires Grätzel cell to achieve internal quantum efficiencies as high as 50-60% [124]. Assembling solid Grätzel solar cells based on semiconductor nanowires have not been extensively explored yet. A self-assembled hybrid P3HT/C$_{60}$ coated-TiO$_2$ nanotube double heterojunction solar cell was recently reported to achieve only about 1% efficiency under AM 1.5 sun illumination [125]. Therefore, research emphasizing on GSSCs need to be significantly pushed forward. We keep very promising expectation to more and more nanostructures that will be designed and created every year to update the Grätzel solar cell efficiency in many aspects.

3. Nanostructured Materials for Organic Solar Cells

In organic solar cells, bonded electron-hole pairs or excitons are created through photon absorption by an electron donor, which is a typical p-type semiconductor organic molecule or polymer. The excitons have to diffuse to the interface that is formed between the connection of electron donor and acceptor (D-A interface) to be dissociated to form free charges for transfer. The electron acceptor typically is n-type materials. This photoinduced electron transfer between donor and acceptor boosts the photogeneration of free charge carriers within the organic photovoltaics [126]. However, the excitions have short lifetime and low mobility. The diffusion length of excitions in organic semiconductors is limited to about 10nm only, less than the optical absorption length. This is a typical characteristic of almost all organic materials used for OPVs, formed an excition diffusion bottleneck, whereby the photogenerated excitions can not reach the D-A interface prior to dissociation into free carries, ultimately limiting the cell efficiency[127].

The exciton diffusion bottleneck imposes an important condition on efficient charge generation, which indicates that anywhere in the active layer, the distance to the interface should be on the order of the exciton diffusion length. However, a double layer of 20nm thin film would not be optically dense, allowing most photons to pass freely, apart from the high absorption coefficients of the donor and acceptor to exceed 10^5 cm^{-1}. Up to today's art, there are several effective strategies to reduce the exciton diffusion bottleneck, so as to improve the cell efficiency, which created different OPV single device nanoarchitectures [127, 128]. Employing a bulk or mixed heterojunction via nanoarchitecutres through creation of ordered bulk-heterojunctions (BHJ) has achieved over 5% PCE, demonstrating a very promising strategy to make the organic solar cells competed with inorganic photovoltaics. The bulk heterojunction nanostructures increase the active material's optical absorption length or the exciton diffusion length to improve the light harvesting efficiency and exciton mobility to the D-A interfaces to enhance the dissociation. The following sections briefly discuss the contribution in creation of the bulk-heterojunciton by nanostructure materials of fullerenes, metal nanoparticles, semiconductor nanorods, and carbon nanotubes.

3.1. Fullerenes

Fullerene families have unique nanostructures in small particle size, spherical-related geometry, as well as their highly electrical conductivity property. The typical fullerene molecule so far being widely used for bulk-polymer heterojunction solar cells is C_{60} derivative PCBM [129]. Simply mixing it with p-type semiconducting polymer as such polythiophene (e.g. P3HT) donor, it can pack tightly to form highly conductive film with excellent orbital overlap between adjacent polymer molecules. The heterojunctions throughout the bulk of the material are created, which ensure quantitative dissociate of photogenerated excitons on a nanometer dimension, irrespective the thickness of the film. This nanoarchitecture improves both electron and exciton diffusion efficiencies, and the intersystem crossing resulting from the large orbital angular momentum inherent in the π-electron system converts all excited states to triplets with their correspondingly long diffusion lengths. Figure 27 shows the close-to-ideal bulk heterojunction nanostructure solar cell model, indicating the phase separation in nanometer scale and the enhanced film thickness feasibility from effectively theoretical 10nm to 100nm for efficiently harvesting sunlight while creating long exciton and electron diffusion lengths [129]. Researchers have extensively explored the different combination ratios and solution-processes to create the nanostructures using PCBM and different p-type semiconductor polymers. Figure 28 gives the representative p-type polymer chemical structures and the corresponding PCBM structures, which were used to build up the heterojunctions. Typical heterojunction nanostructures with PCBM nanoparticles penetrated into the polymer phases are formed as shown in Figure 29 by AFM imaging.

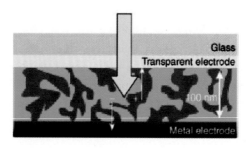

Figure 27. Schematic diagram of a bulk heterojunction (BJH) solar cell, presenting phase separation between donor (red) and the acceptor (blue) materials. (Ref. 129)

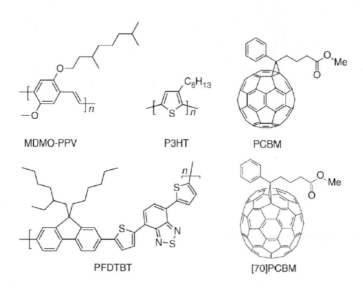

Figure 28. Chemical structures of representative donors and acceptors of materials used in polymer-fullerene bulk heterojunction (BHJ) solar cells. (Ref. 129 Chart 1)

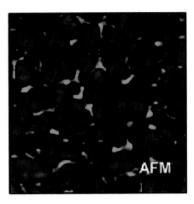

Figure 29. Atomic Force Microscopy phase image ($1 \times 1 \mu m^2$) of nanostructured composite film formed with PCBM fullerene and MDMO-PPV/PCBM (1:4 by wt ratio). (Ref. 130)

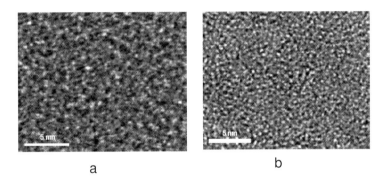

Figure 30. TEM images of not-annealed (a) and annealed (b) PCBM-P3HT blend films. (Ref. 131)

Among the reported p-type semiconduting polymers, P3HT is know to have a high charge carrier mobility and reduced bandgap, as compared with MDMO-PPV. It has been widely used in BHJ solar cells in combination with PCBM [129]. Experiments have demonstrated that thermal post-treatment of the C_{60}-P3HT composite films significantly improves the PCE up to 4.4% [131]. Figure 30 are the TEM images of not-annealed (a) and annealed (b) blend C_{60}-P3HT nanocomposite films reported from the best performance BHJ solar cell by Y. Kim et al.

Organized nanostructures are formed which can be clearly seen from the thermal treatment based on the comparison of the images. The self-organizing properties of P3HT in the BHJ nanostructures are sensitive to its molecular packing, which are varied with processing conditions while directly affecting the optical and electronic properties of the films. High polymer regioregularity (RR) and annealing strongly influence the BHJ cell performance, which can be attributed to the enhanced optical absorption and charge transport resulting from the organization of P3HT chains and domains [131]. Controlling the active layer film growth rate results in an increased hole mobility and balanced charge transport. A PCBM-P3HT BHJ cell film with thickness of over 210nm has been reported to reach the highest PCE of 4.4% [132].

Apart from the polymer-PCBM BHJ cells, using dye to function as the electron donor to form the BHJ PCBM-dye blended film has been carried out.   The reported dyes are as such porphyrins and CuPc photosensitizers. The utilization of double heterojunction, in combination with BHJ, as well as the construction of tandem cells has been reported to achieve a remarkable PCE 5.7% so far [129].

Incorporating a new family of soluble fullerene derivatives into the OPV to form BHJ cells have been reported. Replacing of the PCBM with its new family, such as C_{70} and C_{71}, and efforts focusing on improvement of solubility of the fullerene family in the p-type polymers and enhancement of the optical absorption of the BHJ film are undergoing [133]. Optimization of the nanostructure combinations and the corresponding BHJ cell performances should give a PCE over 6% in the near future to close to the commercialization level.

3.2. Metal Nanoparticles

Metal nanoparticles such as gold and silver have been incorporated into the heterojunction(HJ) cells. Using porphyrin as donor and fullerene C_{60} as the electron acceptor, films with three dimensional(3D) nanostructured arrays can be created by clusterization with gold nanopartilces on nanostructured SnO_2 electrode through electrophoretic deposition [134, 135]. Figure 31 is an organic synthesis strategy to form 3D pattern of a representative optoelectronic film. It is formed by covalent bonded gold nanoparticles and phorphyrin molecules with uniform insertion of C_{60} nanoparticles between the phorphyrin molecules, where the HJ takes place when light illuminating this film. The PCE of this HJ nanocomposite cell has reached as high as 1.5%, which is 45 times higher than that of the reference system consisting of the both single components of porphyrin and fullerene. A broad photocurrent action spectra up to 1000nm of this composite film was observed, which is attributed to a charge transfer type interaction resulting in the long-wavelength absorption of this gold nanoparticle linked 3D nanostructures ($H_2PC_nMPC+C_{60}$, n is the number of CH_2 in the porphyrin linker chain). UV-vis spectroscopy study indicates the formation of the π-complex between porphyrins and C_{60}, and the electron spin resonance (ESR) measurements under photoirradiation confirms the generation of $C_{60}{}^{\bullet}$ anion and porphyrin cation, indicating the porphyrin excited singlet state is quenched by C_{60} via electron transfer in the π-complex rather than by gold nancluster through energy transfer. The gold nanoparticles provide the necessary foundation to organize the donor-acceptor moieties in this new nanoarchitecture for photovoltaics. Comparison of the photocurrent action specta indicates that the higher IPCE and the broader photoresponse are attained with longer chain length of $H_2PCnMPC$. Figure 32(A) shows a representative nanostructure diagram of this 3D assembled architecture with n=15 and (B) gives the IPCE responses of the corresponding nanocomposite film-coated electrodes with different n number under illumination of light >400nm. This is a very interest result in seeking a system to harvest sunlight from visible to near-IR range via changing the bridge linker's chain length and controlling the 3D nanopattern in the composite films.

However, the cited case here utilized electrolyte NaI and I_2 in acetonitrile to regenerate the photosensitizer porphyrin. Replacing the electrolyte NaI and I_2 with P3HT may give us more exciting information in forming solid HJ OPVs using the 3D $H_2PCnMPC+C_{60}$ nanostructures. Also, incorporating other metal nanoparticles into fullerene nanostructures and linking them properly with other chromophores will generate new favorite nanoarchitectures, which would further improve the PCE of HJ OPVs. This indicates organic synthesis plays an important role in the creation of new nonmaterial for high efficient OPVs.

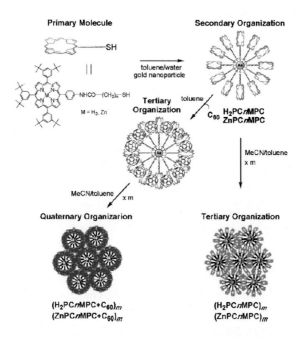

Figure 31. Illustration of high order organization of 3D porphyrin molecules with fullerence C_{60} nanoparticle via clusterization of gold nanoparticles. (Ref. 134)

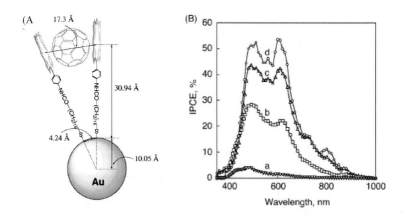

Figure 32. (A) a representative nanostructure of $H_2PCnMPC+C_{60})m$ when n=15; (B) Photocurrent action spectra of the OTE/SnO₂/$H_2PCnMPC+C_{60})m$ electrode ($H_2P=$ 0.19mM): a, n=5, $[C_{60}]$=0.31mM; b, n=11, $[C_{60}]$=0.31mM; c, n=15, $[C_{60}]$=0.31mM; d, n=15, $[C_{60}]$=0.38mM. Electrolyte: 0.5M NaI and 0.01M I_2 in acetonitrile. (Ref. 134)

3.3. Semiconductor Nanocrystal Materials

Semiconductor nanocrystal materials have been explored to be incorporated into conjugated conducting polymers forming hybrid HJ solar cells [136-139]. Typical semiconductor nanomaterials are quantum dots, nanorods, nanowires of the compounds formed by IIIA-VA groups or IIB-VIA groups. They present light absorption starting at the band edge and increases toward higher energy without falling off. The optical absorption can be tuned via the quantum confinement effects through their small particle size control. Light harvesting from visible to near IR range can be realized by different materials choice and size control. In addition, inorganic semiconductor nanocrystals have high electron affinity and high intrinsic charge carrier mobility. To combine the inorganic nanocrystals with conjugated polymers leads to remarkably increasing the charge transfer from polymer donor to nanocrystals and overcoming the low electron mobility problem in the hybrid OPVs. The hybrid organic-inorganic nanocrystal solar cells also present improved mechanical properties and thermal stability [137]. The pioneer group Alivisatos et al blended CdSe nanorods with polymer poly(3-hexylthiophene)(P3HT), created charge transfer junctions with high interfacial area. Figure 33 shows the diagram of the nanostructure solar cell that uses the blended nanocomposite of CdSe nanorids and P3HT. Figure 34 (a) gives the external quantum efficiency (EQE) versus the light wavelength of different nanocrystals with variation of geometry ratios, and Figure 34 (b) and (c) are the TEM images of the CdSe nanorods and corresponding nanocomposite structures prepared with 90% (wt) CdSe nanorods (7nm:60nm) with P3HT, which gave a PCE 1.5% under illumination of A.M. 1.5Global solar conditions in argon gas atmosphere.

Tetrapod-shaped CdTe nanocrystals have been synthesized from solution method, with a wide range of control over arm diameter and length. Creating ordered 3D such nanocrystals-P3HT nanocomposite

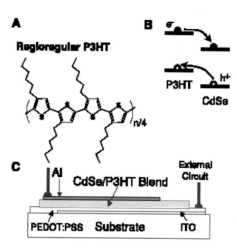

Figure 33. (A) structure of regioregular P3HT. (B) the schematic energy level diagram for CdSe nanorods and P3HT showing the charge transfer of electrons to CdSe and holes to P3HT. (C) the device structure consists of a film about 200nm in thickness sandwiched between an aluminum electrode and a transparent conducting electrode of PEDOT:PSS, which was deposited on an indium tin oxide glass substrate. The active device area is $3mm^2$. The film was spin-cast from a solution of 90% (wt) CdSe nanorods in P3HT in a pyridine-chloroform solvent mixture. (Ref. 136)

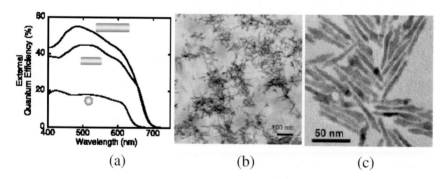

Figure 34. (a) EQE of 7nm-diamter nanorods with length 7, 30, and 60nm. The intensity is at $0.084mW/cm^2$ at 515nm. (b) TEM image of CdSe nanorod with diameter to length of 7nm:60nm. (c) the nanostructures of composite of 90% (wt) CdSe (7nm:60nm) and P3HT. (Ref. 136)

film has been successfully demonstrated by sequential deposition [140]. First, nanocrystal tetrapods are deposited on an electrode surface with a proper linker molecule to touch the substrate surface. The unique tetragonal structure of the nanocrystals gives rise to a natural ordering in the deposited films. Three arms of each tetrapod contact on the substrate at its base, while the fourth arm points up, perpendicularly to the substrate. The ordering evidence can be seen from the creation of polymer composite film by spin-casting P3HT from its optimal solvent to over the nanocrystal film. Figure 35 gives the scanning electron micrographs of the deposited tetrapod nanocrystal film and its composite film with P3HT. The early stage BHJ OPV device has demonstrated a PCE less than 1%.

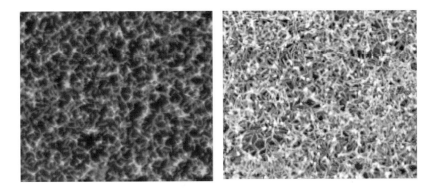

Figure 35. scanning electron micrographs of tetrapod nanocrystal film (left) and its ordered P3HT nanocomposite structures. (Ref. 140)

Compared to CdSe, CdTe as smaller bandgap (1.5eV in bulk), allowing for improved absorption of the solar spectrum. Also, the tetrapods are 3D, allowing for improved electron transport. Their four rod-like arms project symmetrically from a central core, ensuring a transport path across the blend film regardless of their orientation. Therefore, much higher PCE than 1% should be obtained after further optimizing the film morphology and better control the ordered 3D nanoarchitectures.

Bulk heterojunction (BHJ) hybrid solar cells have been developed using hyperbranched CdSe semiconductor nanocrystals and P3HT [141].

The ability to prescribe dispersion and charge percolation characteristics of s composite device through choice of nanocrystal structure may be the advantage of such hyperbranched nanocrystal solar cells over the other hybrid architectures. Figure 36 (a) and (b) give the photovoltaic I~V curve and the corresponding blended composite film. Figure 36 (c) is the TEM image of the CdSe hyperbranched nanocrystal. The cell presents a PCE of 2.18% under a sun AM 1.5G illumination.

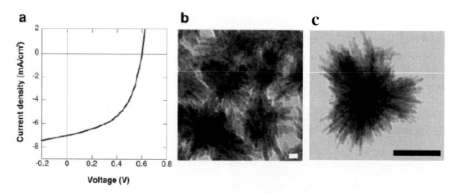

Figure 36. (a) current-voltage characteristics of the blended hyperbranched CdSe nanocrystal with P3HT; (b) the TEM image of this blend (scale bar: 20nm); and (c) the hyperbranched CdSe Nanocrystal (scale bar: 100nm). (Ref. 141)

3.4. Carbon Nanotubes

Carbon nanotubes including the single-walled (SWNTs) and the multi-walled (MWNTs) ones, have been attracted great attention in improving OPV entire performances due to their unique nanostructures and high electrical conductivity, as well as excellent mechanical flexibility. Their high electron affinity makes them function as electron collector and enhance the carrier mobility in the conjugated polymer films. Research using carbon nanotubes to construct nanoscale solar cells is in its early stage. Replacing C_{60} with carbon nanotubes in the BHJ OPVs has been demonstrated a significant enhancement of the open circuit voltage for the photovoltaic performance [142]. In our research, thermal treated SWNTs were uniformly dispersed in P3HT matrix with different weight ratios. Figure 37 (a) presents the diagram of the device structure, and Figure 37 (b) gives the corresponding photocurrent-voltage I~V response. Our finding is that the open circuit photovoltage has been reached over 1.6 volts with a nanocomposite film device formed by thermally treated SWNTs 2% (wt) dispersed in P3HT. This indicates a very promising direction in improving the entire BHJ OPCs using CNTs.

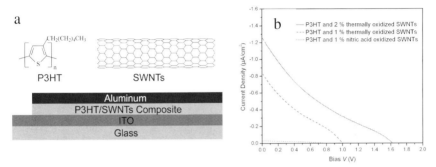

Figure 37. (a) Chemical structure of P3HT, SWNTs and schematic representation of a photovoltaic device, and (b) photovoltaic performance of I~V curve of the nanocomposite films. (Ref. 142)

Figure 38. AFM topography of (a) pristine P3HT film with 2000×2000 nm^2 scanning range, (b) 1% nitric acid purified SWNTs/P3HT composite film with 2000×2000 nm^2 scanning range, (c) 1% thermally oxidized SWNTs/P3HT composite film with 1000×1000 nm^2 scanning range, and (d) 2% thermally oxidized SWNTs/P3HT composite film with 1000×1000 nm^2 scanning range. (Ref. 142)

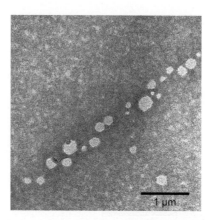

Figure 39. TEM morphology of P3HT microcrystals attached on the sidewall of SWNTs. (Ref. 142)

Our characterizations show that continuous active film with interpenetrating structure formed due to the incorporating of SWNTs into the P3HT matrix (Figure 38), which improves the crystallinity of the resultant film of SWNTs/P3HT composite along with the wall of SWNTs (Figure 39) [142].

Recent report shows that controlled placement of SWNTs monolayer network at different position in polymer-fullerence BHJ solar cells have different impacts to the cell's photovoltaic performance. When SWNTs deposited on the hole-collection side of the active layer lead to an increase in PCE from 4 to 4.9% (under AM 1.5 G, 1.3 suns illumination). When SWNTs deposited on the top of he active layer, it leads to major electro-optical changes in the device functionality, including an increased fluorescence lifetime of P3HT [143]. These new findings really bring about critical questions for how the CNTs impact on the OPVs' performances, which include active layer nanoarchitectures, charge collection and transportation, as well as mechanical properties and so on. These research need to be further performed to make a clear understanding of CNTs's impacts mechanisms to the OPVs.

4. Summary and Discussion

Up to today's art, there are many new nanostructures being created. This review is not able to address them one by one. For the nanoarchitectured solar cells, efforts need to emphasize on Grätzel type solid solar cells since its liquid solar cells have been achieved remarkable PCE over 11%. Organic photovoltaics are competing with Grätzel type solid solar cells in PCE in the currently-reported-highest range of 4.5 to 6%, since the strategy of building up bulk heterojunction significantly improves the photovoltaic responses of the OPVs using fullerenes and carbon nanotubes. A combination of Grätzel type solid solar cells with bulk-heterojunction of OPVs may bring about revolutionary change to the PCE for a single device. Construction tandem stacks and modules would further increase the output power of the cell assembly for commercialization level. Ordered organic-inorganic bulk hetereojunction solar cells and nanowires-based inorganic and organic hybrids may play an important role in corporation with photosensitizing concept to enhance the light harvesting in the sunlight spectrum [144, 145]. To make the nanostructured solar cells to be practically applicable, research on device real lifetime, thermal and photostabilities, as well as the applicable efficiency, and entire cost of the device that includes the manufacturing and materials choices need to be addressed when create new nonmaterial for the desired efficient solar cells. We anticipate the inorganic-organic nanostructure solid solar cells would achieve a PCE over 10% within five to ten years to compete with traditional silicon thin film solar cells.

Acknowledgements

This work was financially supported by grants from National Science Foundation, US Department of Energy, Army Research Office, Washington Technology Center, and Air Force Office of Scientific.

References

1.   A. W. Blakers, A. Wang, A. M. Milne, J. Zhao, and M. A. Green, Applied Physics Letters, 55, 1363 (1989).

2. A. Shah, P. Torres, R. Tscharner, N. Wyrsch, and H. Keppner, Science, 285, 692 (1999).
3. A. Marti and A. Luque, Next Generation Photovoltaics: High Efficiency through Full Spectrum Utilization. Institute of Physics: Bristol, U. K. (2003).
4. M. A. Green, Third Generation Photovoltaics: Advanced Solar Energy Conversion. Springer: Berlin, Germany (2004).
5. M. A. Green, Photovoltaic Principles, Physica E, 14, 11-17 (2002).
6. M.D. Archer and R. Hill, Series on Photoconversion of Solar Energy: Clean Electricity from Photovoltaics, Vol. 1, Imperial College Press, London (2001).
7. Office of Science, Department of Energy, U.S.A, in "Basic Research Needs for Solar Energy Utilization", Report of the Basic Energy Science Workshop on Solar Energy Utilization, San Francisco (April 18-21, 2005).
8. Business-Week Online: "What's Raining on Solar's Parade", http://www.businessweek.com/magazine/content/06_06/b3970108.htm, posted by Feb 6, 2006. Entered the website on September 26, 2006.
9. B. Sun, N. C. Greenham, Physical Chemistry Chemical Physics 8(30), 3557 (2006).
10. H. W. Lee, Solar photovoltaic devices including nanostructured materials. PCT Int. Appl. (2006), 60pp.
11. D. C. Coffey, D. S. Ginger, Time-resolved electrostatic force microscopy of polymer solar cells. Nature Materials 5(9), 735 (2006).
12. L. Tsakalakos, C. S. Korman, A. Ebong, A. M. Srivastava, J. Lee, S. Leboeuf, R. Wojnarowski, O. Sulima, High efficiency inorganic nanorod-enhanced photovoltaic devices. Eur. Pat. Appl. (2006), 38pp.
13. M. Grätzel, Nature 414, 338 (2001).
14. M. Grätzel, "Perspectives for Dye-sensitized Nanocrystalline Solar Cells," Progress in Photovoltaics: Research and Applications 8, 171 (2000).
15. A. Hagfeldt and M. Grätzel, Accounts of Chemical Research 33, 269 (2000).
16. M. Grätzel, Inorg. Chem. 44, 6841 (2005).
17. A. Fujishima and X. T. Zhang, "Solid State Dye-Sensitized Solar Cells", Proc. Japan Acad. 81, Ser. B, 33 (2005).
18. Z. Liang, K. L. Dzienis, J. Xu, Q. Wang, Covalent layer-by-layer assembly of conjugated polymers and CdSe nanoparticles: multilayer structure and photovoltaic properties. Advanced Functional Materials 16(4), 542 (2006).
19. O. F. Yilmaz, S. Chaudhary, M. Ozkan, A hybrid organic -inorganic electrode for enhanced charge injection or collection in organic optoelectronic devices. Nanotechnology 17(15), 3662 (2006).
20. W. U. Huynh, J. J. Dittmer, P. A. Alivisatos, Hybrid Nanorod –polymer Solar Cells, Science, 295 (5564), 0036-8075(2002).
21. Y. Kim, S. Cook, S. M. Tuladhar, S. A. Choulis, J. Nelson, J. R. Durrant, D. D. C. Bradley, M. Giles, I. Mcculloch, C. S. Ha, and M. Ree, Nature Materials, 5, 197 (2006).

22. M.A. Green, "Third Generation Photovoltaics: Ultra-high Conversion Efficiency at Low Cost," Progress in Photovoltaics: Research and Applications 9, 123 (2001).

23. R.H. Bube, Photoelectronic Properties of Semiconductors, Cambridge University Press, Cambridge, 1992.

24. A. J. Nozik, "Spectroscopy and Hot Electron Relaxation Dynamics in Semiconductor Quantum Wells and Quantum Dots," Annual Rev. Phys. Chem., 52, 193 (2001).

25. A. J. Nozik, Quantum dot solar cells. Next Generation Photovoltaics, 196 (2004).

26. A. Marti, N. Lopez, E. Antolin, E. Canovas, C. Stanley, C. Farmer, L. Cuadra, A. Luque, Novel semiconductor solar cell structures: The quantum dot intermediate band solar cell. Thin Solid Films, 511-512, 638-644 (2006).

27. M. Grätzel, "Dye-sensitized solar cells", Journal of Photochemistry and Photobiology C: Photochemistry Review 4, 145-153 (2003).

28. O'Regan , M. A. Grätzel, Nature 353 , 737 (1991).

29. K. Kalyanasundaram, M. Grätzel, Coordinaiton Chemistry Review, 77, 347-414 (1998).

30. A. Hagfeldt and M. Grätzel, Molecular Photovoltaics, Acc. Chem. Res., 33, 269 (2000).

31. U. Bach, D. Lupo, P. Comte, J. E. Moser, F. Weissortel, J. Salbeck, H. Spreitzer, M. Grätzel, Solid-state dye - sensitized mesoporous TiO_2 solar cells with high photon-to-electron conversion efficiencies. Nature, 395 (6702), 583-585 (1998).

32. L. Han, N. Koide, Y. Chiba, A. Islam, R. Komiya, N. Fuke, A. Fukui, and R. Yamanaka, Applied Physics Letters, 86, 213501 (2005).

33. D. Wöhrle and D. Meissner, "Organic Solar Cells," Advanced Materials 3, 129 (1991).

34. G. Yu, J. Gao, J.C. Hummelen, F. Wudl, and A.J. Heeger, "Polymer photovoltaic cells: enhanced efficiencies via a network of internal donor acceptor heterojunctions," Science 270, 1789 (1995).

35. J.J.M. Halls, K. Pickler, R.H. Friend, S.C. Morati, and A.B. Holmes, "Efficient photodiodes from interpenetrating polymer networks," Nature 376, 498 (1995).

36. C.J. Brabec and N.S. Sariciftci, "Polymeric photovoltaic devices," Materials Today, 3 (2000).

37. P. Peumans, A. Yakimov, and S.R. Forrest, "Small molecular weight organic thin-film photodetectors and solar cells," Journal of Applied Physics 93, 3693 (2003).

38. B. Kannan, K. Castelino, and A. Majumdar, "Design of Nanostructured Heterojunction Polymer Photovoltaic Devices," Nanoletters 3, 1729 (2003).

39. H. Hoppe and N.S. Sariciftci, "Organic solar cells: An overview," Journal of Materials Research 19, 1924 (2004).

40. B. A. Gregg, "Exciton Solar Cells", Journal of Physical Chemistry B, 107, 4688 (2003).

41. W.U. Huynh, J.J. Dittmer, and A.P. Alivisatos, "Hybrid Nanorod-Polymer Solar Cells," Science 295, 2425 (2002).

42. C. J. Brabec, "Organic Photovoltaics: Technology and Market", Sol. Energy Mater. Solar Cells, 83, 273-292 (2004).
43. G. Li, V. Shrotriya, J. Hunag, Y. Yao, T. Moriarty, K. Emery, and Y. Yang, Nature materials, 4, 864-868 (2005).
44. D. Gebeyehu, C.J. Brabec, and N.S. Sariciftci, "Solid-state organic/inorganic hybrid solar cells based on conjugated polymers and dye-sensitized TiO_2 electrodes," Thin Solid Films 403-404, 271 (2002).
45. W.U. Huynh, X. Peng, and A.P. Alivisatos, "CdSe Nanocrystal Rods/Poly(3-hexylthiophene) Composite Photovoltaic Devices," Advanced Materials 11, 923 (1999).
46. K. Triyana, T. Yasuda, K. Fujita, and T. Tsutsui, "Tandem-type organic solar cells by stacking different heterojunction materials," Thin Solid Films 477, 198 (2005).
47. J.L. Segura, N. Martin, and D.M. Guldi, "Materials for organic solar cells: the C60/p-conjugated oligomer approach," Chemical Society Reviews 34, 31 (2005).
48. D. M. Adams, L. Brus, C. E. D. Chidsey, S. Creager, C. Creutz, C. R. Kagan, P. V. Kamat, M. Lieberman, S. Lindsay, R. A. Marcus, R. M. Metzger, M. E. Michel-Beyerle, J. R. Miller, M. D. Newton, D. R. Rolison, O. Sankey, K. S. Schanze, J. Yardley, and X. Zhu, "Charge Transfer on the Nanoscale: Current Status", J. Phys. Chem. B, 107, 6668 (2003).
49. K. Tennakone, G. R. R. A. Kumara, I. R. M. Kottegoda, K. G. U. Wijayantta, and V. P. S. Perera, J. Phys. D: Appl. Phys., 31, 1492 (1998).
50. A. Fujishima and X. Zhang, Solid-state dye-sensitized solar cells, Proc. Japan Acad., 81, Ser. B, 33-42 (2005).
51. C. Klein, Md.K. Nazeeruddin, P. Liska, Davide. Di Censo, N. Hirata, E. Palomares, J. R. Durrant, and M. Grätzel, Inorg. Chem. 44, 178 (2005).
52. P. Wang, C. Klein, R. Humphry-Baker, S. M. Zakeeruddin, and M. Grätzel, J. Am. Chem. Soc., 127, 808 (2005).
53. Md. K. Nazeeruddin, P. Pechy, M. Gräztel, Chem. Commun., 1705 (1997).
54. T. Horiuchi, H. Miura, K. Sumioka, and S. Uchida, J. Am. Chem. Soc., 126, 12218 (2004).
55. J. R. Durrant, S. A. Haque, Nat. Mater, 2, 362 (2003).
56. H. Imahort, J. Phys. Chem. B, 108, 6130 (2004).
57. Md. K. Nazeeruddin, S. M. Zakeeruddin, R. Humphry-Baker, M. Jirousek, P. Liska, N. Vlachopoulos, V. Shklover, Christian-H. Fisher, and M. Grätzel, Inorg. Chem., 38, 6298 (1999).
58. M. K. Nazeeruddin, A. Kay, I. Rodicio, R. Humphry-Baker, E. Muller, P. Liska, N. Vlachopoulos, and M. Grätzel, J. Am. Chem. Soc., 115, 6382 (1993).
59. T. Zeng, E. Gladwin, R. O. Claus, "Self-assembled InP Quanutm Dot-TiO_2 Grätzel Solar Cell", MRS proceeding paper, San Francisco, Spring, 2003.
60. I. Robet, V. Subramanian, M. Kuno, and P. V. Kamat, J. Am. Chem. Soc., 128, 2385 (2006).

61. J. L. Blackburn, D.C. Selmarten, R. J. Ellingson, M. Jones, O. Micic, and A. J. Nozik, J. Phys. Chem. B, 109, 2625 (2005).
62. A. Zaban, O. I. Micic, B. A. Gregg, and A. J. Nozik, Langmuir, 14, 3153 (1998).
63. O. I. Micic H. A. Cheong, H. Fu, A. Zunger, J. R. Sprague, A. Mascarenhas, and A. J. Nozik, J. Phys. Chem. B, 101, 4904 (1997).
64. T. J. Bukowski, J. H. Sammons, Critical Review in Solid Sate and Materials Sciences, 27(3/4), 119 (2002).
65. A. J. Nozik, Physica E, 14, 115 (2002).
66. R. Plass, S. Pelet, J. Krueger, and M. Grätzel, J. Phys. Chem. B, 106, 7578 (2002).
67. G. P. Smestad, S. Spiekermann, J. Kowalik, C. D. Grant, A. M. Schwartzberg, J. Zhang, L. M. Tolbert, and E. Moons, Solar Energy Materials and Solar Cells, 76, 85-105 (2003).
68. P. Ravirajian,A. M. Peiro, M. K. Nazeeruddin, M. Gratzel, D. D. C. Bradley, J. R. Durrant, and J. Nelson, J. Physc. Chem. B, 110, 7635 (2006).
69. B. O'Regan and F. Lenzmann, J. Phys. Chem. B, 108, 4342 (2004).
70. J. Jiu, F. Wang, M. Sakamoto, J. Takao, and M. Adachi, "Performance of dye-sensitized solar cell based on nanocrystals of TiO$_2$ film prepared with mixed template method," Solar Energy Materials & Solar Cells 87, 77 (2005).
71. C. Natarajan and G. Nogami, "Electrodeposition of Nanocrystalline Titanium Dioxide Thin Films," Journal of the Electrochemical Society 143, 1547 (1996).
72. S. Karuppuchamy, K. Nonomura, T. Yoshida, T. Sugiura, and H. Minoura, "Cathodic Electrodeposition of oxide semiconductor thin films and their application to dye-sensitized solar cells," Solid State Ionics 151, 19 (2002).
73. T. Miyasaka and Y. Kijitori, "Low-Temperature Fabrication of Dye-Sensitized Plastic Electrodes by Electrophoretic Preparation of Mesoporous TiO$_2$ Layers," Journal of the Electrochemical Society 151, A1767 (2004).
74. A. Hagfeldt and M. Grätzel, "Light-Induced Redox Reactions in Nanocrystalline Systems," Chemical Reviews 95, 49 (1995).
75. M. Thelakkat, C. Schmitz, and H.W. Schmidt, "Fully Vapor-Deposited Thin-Layer Titanium Dioxide Solar Cells," Advanced Materials 14, 577 (2002).
76. F. Pichot, J.R. Pitts, and B.A. Gregg, "Low-Temperature Sintering of TiO$_2$ Colloids: Application to Flexible Dye-Sensitized Solar Cells," Langmuir 16, 5626 (2000).
77. D. Zhang, T. Yoshida, and H. Minoura, "Low Temperature Synthesis of Porous Nanocrystalline TiO$_2$ Thick Film for Dye-Sensitized Solar Cells by Hydrothermal Crystallization," Chemistry Letters, 874 (2002).
78. K.J. Jiang, T. Kitamura, H. Yin, S. Ito, and S. Yanagida, "Dye-sensitized Solar Cells Using Brookite Nanoparticle TiO$_2$ Films as Electrodes," Chemistry Letters, 872 (2002).
79. G. Boschloo, H. Lindström, E. Magnusson, A. Holmberg, and A. Hagfeldt, "Optimization of dye-sensitized solar cells prepared by compression method," Journal of Photochemistry and Photobiology A: Chemistry 148, 11 (2002).

80. M.Y. Song, D.K. Kim, K.J. Ihn, S.M. Jo, and D.Y. Kim, "Electrospun TiO_2 electrodes for dye-sensitized solar cells," Nanotechnology 15, 1861 (2004).

81. A. G. Agrios, I. Cesar, P. Comte, M. K. Nazeeruddin and M. Grätzel,Chem. Mater. Communication, page EST: 2.7, A~C (2006).

82. K. Schwarzburg and F. Willig, "Origin of Photovoltage and Photocurrent in the Nanoporous Dye-Sensitized Electrochemical Solar Cell," Journal of Physical Chemistry B 103, 5743 (1999).

83. J. Krüger, R. Plass, M. Grätzel, P.J. Cameron, and L.M. Peter, "Charge Transport and Back Reaction in Solid-State Dye-Sensitized Solar Cells: A Study Intensity-Modulated Photovoltage and Photocurrent Spectroscopy," Journal of Physical Chemistry B 107, 7536 (2003).

84. J.R. Durrant, S.A. Haque, and E. Palomares, "Towards optimization of electron transfer processes in dye sensitized solar cells," Coordination Chemistry Review 248, 1247 (2004).

85. A.N.M. Green, E. Palomares, S.A. Haque, J.M. Kroon, and J.R. Durrant, "Charge Transport versus Recombination in Dye-Sensitized Solar Cells Employing Nanocrystalline TiO_2 and SnO_2 Films," Journal of Physical Chemistry B 109, 12525 (2005).

86. B.A. Gregg, "Interfacial processes in the dye-sensitized solar cell," Coordination Chemistry Reviews 248, 1215 (2004).

87. S. Rühle and D. Cahen, "Electron Tunneling at the TiO2/Substrate Interface Can Determine Dye-Sensitized Solar Cell Performance," Journal of Physical Chemistry B 108, 17946 (2004).

88. K. Fredin, J. Nissfolk, and A. Hagfeldt, "Brownian dynamics simulations of electrons and ions in mesoporous films," Solar Energy Materials & Solar Cells 86, 283 (2005).

89. D. Cahen and G. Hodes, "Nature of Photovoltaic Action in Dye-Sensitized Solar Cells," Journal of Physical Chemistry B 104, 2053 (2000).

90. A.M. Eppler, I.M. Ballard, and J. Nelson, "Charge transport in porous nanocrystalline titanium oxide," Physica E 14, 197 (2002).

91. J. Nelson, "Continuous-time random-walk model of electron transport in nanocrystalline TiO_2 electrodes," Physical Review B 59, 153 74 (1999).

92. J. Nelson, S.A. Haque, D.R. Klug, and J.R. Durrant, "Trap-limited recombination in dye-sensitized nanocrystalline metal oxide electrodes," Physical Review B 63, 205321-1 (2001).

93. J. Bisquert, A. Zaban, and P. Salvador, "Analysis of the Mechanisms of Electron Recombination in Nanoporous TiO_2 Dye-Sensitized Solar Cells. Nonequilibrium Steady-State Statistics and Interfacial Electron Transfer via Surface States," Journal of Physical Chemistry B 106, 8774 (2002).

94. Y. Tachibana, K. Hara, S. Takano, K. Sayama, and H. Arakawa, "Investigation on anodic photocurrent loss processes in dye sensitized solar cells: comparison

between nanocrystalline SnO_2 and TiO_2 films," Chemical Physics Letters 364, 297 (2002).

95. S. Nakade, M. Matsuda, S. Kambe, Y. Saito, T. Kitamura, T. Sakata, Y. Wada, H. Mori, and S. Yanagida, "Dependence of TiO_2 Nanoparticle Preparation Methods and Annealing Temperature on the Efficiency of Dye-Sensitized Solar Cells," Journal of Physical Chemistry B 106, 10004 (2002).

96. C. Guillard, B. Beaugiraud, C. Dutriez, J.M. Herrmann, H. Jaffrezic, N. Jaffrezic-Renault, and M. Lacroix, "Physicochemical properties and photocatalytic activities of TiO_2-films prepared by sol-gel methods," Applied Catalysis B: Environmental 39, 331 (2002).

97. S. Nakade, Y. Saito, W. Kubo, T. Kitamura, Y. Wada, and S. Yanagida, "Influence of TiO_2 Nanoparticle Size on Electron Diffusion and Recombination in Dye-Sensitized TiO_2 Solar Cells," Journal of Physical Chemistry B 107, 8607 (2003).

98. P.J. Cameron and L.M. Peter, "How Does Back-Reaction at the Conducting Glass Substrate Influence the Dynamic Photovoltage Response of Nanocrystalline Dye-Sensitized Solar Cells?," Journal of Physical Chemistry B 109, 7392 (2005).

99. F. Pichot and B.A. Gregg, "The Photovoltage-Determining Mechanism in Dye-Sensitized Solar Cells," Journal of Physical Chemistry B 104, 6 (2000).

100. X.-T., Zhang, H. W. Taguchi, T. Meng, Q.-B. O. Sato, and A. Fujishima, Solar Energy Mater. Solar Cells, 81, 197-203 (2004).

101. T. Zeng, C. Norris, "Hybrid Grätzel Solid Solar Cells Modified by Atomic Layer Deposition", American Materials Research Society Spring Conference, San Francisco, April 17-21, 2006.

102. E. Stathatos, and P. Lianos, S. M. Zakeeruddin, P. Liska, and M. Grätzel, Chem. Mater. 15, 1825 (2003).

103. P. Wang, S. M. Zakeeruddin, J.-E. Moser, R. Humphry-Baker, and M. Grätzel, J. Am. Chem. Soc., 126, 7164 (2004).

104. K. Keis, C. Bauer, G. Boschloo, A. Hagfeldt, K. Westermark, H. Rensmo, and H. Siegbahn, "Nanostructured ZnO electrodes for dye-sensitized solar cell applications," Journal of Photochemistry and Photobiology A: Chemistry148, 57 (2002).

105. A.B. Kashyout, M. Soliman, M. El Gamal, and M. Fathy, "Preparation and characterization of nanoparticles ZnO films for dye-sensitized solar cells," Materials Chemistry and Physics 90, 230 (2005).

106. S. Chappel, S.G. Chen, and A. Zaban, "TiO_2-coated Nanoporous SnO_2 Electrodes for Dye-Sensitized Solar Cells," Langmuir 18, 3336 (2002).

107. S. Chappel and A. Zaban, "Nanoporous SnO_2 electrodes for dye-sensitized solar cells: improved cell performance by the synthesis of 18 nm SnO_2 colloids," Solar Energy Materials & Solar Cells 71, 141 (2002).

108. S.C. Lee, J.H. Lee, T.S. Oh, and Y.H. Kim, "Fabrication of tin oxide film by sol-gel method for photovoltaic solar cell system," Solar Energy Materials & Solar Cells 75, 481 (2003).

109. K. Tennakone, P.K.M. Bandaranayake, P.V.V. Jayaweera, A. Konno, and G.R.R.A. Kumara, "Dye-sensitized composite semiconductor nanostructures," Physica E 14, 190 (2002).

110. K.M.P Bandaranayake, M.K. Indika Senevirathna, P.M.G.M. Prasad Weligamuwa, and K. Tennakone, "Dye-sensitized solar cells made from nanocrystalline TiO_2 films coated with outer layers of different oxide materials," Coordination Chemistry Reviews 248, 1277 (2004).

111. T. Stergiopoulos, I.M. Arabatzis, M. Kalbac, I. Lukes, and P. Falaras, "Incorporation of innovative compounds in nanostructured photoelectrochemical cells," Journal of Materials Processing Technology 161, 107 (2005).

112. S.G. Chen, S. Chappel, Y. Diamant, and A. Zaban, "Preparation of Nb_2O_5 Coated TiO_2 Nanoporous Electrodes and Their Application to Dye-Sensitized Solar Cells," Chemistry of Materials 13, 4629 (2001).

113. T.S. Kang, S.H. Moon, and K.J. Kim, "Enhanced Photocurrent-Voltage Characteristics of Ru(II)-Dye Sensitized TiO_2 Solar Cells with TiO_2-WO_3 Buffer Layers Prepared by a Sol-Gel Method," Journal of the Electrochemical Society 149, E155 (2002).

114. E. Palomares, J.N. Clifford, S.A. Haque, T. Lutz, and J.R. Durrant, "Control of Charge Recombination Dynamics in Dye Sensitized Solar Cells by the Use of Conformally Deposited Metal Oxide Blocking Layers," Journal of the American Chemical Society 125, 475 (2003).

115. P.K.M. Bandaranayake, P.V.V. Jayaweera, and K. Tennakone, "Dye-sensitization of magnesium-oxide-coated cadmium sulfide," Solar Energy Materials & Solar Cells 76, 57 (2003).

116. J. Bandara, U.W. Pradeep, and R.G.S.J. Bandara, "The role of n-p junction electrodes in minimizing the charge recombination and enhancement of photocurrent and photovoltage in dye sensitized solar cells," Journal of Photochemistry and Photobiology A: Chemistry 170, 273 (2005).

117. E.M. Kaidashev, M. Lorenz, H. von Wenckstern, A. Rahm, H.C. Semmelhack, K.H. Han, G. Benndorf, C. Bundesmann, H. Hochmuth, and M. Grundmann, "High electron mobility of epitaxial ZnO thin films on c-plane sapphire grown by multistep pulsed-laser deposition," Applied Physics Letters 82, 3901 (2003).

118. Th. Dittrich, E.A. Lebedev, and J. Weidmann, "Electron Drift Mobility in Porous TiO_2 (Anatase)," Rapid Research Notes 165, R5 (1998).

119. M. Law, L.E. Greene, J.C. Johnson, R. Saykally, and P. Yang, "Nanowire dye-sensitized solar cells," Nature Materials 4, 455 (2005).

120. M. Adachi, Y. Murata, J. Takao, J. Jiu, M. Sakamoto and F. Wang, J. Am. Chem. Soc., 126, 14943 (2004).

121. G. K. Mor, K. Shankar, M. Paulose, O. K. Varghese and C. A. Grimes, Nano Letters, 6(2), 215 (2006).

122. N. Sakai, Y. Ebina, K. Takada, and T. Sasaki, J. Am. Chem. Soc., 126, 5851 (2004).

123. Tammy Ping-Chun Chou, in Dissertation: "The Study of Titania and Zinc Oxide Nanostructures for the Exploration of Charge Transport Properties in Dye-Sensitized Solar Cells", Department of Materials Science and Engineering, University of Washington (2006).
124. K. S. Leschkies, R. Divkar, J. Basu, E. Enache-Pommer, J. E. Boercker, C. B. Carter, U. R. Kortshagen, D. J. Norris, and E. S. Aydil, Nano Leters, 7(6), 1793 (2007).
125. K. Shankar, G. K. Mor, H. E. Prakasam, O. K. Vrghese, and C. A. Grimes, Langmuir, 10.1021/la7020403 S0743-7463(70)02040-1, published on web10/24/2007.
126. S. Shaheen, D. S. Ginley, and G. E. Jabour, Materials Research Society Bulletin, Vol 30: 10(2005).
127. S. Forrest, Materials Research Society Bulletin, Vol 30: 28 (2005).
128. P. Peumans, V. Bulovic, and S. R. Forrest, Appl. Phy. Lett., 76, 2650 (2000).
129. R. A. J. Janssen, J. C. Hummelen, and N. S. Sariciftci, Materials Bulletin, Vol. 30, 33 (2005).
130. J. K. J. van Duren, X. N. Yang, J. Loos, C. W. T. Bulle-Lieuwma, A. B. Sieval, J. C. Hummelen, and R.A. J. Janssen, Adv. Funct. Mater., 14: 425 (2004).
131. Y. Kim, S. Cook, S. M. Tuladhar, S. A. Choulis, J. Nelson, J. R. Durrant, D. D. C. Bradley, M. Giles, I. Mcculloch, C. S. HA, and M. Ree, Nature, 5, 197 (2006).
132. G. Li, V. Shrotriya, J. Huang, Y. Yao, T. Moriarty, K. Emery, and Y. Yang, Nature materials, 4, 865 (2005).
133. S. A. Backer, K. Sivula, D. F. Kavulak, and J. M. J. Frecher, Chem. Mater., 19, 2927 (2007).
134. T. Hasobe, H. Imahori, P. V. Kamat, T. K. Ahn, S. K. Kim, D. kim, A. Fujimoto, T. Hirakawa, and S. Fukuzumi, J. Am. Chem. Soc., 127, 1216 (2005).
135. P. V. Kamat, J. Phys. Chem. C, 111, 2834 (2007).
136. W. U. Huynh, J. J. Dittmer, A. P. Alivisatos, Science, 295, 2425 (2002).
137. D. J. Milliron, I. Gur, and A. P. Alivisatos, MRS bulletin, 30, 41 (2005).
138. Y. Yin and A. P. Alivisatos, Nature, 437, 664 (2005).
139. M. Pientka, V. Dyakonov, D. Meissner, A. Rogach, D. Talapin, H. Weller, L. Lutsen, and D. Vanderzande, Nanotechnology, 15, 163 (2004).
140. I. Gur, N. A. Fromer, and A. P. Alivisatos, J. Phys. Chem. B, 110, 25543 (2006).
141. I. Gur, N. A. Fromer, C. P. Chen, A. G. Kanaras, and A. P. Alivistos, Nano Letters, 7(2), 409 (2007).
142. J. Geng and T. Zeng, J. Am. Chem. Soc., 128, 16827 (2006).
143. S. Chaudhary, H. Lu, A. M. Müller, C. J. Bardeen, and M. Ozkan, Nano Letters, 7(7), 1973 (2007).
144. K. M. Coakley, Y. Liu, C. Goh, and M. McGehee, MRS Bulletin, 30, 37 (2005).
145. C. J. Brabec, J. A. Hauch, P. Schilinsky, and C. Waldauf, MRS Bulletin, 30, 50 (2005).